Complete Guide to PEST CONTROL — WITH AND WITHOUT CHEMICALS

FOURTH EDITION

Complete Guide To

PEST CONTROL

— WITH AND WITHOUT CHEMICALS

FOURTH EDITION

GEORGE W. WARE
UNIVERSITY OF ARIZONA

Ware, George Whitaker, 1927 --

Printed in United States of America
Library of Congress Number 2005921853
ISBN 1-892829-13-4

Printed in the United States of America.

To my wife,	*my children,*	*and grandchildren*
Doris	*Cindy*	*John*
	Lynn	*Karyn*
	Sam	*Ashleigh*
	Melanie	*Lauren*
	Julie	*Matthew*
	Randy	*Taylor*

DISCLAIMER CLAUSE

The author has made every attempt to provide up-to-date, scientific information for the novice to solve every variety of home pest control problems. Many sources of information have been gleaned to acquire these suggestions for both chemical and non-chemical controls. The pesticides recommended herein were registered for the uses described at the time of this writing and are known to be safe and effective if used according to the directions on the label. However, due to constantly changing laws and regulations, the author can assume no liability for these suggestions. The user assumes all such responsibility in following these as well as all other pest control suggestions involving pesticides. Those pesticides recommended in this book are registered with the Office of Pesticide Programs of the Environmental Protection Agency (EPA) and are safe for the uses indicated when following the directions for their use printed on the label or printed material accompanying the product. Suggestions for insect control are limited to uses in and around the home and do not apply to food-handling establishments.

The author and publisher are in no way responsible for the application or use of the chemicals mentioned or described herein. They make no warranties, express or implied, as to the accuracy or adequacy of any of the information presented in this book, nor do they guarantee the current status of registered uses of any of the chemicals with the U.S. Environmental Protection Agency. Also, by the omission, either unintentionally or from lack of space, of certain trade names and of some of the formulated products available to the public, the author is not endorsing those companies whose brand names or products are listed.

Contents

Preface

The title, *COMPLETE GUIDE TO PEST CONTROL—"with and without chemicals"* is a misnomer. Rather, it should be titled *COMPLETE GUIDE TO PEST CONTROL—with and without conventional pesticides*, because essentially everything tangible is chemical in composition. Consequently, we would not be able to control pests "without chemicals"! When this book first appeared some 25 years ago, the term *pesticide* identified those materials in somewhat bad repute. The word *chemicals* was not exactly popular, but carried less of an ominous image than *pesticides*, thus determining the choice of the secondary title, *with and without chemicals*.

It has been an immensely popular book because of its very practical nature and easy-to-follow approach. I know that you, as a conscientious homeowner or gardener, will find it quite useful in solving your day-to-day annoying pest problems, so place it in a prominent place for easy reference.

Homeowners and apartment dwellers spend $3.66 billion annually on professional pest control when they could easily handle most of it themselves. Why? Because they have been led to believe that it is too complicated and difficult to do themselves. Because they don't know where to go to get dependable information for do-it-yourself pest control. Because they believe that only "professionals" can do an effective job. Some are simply frightened by the thought of working with pesticides or around those "creepy, crawling things."

In your hands is the best possible source of reliable information on do-it-yourself pest control, the Fourth Edition of *COMPLETE GUIDE TO PEST CONTROL — WITH AND WITHOUT CHEMICALS*. It contains good, reliable information on a choice of ways to control insects, plant diseases, weeds, rodents and other small animals, pest birds, snails, nematodes, algae and even your neighbors' straying and annoying pets, both economically and with thoughtful consideration for the environment.

This book was written to meet the needs of urban and suburban residents. It describes pest problems that the average household or apartment dweller might encounter anywhere in the United States, with accompanying safe and simple methods for their control in and around the home. Every effort has been made to include non-chemical (non-pesticide) methods, along with traditional chemical or pesticide controls, when they were known from reliable sources.

This book presents how-to-do-it information for the layperson in a simple, understandable way, that provides an appreciation for the very important chemical and non-chemical tools and a knowledge of the biology and ecology of the pests they are intended to control. Discussions of individual pests or groups of pests are not restricted to bare definitions, but rather adequate information is given to convey something of their importance in the home, garden, lawn or orchard.

Here is a catalog of selected pests and their intelligent manipulation with and without the use of pesticides. The individual pests used for illustrating certain points were not selected to the exclusion of others, but were chosen based on their frequency of appearance as pests in and around homes in North America. Such memorable characteristics as a notorious history, catchy name or unique biology make them even more interesting. The selection of the individual pesticides for control purposes indicates that they will do the best job, according to the authorities whose reports and research were used.

While writing this fundamental book on pests and pest control and their important position in today's home life, I have attempted to use an everyday, factual, yet unemotional approach to a subject that is more often than not presented highly emotionally charged. Also, an effort has been made to avoid the use of technical and scientific terms prior to their introduction and discussion. In trying to avoid the natural tendency of oversimplification, however, it was necessary to compromise between the too simple and the incomprehensible.

No book on this subject could be complete because of the tremendous amount of material to be covered. It is intended to present a comprehensive picture of major pests and their management and is not intended as an exhaustive study of every possible pest found in all regions. It will be of interest, not only to homeowner, but also to those who work small farms, professional gardeners, grounds keepers, golf course superintendents, landscape maintenance persons, structural pest control operators, apartment managers and to students enrolled in various pest control, ecological and environmental courses.

The Third Edition of COMPLE*TE GUIDE TO PEST CONTROL — WITH AND WITHOUT CHEMICALS* was adopted in numerous junior- and four-year colleges as a textbook and used in large numbers by the U. S. Peace Corps in foreign countries where traditional control materials were not available.

A great many ideas and some data are presented in various sections of the book without direct citations of sources. At the end is the *Bibliography,* which lists the sources of information and contributors whose work or writings were used.

During the months required to make this revision, I sorted through hundreds of scientific papers, state and federal pest control recommendations, commercial ads and magazine clippings. Everything I had collected and carefully catalogued had a place in this Fourth Edition. There was but one problem — there simply wasn't room to include everything useful to the universal reader. Much of the research collection carefully sorted to manila folders is still there, unused.

To Doris, Cindy, Lynn, Sam, Melanie, Julie and Randy, my wife and children and to John, Karyn, Matthew, Ashleigh, Lauren and Taylor, my grandchildren (who now know that butterflies come from caterpillars and caterpillars come from eggs and eggs come from butterflies), I owe a great debt of appreciation for permitting me the isolation and time to see this Fourth Edition completed.

George W. Ware

Tucson, Arizona
November 2004

CHAPTER 1

PESTS — An Overview

...the locusts . . invaded all Egypt and settled down in every area of the country in great numbers. Never before had there been such a plague of locusts.

Exodus, 10:13-14

WHY DO IT YOURSELF?

Pest control in the U.S. is big, big business! There were 103.9 million households in the U.S. in 1999, of which 77 million, or 74%, used pesticides. These consumers used 80 million pounds of conventional pesticide active ingredients (a.i.) in their homes, on their lawns and gardens and 60 million pounds of other pesticides (sulfur, oils, insect repellents, moth control products and others) in 1999, according to the latest report from the U.S. Environmental Protection Agency. For all of this they paid $1.984 billion, or 18% of the total spent for pesticides in the U.S. (Donaldson, et al., 2002)

Considering all types of pesticides, total usage, including agricultural, amounted to 1.244 billion pounds, or 11.97 pounds per person. A portion of this figure is residential pesticide use for home, lawn and garden of 140 million pounds, or 0.51 pounds per person, or 1.35 pounds per household. The average total expenditure per household for pesticides in 1999 was $19.09.

This is small change when compared with professional pest control. In 1999, there were 33,100 commercial pest control firms registered in the U.S. Their sales amounted to approximately $6.1 billion, with about 60% coming from the residential market. This averages approximately $36.00 per household for commercial pest control, compared to about one half that spent for do-it-yourself pest control!

WHAT ARE PESTS?

Pests are any unwanted plants, animals or microorganisms. Every homeowner and apartment dweller has pest problems. In fact, we spend more than $3.66 billion annually for professional pest control services in our homes and apartments. These pests may include, among other things, filthy, annoying and disease-transmitting flies, mosquitoes and cockroaches; moths that eat woolens; beetles that feed on leather goods and infest packaged foods; slugs, snails, aphids, mites, beetles, caterpillars and bugs feeding on our lawns, garden, trees and ornamentals; termites that nibble away at our wooden buildings, books and cellulose products; diseases that mar and destroy our fruits, vegetables and plants; algae growing on the walls or greening the water of our swimming pools; slimes and mildews that grow on shower curtains and stalls and under the rims of sinks; rats and mice that leave their fecal pellets scattered around in exchange for the food they eat; dogs that designate their territories by urinating on automobile wheels, shrubs and favorite flowers or defecate on lawns and driveways; cats that yowl and urinate on our auto windshields at night or catch song birds; and annoying birds that leave their feces on widow ledges, sidewalks and statues of yesterday's heroes, or flock noisily together in the autumn by the thousands, leaving their tree roosts literally white-washed with their feces.

To the commercial grower pests could include insects and mites that damage crops; weeds that compete with field crops for nutrients and moisture; aquatic plants that clog irrigation and drainage ditches; diseases of plants caused by fungi, bacteria and viruses; nematodes, snails and slugs; rodents that feed on grain, young plants and the bark of fruit trees; and birds of every imaginable species that eat their weight every day in young plant seedlings and grain from fields and animal feedlots as well as from storage.

IMPORTANCE OF PESTS AND THEIR DAMAGE

That sum of $3.66 billion plus spent every year for pest control services from professional operators does not include an equal amount of do-it-yourself applications, the subject of this book. We no longer must tolerate carpet beetles in our wall-to-wall floor covering, moths in our closets, or cockroaches in our kitchen. We now consider the presence of these pests in our castles a matter of social indignation as well as a health hazard and general nuisance.

It is estimated that in the United States agriculture sector alone, insects, weeds, plant diseases and nematodes account for losses up to $35 billion annually. This is for just agriculture. There is no way of estimating how much more than the $3.66 billion urban pest control costs the homeowner, for in addition to his own efforts and expenditures, many tax dollars are involved in municipal, county and state pest control projects for flies, mosquitoes, cockroaches and rats. It is because of the economic implications of such losses and savings that pests and their control have assumed their importance, both to the homeowner and the apartment dweller, whose only claim to a green thumb may be a window-box 32 floors above the street.

Plants are the world's main source of food. They compete with about 100,000 plant diseases caused by viruses, bacteria, microplasma-like organisms, rickettsias, fungi, algae and parasitic seed plants; 30,000 species of weeds the world over, with approximately 1,800 species causing serious economic loss; 3,000 species of nematodes that attack crop plants with more than 1,000 that cause damage; and over 1 million species of insects of which 10,000 species add to the devastating loss of crops throughout the world.

An astounding one-third of the world's food crops are destroyed by these pests during growth, harvesting and storage. Losses are even higher in emerging countries: Latin America loses to pests approximately 40% of everything produced. The Food and Agriculture Organization (FAO) has estimated that 50% of cotton production in developing countries would be destroyed without proper insect control. The home gardener will lose approximately 25% of his products in the same fashion if pests are left uncontrolled.

Here are a few good examples of specific increases in yields resulting from the chemical control of insects in the U.S.: cotton, 100%; corn, 25%; potatoes, 35%; onions, 140%; tobacco, 125%; beet seed, 180%; alfalfa seed, 160%; and milk production, 15%.

Equally important are the agricultural losses from weeds. They deprive crop plants of moisture and nutritive substances in the soil. They shade crop plants and hinder their normal growth. They contaminate harvested grain with seeds that may be poisonous to man and animals. In some instances, complete loss of the crop results from disastrous competitive effects of weeds. For the home lawn, if not properly managed, weeds can become the dominant species instead of the golf-green dream of the owner.

PESTS IN HISTORY

America's history contains innumerable influences resulting from the mass destruction of crops by diseases and insects. In 1845-1851, the potato famine in Ireland occurred as a result of a massive infection of potatoes by a fungus, *Phytophthora infestans*, now commonly referred to as late blight. (maneb or zineb would control that handily today, with two or three applications.) This resulted in the loss of about one million lives and the cultural invasion of America by Irish refugees. The sad epilog, however, is that the infected potatoes were edible and nutritious, but a superstitious population refused to use diseased tubers. Even now, late blight still causes the annual loss of over 22 million tons of potatoes worldwide. In 1930, 30% of the U.S. wheat crop was lost to stem rust, the same disease that destroyed three million tons of wheat in Western Canada in 1954.

In the summer of 1986, locusts caused extensive damage to wheat and range grasses in the West and Northwest U. S., causing millions of dollars in losses and control costs. In 1987 and more recently in 1995, locusts were once again a serious plague in the Sahel countries of Africa.

In addition to plant epidemics of insects and diseases, there are devastating human and animal diseases whose causal organisms are transmitted by insects. In 1991, five people contracted horse sleeping sickness (Eastern equine encephalitis) in Florida—the most since 1978—two of the five died. At the same time more than 100 horses died. This disease is transmitted by a mosquito, the Asian Tiger (*Aedes albopictus)* (Kunerth, 1992). In 1971, Venezuelan equine encephalitis appeared in southern Texas, moving in from Mexico. However, through a very concerted suppression effort involving horse vaccination, a quarantine on horse movement and extensive mosquito control measures, the reported cases were limited to 88 humans and 192 horses.

In 1995, three insect-borne diseases that normally occur in Africa were reported in Southern Mexico. These were dengue fever, yellow fever and Venezuelan equine encephalitis. In the case of dengue hemorrhagic fever, 168 cases were reported in Mexico, in 1995, resulting in 16 deaths (Pinheiro 1995). The most recent outbreak of Venezuelan equine encephalitis was in Venezuela, 1995. There were reported 11,390 human illnesses resulting in 16 deaths and 475 dead animals, including horses, mules and donkeys (Ruiz 1995).

The West Nile Virus first appeared in New York State in 1999 and now has spread over large portions of the U.S. In 2002, there were 3900 human cases in the U.S., resulting in 274 deaths. This, like most of the insect-borne viruses, is spread by several species of mosquitoes, with birds, horses and several different wildlife species serving as the virus reservoir. West Nile Virus is the most recent of a long list of diseases transmitted to humans by insects (Illinois Agrinews, 2003).

As late as 1955, malaria (transmitted from person to person only by female mosquitoes of the genus *Anopheles*) infected more than 200 million persons throughout the world. The annual death from this debilitating disease has been reduced, through the selective application of insecticides, from 6 million in 1939, to 2.5 million in 1965, down to about 1 million in 1991. It is estimated by the Centers for Disease Control in Atlanta, that about 1,000 cases of malaria are imported into the U.S. annually, usually from travelers to Africa, Southeast Asia and South America (Rasche, 1992). With the judicious use of insecticides, progress has been made in the control of other important tropical diseases, e.g. yellow fever (transmitted by *Aedes aegypti* mosquitoes), sleeping sickness (transmitted by tsetse flies of the genus *Glossina*) and Chagas' disease (carried by "kissing bugs" of the genus *Triatoma*).

Construction of the Panama Canal was abandoned in the 19th Century by the French because more than 30,000 of their laborers died from yellow fever and other diseases. Think of it—30,000!

Since the first recorded epidemic of the Black Death or Bubonic Plague, it is estimated that more than 65 million persons have died from this disease transmitted by the rat flea (*Nosopsyllus fasciatus*).

Bubonic Plague is endemic to northern Arizona, northeastern California, southern Colorado and northern New Mexico, carried by fleas on prairie dogs that colonize these areas. Infected fleas bite humans that come in contact with these colonies, transmitting the disease, which is usually fatal when not appropriately treated with antibiotics. Some 2-3 deaths occur each year from the plague in these areas. From 1947, there have been 390 cases of plague in the U.S., causing 60 deaths (Centers for Disease Control and Prevention, 1997).

The number of deaths resulting from all wars appears paltry beside the morbid toll taken by insect-borne diseases. Currently there is the ever-lurking danger to humans from such diseases as encephalitis, relapsing fever, sleeping sickness, typhus, malaria and West Nile Virus.

Controlling our pests is neither a luxury nor mark of affluence—it is essential.

CHAPTER 2

You pays your money,
you takes your choice.

Caption to cartoon by John Leech in *PUNCH*, Jan 3, 1846.

CONTROLS — You Have Choices

CHOICES

For more than 3 million years, humanoids have been on earth and when early man moved out of his tree into a cave and later into a shelter of his own creation, he brought with him old and unwanted guests and encountered new ones. Today mankind has a magnificent variety of these pesky insect pests with which to compete. Even the Bible mentions ants, fleas and moths as early pests of man. The apostle James (5:2) wrote "Your riches have rotted and your garments are moth-eaten". Similarly, Matthew (6:10) recorded for us, "Do not lay up for yourselves treasures on earth, where moth and rust consume . . ." (It is highly probable that even then they practiced a form of moth control for their woolens, found in the chapter on household insects, by exposing their garments to the noonday sun.)

Man hasn't taken all these pest problems sitting down. Undoubtedly in his early encounters with pests he learned that the easiest method of dealing with his competitors was to avoid them, live and move when and where the pests were not, by carefully selecting his habitat. When he was compelled to deal with the pests, regardless of the habitat, he may have covered himself with dust or mud and eventually some forms of clothing, as barriers. Other methods included removing them by hand or killing them outright, forms of physical and mechanical control.

These were non-chemical methods of control. Man had only the sum of his skills to fight pests.

Chemical controls were slow in coming and often as not, accidental more than intentional and were based more on superstition than fact. The Greeks burned brimstone (sulfur) as an insecticide or purifier. The Bible refers to the use of ashes and salt as herbicides. That's about the way it remained until after the Civil War in the United States.

Historians have traced the use of pesticides to the time of Homer, around 1000 B.C. By 900 A.D., the Chinese were using arsenic as an insecticide in their gardens (Fletcher 1974). Pliny the Elder (A.D. 23-79) recorded most of the early insecticide uses in his Natural History, collected largely from the folklore and Greek writings of the previous two or three centuries. Among these were the use of gall from green lizards to protect apples from worms and rot. Since then, a variety of materials have been used with dubious results: extracts of pepper, whitewash, vinegar, turpentine, fish oil, brine, lye and many others.

Really, nothing much happened in the way of chemical control, at least in this country, until 1868, when kerosene emulsions were developed for the control of various scale insects on citrus. Much of the rest you may know.

Today, unlike our ancestors, we have a choice of chemical or non-chemical control methods. It is the purpose of this book to present the best of both.

NON-CHEMICAL CONTROL — Natural Alternatives

The enthusiasm for *natural* or *organic* gardening has grown amazingly since the second edition of this book in 1988. Included in this pattern of gardening is the enrichment of soil with naturally occurring products, including mulches, composts and animal manure and controlling insects, diseases and weeds with other than chemical methods. These methods stress cultural, mechanical and biological controls along with strict avoidance or limited use of insecticides, fungicides and herbicides, unless they are naturally occurring or derived from natural sources.

Organic gardening obviously requires more personal attention and human energy than the standard integrated techniques and generally results in greater damage to the produce. For persons with the interest and time, this natural scheme of gardening is highly appealing. After all, our forefathers managed their gardens and field crops in this manner until a little over a generation ago.

A combination of several methods is usually needed for satisfactory control of the many garden pests. Since insect and disease control methods vary in their effectiveness, you may wish to select alternative methods to correspond with differences in plant development and productivity, insect damage, weather conditions and cultural practices.

CULTURAL METHODS

Cultural control alters routine production practices in ways that are detrimental to the biological success of insects, diseases and weeds. For instance, spading or plowing and cultivating the soil kills weeds, buries disease organisms and exposes soil insects to adverse weather conditions, birds and other predators. Additionally, deep spading or plowing buries some insects, preventing their emergence.

Crop rotation is effective against insects and diseases that are fairly specific in their range of affected plants and especially against insects with short migration ranges. The movement of crops to different locations will isolate these pests from their food source. If additional land is not available for an alternate site, change the sequence of plants grown in the garden. Don't plant members of the same family in the same location in consecutive seasons, such as melons, squash and cucumbers following each other.

Intelligent use of fertilizers and water induces normal, healthy plant growth and enhances the plant's capability of tolerating insect and disease damage. Along this same line, zealous use of compost or manure in gardens makes them highly attractive to millipedes, white grubs and certain mulch-inhabiting beetles and promotes mildew and some stem blights.

Planting or harvesting times can be changed to reduce disease vulnerability or keep insect pests separated from susceptible stages of the host plant. Corn and bean seed damage by seed maggots can be greatly reduced by delaying planting until the soil is warm enough to result in rapid germination. This same technique reduces root and seedling diseases that attack slow-growing plants. Placing protective coverings ("hot caps") over seedlings during early season not only preserves heat, but also protects plants from wind, hail, some insects and certain cool weather diseases. In some situations a healthy transplant overcomes insect and disease damage more readily than plants developing from seed in the garden.

The removal of crop remains and disposal of weeds and other volunteer plants eliminates disease organism sources and food and shelter for many arthropod pests including cutworms, webworms, white grubs, aphids, millipedes and spider mites. When garden plants stop producing, spade them into the soil to convert them to compost. The exception is when root-knot nematode is present, for this merely increases its distribution in the garden.

INTERPLANTING — Avoiding Insect Pests

Interplanting is an orderly mixing of crop plants aimed at diversifying populations. There are many claims made about the ability of certain plants to protect other species from insect damage. Unfortunately, I have no hard research data to support the claims, but rather many literature sources and testimonials. Quite likely they work to some extent and several are included as trial combinations:

1. Alternate plants of cucumbers with radishes or nasturtiums to reduce numbers of cucumber beetles.
2. Nasturtium is reported to repel aphids and squash bugs.
3. Interplant basil or borage with tomatoes to discourage tomato hornworms.
4. Mix planting of beans with marigolds or summer savory to reduce attraction to Mexican bean beetles. Watch for increased spider mites.
5. Interplant catnip with eggplant, tomato, or potato to repel flea beetles.
6. Interplanting datura or rue with plants to be protected is reported to repel Japanese beetles.
7. The weed, deadnettle, or interplanted beans are purported to prevent Colorado potato beetle from invading potatoes.

8. Chives or garlic interplanted with lettuce or peas may reduce aphids.
9. Flax or horseradish also have reputations for repelling potato beetle.
10. Geraniums planted among roses or grapes are reported to repel Japanese beetles.
11. Garlic planted among various vegetables is reported to repel Japanese beetles, aphids, the vegetable weevil and spider mites.
12. Marigolds interplanted with curcubits may help avoid cucumber beetles.
13. Mint, thyme or hyssop planted adjacent to most cole crops may reduce cabbageworms.
14. Potatoes interplanted with beans may reduce Mexican bean beetles.
15. Castor bean (as is Castor Oil) is repellent to moles and is said to repel aphids.
16. Pennyroyal is avoided by ants and aphids, when interplanted with susceptible hosts.
17. Rosemary and sage are reported to be repellent against cabbage butterfly, bean beetles and carrot flies.
18. Sassafras as an interplant is supposed to repel aphids, as does stinging nettle, a weed.
19. Tansy as an interplant is reported to ward off Japanese beetles, striped cucumber beetles, squash bugs and ants.
20. Sweet alyssum lures beneficial insects to gardens and is especially useful in controlling aphid populations.

The principle involved here is to avoid planting all plants of one kind together. This is comparable to monoculture in agriculture, large fields planted to one crop. A patch or row of plants in a garden is much more attractive than one or two plants mingled in with other species. The interplanting principle literally causes vulnerable plants to lose their identity in the forest of other garden plants, thus losing their attractiveness to egg-laying by adult insects. It becomes a simple and logical way to avoid infestation, which surpasses controlling the pests after they nibble on the beautiful results of your hard work.

I know of no recommendations for controlling plant diseases by interplanting.

RESISTANT VARIETIES

Conscientious gardeners should make every effort to plant species or varieties that are resistant or tolerant to disease and insect damage. Resistance in plants is likely to be interpreted by the layman as meaning immune to damage. In reality, it distinguishes plant varieties that exhibit less insect or disease damage when compared to other varieties under similar circumstances. Some varieties may not taste as good to the pest, may not support disease organisms or may possess certain physical or chemical properties which repel or discourage insect feeding or egg-laying, or may be able to support insect populations with no appreciable damage or alteration in quality or yield.

When buying seeds or plants, check seed catalogs for information on resistant varieties that grow well in your area. Inquire as well at your County Agent's or seed dealers and nurserymen in your area. (The County Agent can be found by looking in the white pages of the telephone directory for Cooperative Extension Service listed under the local County Government). Vegetable varieties that grow well and are resistant to certain pests in Ohio may be a flop in California or Montana. Experienced gardening neighbors and local garden clubs may be the best information sources of all! Resistant varieties are included in the table of insect and disease control when known.

MECHANICAL AND PHYSICAL METHODS

Mechanical control actually aims activities directly toward diseased tissues and insects in various stages of development and is usually more practical for small gardens than for large. These methods can be used in combinations or singly as seems fitting.

Removing by hand or handpicking of diseased stems, leaves and fruit, or insects and insect egg masses insures immediate and positive control. This is especially effective with anthracnose and leaf spots and foliage-consuming insects as potato beetles, hornworms, or bean beetles.

Plant guards or preventive devices are easy to use against insects, although sometimes more ornamental than effective. These include: (1) cheesecloth screens for cold frames and hot beds to prevent insect egg-laying; (2) paper collars, paper cups or tin cans with the end removed to prevent cutworm damage; (3) sticky barriers on the trunks of trees and woody shrubs to prevent damage by crawling insects; (4) mesh covers for small fruit trees and berry bushes to screen out larger insects and birds; and (5) aluminum foil on the soil beneath plants to repel aphids and leafhoppers, referred to as aluminum foil mulch.

A stiff stream of water removes insects without injuring the plants or drenching the soil. Simple water sprays from a hose or pressure sprayer, sometimes with a small amount of soap or detergent added, will dislodge certain pests as aphids, mealybugs and spider mites.

Heavy mulching or plastic mulches work well in preventing weed seed germination.

TRAPS

Traps of various types have always been popular because of the visible results. Examples: (1) slugs and pillbugs can be trapped under boards on the ground; (2) earwigs are trapped in rolled up newspaper placed among the plants or other locations where they gather; (3) a 2-quart container half-filled with a 9:1 dilution of water to molasses will collect grasshoppers, moths and certain beetles and (4) the famous beer pan trap for slugs and snails is made by placing a small can or pan flush with the soil and half-filling with beer to attract and drown them. Generally speaking, simple types are the easiest to set up and the most effective. Commercial traps are seldom better than those made at home.

Light traps, particularly blacklight or bluelight traps (special bulbs that emit a higher proportion of ultraviolet light that is highly attractive to nocturnal insects) are a good insect monitoring tool but provide little or no protection for the garden. True, they usually capture a tremendous number of insects, however, a close examination of light trap collections shows that they attract both beneficial and harmful insects that would ordinarily not be found in that area. Those insects attracted but not captured remain in the area and the destructive ones may cause damage later. Also some species that are wingless and those active only during the day (diurnal, as opposed to nocturnal) are not caught in these traps. Consequently, the value of blacklight or simple light traps in protecting the home garden is generally of no benefit and in some instances detrimental.

Sex lure or pheromone traps are quite effective in the home orchard and garden. These are discussed in detail in Chap. 4, BIORATIONALS.

BIOLOGICAL METHODS

Biological control is the use of parasites, predators or disease pathogens (bacteria, fungi, viruses and others) to suppress pest populations to low enough levels to avoid economic losses. There are 3 categories of biological control: (1) introduction of natural enemies which are not native to the area and which will have to establish and perpetuate themselves; (2) expanding existing populations of natural enemies by collecting, rearing and releasing additional parasites or predators, (inundative releases); and (3) conserving resident beneficial insects by the judicious use of insecticides and the maintenance of alternate host insects so that the beneficials can continue to reproduce and be available when needed. Figure 1 shows some of the beneficial insects found in the home garden and orchard.

The introduction of a parasite or predator does not guarantee its success as a biological control. However, certain conditions can indicate the potential value of such a natural enemy. The effectiveness of a predator or parasite is usually dependent on (1) its ability to find the host when host populations are small, (2) its ability to survive under all host-inhabited conditions, (3) its ability to utilize alternate hosts when the primary host supply runs short, (4) a high reproductive capacity and short life cycle (high biological potential) and (5) close synchrony of its life cycle with the host so the desirable host stage is available when needed by its natural enemy.

The garden and immediate surroundings are alive with many beneficial organisms that are there naturally; however, they may not be numerous enough to control a pest before its damage is done. Actually, parasites and predators are most effective when pest populations have stabilized or are relatively low. Their influence on increasing pest populations is usually minimal since any increase in parasite and predator numbers depends on an even greater increase in pest numbers. Disease pathogens, however, seem to be most effective when pest populations are large. Thus the nature of the host-enemy relationship makes it impossible to have an insect-free garden and simultaneously retain sizable populations of beneficials.

Several biological control agents are available to gardeners. There is a list of sources in Appendix G, or your County Extension Agent or Extension Entomologist can supply a list of sources.

FIGURE 1. Some of the beneficial insect predators found in the home garden and orchard.

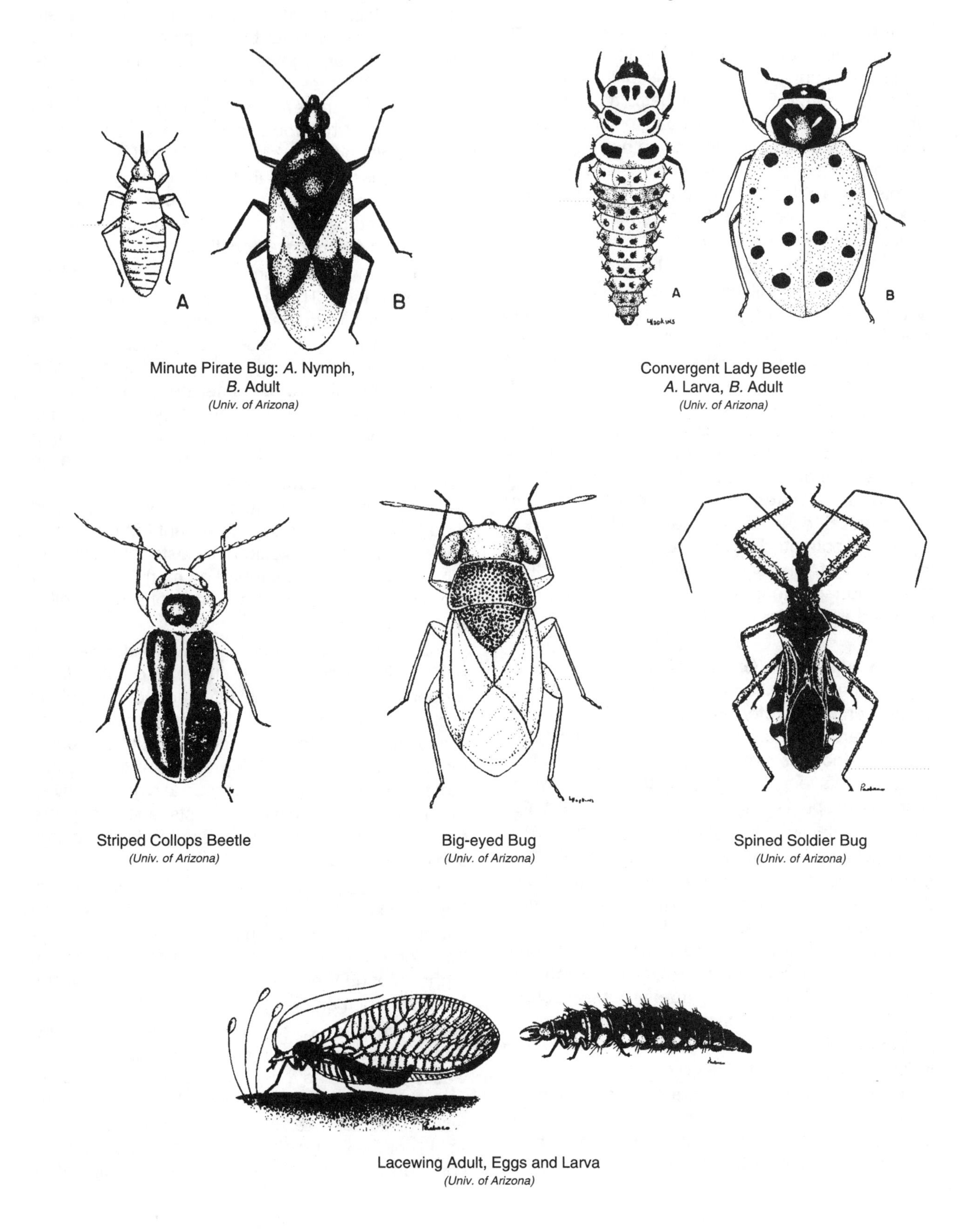

Minute Pirate Bug: *A.* Nymph, *B.* Adult
(Univ. of Arizona)

Convergent Lady Beetle *A.* Larva, *B.* Adult
(Univ. of Arizona)

Striped Collops Beetle
(Univ. of Arizona)

Big-eyed Bug
(Univ. of Arizona)

Spined Soldier Bug
(Univ. of Arizona)

Lacewing Adult, Eggs and Larva
(Univ. of Arizona)

Predators

Praying mantids are usually sold in the form of egg cases and are readily available. Mantids are hungry hunters, are cannibalistic, and will eat their brothers and sisters immediately after hatching; thus, few survive. They really don't live up to their reputations because they usually wait for prey rather than search them out. Naturally this determines the kinds of insects it captures. They prefer grasshoppers, crickets, wasps, bees, and flies, none of which are particularly important garden pests.

Lady beetles or ladybird beetles are also readily available commercially since they can be collected by the millions in the California mountains during their summer hibernation. Adult lady beetles and larvae prefer aphids, though they will feed readily on mealybugs, spider mites and certain other soft-bodied pests and eggs. They do not feed on caterpillars, grubs and other beetles. If an adequate food supply of aphids or other hosts is not available at the release point, lady beetles will move out until they find sufficient food. Because of their physiological conditions under which they were originally collected, they may leave the area regardless of the food supply, to complete their hibernation or aestivation cycle. So, don't be surprised if today's release is absent from rollcall tomorrow.

Green lacewings (Chrysopa) are usually sold as eggs, since the adults are difficult to manage. The larvae, commonly referred to as *aphid lions*, feed on several garden pests including aphids, spider mites, thrips, leafhopper nymphs, moth eggs and small larvae. Adults feed on pollen, nectar and honeydew secreted by aphids. Introduced lacewings must also have an available food supply or they too will leave.

There are several other beneficial insects that are resident helpers and are not available commercially. With the ecological mistakes of man now becoming more apparent, it is reassuring to know that nature can establish certain controls which prevent some destructive insects from overpopulating the environment. The attentive gardener can encourage and prepare the conditions for an increase in insect predators by not using insecticides except where and when absolutely necessary.

True bugs occur naturally, are seen around the yard and garden and are among the predaceous beneficial insects, for they feed on insect pests. Bugs stab their hosts with piercing sucking mouthparts and inject digestive enzymes, then remove the liquid contents. Included in the true bug predators are assassin bugs, damsel bugs, spined soldier bugs and the smaller bigeyed bugs and minute pirate bugs. Together they prey on caterpillars, lygus bugs, fleahoppers, spider mites, aphids, thrips and leafhoppers. When food becomes scarce they may occasionally become cannibalistic and attack their own species. As adults, the bugs are winged and fly readily. The nymphs, or immature stages, are wingless and cannot fly, though they resemble the adults in all other characteristics.

Common wasps haunt your trees and gardens looking for food. They may be the common paper wasp, hornet, yellow jacket, mud dauber, or wasps so tiny that they can be seen only after very close observation. Wasps vary widely in size, color and general body structure. Some are parasites while others are predators. Tiny parasitic wasps lay eggs in bodies of insects and their developing offspring devour and kill their hosts from within. Large predaceous wasps—well known to every gardener—sting caterpillars to paralyze them, and feed them to their young. Mud daubers commonly seek out spiders to stock their nests for feeding of larvae. The author has found as many as 10 black widow spiders, male and female, in one larval cell of a mud dauber.

Ground beetles include hundreds of species and exhibit differences in size, shape and color. Most ground beetles are somewhat flattened, dark, shiny and have visible mandibles or jaws with which they grasp their prey. They may be found under stones, logs, bark, debris or running about on the ground. Most of them hide by day and feed at night. Nearly all are predaceous on other insects and many are beneficial. There are several that feed primarily on snails and slugs.

Dragonflies are those large, biplane-like insects that soar and dart about near and over ponds and streams. They depend entirely on other insects for their food, both catching and eating their prey in flight. Mosquitoes and flies make up a large proportion of their diets.

Syrphid flies are commonly called *flower flies*. They may be brightly colored and several resemble wasps and bees hovering over flowers. They do not sting because they are true flies. The larvae are the active predators, resemble maggots and feed on aphids or the young of termites, ants, or bees. Most commonly they can be found in dense colonies of aphids.

Antlions, or "doodlebugs", are larvae of large, clear-winged predatory insects and are strange looking creatures with long sickle-shaped mouthparts. They are really first cousins to the lacewings or aphid lions mentioned earlier. Antlions are more commonly found in the South and Southwest, but there are a few species found throughout most of the U.S. The larvae hide in cone-shaped burrows in dry soil waiting for an ant or other unlucky insect to stumble into its lair. Once inside, the insect is quickly dispatched.

Lightning bugs occur at evening during the early summer and late spring and are conspicuous by their blinking yellow light. Most of the larvae are luminescent and are given the name, glow-worms. The larvae feed on various smaller insects and on snails.

Spiders make their contribution to insect control not only in the garden but on porches, roof overhangs, in trees and shrubs and inside the home. Basically there are two types of spiders, the web-spinning varieties that capture flying insects and the crab spiders that do not make webs but rather lie in wait for their unsuspecting prey. Spiders, unlike previously mentioned beneficial insects, are not very selective and will capture and feed on any insect within their size range, including the beneficial species.

Parasites

Trichogramma wasps are also available commercially "in season". These tiny wasps deposit their eggs in the eggs of other insects, namely moths, including armyworms, cutworms, fruitworms and many others found in orchards and gardens. These egg-parasitic wasps should be released when the moths are laying eggs, but a sequence of releases throughout the season is preferable to a single large release. The results of these releases will depend on timing, selection of *Trichogramma* species and placement of wasps near host egg masses. A list of caterpillars whose eggs are parasitized by *Trichogramma* wasps is presented in Table 1.

Braconid wasps and other tiny wasps are seldom seen except as they form their little white cocoons on the outside of their host after killing it. There are many species of parasitic wasps most of which are quite small. Like the related Ichneumen wasps, they feed on the inner body fluids of their hosts. The most common ones are parasitic on sphinx moth caterpillars like the tobacco and tomato hornworms.

Parasitic flies are another important group of insects that help hold down populations of caterpillars. These are known as tachinid flies, are about the size of house flies but are darker and covered with heavy

TABLE 1. Some of the important caterpillars whose eggs are parasitized by *Trichogramma* sps. wasps.

There are several species of the tiny, parasitic trichogramma wasps, all of which lay their eggs in the eggs of other insects. There are more than 200 known caterpillar pests whose moth eggs are attacked by Trichogramma.

alfalfa caterpillar	European corn borer	peachtree borer
Angoumois grain moth	fall cankerworm	pecan nutcase bearer
armyworm	fall webworm	pink bollworm
bagworm	forest tent caterpillar	plum moth
bean leaf roller	giant silkworm	polyhemus moth
browntail moth	grassworm	promethea moth
cabbage looper	gypsy moth	prominents
California oakworm	hawk moth	regal moth
carpenterworm	hummingbird moth	rosy maple moth
codling moth	imported cabbageworm	royal moth
corn earworm	inchworm	skippers
(cotton bollwom)	(measuring worm)	squash vine borer
(tomato fruitworm)	(spanworm)	swallowtail butterfly
cotton leafworm	io moth	tobacco budworm
cutworm	luna moth	tobacco hornworm
dagger moth	nymphalid butterfly	tomato hornworm
datanas	Oriental fruit moth	tussock moth
eastern tent caterpillar	overflow worm	wax moth

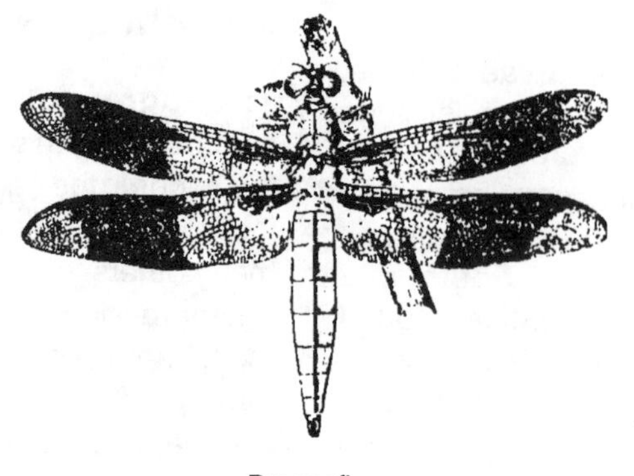

Dragonfly
(U.S.D.A.)

Parasitic Wasp
(U.S.D.A.)

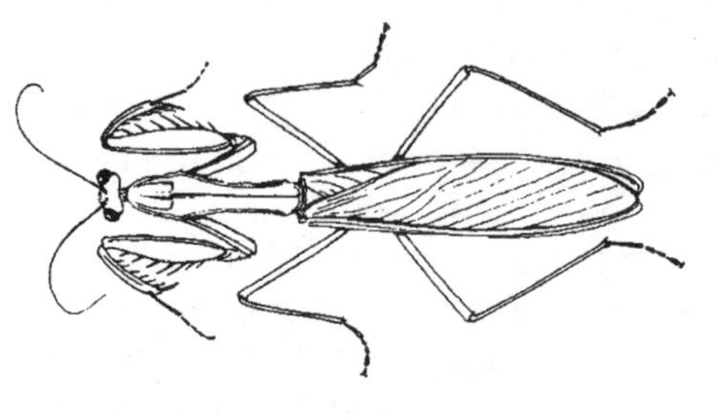

Praying Mantid
(U.S.D.A.)

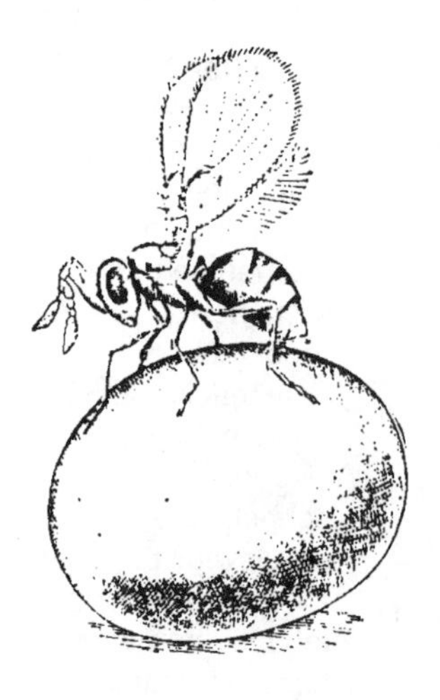

Trichogramma female stinging a moth egg and placing its own egg inside
(U.S.D.A.)

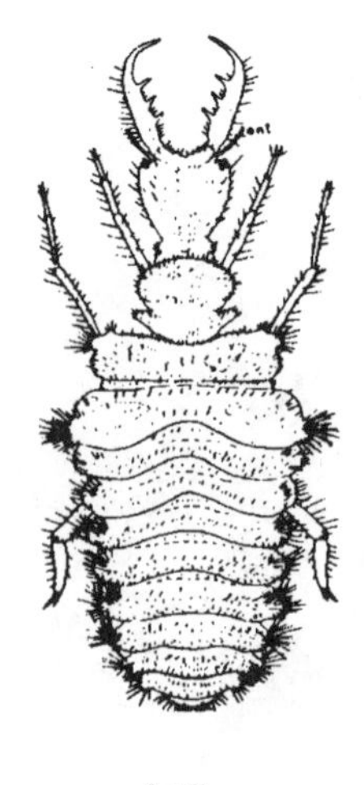

Antlion
(U.S.D.A.)

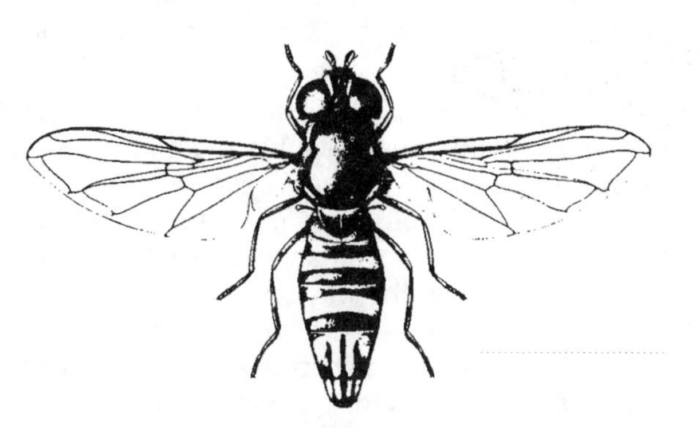

Syrphid or Flower Fly
(U.S.D.A.)

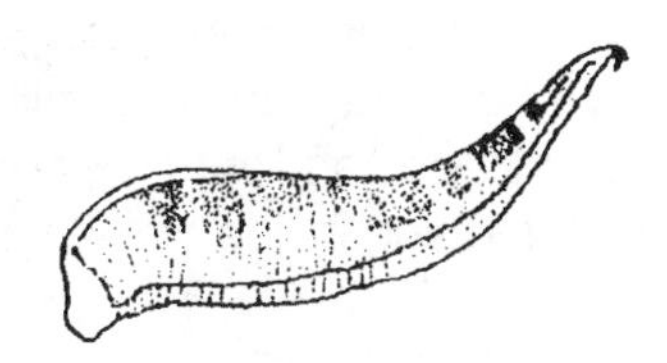

Syrphid Fly Maggot
(Univ. of Arizona)

Tachinid Fly
(Univ. of Arizona)

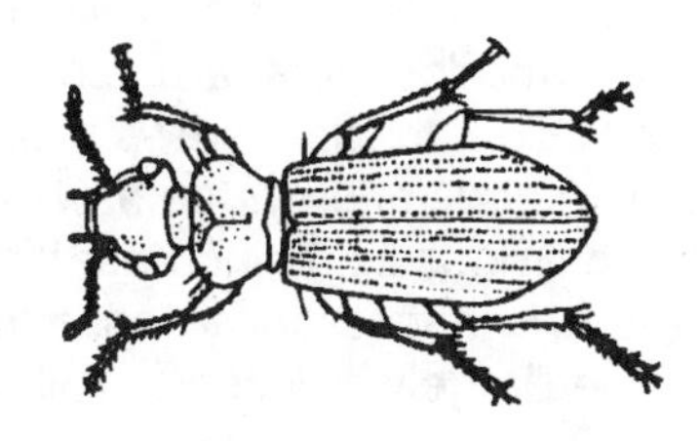

Ground Beetle
(U.S.D.A.)

Dark Paper Wasp
(U.S. Public Health Service)

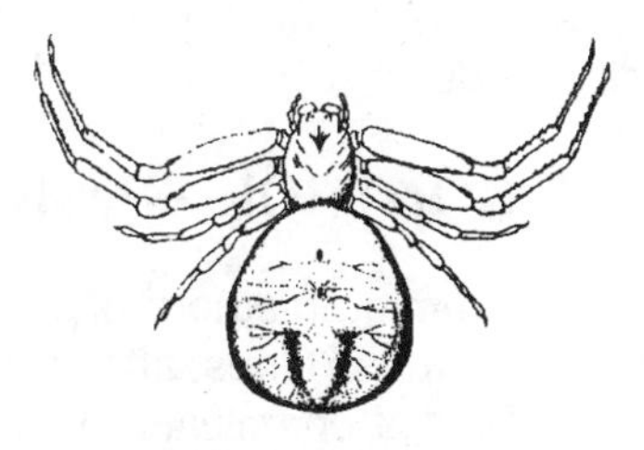

Crab Spider
(Univ. of Arizona)

bristles. They lay their eggs on, in or near caterpillars and in a day or two the eggs hatch. The parasitic fly larva lives inside and feeds on the caterpillar until it pupates. Usually at this stage the caterpillar will die, though it continues to feed and appears normal while the parasite is developing inside.

Disease Organisms—Microbial Insecticides

These are discussed in detail in Chapter 4, BIORATIONALS.

Birds

Far more important in preventing insect outbreaks than in controlling them are birds, birds of all kinds. All bird species feed on insects to some extent. The creepers, flycatchers, nuthatches, swallows, vireos, warblers and woodpeckers are almost entirely insectivorous, while blackbirds, crows, gulls, magpies, robins and even birds of prey, the hawks and owls, commonly feed on insects. To increase bird numbers near gardens it is necessary to encourage those species that feed largely on insects. If all species are encouraged, including those that damage gardens, such as blackbirds, robins and starlings, trouble will develop. Insect-feeding birds can best be encouraged by providing cover, supplementary feed and prevention of predation from cats.

Many insectivorous birds can be attracted to the home area by planting ornamentals that provide suitable bird cover and food. Especially valuable are the following trees and shrubs: apple, bittersweet, cherry, cotoneaster, crabapple, dogwood, elderberry, firethorn, gray birch, hawthorn, highbush cranberry, holly, male mulberry, mountain ash, plum, red cedar, Russian olive, sumac (aromatic and staghorn), tupelo, white oak and wild plum or cherry. Additionally, sunflowers are especially attractive to insect-feeding birds.

CHEMICAL CONTROL—Pesticides

Frequently non-chemical methods of pest control will fail to do the job. At these times, after everything has been tried, the use of chemicals may be the only alternative. These chemicals are classed under the broad heading of *pesticides.* (A pesticide is anything which kills pests, thus *Bacillus thuringiensis,* found under Microbials in Chap. 4, BIORATIONALS, is a pesticide. Because it occurs naturally or is not synthesized in a chemical factory does not remove it from the pesticide classification.)

Because this book is about pests and their control, with and without chemicals, we must now discuss several aspects of the chemical tools. Pesticides, those remarkable products used to control pests around the home as well as in agriculture, have a prominent place in the day-to-day activities of our technologically-advanced society. Pesticides can't be ignored and they won't go away. Like it or not, we find ourselves well into the 21st century, just the beginning of what historians will refer to as "The Chemical Age." Time, population density and technology have isolated us from the "good old days" (whenever they were) and we could no more return to the non-chemical, back-to-nature way of life, than we could park our automobiles and walk where we normally drive.

Because pesticides are now a way of life and are essential ingredients in our affluent lives both here and abroad, it becomes necessary for every educated person, every conscientious citizen, to know something about these valuable tools. Practically speaking, there are two classes of pesticides: Those naturally occurring micro-organisims and minerals, chemical compounds and plant parts or plant extracts that we use as pesticides and those man-made chemicals that are designed and used only as pesticides.

Synthetic pesticides, by their simplest definition, are those made or synthesized by man in laboratories or factories. Examples of these include the insecticides such as malathion and permethrin, herbicides such as 2,4-D and Roundup®, or the fungicides maneb and captan.

The first synthetic pesticide was probably kerosene emulsion sprayed on citrus to control scale insects in California in 1868. Several more complex synthetic pesticides were discovered in the 1920's and 1930's, some of which are still in use. The real surge of development, however, came with World War II, beginning with the discovery of DDT.

CHAPTER 3

PESTICIDES

The proposed decoctions and washes we are well satisfied, in the majority of instances, are as useless in application as they are ridiculous in composition. . .
Editorial, *Practical Entomologist,* Oct. 30, 1865.

PESTICIDES – Big Business

More than 1.24 billion pounds of pesticides were used in the U. S. in 1999, valued in excess of $11.2 billion at the retail level. Of this, agriculture used 77%, industry and government utilized 14%, while home and garden use amounted to 9% of the volume, or about 80 million pounds. (Donaldson et al., 2002).

Most pesticides are synthetic, though a few are produced naturally by plants. The U.S. Environmental Protection Agency (EPA) had more than 900 pesticides registered in 2002. Of these 252 were herbicides, 225 insecticides, 180 fungicides, 36 rodenticides and 207 were biocides (disinfectants). These are sold in the form of some 20,000 products or formulations and 11.97 pounds of pesticides are used each year to feed, clothe and protect every man, woman and child in the United States alone. Of these 11.97 pounds, 1.35 pounds is used at home by the homeowner as do-it-yourself pest control and that is what part of this book is about. (Donaldson et al. 2002).

Pesticides have become extremely beneficial tools to the urbanite, the home gardener. We depend on pesticides, perhaps more than we realize. For algae control in the swimming pool, weed control in our lawns, flea collars and powders for pets, sprays for controlling a myriad of garden and lawn insects and diseases, household sprays for ants and roaches, aerosols for flies and mosquitoes, soil and wood treatment for termite protection by professional exterminators, baits for the control of mice and rats, woolen treatment at the dry cleaners for clothes moth protection and repellents to keep off biting flies, chiggers and mosquitoes when camping or fishing.

These chemical tools are used by the homeowner as intentional additions to our home and garden environment in order to improve environmental quality for ourselves, our animals and our plants, giving us the advantage over our pest competitors. Pesticides are used in agriculture to increase the ratio of cost/benefit in favor of the grower and ultimately the consumer of food and fiber products—the public. Pesticides have contributed significantly to the increased productive capacity of the U.S. farmer, each of whom produced food and fiber for 3 persons in 1776, 73 in 1970 and more than 130 persons in 2005. Where can we find a more successful technology story?

PESTICIDE LANGUAGE

To one person the word pesticide may suggest the insecticide malathion. To another it may conjure up the herbicide Roundup® and still another the garden fungicide maneb. All are correct, however only in part, for their uses and effects are totally unrelated.

"Pesticide" is an all-inclusive, but nondescript word meaning "killer of pests." The various generic words ending in "-cide" (from the Latin-cida, to kill) are classes of pesticides, such as fungicides and insecticides. In the table below are listed the various pesticides and other classes of chemical compounds not commonly considered pesticides. These others, however, are included among the pesticides as defined by federal and state laws.

Pesticides are legally classed or defined in most state and federal laws as "any substance used for controlling, preventing, destroying, repelling, or mitigating any pest." Should you ever pursue the

TABLE 2. A list of pesticide classes, their use and derivation.

Pesticide class	***Function***	***Root-word derivation*** [a]
Acaricide	Kills mites	Gr. *akari,* "mite or tick"
Algaecide	Kills algae	L. *alga,* "seaweed"
Avicide	Kills or repels birds	L. *avis,* "bird"
Bactericide	Kills bacteria	L. *bacterium;* Gr. *baktron,* "a staff"
Biocide	Kills microorganisms	Gr. *bios* "life or living"
Fungicide	Kills fungi	L. *fungus,* Gr. *spongos,* "mushroom"
Herbicide	Kills weeds	L. *herba,* "an annual plant"
Insecticide	Kills insects	L. *insectum,* "cut or divided into segments"
Larvicide	Kills larvae (usually mosquito)	L. *lar,* "mask or evil spirit"
Miticide	Kills mites	Synonymous with Acaricide
Molluscicide	Kills snails and slugs (may include barnacles, clams, mussels)	L. *molluscus,* "soft- or thin-shelled"
Nematicide	Kills nematodes	L. *nematoda,* Gr. *nema,* "thread"
Ovicide	Destroys eggs	L. *ovum,* "egg"
Pediculicide	Kills lice (head, body, crab)	L. *pedis,* "louse"
Piscicide	Kills fish	L. *piscis,* "a fish"
Predicide	Kills predators (coyotes, usually)	L. *praeda,* "prey"
Rodenticide	Kills rodents	L. *rodere,* "to gnaw"
Silvicide	Kills trees and brush	L. *silva,* "forest"
Slimicide	Kills slimes	Anglo-Saxon, *slim*
Termiticide	Kills termites	L. *termes,* "wood-boring worm"
Viricide	Kills viruses	L. "slimy liquid"
	Chemicals classed as pesticides not bearing the -cide suffix	
Antimicrobials	Control microorganisms	
Attractants	Attract insects	
Chemosterilants	Sterilize insects or pest vertebrates (birds, rodents)	
Defoliants	Remove leaves	
Desiccants	Speed drying of plants	
Disinfectants	Destroy or inactivate harmful microorganisms	
Growth regulators	Stimulate or retard growth of plants or insects	
Pheromones	Attract insects or vertebrates	
Repellents	Repel insects, mites and ticks, or pest vertebrates (dogs, rabbits, deer, birds)	

[a] *Gr.* indicates Greek origin; *L.* indicates Latin origin

subject of pesticides from a legal viewpoint they would be discussed as Pesticides or occasionally as economic poisons.

The pesticide vocabulary not only includes all of the commonly-used pesticides, but also includes groups of chemicals which do not actually kill pests, as shown in Table 2. However, because they fit logically as well as legally into this umbrella word, pesticides, they are included.

NAMING OF PESTICIDES

A passing knowledge of pesticides involves, among other things, learning something of their names or nomenclature. For example, let us look at malathion, a commonly known household insecticide.

Malathion is the *common* name for the compound. Common names are selected officially by the appropriate professional scientific society and approved by the American National Standards Institute and the International Organization for Standardization. Common names of insecticides are selected by the Entomological Society of America; herbicides by the Weed Science Society of America; and fungicides by the American Phytopathological Society. The *proprietary* name, Malixol®, trade name or brand name, for the pesticide is given to a particular pesticide on various formulations by their formulators. To illustrate, Malixol® is also known as Malate®, Malathate®, Mallet®, etc.

Common names are assigned to avoid the confusion resulting from the use of several trade names, as just illustrated.

The long chemical name, diethyl mercaptosuccinate, S-ester with 0,0-dimethyl phosphorodithioate, is the scientific or chemical name. It is usually presented according to the principles of nomenclature used in Chemical Abstracts, a scientific abstracting journal which is generally accepted as the world's standard for chemical names, or IUPAC, an international organization.

The chemical name is required on the label of all pesticide containers.

FORMULATIONS

Pesticides are formulated to improve their properties of handling, application, effectiveness, safety and storage. After a pesticide is manufactured in its relatively pure form, the *technical grade material*, whether herbicide, insecticide, fungicide, or other classification, the next step is formulation. It is processed into a usable form for direct application, or for dilution followed by application. The formulation is the final physical condition in which the pesticide is sold for use. The technical grade material may be formulated by its basic manufacturer or sold to a formulator. The formulated pesticide will be sold under the formulator's brand name or it may be custom-formulated for another firm.

Formulation is the processing of a pesticidal compound by any method that will improve its properties of storage, handling, application, effectiveness, or safety. The term is usually reserved for commercial preparation prior to actual use and does not include the final dilution in application equipment.

The real test for a pesticide is acceptance by the user. To be accepted for use by the home gardener or pest control operator, a pesticide must be effective, safe and easy to apply, but not necessarily economical, especially from the home gardener's viewpoint. The urbanite commonly pays 5 to 25 times the price that a grower may pay for a given weight of a particular pesticide, depending to a great extent on the formulation. For instance, the most expensive formulation of an insecticide is the pressurized aerosol.

Pesticides, then, are formulated into many usable forms for satisfactory storage, for effective application, for safety to the applicator and the environment, for ease of application with readily available equipment and for economy. This is not always simply accomplished, due to the chemical and physical characteristics of the technical grade pesticide. For example, some materials in their "raw" or technical condition are liquids, others solids; some are stable to air and sunlight, whereas others, are not; some are volatile, others not; some are water soluble, some oil soluble and others for example, the insecticide carbaryl (Sevin®), is insoluble in both water and oil. These characteristics pose problems to the formulator, since the final formulated product must meet the standards of acceptability by the user.

About 98% of all pesticides used in the United States in 2005 are manufactured in the formulations appearing in the simplified classification presented in Table 3. Familiarity with the more important formulations is essential to the well-informed home gardener. We will now examine the major formulations used for the home, lawn and garden, as well as those employed in structural pest control and agriculture.

Sprays

Emulsifiable Concentrates. (EC). Formulation trends shift with time and need. Traditionally, pesticides have been applied as water sprays, water suspensions, oil sprays, dusts and granules. Spray formulations are prepared for insecticides, herbicides, miticides, fungicides, algaecides, growth regulators, disinfectants, repellents and molluscicides. Consequently, more than 85% of all pesticides are applied as sprays. The bulk of these are currently applied as water emulsions made from emulsifiable concentrates.

Emulsifiable concentrates are concentrated solutions of the technical grade material with enough emulsifier added to make the concentrate mix (emulsify) readily with water for spraying. The emulsifier is a detergent-like material that makes possible the suspension of microscopically small oil droplets in water to form an emulsion.

When an emulsifiable concentrate is added to water, the emulsifier causes the oil to disperse

TABLE 3. Common formulations of pesticides [a]

1. Liquids/Sprays (insecticides, herbicides, fungicides)
   a. Emulsifiable concentrates (also emulsible concentrates)
   b. Water-miscible liquids, (sometimes referred to as *liquids*)
   c. Wettable powders and wettable granules
   d. Water-soluble powders, e.g., prepackaged, tank drop-ins, for agricultural and pest control operator use
   e. Gels, packaged in water-soluble bags, e.g., Buctril® Gel
   f. Oil solutions, e.g., barn and corral ready-to-use sprays, and mosquito larvicides
   g. Liquid concentrates for water-hose sprayers, lawn/garden
   h. Soluble pellets for water-hose attachments
   i. Flowable or sprayable suspensions
   j. Flowable microencapsulated suspensions, e.g., Penncap M®, Dursban ME®
   k. Ultralow-volume (ULV) concentrates (agricultural and forestry use only)
   l. Fogging concentrates, e.g., public health mosquito and fly abatement foggers
   m. Foam concentrates (herbicides; insecticides in structural applications)
   n. Shampoos (human head lice [RID®]; pets for fleas)
   o. Mousse gels (human head lice [RID®])
2. Dusts (insecticides, fungicides)
   a. Undiluted toxic agent
   b. Toxic agent with active diluent, e.g., sulfur, diatomaceous earth
   c. Toxic agents with inert diluent, e.g., home garden insecticide-fungicide combination in pyrophyllite carrier
   d. Aerosol dust, e.g., silica aerogel in aerosol form
3. Aerosols (insecticides, repellents, disinfectants)
   a. Pushbutton
   b. Total release
4. Granulars (insecticides, herbicides, algaecides)
   a. Inert carrier impregnated with pesticide
   b. Soluble granules, e.g., dry flowable herbicides
   c. Water dispersable granules
5. Fumigants (insecticides, nematicides, herbicides)
   a. Stored products and space treatment, e.g., liquids, gases, moth crystals
   b. Soil treatment liquids that vaporize
   c. Greenhouse smoke generators, e.g., Nico-Fume® (discontinued)
6. Impregnates (insecticides, fungicides, herbicides)
   a. Polymeric materials containing a volatile insecticide, e.g., No-Pest Strips®, pet collars
   b. Polymeric materials containing non-volatile insecticides, e.g., pet collars, adhesive tapes, pet tags, livestock eartags
   c. Mothproofing agents for woolens
   d. Wood preservatives
   e. Wax bars (herbicides)
   f. Insecticide soaps for pets
7. Fertilizer combinations with herbicides, insecticides or fungicides
8. Baits (insecticides, molluscicides, rodenticides and avicides)
9. Slow-release pesticides
   a. Microencapsulated materials for agriculture, mosquito abatement, and household, e.g., Penncap M®, Knox Out 2FM®
   b. Paint-on lacquers for pest control operators, e.g., Killmaster® II (discontinued)
   c. Interior latex house paints for home use
   d. Adhesive tapes for pest control operators and homeowners, e.g., Hercon®, Vaportape II®
   e. Resin strips containing volatile organophosphate fumigant, e.g., No-Pest Strips® or pyrethroids used in livestock eartags
10. Insect repellents
    a. Aerosols
    b. Rub-ons (liquids, lotions, paper wipes and sticks)
    c. Vapor-producing candles, torch fuels, smoldering "punk" or coils.
11. Insect attractants
    a. Food, e.g., Japanese beetle traps, ant and grasshopper, hornet and wasp and Mediterranean fruit fly baits
    b. Sex lures, e.g., pheromones for agricultural and forest pests (gypsy moth), household (cockroach traps)
12. Animal systemics (insecticides, parasiticides)
    a. Oral (premeasured capsules or liquids)
    b. Dermal (pour-on or sprays)
    c. Feed-additive, e.g., impregnated salt block and feed concentrates

[a] This list is incomplete, but does contain most of the common formulations.

immediately and uniformly throughout the water, with agitation, giving it an opaque or milky appearance. This oil-in-water suspension is a normal emulsion. There are a few rare formulations of invert emulsions, which are water-in-oil suspensions and are opaque in the concentrated forms, resembling salad dressing or face cream. These are employed almost exclusively as herbicide formulations for agricultural and industrial use. The thickened sprays result in reduced drift and can be applied in sensitive situations.

If properly formulated, emulsifiable concentrates should remain suspended without further agitation for several days after dilution with water. A pesticide concentrate that has been held over by the home gardener from last year can be easily tested for its emulsifiable quality by adding one tablespoon to a quart of water and allowing the emulsion to stand after shaking. The material should remain uniformly suspended for at least 24 hours with no precipitate. If a precipitate does form, the same condition may occur in your spray tank, resulting in a clogged nozzle and uneven application. This can be remedied by adding two tablespoons of a quality liquid dish washing detergent to each pint of concentrate and mixing thoroughly. The bulk of pesticides available to the homeowner are formulated as emulsifiable concentrates and generally have a shelf-life of about 3 years. Emulsifiable concentrates should be stored where they will not freeze, which usually causes separation of ingredients and failure to remain emulsified when diluted.

Water-Miscible Liquids readily mix with water. The technical grade material may be water-miscible initially or it may be alcohol-miscible and formulated with an alcohol to become water-miscible. These formulations resemble the emulsible concentrates in viscosity and color but do not become milky when diluted with water. Few of the home and garden pesticides are sold as water miscibles since few of the pesticides that are safe for home use have these physical characteristics.

Wettable Powders, (WP) are essentially concentrated dusts containing a wetting agent to facilitate the mixing of the powder with water before spraying. The technical material is added to the inert diluent, in this case a finely ground talc or clay, in addition to a wetting agent, similar to a dry soap or detergent and thoroughly ground together in a ball mill. Without the wetting agent, the powder would float when added to water and the two would be almost impossible to mix. Because wettable powders usually contain from 50% to 75% clay or talc, they sink rather quickly to the bottom of spray tanks unless the tank is shaken repeatedly during use. Many of the insecticides sold for garden use are in the form of wettable powders because there is very little chance that this formulation will be *phytotoxic*, that is burn foliage, even at high concentrations. This is not true for emulsible concentrates, since the original carrier is usually a solvent, which alone in relatively moderate concentrations can cause foliage burning at high temperatures (90°F. and above).

Water Soluble Powders (SP) are appropriately named and self explanatory. Here, the technical grade material is a finely ground water-soluble solid and contains nothing else to assist its solution in water. It is merely added to the proper amount of water in the spray tank where it dissolves immediately. Unlike the wettable powders and flowables, these formulations do not require repeated agitation; they are true solutions and do not settle to the bottom.

Oil Solutions in their commonest form are the ready-to-use, household and garden insecticide sprays, sold in an array of bottles, cans and plastic containers, all usually equipped with a handy spray atomizer. Not to be confused with aerosols, these sprays are intended to be used directly on pests or where they frequent. Oil solutions may be used as weed sprays, for fly control around stalls and corrals, for standing pools to control mosquito larvae, in fogging machines for mosquito and fly abatement programs, or for household insect sprays purchased in supermarkets, hardware and garden supply stores. For commercial use they may be sold as oil concentrates to be diluted with kerosene or diesel fuel before application, or as the dilute, ready-to-use form. In either case, the compound is dissolved in a light-weight oil and is applied as an oil spray; it contains no emulsifier or wetting agent.

Soluble Pellets, despite their seeming convenience and ease of handling with a water hose, are not very effective. They are sold in kits, including the water hose attachment, fertilizer, fungicide, insecticide and even a car wash detergent and wax pellets. The actual amount of active ingredient is very small and uniform distribution with a watering hose is difficult.

Flowable or Sprayable Suspensions are ingenious solutions to formulation problems.

Generally speaking, these are suspensions of chemicals in granular, microencapsulated, powder or liquid form, suspended in water or in an oil-water emulsion. This usually results in a thick, concentrated form which is diluted with water before spraying, somewhat like an emulsifiable concentrate. In these instances, however, the suspended particles, whether liquid or solid, were suspended in the formulated product as sold, rather than forming an emulsion or suspension when diluted with water. Most of the unique pesticide formulations in recent years are found in this group.

Fogging Concentrates are the formulations sold strictly for public health use in the control of nuisance or disease vectors, such as flies and mosquitoes and to pest control operators. Fogging machines generate droplets whose diameters are usually less than 10 microns but greater than 1 micron. They are of two types. The thermal fogging device utilizes a flash heating of the oil solvent to produce a visible vapor or smoke. Unheated foggers atomize a tiny jet of liquid in a venturi tube through which passes an ultra-high velocity air stream. The materials used in fogging machines depend on the type of fogger. Thermal foggers use oil only, whereas un-heated generators use water, emulsions, or oils.

Dusts

Historically, dusts (D) have been the simplest formulations of pesticides and the easiest to apply. Examples of the undiluted toxic agent are sulfur dust used on ornamentals for some disease and mite control and the household roach dusts, boric acid. An example of the toxic agent with active diluent would be one of the garden insecticides or fungicides having sulfur dust as its carrier or diluent. A toxic agent with an inert diluent is the most common type of dust formulation in use today in the home garden. Insecticide-fungicide combinations are applied in this manner, with the carrier being an inert clay such as talc or pyrophyllite. The last type, the aerosol dust, is a powdered silica aerogel in a liquefied gas propellant that can be directed into crevices of homes and commercial structures for insect control. The aerosol dust was never popular for home use.

Despite their ease in handling, formulation and application, dusts are the least economical of the pesticide formulations. The reason is that dusts have a poor rate of deposit on foliage, unless it is wet from dew or rain. The remainder drifts upward and downwind. Which may be annoying to the downwind neighbor. Under similar circumstances, a garden hose or hand sprayer application of water emulsion spray will deposit 70% to 80% of the pesticide on target plants or turf.

Aerosols

Most of us have been raised in the aerosol culture. bug bombs, hair sprays, underarm deodorants, home deodorizers, oven sprays, window sprays, repellents, paints, garbage can and shower-tub disinfectants and supremely, the anti-itch remedies for athletes foot. Among aerosol pesticides the insecticides are dominant. Developed during World War II for use by military personnel living in tents, the push-button variety was used as space sprays to knock down mosquitoes. More recently the total-release aerosol has been designed to discharge its entire contents in a single application. They are now available for home owners as well as commercial pest control operators. In either case, the nozzle is depressed and locked into place, permitting the aerosol total emission while the occupants leave and remain away for a few hours. Aerosols are effective only against resident flying and crawling insects and provide little or no residual effect as do conscientiously-applied sprays. How do aerosols work? Essentially the active ingredients must be soluble in the volatile petroleum solvent, which is pressurized by carbon dioxide, nitrogen or nitrous oxide. When the solvent is released it evaporates almost instantly, leaving the micro-sized droplets of toxicant suspended in air. I should point out that aerosols commonly produce droplets well below 10 microns in diameter, which are respirable, meaning that they will be absorbed by alveolar tissue in the lungs rather than impinging in the bronchioles, as do larger droplets. Consequently, aerosols of every type should be handled with discretion and breathed as little as possible, including hair spray aerosols!

Fertilizer Combinations

Fertilizer-pesticide combinations are fairly common formulations to the home gardener who has purchased a lawn or turf fertilizer containing a herbicide for crabgrass control, insecticide for grubs

..., and/or a fungicide for numerous ...seases.

Granulars

Granular pesticides (G) overcome the disadvantages of dusts in their handling characteristics. The granules are small pellets formed from inert clays and sprayed with a solution of the toxicant to give the desired content. After the solvent has evaporated, the granules are packaged for use. Granular materials range in size from 20 to 80 mesh, which refers to the number of grids per linear inch of screen through which they will pass. Only insecticides and a few herbicides, are formulated as granules. They range from 1% to 15% active ingredient, including some systemic insecticides as granules available for garden, lawn and ornamentals. Granular materials may be applied at virtually any time of day in winds up to 20 mph without problems of drift, an impossible task with sprays or dusts. They also lend themselves to soil application in the drill at planting time to protect the roots from insects or to introduce a systemic to the roots for transport to above-ground parts in certain garden vegetables, lawns and ornamentals.

Fumigants

Fumigants are a rather loosely defined group of formulations. Plastic insecticide-impregnated pest strips and pet collars of similar materials are really a slow-release formulation, permitting the insecticide to work its way slowly to the surface and volatilize. Moth crystals and moth balls (paradichlorobenzene and naphthalene) are crystalline solids that evaporate slowly at room temperatures, exerting both a repellent as well as an insecticidal effect. (They can also be used in small quantity to keep cats and dogs off of or away from their favorite parking places.) Soil fumigants are used in horticultural nurseries, greenhouses and on high-value cropland to control nematodes, insect larvae and adults and sometimes diseases. Depending on the fumigant, the treated soils may require covering with plastic sheets for several days to retain the volatile chemical, allowing it to exert its maximum effect. The latter are not likely to be used in a home garden situation and are not recommended in this book.

Impregnates

Impregnating materials mentioned here will include only treatment of woolens for mothproofing and timbers against wood-destroying organisms. For several years woolens and occasionally leather garments have been mothproofed in the final stage of dry-cleaning (using chlorinated solvents). The last solvent rinse contains a weak concentration of the insecticides Permethrin or resmethrin which have long residual qualities against moths and leather-eating beetle larvae. Railroad ties, telephone and light poles, fence posts and other wooden objects that have close contact with or are actually buried in the ground, soon begin to deteriorate as a result of attacks from fungal decay microorganisms and insects, particularly termites, unless treated with fungicides and insecticides. Such treatments permit poles to stand for 40 to 60 years, that would otherwise have been replaced in 5 to 10 years. The insecticides of choice for wood exposed to potential termite damage were dieldrin and chlordane. However, because EPA has canceled the use of these as termiticides, less effective materials, mostly pyrethroids, are now used.

Baits

Several baits can be purchased for home and garden. These include baits for snails and slugs, ants, wasps and hornets, crickets and cockroaches and rats and mice. They contain a toxicant incorporated into materials that are relished by the target pests. Here is another example where spot application, placing the bait in selected places accessible only to the target species, permits the use of very small quantities of often times highly toxic materials in a totally safe manner, with no environmental disruption.

Slow-Release

Slow-release insecticides are recently available to the homeowner and few in number. They involve the incorporation of the toxicant in a permeable covering, which permits its escape at a reduced, but effective rate. Several insecticides for pest control operator use have been microencapsulated into tiny plastic spheres. These are sprayed into areas where insects crawl or hide. The insecticide escapes through the sphere wall over an extended period, preserving its effectiveness much longer than if applied as an emulsion spray. Two of these products are Penncapthrin® (permethrin) and Dursban ME®.

A new insecticidal adhesive tape works essentially by the same principle but is perhaps related more to the plastic strips mentioned under fumigants. The adhesive back is exposed by removing a protective strip and the tape is attached beneath counters, under shelves and in other protected places. This new formulation of propoxur has become available to homeowners as Hercon Insectape®.

In closing this section of pesticides, it might appear that there are no limits to the different forms in which a pesticide can be prepared. This is almost the case. While reading the last entry on slow-release formulations, your imagination may have been stimulated to think about formulations of the future. If not through economy, then by efforts of the Environmental Protection Agency, we will learn to formulate and apply pesticides in extremely conservative ways, to preserve our health, our resources and the environment that at times appears to be in some jeopardy. In summary, pesticides are formulated to improve their properties of handling, application, effectiveness, safety and storage.

INSECTICIDES

As the facts fall into place, man has been on earth somewhere between 1 and 2 million years, a figure that most of us have difficulty in conceiving. In contrast, however, try to visualize 250,000,000 years, the period insects are known to have existed. Despite their head start, man has been able to carve his niche and to forge a path through their devastations. He has learned to live and to compete with them. There is no way to determine when insecticides became a tool, but we can guess that the first materials used by our primitive ancestors were mud and dust spread over their skin to repel biting and tickling insects, a habit resembling those of water buffalo, pigs and elephants. In fact, we could speculate that their first use was around the home tree or cave, making them domestic or urban repellents. Fictitious as it may seem, this may have been the origin of urban pest control.

It was not until 1690 that the first truly insecticidal material was used—water extracts of tobacco sprinkled on garden plants to kill sucking insects. Then, about 1800, a louse powder was being used in the Napoleonic Wars which was made by grinding the flower heads of a chrysanthemum. You know the active ingredients as pyrethrins. These naturally occurring or botanical insecticides were the first real step forward in man's perpetual war against insects.

A complete listing of insecticides is presented in Appendix B.

Botanicals

The botanical insecticides are of great interest to many gardeners, especially organic gardeners because they are "natural" insecticides, toxicants derived from plants. Historically, the plant materials have been in use longer than any other group of insecticides with the possible exception of sulfur. Tobacco, pyrethrum, derris, hellebore, quassia, camphor and turpentine were some of the more important plant products in use before the organized search for synthetic insecticides had begun.

Some of the most widely used insecticides have come from plants. The flowers, leaves, or roots have been finely ground and used in this form, or the toxic ingredients have been extracted and used alone or in mixtures with other toxicants.

There are 10 natural or botanically-derived insecticides that will be of interest to gardeners in general but especially to the organic gardener. Limonene (or *d*-limonene), neem, pyrethrins, rotenone, sabadilla, ryania, cinnamaldehyde, eugenol, jojoba oil and Rosemary oil. All except limonene and neem are exempt from the requirement of a tolerance when applied to growing fruit and vegetables. That is, they can be eaten anytime after application, but these botanical insecticides must be used according to label directions. (See Appendix F for pesticides that can be made at home).

Limonene. Limonene is the latest addition to the botanical insecticides. Extracted from citrus peels, it is effective against all external pests of pets, including fleas, lice, mites and ticks and is virtually non-toxic to warm blooded animals. Several insecticidal substances occur in citrus oil, but the most important appears to be limonene, which constitutes about 98% of orange peel oil by weight.

The insecticidal quality of citrus has been known for a long time. The juice of lemons was used as a remedy for mosquitoes centuries ago on Sir Francis Drake's third voyage to the New World (1572). In 1915, research narrowed the site of insecticidal activity to material from the oil of citrus peel. Not yet identified, it was toxic to the eggs and larvae of the

Mediterranean fruit fly. It is available as ready-to-use sprays, aerosols, shampoos and dips for pets.

Neem. Oil extracts from seeds of the neem tree *(Azadirachta indica)* contain the active ingredient azadirachtin, a nortriterpenoid belonging to the lemonoids. Azadirachtin, a light green powder with a garlic-like odor, has shown some rather sensational insecticidal, fungicidal and bactericidal properties, including insect growth regulator qualities. It disrupts molting by inhibiting biosynthesis or metabolism of ecdysone, the juvenile molting hormone. Azatin® is marketed as an insect growth regulator and Amazin® or Ornazin® as a stomach/contact insecticide for greenhouse and ornamentals.

Pyrethrum. Pyrethrum is extracted from the flowers of a chrysanthemum grown in Kenya, Africa and Ecuador, South America. It has an Oral LD_{50} of approximately 1,500 mg/kg and is one of the oldest household insecticides available (LD_{50} means a dose lethal to 50% of the organisms treated, usually the white laboratory rat and the higher the figure the safer the pesticide). (See p. 266 for more details regarding the meaning of LD_{50}) The ground, dried flower heads were used back in the 19th century as the original louse powder to control body lice in the Napoleonic Wars. Pyrethrins act on insects with phenomenal speed causing immediate paralysis, thus its popularity in fast knock-down household aerosol sprays. However, unless it is formulated with one of the synergists, most of the paralyzed insects recover to once again become pests. Pyrethrins are formulated as household sprays and aerosols and are available as spray concentrates and dusts for use on vegetables, fruit trees, ornamental shrubs and flowering plants at any stage of growth. Vegetables and fruit sprayed or dusted with pyrethrins may be harvested or eaten immediately. In other words, there is no waiting interval required between application and harvest of the food crop.

Because of its general safety to man and his domestic animals and its effectiveness against practically every known crawling and flying insect pest, pyrethrins have a greater range of uses than any other insecticide, numbering literally in the thousands!

Rotenone. Rotenoids, the rotenone-related materials, have been used as crop insecticides since 1848, when they were applied to plants to control leaf-eating caterpillars. However, they have been used for centuries before that (at least since 1649) in South America to paralyze fish, causing them to surface.

Rotenoids are produced in the roots of two genera of the legume (bean) family, *Derris* grown in Malaya and the East Indies and *Lonchocarpus* (also called cubé) grown in South America. Rotenone has an Oral LD_{50} of approximately 350 mg/kg and has been used for generations as the ideal general garden insecticide. It is harmless to plants, highly toxic to fish and many insects, especially caterpillars, moderately toxic to warm-blooded animals and leaves no harmful residues on vegetables. There is no waiting interval between application and harvest of a food crop.

It is both a contact as well as a stomach poison to insects and is sold as spray concentrates and ready-to-use dust. It kills insects slowly, but causes them to stop their feeding almost immediately. Like all the other botanical insecticides its life in the sun is short, 1 to 3 days. It is useful against caterpillars, aphids, beetles, true bugs, leafhoppers, thrips, spider mites, ants, rose slugs, whiteflies, sawflies, bagworms, armyworms, cutworms, leafrollers, midges and a host of other pests. Rotenone is not for use in the home against household pests.

Rotenone is the most useful piscicide (fish control chemical) available for reclaiming lakes for game fishing. It eliminates all fish, closing the lake to reintroduction of rough species. After treatment, the lake can be restocked with the desired species. Rotenone is a selective piscicide in that it kills all fish at dosages that are relatively non-toxic to fish food organisms. It also breaks down quickly leaving no residues harmful to fish used for restocking. The recommended rate is 0.5 part of rotenone to one million parts of water (ppm), or 1.36 pounds per acre-foot of water.

Sabadilla. Sabadilla is extracted from the seeds of a member of the lily family. Its Oral LD_{50} is approximately 5,000 mg/kg, making it the least toxic to warm blooded animals of the six botanical insecticides discussed. It acts as both a contact and stomach poison for insects. It is irritating to the eyes of humans and causes violent sneezing in some sensitive individuals. It deteriorates rapidly in sunlight and can be used safely on food crops with no waiting interval required by the Environmental Protection Agency. Sabadilla will probably be the most difficult of the 10 botanical insecticides to purchase, simply because there was a period of about 15 years during which there was hardly any demand for it.

Sabadilla controls pests on most commonly grown vegetables – – caterpillars, grasshoppers, beetles, leafhoppers, thrips, chinch bugs, stink bugs, harlequin and squash bugs, other true bugs and potato psyllids. It is not very useful against aphids and will not control spider mites. For commercial agriculture, sabadilla is registered only for the control of citrus thrips.

Ryania. Ryania is another botanical or plant-derived insecticide that is quite safe for man and his animals—so safe, that no waiting is required between the time of application to food crops and harvest, as there is for most other insecticides. Ryania is made from the ground roots of the ryania shrub grown in Trinidad and like nicotine, belongs to the chemical class of alkaloids. It has an Oral LD_{50} of approximately 750 mg/kg. It is a slow-acting insecticide, requiring as much as 24 hours to kill. Insects exposed to ryania usually stop their feeding almost immediately, making it particularly useful for caterpillars.

The preferred uses for ryania are against fruit and foliage-eating caterpillars on fruit trees, especially the codling moth on apple trees. However, it is useful against almost all plant-feeding insects, making it an ideal material for small orchards of deciduous fruits. It is not effective against spider mites.

Ryania is exempt from a waiting period between application and harvest and is registered by the EPA for use on citrus, apples, pears, walnuts and corn to control citrus thrips, codling moth and the European corn borer. It is an ideal vegetable garden insecticide for the control of aphids, cabbage loopers, Colorado potato beetle, corn borers, cucumber beetles, diamond back moth, flea beetles, leafhoppers, Mexican bean beetle, spittle bugs and tomato hornworms. Though not very effective, ryania can be used in the home to control ants, silverfish, cockroaches, spiders and crickets.

On ornamentals it will control aphids and lace bugs; aphids, Japanese beetle, thrips and whiteflies on roses; aphids, raspberry fruitworm and sawfly on brambles and aphids (except the woolly aphid) and Japanese beetle on grapes.

Ryania may be difficult to obtain. See Appendix G for sources.

Cinnamaldehyde. Cinnamaldehyde, sold as Cinnamite® and Valero®, is derived from Ceylon and Chinese cinnamon oils. It is used in greenhouses and on horticultural crops as an insecticide and fungicide for aphids, mites and powdery mildew. It attracts corn rootworms and their adult beetles and is known to repel dogs, cats and most other mammals.

Eugenol. Eugenol, or oil of cloves, is sold as Matran 33® and used as an insecticide on many crops and ornamentals. It is the major attractant used in Japanese beetle traps.

Jojoba oil. Jojoba oil is extracted from beans of the southwestern desert jojoba plant and sold as Detur®, which kills or repels whiteflies on all crops and is also fungicidal. Eco E-Rase® controls powdery mildew on grapes and ornamentals.

Rosemary oil. Rosemary oil, (Hexacide® and EcoTrol®) is both insecticidal and miticidal and also has some antifungal action. It is used on several vegetables, ornamentals, fruits and nuts.

Nicotine and nicotine sulfate. These products have been used literally for centuries, but are extremely toxic to humans and our domestic animals. Consequently, nicotine and nicotine sulfate are no longer registered for use by the EPA. Replacing nicotine are the synthetic nicotinoids. (See *Nicotinoids* under the SYNTHETICS section.)

In closing, let me point out that these botanical insecticides are chemicals and no safer than most of the currently available synthetic insecticides. Their only distinction is that they are produced by plants and extracted for use by man. (See Appendix F for list of pesticides that can be made at home).

Soaps

Soap dilutions have been used for control of soft-bodied plant pests, such as aphids, spider mites and mealy bugs, since 1787, when this control method first appeared in writing. Undoubtedly, soap had been used long before that. Most often these were derived from either plant oils (cottonseed, olive, palm, or coconut) or from animal fat, e.g. lard, whale oil, or fish oil. The principal value of soap lies in its capacity to disrupt the cuticle and break down cell membranes resulting in rapid death of insects and mites. With its surface tension much reduced, water readily penetrates insect spiracles, reducing oxygen availability. Thus, a part of soap's mode of action is the "drowning" of exposed insects.

Soaps are the alkali salts of fatty acids. Both soft

soaps (potassium salts) and the hard soaps (sodium salts) are water soluble, with the soft soaps being the most effective. Overall, the most efficacious fatty acid salts are those near the carbon chain length of lauric acid (C_{12}), which include caproic (C_{10}), myristic (C_{14}), palmitic (C_{16}) and stearic acids (C_{18}). Thus the molecular formula of lauric acid potassium soap (potassium laurate) would be:

$$C_{11}\ H_{23}COOK$$

Only since the mid 1990's have the soft soaps come into vogue. Some of the insecticidal soaps are M-Pede® (Dow AgroSciences), Concern® (St. Gabriel Labs), K-Neem® (Peaceful Valley Supply), while the herbicidal soaps are Scythe® (Dow AgroSciences), QuiK Weed® (Monterey Lawn & Garden Products) and WeedAside® (Gardens Alive!).

Synthetics

ORGANOCHLORINES

As recently as 1940, our insecticide supply was limited to lead or calcium arsenate, petroleum oils, nicotine, pyrethrum, rotenone, sulfur, hydrogen cyanide gas and cryolite. World War II opened the Chemical Era with the introduction of a totally new concept of insecticide control chemicals—synthetic organic insecticides, the first of which was DDT.

The advantages of the synthetic insecticides over the botanical or natural insecticides were efficacy, cost and persistence. Much less of the synthetic materials was needed to kill, the cost was usually much less and their effectiveness lasted longer, sometimes as much as months. These synthetics were so easy to use and so economical that they soon replaced the older materials.

The organochlorine insecticides should be familiar because of the notoriety given them in recent years by the press. The organochlorines are insecticides that contain carbon (thus the name organo-), chlorine and hydrogen. They are also referred to by other names. "chlorinated hydrocarbons," "chlorinated organics," "chlorinated insecticides," "chlorinated synthetics," and perhaps others. They included chlordane, heptachlor, lindane, aldrin, endrin, mirex, TDE, dicofol (Kelthane®), methoxychlor, endosulfan, (Thiodan®), DDT and dieldrin.

DDT and Relatives. DDT was first known chemically as dichloro diphenyl trichloroethane, thus ddt or DDT. It is available only as an insecticide for louse control on humans through a physician's prescription.

DDT is a very persistent insecticide, which is one reason that its registrations were cancelled by EPA. Persistence, as used here, implies a chemical stability giving the products long lives in soil and aquatic environments and in animal and plant tissues. They are not readily broken down by microorganisms, enzymes, heat or ultraviolet light. From the insecticidal viewpoint these are good characteristics. From the environmental viewpoint they are not. Using these qualities, the other relatives of DDT are considered nonpersistent.

How does DDT kill? The mode of action, or type of biological activity, has never been clearly worked out for DDT or any of its relatives. It does affect the neurons or nerve fibers in a way that prevents normal transmission of nerve impulses, both in insects and mammals. Eventually the neurons fire impulses spontaneously, causing the muscles to twitch; this may lead to convulsions and death. There are several valid theories for DDT's mode of action, but none have been clearly proved.

DDT is relatively stable to the ultraviolet of sunlight, not readily broken down by microorganisms in the soil or elsewhere, stable to heat, acids and unyielding in the presence of almost all enzymes. In other words, it is poorly biodegradable. Next, it has practically zero water solubility. DDT has been reported in the chemical literature to be probably the most water insoluble compound ever synthesized. Its water solubility is actually somewhere in the neighborhood of 6 parts per billion parts of water (ppb). On the other hand, it is quite soluble in fatty tissue and as a consequence of its resistance to metabolism, it is readily stored in fatty tissue of any animal ingesting DDT alone or DDT dissolved in the food it eats, even when it is part of another animal.

If it is not readily metabolized and thus not excreted and if it is freely stored in body fat, it should come as no surprise that it accumulates in every animal that preys on other animals. It also accumulates in animals that eat plant tissue bearing even traces of DDT. Here we aim at the dairy and beef cow. The dairy cow excretes (or secretes) a large share of the ingested DDT in its milk fat. Man drinks milk and eats the fatted calf. Guess where the DDT is now.

The same story is repeated time and again in food chains ending in the osprey, falcon, golden eagle, sea gull, pelican and so on.

The explanation of these food chain oddities is this. Any chemical that possesses the characteristics of stability and fat solubility will follow the same biological magnification (condensed to biomagnification) as DDT. Other insecticides incriminated to some extent in biomagnification, belonging to the organochlorine group are TDE, DDE (a major metabolite of DDT), dieldrin, aldrin, several isomers of BHC, endrin, heptachlor and mirex. Of course, they all possess these two crucial prerequisites.

Lindane. HCH (BHC), Hexachlorocyclohexane, was first discovered in 1825. But like DDT, was not known to have insecticidal properties until 1940, when French and British entomologists found the material to be active against all insects tested.

Lindane has a higher vapor action than most insecticides and is recommended in instances where the vaporized insecticide can reach the insect, such as with borers in the trunks of fruit trees and woody ornamentals.

Since the Gamma isomer was the only active ingredient, methods were developed to manufacture a product containing 99% Gamma isomer, Lindane, which was effective against most insects, but also quite expensive, making it impractical for crop use.

Lindane is available as a pharmaceutical for the control of human lice through a physician's prescription and for selected uses such as seed treatment.

Cyclodienes. The cyclodienes were a prominent and extremely useful group of insecticides, also known as the diene-organochlorine insecticides.

They were developed after World War II and are, therefore, of more recent origin than DDT (1939) and HCH (1940). The 3 compounds listed below were first described in the scientific literature or patented in the year indicated. chlordane, 1945; dieldrin, 1948; and endosulfan (Thiodan®), 1956.

The cyclodienes are persistent insecticides, except endosulfan, and are stable in soil and relatively stable to the ultraviolet action of sunlight. Consequently, chlordane and dieldrin were used only as soil insecticides for the control of termites. Because of their persistence, the use of cyclodienes on crops was restricted; undesirable residues remained beyond the time for harvest. To suggest the effectiveness of cyclodienes as termite control agents, structures treated with chlordane and dieldrin in the year of their development are still protected from damage. This is 60 and 57 years, respectively. It would be elementary to say that these insecticides are the most effective, long-lasting, economical and safest termite control agents known. Unfortunately, these superb insecticides have been phased out as wood protectants and replaced with less effective organophosphate and pyrethroid insecticides.

All of the organochlorine insecticides lost their registrations in the 1970's and are no longer used in the U.S., except lindane.

ORGANOPHOSPHATES

The chemically unstable organophosphate (OP) insecticides replaced the persistent organochlorine compounds. This is especially true with regard to their use around the home and garden.

The OPs have several commonly used names, any of which are correct. Organic phosphates, phosphorus insecticides, nerve gas relatives, phosphates, phosphate insecticides and phosphorus esters. They are all derived from phosphoric acid and are generally the most toxic of all pesticides to vertebrate animals, except those recommended for the home, which are quite safe to use.

The OPs have two distinctive features. First, they are generally more toxic to vertebrates than most of the other insecticides and second, they are chemically unstable or nonpersistent and break down rather quickly and easily. It is this latter quality that brings them into general agricultural use as substitutes for the persistent organochlorines.

Malathion, one of the oldest and safest organophosphate insecticides, appeared in 1949. It is highly recommended and commonly used in and around the home with little or no hazard either to man or his pets. It controls a wide variety of pests including aphids, spider mites, scale insects, mosquitoes, house flies and a broad spectrum of other sucking and chewing insects attacking fruit, vegetables, ornamentals and stored products. It is safe to use on most pets as a dust. Malathion is sold to the homeowner as emulsifiable concentrates, wettable powders and dusts.

Phosmet (Imidan®, Prolate®) is used on fruit trees for the control of curculio, leafroller, cherry fruit flies and fruitworm, rose chafer, leafhoppers, codling moth, aphids and oystershell scale. It appears in many state home orchard recommendations, but is not recommended for the vegetable garden.

Trichlorfon (Dylox®) is a chlorinated OP, which has been useful for crop pest control and fly control around barns and other farm buildings. It is registered for several vegetables and ornamentals but will not be found often as a recommended insecticide in this book.

Naled (Dibrom®) is an organophosphate that contains bromine in addition to chlorine, making it somewhat unique. Even though it is registered with the EPA for several vegetable crops, it is recommended only for use on ornamentals in this book. It has a broad spectrum of activity, much as malathion does and can be particularly useful in fly control around barns, stables, poultry houses and kennels.

One of the old standbys is diazinon which appeared first in 1952. With the exception of DDT, more diazinon has been used in and around homes than any other insecticide. It is no longer registered for use.

Chlorpyrifos (Dursban®) was the most frequently used insecticide by pest control operators in homes and restaurants for controlling cockroaches and other household insects. Chlorpyrifos is no longer registered for homeowner use.

Dichlorvos (Vapona®, DDVP) is not only a contact and stomach poison but also acts as a fumigant. It has been incorporated into vinyl plastic pet collars and pest strips from which it escapes slowly as a fumigant. It may last up to several months and is very effective for insect control in the home and other closed areas.

Isofenphos (Oftanol®) is registered only for turf and non-crop areas, ornamentals and nursery stock and not for use in the home garden or orchard. Around the home, it is most useful as a turf insecticide, in the control of chinch bug, flea beetles, mole crickets, sod webworm, white grubs, wireworms and other soil insects. The only formulation available will probably be the 1.5% granules.

Isazofos is both an insecticide and nematicide, formulated as 1% and 2% granules. It is registered for and used around the home, only on turf, for the control of chinch bugs, cutworms, sod webworms, Japanese beetles, leafhoppers, mole crickets, nematodes, white grubs and other soil insects.

Systemics. Systemic insecticides are those that are taken into the roots of plants and translocated to the above ground parts, where they are toxic to any sucking insects feeding on the plant juices. Normally caterpillars and other plant tissue-feeding insects are not controlled, because they do not ingest enough of the systemic-containing juices to be affected.

Contained among the several plant systemics are dimethoate (Roxion®) and acephate (Orthene®, Pinpoint®) both of which can be used safely by the homeowner and are the only ones recommended in this book.

Dimethoate is used as a residual wall spray in farm buildings for fly control and for control of a wide range of mites and insects on ornamentals, vegetables, apples and pears, citrus and melons.

Disulfoton (Di-Syston®) is a systemic insecticide/ miticide. It can give up to 5-6 weeks of control from seed treatment or application in the seed furrow or as a side dressing, controlling many species of insects and mites and especially sucking insects, such as aphids only for ornamentals.

Acephate (Orthene®) is one of the systemic insecticides available for home use, though it also acts as a contact insecticide. It has moderate persistence with 10 to 15 days of systemic activity. Around the home it controls aphids, cabbage loopers, bagworms, tent caterpillars, gypsy moths, lace bugs, leaf miners and rollers, leafhoppers, thrips and webworms. It has a few garden vegetable uses and more on ornamentals. Check the label.

ORGANOSULFURS

The organosulfurs, as the name suggests, are synthetic organic insecticides built around sulfur. Dusting sulfur alone is a good acaricide (miticide), particularly in hot weather. The organosulfurs, however, are far superior, requiring much less material to achieve control. Of greater interest, however, is that the organosulfurs, even though toxic to mites, have very low toxicity to insects. As a result they are normally used only for mite control.

This group has one other valuable property. They are usually ovicidal (kill the eggs) as well as being toxic to the young and adult mites.

Propargite (Omite®, Comite®, Ornamite®) is an established acaricide, effective against a broad range of mites on a variety of crop hosts. It can be used on most fruit trees, garden vegetables and ornamentals with complete safety. It is phytotoxic to a few plants, leaving blemishes on the leaves and sometimes fruit. The label should be checked for these sensitive plants before using.

Oxythioquinox (Morestan®) is an amazing chemical. It has the distinction of being all things to all pests—an insecticide, acaricide, fungicide and ovicide! It is found frequently as a recommended material for control of pests on ornamentals, but not for the vegetable garden. It gives residual control of mites, mite eggs, powdery mildew and pear psylla. It is registered only for citrus and ornamentals.

CARBAMATES

Considering that the organophosphate insecticides are derivatives of phosphoric acid, then the carbamates must be derivatives of carbamic acid. The carbamates kill insects in ways similar to the organophosphates.

Carbaryl (Sevin®), the first successful carbamate, was introduced in 1956. More of it has been used world over than all the remaining carbamates combined. Two distinctive qualities have resulted in its popularity. Very low mammalian oral and dermal toxicity and a rather broad spectrum of insect control. This has led to its wide use as a lawn and garden insecticide.

Carbaryl is registered for insect control on more than 100 different crops. These include vegetables, citrus, deciduous fruit, forage crops, field crops, forests, lawn and turf, nuts, ornamentals, rangeland, shade trees, poultry and most pets.

Another carbamate, propoxur (Baygon®), is highly effective against cockroaches that have developed resistance to the organophosphates. Propoxur is used by most structural pest control operators for roaches and other household insects in restaurants, kitchens and homes. It is also formulated as bottled sprays for home use.

Bendiocarb (Garvox®, Multamat®) is a carbamate insecticide used mostly for soil insects, but is also effective for ornamentals and some turf pests. It cannot be used for fruit and vegetable pest control.

PYRETHROIDS

Pyrethrum was almost never used for agriculture because of its cost and instability in sunlight. Beginning in the 1980's, however, several synthetic pyrethrin-like materials have become available for agriculture and can now be used around the home. They are referred to as synthetic pyrethroids or simply pyrethroids. They are fairly stable to sunlight and are generally effective at very low rates. Examples of pyrethroids recommended for use on livestock and pets are permethrin, cypermethrin and resmethrin as sprays and ear tags. In other instances tetramethrin, phenothrin, allethrin, deltamethrin, bifenthrin and cyfluthrin are recommended for household pest control and on ornamentals. Most are quite safe to use around pets and humans and are effective across a broad range of insect and mite pests. Check the label.

NICOTINOIDS

A new class of insecticides with a new mode of action are the nicotinoids. Earlier names given them were nitro-quanidines, neonicotinyls and chloronicotinyls. Nicotinoids are similar to and modeled after the natural nicotine, just as the synthetic pyrethroids are similar to and modeled after the natural pyrethrins. Imidicloprid was the first of the nicotinoids and marketed worldwide under several proprietary names (Admire®, Confidor®, Gaucho®, Merit®, Premier® and Provado®). Surprisingly, it likely is used in the greatest volume worldwide of all insecticides.

Imidacloprid is systemic, having good root-systemic qualities and outstanding contact and stomach action. It is used as a soil, seed or foliar treatment, on fruits and vegetables and on turf, for the control of sucking insects. It also offers promise as a termiticide.

Other recent nicotinoids are acetamiprid (Assail®, Intruder®, TriStar® and Spear®), thiamethoxam (Actara®, Cruiser®, Platinum®) and nitenpyram (Bestguard®). Currently, imidacloprid is the only one of this new group available to the homeowner/gardener.

ANTIBIOTICS

Antibiotics are usually identified with penicillin, tetracycline and chloramphenicol, used against bacterial diseases of man and domestic animals. However, there is a group of compounds known as avermectins, insecticidal, acaricidal and anthelminthic (for control of internal parasites) agents that have been isolated from the fermentation products of *Streptomyces avermitilis*, of the Actinomycete family.

Abamectin is the common name assigned to Avermectin®, a mixture of avermectins B1a and B1b. Avid® and Agri-Mek® are used as a miticide and insecticide and Zephyr® in various cockroach ant baits. Abamectin's greatest promise is in the control of the citrus rust mite and the two spotted mite and leafminers on greenhouse crops and some field vegetables, e.g. celery, tomatoes. Abamectin has certain local systemic qualities, permitting it to kill mites on a leaf's underside when only the upper surface was treated and leaf miners that feed between the upper and lower surfaces.

SYNERGISTS OR ACTIVATORS

Synergists are not in themselves considered toxic or insecticidal but are materials used with insecticides to synergize or enhance the activity of the insecticides. They are added to certain insecticides in the ratio of 8:1 or 10:1, synergist. insecticide. The first synergist was introduced in 1940 to increase the effectiveness of the plant-derived insecticide, pyrethrin. Since then many materials have been introduced, but only a few have survived because of cost and ineffectiveness. Synergists are found in practically all of the "bug-bomb" aerosols to enhance the action of the fast knock-down insecticides, pyrethrin and pyrethroids, against flying insects.

The synergists are usually used in spray prepared for the home and garden, stored grain and on livestock, particularly in dairy barns. Synergists and the insecticides they synergize, such as natural pyrethrins, are quite expensive, thus seldom if ever used on crops. The new synthetic pyrethrins (or pyrethroids) however, are stable to sunlight and weather and are being used on field crops and some vegetables.

The only synergist commonly used today in sprays and aerosols is piperonyl butoxide.

Synergists are commonly found in aerosols and ready-to-use sprays for ornamental plants outside as well as inside and in aerosols intended for household pests that contain pyrethrum and the pyrethroids as the active ingredient(s).

MICROBIALS

The insect disease pathogens, known as microbial insecticides, are discussed in Chapter 4, BIORATIONALS.

INSECT REPELLENTS

Historically repellents included smokes, plants hung in dwellings or rubbed on the skin as the fresh plant or brews, oils, pitches, tars and various earths applied to the body. Camel urine sprinkled on the clothing has been of value, though questionable, in certain locales. Camphor crystals sprinkled among woolens has been used for decades to repel clothes moths. Prior to man's more edified approach to insect olefaction and behavior it was assumed that if a substance was repugnant to man it would likewise be repellent to annoying insects.

Prior to World War II there were only 4 principal repellents. (1) oil of citronella, discovered in 1901, used also as a hair-dressing fragrance by certain Eastern cultures; (2) dimethyl phthalate, discovered in 1929; (3) Indalone, introduced in 1937 and (4) Rutgers 6-12, which became available in 1939.

Beginning with World War II and the introduction of American military personnel into the tropics, it became necessary to find new repellents which would survive both time and dilution by perspiration. The ideal repellent would be nontoxic and nonirritating to man, nonplasticizing and long lasting (12 hours) against mosquitoes, biting flies, ticks, fleas and chiggers.

Insect repellents come in every conceivable formulation—undiluted, diluted in a cosmetic solvent with added fragrance, aerosols, creams, lotions, treated cloths to be rubbed on the skin, grease sticks, powders, suntan oils and clothes-impregnating laundry emulsions. Despite the formulation, their periods of protection vary with the chemical, the individual, the general environment, insect species and avidity of the insect.

What happens to repellents once applied? Why are they not effective longer than 1 or 2 hours? No single answer is satisfactory but generally they evaporate, are absorbed by the skin, are lost by abrasion of clothing or other surfaces and diluted by perspiration. Usually it is a combination of them all and the only solution is a fresh application.

Deet (Delphene®, Off®) and dimethyl phthalate are those most commonly found in today's repellents. Of these, deet is by far superior to all others against biting flies and mosquitoes.

FUMIGANTS

The fumigants are small, volatile, organic molecules that become gases at temperatures above 40°F. They are usually heavier than air and commonly contain one or more of the halogens (chlorine, bromine or fluorine). Fumigants are fat-soluble and their effects are reversible.

Fat solubility appears to be an important factor in the action of fumigants, since these narcotics lodge in fat-containing tissues, which are found in the nervous system. Only two materials are recommended for home use, naphthalene and paradichlorobenzene (PDB). You know them as moth balls and moth crystals.

Dichlorvos, mentioned in the organophosphates, acts as a fumigant when used as No-Pest® strips in enclosed spaces.

INORGANICS

Inorganic insecticides are those that do not contain carbon. Usually they are white and crystalline resembling the salts. They are stable chemicals, do not evaporate and are frequently soluble in water.

Sulfur. Sulfur dusts are especially toxic to mites, such as chiggers and spider mites of every variety, thrips, newly-hatched scale insects and as a stomach poison for some caterpillars. Sulfur dusts and sprays, made from wettable sulfur, are fungicidal, especially against powdery mildews.

Lime-sulfur is an old dormant-spray material, used on fruit trees for fungus diseases, thrips and various mites. Lime-sulfur solutions can be made at home by cooking together sulfur and unslaked lime (quicklime) in water. It has a disagreeable odor and is caustic to the skin. As with any other insecticide, it can be used safely and effectively if handled with the usual element of caution and common sense. (See Recipes for Home-made Insecticides in Appendix F).

Cryolite is a naturally occurring mineral that is mined in its pure form and is used as a stomach poison insecticide for chewing insects, especially caterpillars. Its low toxicity hazard and lack of effect on most predators make this a very useful material for the home garden, orchard and ornamentals. The insecticidal properties of cryolite were discovered in the 1920's and it was used heavily until the mid 1950's when it was displaced by the more effective synthetic organic insecticides. It is registered for use on most garden crops, fruit crops and ornamentals and provides control for several days since it does not evaporate or deteriorate under sunlight. I am acquainted with only two products available to the homeowner, Kryocide® and Prokil® Cryolite. These are available as dusts, wettable powder and aqueous suspensions.

Boric Acid powder used indoors is especially effective against all species of cockroaches and in baits for most species of ants.

A complete listing of insecticides is presented in Appendix B.

FUNGICIDES

Fungicides, strictly speaking, are chemicals used to kill or halt the development of fungi. Most plant diseases can be controlled to some extent with today's fungicides. Among those that are very difficult to control with chemicals are *Phytophthora* and *Rhizoctonia* root rots, *Fusarium, Verticillium* and bacterial wilts and the viruses. The difficulties with these diseases are that they occur either below ground and beyond the reach of fungicides, or they are systemic within the plant. Some of these can be controlled with the new systemic fungicides, discussed later.

Other diseases are controlled by seed treatment with selective materials. Others are controlled with resistant varieties. Diseases of fruit and vegetables are often controlled by spray or dusts of fungicides.

Historically, fungicides have centered around sulfur, copper and mercury compounds and even today, most of our plant diseases could be controlled by these groups. However, the sulfur and copper compounds can retard growth in sensitive plants, and as a result the organic fungicides were developed. These sometimes have greater fungicidal activity and usually have less phytoxicity.

The general-purpose fungicides for home use include inorganic forms of copper, sulfur, a variety of organic compounds and two or three systemic fungicides. The general purpose lawn and garden fungicides are few in number and are organic compounds.

There are 180-190 fungicidal materials in our present arsenal, most of which are recently discovered organic compounds. Most of these act as protectants, preventing spore germination and subsequent fungal penetration of plant tissues. Protectants are applied repeatedly to cover new plant growth and to replenish the fungicide that has deteriorated or has been washed off by rain.

There are two basic kinds of fungicides, the protective fungicides (protectants) and the systemic fungicides.

A complete listing of fungicides is presented in Appendix D.

PROTECTANTS

It is necessary that the protectants be applied to protect plants during stages when they are vulnerable to inoculation by pathogens, before there is any evidence of disease.

The application principle for protectants differs from that of herbicides and insecticides. Only that portion of the plant which has a coating of dust or spray film of fungicide is protected from disease. Thus a good, uniform coverage is essential. They are applied as sprays or dusts, but sprays are preferable since the films stick more readily, remain longer, can be applied during any time of the day and result in less off-target drift.

Protectants can help to control certain diseases after the symptoms appear, referred to as chemotherapeutants. Also, protectants are commonly used even after symptoms of disease have appeared. Eradicant fungicides are usually applied directly to the pathogen during its "overwintering" stage, e.g. Bordeaux mixture on fruit trees, long before disease has begun and symptoms have appeared. In the case of crops, however, if their sale depends on appearance, such as lettuce and celery, then the fungicide must be applied as a protectant spray in advance of the pathogens to prevent the disease.

INORGANICS

Sulfur. Sulfur in several forms is probably the oldest effective fungicide known to man and still a very useful home-garden fungicide. There are three physical forms or formulations of sulfur used as fungicides. The first is finely ground sulfur dust which contains 1% to 5% clay or talc to assist in dusting qualities. The sulfur in this form may be used as a carrier for another fungicide or an insecticide. The second is flotation or colloidal sulfur, which is so very fine that it must be formulated as a wet paste in order to be mixed with water. In its original microparticle size, it would be impossible to mix with water and would merely float. Wettable sulfur is the third form; it is finely ground with a wetting agent so that it mixes readily with water for spraying. The easiest to use, of course, is dusting sulfur when plants are slightly moist with the morning dew.

Some pathogenic organisms and most mites are killed by direct contact with sulfur and also by its fumigant action at temperatures above 70°F. The fumigant effect is, however, somewhat secondary at marginal temperatures and under windy conditions. It is quite effective in controlling powdery mildews of plants that are not unduly sensitive to sulfur. Unlike those of any other fungus, spores of powdery mildews will germinate in the absence of a film of water and penetrate the leaf tissue. Its fumigant effect—acting at a distance— is undoubtedly important in killing spores of powdery mildews. In its fungicidal action sulfur is eventually converted to hydrogen sulfide (H_2S), which is a toxic entity to the cellular proteins of the fungus.

Copper. Most inorganic copper fungicides are practically insoluble in water and are the pretty blue, green, red, or yellow powders sold for plant disease control. Included in these is Bordeaux mixture, named after the Bordeaux region in France where it originated. Bordeaux is a chemically undefined mixture of copper sulfate and hydrated lime, which was accidentally discovered when sprayed on grapes in Bordeaux to scare off "freeloaders". It was soon observed that downy mildew, a disease of grapes, disappeared from the treated plants. From this unique origin began the commercialization of fungicides.

Bordeaux mixture is the oldest recognized fungicide and is readily available to home gardeners in the apple and grape producing districts of the country.

Bordeaux is one of two do-it-yourself fungicides that can be made at home. The other is lime-sulfur and both recipes can be found in Appendix F.

TABLE 4. Some of the inorganic copper compounds used as fungicides.

NAME	USES
Cupric sulfate	Seed treatment and preparation of Bordeaux mixture
Copper oxychloride	Powdery mildews
Copper oxychloride sulfate	Many fungal diseases
Copper zinc chromates	Diseases of potato, tomato, cucurbits, peanuts and citrus.
Cuprous oxide	Powdery mildews
Basic copper sulfate	Seed treatment and preparation of Bordeaux mixture
Cupric carbonate	Many fungal diseases

While primarily a fungicide, Bordeaux is also repellent to many insects, such as flea beetles, leafhoppers and potato psyllid, when sprayed over the leaves of plants.

The copper ion, which becomes available from both the highly soluble and relatively insoluble copper salts, provides the fungicidal as well as phytotoxic and poisonous properties. A few of the many inorganic copper compounds used over the years are presented in Table 4.

Protective fungicides have very low solubilities but, in water, some toxicant does go into solution. That small quantity absorbed by the fungal spore is then replaced in solution from the residue. The spore accumulates the toxic ion and "commits suicide", so to speak. Except for powdery mildews, water permits spore germination and solubilizes the toxic portion of the fungicidal residue.

The copper ion is toxic to all plant cells and must be used in discrete dosages or in relatively insoluble forms to prevent killing all or portions of the host plant. This is the basis for the use of relatively insoluble or "fixed" copper fungicides, which release only very low levels of copper, adequate for fungicidal activity, but not enough to become toxic to the host plant.

Copper compounds are not easily washed from leaves by rain since they are relatively insoluble in water, thus give longer protection against disease than do most of the organic materials. They are relatively safe to use and require no special precautions during spraying. Despite the fact that copper is an essential element for plants, there is some danger in an accumulation of copper in agricultural soils resulting from frequent and prolonged use. In fact, certain citrus growers in Florida have experienced a serious problem of copper toxicity after using fixed copper for disease control for many years.

ORGANICS

A host of synthetic sulfur and other organic fungicides have been developed over the past 50 yeas to replace the more harsh, less selective inorganic materials. Most of them have had no measurable build-up effect on the environment after many years of use. The first of the organic sulfur fungicides was thiram, discovered in 1931. This was followed by many others. Then came other new classes, the dithiocarbamates and dicarboximides (zineb and captan) introduced in 1943 and 1949, respectively. Since then, the organic synthesists have literally opened the doors with now more than 170 fungicides of all classes in use and in various stages of development.

The newer organic fungicides possess several outstanding qualities. They are extremely efficient, that is, smaller quantities are required than of those used in the past; they usually last longer; and they are safer for crops, animals and the environment. Too, most of the newer fungicides have very low phytotoxicity, many showing at least a 10-fold safety factor over the copper materials. Most of them are readily degraded by soil microorganisms, thus preventing their accumulation in soils.

Dithiocarbamates. These are the "old reliables", thiram, nabam, maneb, mancozeb, (Dithane®, Fore®) ferbam, ziram, metiram, metam-sodium, (Vapam®) and zineb, all developed in the early 1930's and 1940's. They probably have greater popularity, including home garden use, than all other fungicides combined. Except for systemic action, they are employed collectively in every use known for fungicides.

Thiram is used as a seed protectant and for certain fungus diseases of apples, peaches, strawberries and tomatoes. As a turf fungicide it is applied for control of large brown patch and dollar spot of turf. It acts as a repellent for rabbit and deer, when used on fruit trees, shrubs, ornamentals and nursery stock.

Maneb is used for the control of early and late blights on potatoes, tomatoes and many other diseases of fruits and vegetables.

The principal uses of ferbam are in the control of apple scab and cedar apple rust and tobacco blue mold. It is also applied as a protective fungicide to other crops against many fungus diseases.

Zineb and nabam were used on a variety of fruits and vegetables, especially on potato seed pieces only and on tomatoes for blight, but are no longer registered for use in the U.S.

Substituted Aromatics. This is a somewhat arbitrary classification assigned to the benzene derivatives that possess long-recognized fungicidal properties. Pentachloro-nitrobenzene (PCNB) (Terraclor®, Turfside®) was introduced in the 1930's as a fungicide for seed treatment and selected foliage applications. Chlorothalonil (Bravo®, Daconil®) first became available in 1964 and has proved to be a very useful, broad-spectrum foliage protectant fungicide. It has wide use in the garden, including beans, cole crops, carrot, sweet corn, cucumber, cantaloupe, musk-melon, honeydew, watermelon, squash, pumpkin, onion, potato, tomato and turf.

Thiadiazoles. The thiazoles contain only the ever-popular etridiazole (Terrazole®, Koban®, Truban®) which is used as a soil fungicide. Terrazole® is recommended as a turf fungicide for the control of Pythium diseases (Pythium blight, cottony blight, grease spot, spot blight and damping off). It is also effective against the seedling disease complex (Pythium, Fusarium and Rhizoctonia) of several garden and field crops. Application can be either by seed treatment or in-furrow at time of planting.

Phthalimides. Two extremely useful foliage protectant fungicides belong to this group. captan appeared in 1949 and was undoubtedly the most heavily used fungicide around the home of all classes; folpet appeared in 1962. They are used primarily as foliage dusts and sprays on fruits, vegetables and ornamentals.

Captan controls scabs, blotches, rots, mildew and other diseases on fruit, vegetables and flowers. It is used also as a dust or slurry seed treatment. Folpet controls apple scab, cherry leaf spot, rose black spot and rose mildew. It is also useful as seed and plant bed treatments. Folpet is no longer registered in the U.S.

Captan is one of the safest of all pesticides available and is recommended for lawn and garden use as seed treatments and as a protectant for mildews, late blight and other diseases. Remember the old garden center adage, "When in doubt use captan."

SYSTEMICS

Systemic fungicides, those that penetrate the plant cuticle and are translocated to growing points, have become available only in recent years and very few are yet available. Most systemic fungicides have eradicant properties which stop the progress of existing infections. They are therapeutic in that they can be used to cure plant diseases. A few of the systemics can be applied as soil treatments and are slowly absorbed through the roots to give prolonged disease control.

Systemics offer much better control of diseases than is possible with a protectant fungicide that requires uniform application and remains essentially where it is sprayed onto the plant surfaces. There is, however, some redistribution of protective fungicidal residues on the surfaces of sprayed or dusted plants giving them longer residual activity than would be expected.

Benzimidazoles. The benzimidazoles, represented by thiophanate-methyl which was introduced in 1967, have received wide acceptance as systemic fungicides against a broad spectrum of diseases. Thiophanate has the widest spectrum of fungitoxic activity of all the newer systemics, including control of Sclerotinia, Botrytis, Rhizoctonia, powdery mildews and apple scab. Thiophanate (Topspin®, Fungo®, Mildothane®), although not a benzimidazole initially, is converted to one by the host plant and the fungus through their metabolism. Thiophanate has been used in foliar applications, seed treatment, dipping of fruit or roots and soil application. It also controls brown patch, Fusarium blight, dollar spot, red thread, stripe smut and powdery mildew on turf.

Thiabendazole (Arbotect®, Mertect®) is effective against a broad spectrum of diseases similar in range to benomyl, e.g., Sclerotinia, Botrytis and Rhizoctonia species and powdery mildews.

Systemic fungicides cover susceptible foliage and flower parts more efficiently than protectant fungicides because of their ability to translocate through the cuticle and across leaves. They bring into play the perfect method of disease control by attacking the pathogen at its site of entry or activity and reduce the risk of contaminating the environment by frequent broadcast fungicidal treatments. Undoubtedly, as new and more selective systemic molecules are synthesized, they will gradually replace the protectants which comprise the bulk of the fungicidal arsenal.

Some, but not all, of the more recent systemic fungicides are now available to the home gardener. Here are a few of those you may want to look for in your local garden supply center.

Oxathiins—These are represented by carboxin (Vitavax®) and oxycarboxin (Plantvax®), the earliest systemic fungicides. Carboxin is used as a seed treatment for the control of smuts, bunts, Rhizoctonia damping-off, loose smut, sore-skin and flag smut. Oxycarboxin is applied to the foliage of carnations and geraniums in greenhouses for the control of rust.

Dicarboximides—Iprodione (Rovral®) and vinclozolin (Ronilan®) are contact fungicides with both preventive and curative effects, acting somewhat like systemics. They are particularly effective against Botrytis, Monilinia and Sclerotinia. Iprodione is also active against Alternaria, Helminthosporium, Rhizoctonia, Corticium, Typhula and Fusarium.

Pyrimidinols—Cyprodinil (Vangard®) is a broad-spectrum systemic used on grapes, pome and stone fruits, strawberries and vegetables. Pyrimethanil (Scala®) is for gray mold on fruit, vegetables, vines and ornamentals and leaf scab on pome fruit.

Phenylamides—Metalaxyl (Metax®, Rampart®) is the only one of this broad group available to the home gardener. It is used as a seed treatment for damping off and root rot, soil-borne diseases caused by Pythium and Phytophthora and the downy mildews, foliar diseases. There are other phenylamide fungicides, but not available to the home gardener.

Triazoles—The systemic triazole fungicides have both protective and curative properties. They are broad spectrum, effective against mildews and rusts on vegetables, cereals, deciduous fruit, grapes and ornamentals. These include triadimefon (Bayleton®), propiconazole (Tilt®, Orbit®), hexaconazole (Anvil®), penconazole (Topas®), tebuconazole (Folicur®), myclobutanil (Systhane®, Rally®) plus several others. Check the label for crops listed.

Piperazines—Triforine (Funginex®) is the only important member of this group. It is amazingly effective against powdery mildew on any host and against rust and scab on apples, anthracnose, mildew and rust on ornamentals, brown rot, rusts, leaf spot and black spot. It is registered for apples, asparagus, cherries, apricots, plums, nectarines, ornamentals, peaches and roses.

Imidazoles—These are related to the triazoles and include imazalil (Freshgard®), fenamidone (Reason®) and cyazofamid (Ranman®). All are active against early and late blights on potato, tomato and downy mildew on vegetables, cucurbits and grapes.

Organophosphates—Surprisingly, there are fungicides classed as organophosphates. Fosetyl (Alliette®), is used for downy mildews and *Phytophthora* on non-bearing citrus, ornamentals, grapes, bush berries and turf. Tolclofos-methyl (Rizolex®) is for seedling diseases on many crops and most fungal diseases of turf.

Morpholines—These are new to the U.S. and include dodemorph (Meltatox®), tridemorph (Calixin®), fenpropimorph (Corbel®) and dimethomorph (Acrobat®, Forum®). Morpholines are all systemic with curative and preventative qualities, for powdery mildew, rust and black spot on greenhouse ornamentals, fruits and vegetables, downy mildews, late blights, root rots for grapes, potatoes and tomatoes.

Strobilurins—Are the newest of the systemics and very different in origin. They are based on natural fungicides produced by certain edible mushrooms and wood-decaying mushrooms. Azoxystrobin (Abound®, Heritage®) and kresoxim-methyl (Stroby®, Cygnus®) are broad spectrum preventive fungicides with systemic and curative properties for foliar and soil-borne diseases including downy and powdery mildews, leaf spots and rusts on a variety of crops. Other strobilurins being developed but not available to the layman are famoxadone, trifloxystrobin, fluoxastrobin, orysastrobin and pyraclostrobin.

Acetamides—There is but one fungicide registered in this group, cymoxanil (Curzate®). It is a preventative, post-infection-curative for potatoes and tomatoes to control downy mildew, deadarm and late blights.

Quinolines—Quinoxyfen (Fortress®) is also the only

registered fungicide in this group. It is foliar applied to cereals, tree crops, grapes and vegetables, for powdery mildew, scab and other diseases.

There are several other very recent fungicide groups, not available to the homeowner, but look quite promising for agriculture. They are listed here only as a matter of information. anilides, benzamids, carboxamids, propionamids, sulfamids and valinamides.

Dinitrophenols—The basic dinitrophenol molecule has a broad range of toxicities. Compounds derived from it are used as herbicides, insecticides, ovicides and fungicides. They act by preventing the utilization of nutritional energy.

One dinitrophenol has been used since the late 1930's, dinocap (Mildane®), both as an acaricide and for powdery mildew on a number of fruit and vegetable crops and ornamentals. Dinocap undoubtedly acts in the vapor phase since it is quite effective against powdery mildews whose spores germinate in the absence of water. This has been a popular home fungicide.

Guanidines—Dodine (Melprex®, Syllit®), a fungicide introduced in the middle 1950's has proved effective in controlling certain diseases such as apple, pear and pecan scab and cherry leaf spot, foliar diseases of strawberries, blossom brown rot on peaches and cherries (Western states only), peach leaf curl (western states only), bacterial leafspot on peaches, leaf blight of sycamore and black walnut. It has disease-specificity and slight systemic qualities and is taken up rapidly by fungal cells. It offers some control of powdery mildew.

ANTIBIOTICS—The antibiotic fungicides are substances produced by microorganisms, which in very dilute concentration inhibit growth and even destroy other microorganisms. The largest source of antifungal antibiotics is the actinomycetes, a group of the lower plants. Within this group is one amazing species, *Streptomyces griseus,* from which are derived streptomycin and cycloheximide.

Streptomycin (Agri-Mycin®, Streptrol®) is the only antibiotic recommended to the home gardener and is used as dusts, sprays and seed treatment. It will control bacterial diseases such as blight on apples and pears, soft rot on leafy vegetables and some seedling diseases. It is also effective against a few fungal diseases.

A complete listing of fungicides is presented in Appendix D.

DISEASE RESISTANCE—Disease resistance to chemicals other than the heavy metals occurs commonly in fungal and on rare occasions in bacterial plant disease pathogens. Several growing seasons after a new fungicide appears, it becomes noticeably less effective against a particular disease. As our fungicides become more specific for selected diseases, we can expect the pathogens to become resistant. This can be attributed to the singular mode of action of a particular fungicide which disrupts only one genetically controlled process in the metabolism of the pathogen. The result is that resistant populations appear suddenly, either by selection of resistant individuals in a population or by a single gene mutation. Generally, the more specific the site and mode of fungicidal action, the greater the likelihood for a pathogen to develop a tolerance to that chemical.

HERBICIDES

The first chemicals utilized in weed control were inorganic compounds. These were brine and a mixture of salt and ashes, both of which were used as early as Biblical times by the Romans to sterilize the land. In 1896, copper sulfate was used selectively to kill weeds in grain fields.

Herbicides, or chemical weed killers, have largely replaced mechanical methods of agricultural and industrial weed control in the past 50 years, especially in intensive and highly mechanized agriculture. Herbicides provide a more effective and economical means of weed control than cultivation, hoeing and hand pulling (except in the home garden). Without the use of herbicides in agriculture, it would have been impossible to mechanize fully the production of cotton, sugar beets, grains, potatoes and corn.

Other locations where herbicides are used extensively include areas such as industrial sites, roadsides, ditch banks, irrigation canals, fence lines, recreational areas, railroad embankments and power lines. Herbicides remove undesirable plants that might cause damage, present fire hazards, or impede work crews. They also reduce costs of labor for mowing.

TABLE 5. Herbicide classification chart.

Herbicide (Common Name)	Selective			Nonselective		
		Foliage	Soil		Foliage	Soil
	Contact	Translocated	Residual	Contact	Translocated	Residual
Benefin			X			
Bensulide			X			
Bentazone	X	X				
Bromoxynil				X		
Cacodylic acid				X		
DCPA (Dacthal®)						X
Dicamba		X	X			
Dichlobenil						X
Diquat				X		
DSMA		X			X	
EPTC			X			
Fenoxaprop-P-ethyl	X	X				
Fluazifop-P-butyl		X				
Glyphosate		X				
MSMA		X			X	
Oryzalin			X			
Oxadiazon	X	X	X			
Petroleun Oils				X		
Sethoxydim		X				
Siduron	X		X			
Simazine				X		X
Trifluralin			X			
2,4-D amine		X				

Herbicides are classifed as selective when they are used to kill weeds without harming the crop and nonselective when the purpose is to kill all vegetation.

Selective and nonselective materials can be applied to weed foliage or to soil containing weed seeds and seedlings depending on the mode of action. True selectivity refers to the capacity of an herbicide, when applied at the proper dosage and time, to be active only against certain species of plants but not against others. But selectivity can also be achieved by placement, as when a nonselective herbicide is applied in such a way that it contacts the weeds but not the crop.

The classification of herbicides would be a simple matter if only the selective and nonselective categories existed. However there are multiple-classification schemes which may be based on selectivity, contact vs. translocated, timing, area covered and chemical classification. One method of classifying the herbicides is presented in Table 5.

Each herbicide affects either by contact or translocation and some by both. Contact herbicides kill the plant parts to which the chemical is applied and are most effective against annuals, those weeds that germinate from seeds and grow to maturity each year. Complete coverage is essential in weed control with contact materials.

The translocated herbicides are absorbed either by roots or above-ground parts of plants and are then moved within the plant system to distant tissues. Translocated herbicides may be effective against all weed types, however their greatest advantage is seen when used to control established perennials, those weeds that continue their growth from year to year. Uniform application is needed for the translocated materials, whereas complete coverage is not required.

Another method of classification is the timing of herbicide application with regard to the stage of crop or weed development. The three categories of timing are preplanting, preemergence and postemergence.

Preplanting applications for control of annual weeds are made to an area before the crop is planted, within a few days or weeks of planting. Preemergence applications are completed prior to emergence of the crop or weeds, depending on definition, after planting. Postemergence applications are made after the crop or weed emerges from the soil.

Broadcast applications cover the entire area, including the crop. Spot treatments are confined to small areas of weeds. Directed sprays are applied to selected weeds or to the soil to avoid contact with the crop.

A complete listing of herbicides is presented in Appendix C.

INORGANICS

Intensive EPA restrictions have been placed on some of the inorganics because of their persistence in soils. The inorganic herbicides are not wise choices for use around the home except with great care when removing all vegetation from an area.

ORGANICS

Arsenicals. Widely used as agricultural herbicides and in some instances on turf, are the organic arsenicals, namely the arsenic and arsonic acid derivatives. Cacodylic acid (dimethylarsinic acid), a derivative of arsenic acid, is used for the seasonal killing of lawns for their conversion to winter or summer lawns. Salts of arsonic acid, disodium methanearsonate (DSMA) and monosodium acid methanearsonate (MSMA), can be used for perennial weed control in Bermuda grass turf. The organic arsenicals are much less toxic to mammals than the inorganic forms, are crystalline solids and are relatively soluble in water.

Arsonates are absorbed and translocated to underground tubers and rhizomes, making them extremely useful against johnsongrass and nut sedges, when applied as spot treatments.

Phenoxys. An organic herbicide introduced in 1944, later to be known as 2,4-D, was the first of the "phenoxy herbicides," "phenoxyacetic acid derivatives," or "hormone" weed killers. These were highly selective for broadleaf weeds and were translocated throughout the plant. 2,4-D provided most of the impetus in the commercial search for other organic herbicides in the 1940's. There are several compounds belonging to this group, of which 2,4-D, 2,4,5-T and silvex are the most familiar.

The phenoxy herbicides cause responses in broadleaf plants resembling those of auxins (growth hormones). They affect cellular division and result in unusual and rapid growth in treated plants.

2,4-D and 2,4,5-T have been used for years in gargantuan volume world wide. 2,4,5-T, used mainly for control of woody perennials, became the subject of heavy investigation, particularly on its use in Vietnam in the form of "Agent Orange". Silvex and 2,4,5-T were banned by the EPA in 1979.

The 2,4-D amine is probably the most useful of all herbicides to the happy lawn owner, in the control of broadleaved lawn weeds.

Diphenyl Ethers. The only member of this group recommended for home garden and orchard is fluazifop-P-butyl (Ornamec®, Fusilade®). It is a selective herbicide for the control of grass weeds in broadleaved crops, ornamentals, vines and remains active in the soil up to four months.

Dinitroanilines. Benefin, oryzalin, pendimethalin and trifluralin belong to this group. Two new additions are prodiamine and ethalfluralin. Benefin is used against annual grasses and broadleaf weeds in established turf. Oryzalin is good for grass weeds in fruit, nut, vine crops and ornamentals and should be watered into soil to reach the weed germination zone. Pendimethalin is good against foxtails, barnyard grass and crabgrass. Trifluralin is applied in granular form, followed by water incorporation, to ornamental trees, flowers and shrubs. Prodiamine is for turf and ornamentals.

Ureas. The substituted ureas are a group of compounds used primarily as selective preemergence herbicides. The only one of this group recommended for use around the home is siduron (Tupersan®). The ureas are strongly absorbed by the soil, then absorbed by roots. They inhibit photosynthesis, the production of plant sugars with the energy from light.

Siduron is used only for the establishment of lawns. It is applied as a preemergence herbicide for the control of annual weed grasses, such as crabgrass, foxtail and barnyard grass, in newly seeded or established planting's of bluegrass, fescue, redtop, smooth brome, perennial ryegrass, orchard grass and certain strains of bentgrass.

Thiocarbamates. The esters of carbamic acid are physiologically quite active. As we have seen, some carbamates are insecticidal, while others are fungicidal. Discovered in 1945, the thiocarbamates are used primarily as selective preemergence herbicides, but some are also effective for post-emergent control.

EPTC belongs to the thiocarbamates and is the only one of this category recommended for use around the home. In this instance, EPTC is recommended only in the granular formulation for nutsedge, johnsongrass and quackgrass control in and around woody ornamentals.

Triazines. The triazines are strong inhibitors of photosynthesis. Their selectivity depends on the ability of tolerant plants to degrade or metabolize the herbicide, whereas the susceptible plants do not. Triazines are applied to the soil primarily for their postemergence activity against broadleaf weeds.

There are many triazines on the market today, only one of which is used around the home. Simazine (Princep®) is used where all vegetation needs to be controlled, for example, around patios, walks and driveways and for non-selective weed control in lawns, orchards and vineyards.

Phthalic Acids. A number of these materials are employed as herbicides and are applied to the soil against germinating seeds and seedlings.

DCPA (Dacthal®) is placed on top of soil before weed seed germination. Widely used for crabgrass control on turf, it is approved for use on turf and ornamentals and is effective against smooth and hairy crabgrass, green and yellow foxtails, fall panicum and other annual grasses in addition to the perennial witchgrass. It is also useful against certain broadleaf weeds such as carpet weed, dodder, purslane and common chickweed.

Nitriles. Nitriles are organic compounds containing a cyanide group. There are several substituted nitrile herbicides that have a wide spectrum of uses against grasses and broadleaf weeds. Their actions are broad involving seedling growth and potato sprout inhibition. They are fast acting which is attributed primarily to rapid permeation.

Dichlobenil (Casoron®) is used for weed control around woody ornamentals, nurseries and in fruit orchards. Bromoxynil (Pardner®, Buctril®) is used on industrial sites, vacant lots, driveways and other areas to be freed from weeds for post-emergent control of weeds such as blue mustard, fiddleneck, corn gromwell, cowcockle, field pennycress, green smartweed, groundsel, lambsquarters, London rocket, shepherd's-purse, silverleaf nightshade, tartary buckwheat, tarweed, tumble mustard, wild buckwheat and wild mustard.

Bipyridyliums. There are two important herbicides in this group, diquat (Reward®) and paraquat. Both are contact herbicides that damage plant tissues quickly, causing the plants to appear frostbitten because of cell membrane destruction. Both materials reduce photosynthesis and are more effective in the light than in the dark. Neither material is active in soils. Diquat is the only one recommended and available to the amateur.

Amides. One of the better turf herbicides, especially for the control of crabgrass, is bensulide (Prefar®). It is an organophosphate but is considered one of the less toxic herbicidal materials. Bensulide acts by inhibiting cell division in root tips. It is used as a preemergence herbicide in lawns to control certain grasses and broadleaf weeds, but it fails as a foliar spray because it is not translocated.

Phosphono Amino Acids. Glyphosate (Roundup®) belongs to this classification, a very popular herbicide. It is useful for the control of many annual and perennial grasses and broadleaf weeds. A foliarly applied, translocated herbicide, it may be applied in spring, summer, or fall. It is ideal for general nonselective weed control in noncrop areas such as in industrial areas and in turfgrass establishment or renovation.

Many chemical and use classes of herbicides are available. Those discussed here are but a cross-section of the existing herbicides since they are available for use around the home (except paraquat). We can expect to see new and different classes develop in the future just as in the past.

The home gardener is urged to review the container label carefully to gain an understanding of why a certain herbicide is used in a particular way and of how to use it in ways that will not injure or kill wanted plants.

A complete listing of herbicides is presented in Appendix C.

MOLLUSCICIDES

Metaldehyde has been used in baits for the control of slugs and snails commercially and around the home since its discovery in 1936. Its continued use and success can be attributed both to its attractant and toxicant qualities. Many other materials have been used as baits, sprays, fumigants and contact toxicants, with but little success.

The other two toxicants are carbamate insecticides that have proved very successful as bait formulations. Carbaryl (Sevin®) was the first successful carbamate insecticide and has very low mammalian oral and dermal toxicity.

Methiocarb (Mesurol®) is the most effective of the insecticides registered for use against snails and slugs on ornamentals. It is highly effective against these pests and has also demonstrated repellency to several bird species.

Sodium fluosilicate is formulated commercially with metaldehyde as a slug and snail bait.

NEMATICIDES

Nematodes are covered with an impermeable cuticle that provides them with considerable protection. Chemicals with outstanding penetration characteristics are required, therefore, for their control.

Nematicides are seldom used by the homeowner unless he has a greenhouse or cold frames. For the most part we can say that nematicides are not and should not be used by the layman due mainly to their hazard.

Those that are available commercially fall into four groups. (1) halogenated hydrocarbons (some of which were described earlier), (2) isothiocyanates, (3) organophosphate insecticides and (4) carbamate or oxime insecticides. Only isothiocyanates are available to the homeowner.

Most of today's nematicides are soil fumigants, volatile halogenated hydrocarbons. To be successful they must have a high vapor pressure to spread through the soil and contact nematodes in the water film surrounding soil particles.

Isothiocyanates. In this classification belongs metam-sodium (Vapam®). It is a dithiocarbamate mentioned under the fungicides, but it is activated in the soil and is effective against all living matter in the soil, including nematodes and their eggs, fungi, weeds and weed seeds.

RODENTICIDES

Most of the methods used to control rodents are aimed at destroying them. Poisoning, shooting, trapping and fumigation are among the methods preferred. Of these, poisoning is most widely used and probably the most effective and economical. Because rodent control is in itself a diverse and complicated subject, we will mention only those more commonly used rodenticides.

Rodenticides differ widely in their chemical nature. Strange to say, they also differ widely in the hazard they present under practical conditions, even though all of them are used to kill animals that are physiologically similar to man.

A complete listing of rodenticides is presented in Appendix E.

Coumarins (anticoagulants). The most successful group of rodenticides are the coumarins, represented classically by Warfarin®. There are five compounds belonging to this classification, all of which have been very successful rodenticides. Their mode of action is twofold. (1) inhibition of prothrombin, the material in blood responsible for clotting and (2) capillary damage, resulting in internal bleeding. The coumarins require repeated ingestion over a period of several days, leaving the unsuspecting rodents growing weaker daily. The coumarins are thus considered relatively safe, since repeated accidental ingestion would be required to produce serious illness. In the case of most other rodenticides a single accidental ingestion could be fatal to man.

Of the several types of rodenticides available, only those with anticoagulant properties are safe to use around the home.

Warfarin was released in 1950 by Wisconsin Alumni Research Foundation (thus its name WARF coumarin, or WARFarin). It was immediately successful as a rat poison because rats did not develop "bait shyness" as they did with other baits during the required ingestion period of several days.

Two very effective coumarin rodenticides are available, brodifacoum (Talon® and Ratak Plus®) and bromadiolone (Maki®, Boothill® and Hawk®). These differ from the earlier coumarins in that, although they are anticoagulant in their mode of action, they require but a single feeding for rodent death to occur within 4 to 7 days. They are both effective against rodents that are resistant to conventional anticoagulants.

To protect children and pets from accidental ingestion of these highly toxic anticoagulants, the treated baits must be placed in tamper-proof boxes or in locations not accessible to children.

These rodenticides are relatively specific for rodents in that they are offered as rodent- attractive, premixed baits, inaccessible to pets and domestic animals and their relative toxicity to other warm-blooded animals is low. For instance, the rat is 2 times more sensitive to brodifacoum than the pig and dog, 36 times more than the chicken and 90 times more sensitive than the cat.

There are several baits of both brodifacoum and bromadiolone that can be purchased for home use at the hardware or garden store and which are highly effective against mice, rats and most other rodents.

Indandiones. Diphacinone (Promar®, Tomcat®, Ramik®) belongs to this class, has the same remarkable anticoagulant property as Warfarin and has replaced Warfarin where rodent avoidance behavior (bait shyness) made it ineffective. Sold as baits containing up to 0.0075% of the toxicant, it must be ingested for several consecutive days before it becomes effective. Because of this characteristic, the anticoagulants provide a definite safety factor for children and pets.

A complete listing of rodenticides is presented in Appendix E.

Rodent Repellents. "Rodent" is rather all-inclusive and perhaps a bit deceptive, because all rodent repellents are not registered for all rodents. This requires the accurate identification of the particular pest and selection of one of the following materials or combinations which clearly indicates that pest on its label. Naphthalene, paradichlorobenzene, polybutenes, polyethylene and thiram. R-55 (tert-butyl dimethyltrithioper-oxycarbamate), which is used to impregnate jacketed cables to prevent gophers and other rodents from chewing them. Ro-Pel® is perhaps the best rodent repellent available. It is repellent to virtually all mammals and birds, as well.

AVICIDES

The old, general purpose poison, strychnine was registered by EPA as an avicide for the control of English sparrows and pigeons when used as a 0.5% grain bait. Strychnine acts very quickly leaving the treated area strewn with dead birds which should be removed at regular intervals. Pre-baiting for several days is necessary before distributing the treated bait. Strychnine is no longer registered for birds – only for gopher and mole control.

Starlicide® is a chlorinated compound used as a slow-acting avicide against starlings and blackbirds. It is not effective against house sparrows. Because the material kills slowly, requiring from 1 to 3 days, large numbers of dead birds do not appear in the treated area but rather die in flight or at their roosts. It is formulated at 0.1% and 1% baits and is available only to pest control operators trained in bird control.

BIRD REPELLENTS

Bird repellents can be divided into three categories. (1) olfactory (odor), (2) tactile (touch) and (3) gustatory (taste). In the first category, only naphthalene granules or flakes are registered by the EPA. It should come as no surprise that naphthalene is repellent to all domestic animals as well.

Tactile repellents are made of various gooey combinations of castor oil, petrolatum, polybutene, resins, diphenylamine, pentachlorophenol, quinone, zinc oxide and aromatic solvents applied as thin strips or beads to roosts, window ledges and other favorite resting places.

The taste repellents are varied and somewhat surprising in certain instances, since they have other uses. The fungicides captan and copper oxalate are examples and are used as seed treatments to repel seed-pulling birds. Two other popular seed treatments are anthraquinone and chloralose (Alfamat®). Turpentine, an old standby with multiple uses, can also be used as a seed treatment. Methiocarb (Mesurol®) is registered only as a seed dressing bird repellent.

Probably the most successful of the avicides is Avitrol® (4-aminopyridine), used as a repellent. The repellency results from the distress calls made by affected birds after eating treated seeds. The material has a relatively low LD_{50}, but causes some mortality. Birds that have not eaten the treated seeds are immediately warned and do not return. Avitrol® is to be used only by licensed pest control operators for driving away flocks of nuisance or feed-consuming birds from feedlots, fields, airports, warehouse premises, public buildings and grain-processing plants.

ALGAECIDES

Copper Compounds. Rather for industrial, public water systems and agriculture but not for the swimming pools are the organic copper complexes. Most commonly used among these are the copper triethanolamine materials. Generally they can be used as surface sprays for filamentous and planktonic forms of algae in potable water reservoirs, irrigation water storage and supply systems, farm, fish and fire ponds, lake and fish hatcheries and aquaria. The treated water can be used immediately for its intended purpose. The same generalities apply also to another organic copper compound, copper ethylenediamine complex.

Quaternary Ammonium Halides. The quaternary ammonium (QA) compounds include a host of algaecides. The QA's are characterized as containing a chlorine or bromine and nitrogen.

They are general purpose antiseptics, germicides and disinfectants, ideal for algae control in the greenhouse as pot dips and wall, bench and floor sprays, swimming pool and recirculation water systems. Algae control lasts up to several months.

The alkyldimethylbenzylammonium chlorides can be used to give long-lasting control of algae and bacteria in swimming pools, cooling systems, air conditioning systems and glass houses and compose the bulk of the quaternary ammonium halides. They are toxic to fish and consequently cannot be used in fish ponds or aquaria. Dimanin C® (sodium dichloroisocyanurate) also gives long-lasting control of algae in swimming pools and can be used as a disinfectant.

Hypochlorites. Algaecides for the home swimming pool and public pools are usually the ever-popular calcium- and sodium-hypochlorites. They are not only good algaecides but also excellent disinfectants. Calcium hypochlorite is the most commonly used pool algicide and contains 70% available chlorine. It is formulated as aqueous solutions, powder, granules and compressed cakes up to one-quarter pound in size for slow release. Calcium hypochlorite is also the source of most bottled laundry bleaches (5 to 12% $CaOCl_2$).

Endothall (Aquathol®, Hydrothol®) an agricultural herbicide, is also registered as an algicide for use against blue-green algae in lakes and ponds and for algae on lawns. It is not a disinfectant and is quite toxic to fish.

BIOCIDES (Disinfectants)

Disinfectants now designated as *biocides* by the EPA, are an almost limitless number of chemical agents for controlling microorganisms and new ones appear on the market regularly. A common problem confronting persons who must utilize disinfectants or antiseptics is which one to select and how to use it. There is no single ideal or all-purpose disinfectant, thus the compound to choose is the one that will kill the organisms present in the shortest time, with no damage to the contaminated substrate.

The EPA had registered in 2003 approximately 225 active ingredients as biocides made into more than 5,000 products. There are 23 major use categories further divided into 200 use sites. Thus, the reader quickly sees the difficulty in making specific, precise recommendations.

Before exploring the various chemicals referred to as biocides, it is necessary to make a quick distinction between two confusing words, *antisepsis* and *sanitation. Antisepsis* is the disinfection of skin and mucous membranes, while *sanitation* is the disinfection of inanimate surfaces. Consequently, much more severe treatment can be used for sanitation than for antisepsis. We will direct our presentation only to the use of biocides used in sanitation.

Phenols. Phenol is the oldest recognized disinfectant, because Dr. Joseph Lister used carbolic acid (phenol) in the 1860's as a germicide in the operating room. When greatly diluted, its deadly effect is due to protein precipitation. It is used as the standard for comparison of the activities of other disinfectants in terms of phenol coeffecients. Phenol and the cresols have very distinct odors, which change little with modification of their chemical structures. Lysol® and its mimics belong to this group. The addition of chlorine or a short chain organic compound increase the activity of the phenols. The most commonly used biocide, related to the phenolic and chlorinated antimicrobials, is triclosan. It is the active ingredient in a huge number of germ-killing household and personal care products – bar soap, liquid soap, toothpaste and cleaning/ disinfecting agents applied to kitchen and bathroom surfaces.

Halogens. These are compounds containing chlorine, iodine, bromine or fluorine, both organic and inorganic. Generally, the inorganic halogens are deadly to all living cells.

Chlorine. Chlorine was first used as a deodorant and later as a disinfecting agent. It is a standard treatment for drinking water in all communities of the U.S.. Hypochlorites are those most commonly used in disinfecting and deodorizing procedures because they are relatively safe to handle, colorless, good bleaches and do not stain. Several organic chlorine derivatives are used for the disinfection of water, particularly for campers, hikers and the military. The most common of these are halazone and succinchlorimide. Chlorine is the dominant element in disinfectants, in that roughly 25% of those registered with EPA contain one or more atoms of this important halogen.

Hypochlorites. Calcium hypochlorite and sodium hypochlorite mentioned earlier as algaecides, are popular compounds, widely used both domestically and industrially. They are available as powders or liquid solutions and in varying concentrations depending on the use. Products containing 5% to 70% calcium hypochlorite are used for sanitizing dairy equipment and eating utensils in restaurants. Solutions of sodium hypochlorite are used as a household disinfectant; higher concentrations of 5% to 12% are also used as household bleaches and disinfectants and for use as sanitizing agents in dairy and food-processing establishments.

Chloramines. Another category of chlorine compounds used as disinfectants, or sanitizing agents, are the chloramines. Three chemicals in this group are monochloramine, Chloramine-T® and azochloramide. One of the advantages of the chloramines over the hypochlorites is their stability and prolonged chlorine release.

The germicidal action of chlorine and its compounds comes through the formation of hypochlorous acid when free chlorine reacts with water. Similarly, hypochlorites and chloramines also form hypochlorous acid. The hypochlorous acid formed in each instance is further decomposed releasing oxygen. The oxygen released in this reaction (nascent oxygen) is a strong oxidizing agent and through its action on cellular constituents, microorganisms are destroyed. The killing of cells by chlorine and its compounds is also due in part to the direct combination of chlorine with cellular materials.

Iodine. Water or alcohol solutions of iodine are highly antiseptic and have been used for decades before surgical procedures. Several metallic salts, such as sodium and potassium iodide, are registered as disinfectants, but the number of compounds containing iodine nowhere approaches those containing chlorine.

Peroxides. Hydrogen peroxide (H_2O_2) is a highly effective and nontoxic antiseptic. The molecule is unstable and when warmed degrades into water and free oxygen. At concentrations of 0.3% to 6.0%, hydrogen peroxide is used in disinfection and at concentrations of 6% to 25%, in sterilization

Alcohols. Alcohols denature proteins, possibly by dehydration and they also act as solvents for lipids. Consequently, membranes are likely to be disrupted and enzymes inactivated in the presence of alcohols. Three alcohols are used. ethanol or grain alcohol (CH_3CH_2OH) ; methanol or wood alcohol (CH_3OH); and isopropanol or rubbing alcohol ($[CH_3]_2CHOH$). As a rule of thumb, the bactericidal value increases as the molecular weight increases. Isopropyl alcohol is therefore the most widely used of the three. In practice, a solution of 70-80% alcohol in water is used. Percentages above 90 and below 50 are usually less effective, except for isopropyl alcohol, which is effective even in 99% solutions. A 10-minute exposure is sufficient to kill vegetative cells but not spores.

Quaternary Ammonium Compounds —These are organic compounds that have two chemical ends or poles. One end is hydrophilic and mixes well with water while the other is hydrophobic and does not. As a result, the compounds orient themselves on the surfaces of objects with their hydrophilic poles toward the water. Basically, these may be classed as ionic or non-ionic detergents. The ionic are either anionic (negatively charged) or cationic (positively charged). The anionic detergents are only mildly bactericidal.

The cationic materials, which are the quaternary ammonium compounds, are extremely bactericidal especially for *Staphylococcus,* but do not affect spores. Hard water, containing calcium or magnesium ions, will interfere with their action and they also rust metallic objects. Even with these disadvantages, the cationic detergents are among the most widely used disinfecting chemicals, since they are easily handled and are not irritating to the skin in concentrations ordinarily used.

Aldehydes. Combinations of formaldehyde and alcohol are outstanding sterilizing agents, with the exception of the residue remaining after its use. A related compound, glutaraldehyde, in solution is as effective as formaldehyde. Most organisms are killed in 5 minutes exposure to glutaraldehyde, while bacterial spores succumb in 3 to 12 hours.

There are other classes of disinfectants including a number of dyes, acids and alkalies and fumigants (ethylene oxide and methyl bromide) which are very effective under certain conditions. However, due to the scope of this book, they are only mentioned in passing.

Remember that no single chemical antimicrobial agent is best for any and all purposes. This is not surprising in view of the variety of conditions under which agents may be used, differences in modes of action and the many types of microbial cells to be destroyed.

DEODORANTS

There are frequent occasions following the control of small mammal pests when there arises the need for a deodorant, for example, when a mouse or rat dies in some unreachable void within the walls or in a crawl space, or bat guano becomes pungent. The mere presence of rodents in large numbers will result in a musky, offensive odor. The odors from small home fires, cigarette and cigar smoke and cooking odors can be equally offensive. Deodorants can provide partial or complete relief.

To handle unpleasant odors, isobornyl acetate, found in several deodorizers, (see Skunks), quaternary ammonium compounds (above) Styamine 1622, Zephiran chloride, Meelium® and Bactine® can be used as an aerosol, mist spray, or in a bowl or bottle with a tissue or cotton wick. These are more than masking agents. They are deodorants, in that they react chemically with the odors. These are the materials that the professionals use.

Several of the old familiar masking agents also can be used. These are oil of pine, oil of peppermint, oil of wintergreen, formalin and anise. For example, 10 drops of pine oil in one gallon of water can be applied with an atomizer or with a mist sprayer to provide relief. Unless you like the fragrance of the men's gym, I do not recommend using oil of wintergreen.

CHAPTER 4

We have to find in the next 25 years, food for as many people again as we have been able to develop in the whole history of man.

Jean Mayer (1975)

BIORATIONALS

Biorational pesticides are ideally those control substances that affect only the target pest and have no undesirable side effects. *Biorationals* refer to those control substances of natural origin, or man-made substances resembling those of natural origin. They have a detrimental or lethal effect only on specific target pests; they possess a unique mode of action; they are non-toxic to man and his domestic plants and animals and they have no adverse effects on wildlife and the environment.

Biorational pesticides are conveniently classified for our purposes into two distinct groups: (1) microbial (bacteria, fungi, viruses, protozoa and nematodes) and (2) biochemical (hormones, enzymes, pheromones and natural insect and plant growth regulators)

INSECT CONTROL

The first generation of insecticides were the stomach poisons, such as the arsenicals, heavy metals and fluorine compounds. The second generation includes the familiar contact insecticides: organochlorines, organophosphates, carbamates and pyrethroids.

The third generation is the biorationals. These chemicals and microorganisms are environmentally sound, closely resembling or identical to chemicals produced by insects and plants and those naturally occurring insect and plant disease microorganisms.

Microbials

Microbial insecticides obtain their name from microbes, or microorganisms, which are sometimes used by man to control certain insect pests. Insects, like mammals, also have diseases caused by bacteria, fungi, viruses, protozoa and nematodes.

These insect and mite disease microorganisms do not harm other animals or plants. The reverse of this is also true. Vertebrate disease microorganisms do not harm insects. This method of insect control is ideal in that the diseases are usually rather specific to targeted pests. Undoubtedly, the future holds many such materials in the arsenal of biorational insecticides, since several new insect pathogens are identified each year. Now there are many produced commercially and approved by the U.S. Environmental Protection Agency for use on food and feed crops. There is always the concern regarding the very remote possibility of man's susceptibility to these diseases, thus the slow advances and exceptional precautionary testing before an organism is registered. A list of those registered by the EPA is shown in Table 6.

Bacteria—*Bacillus thuringiensis* (Bt), the insecticidal bacterium, was discovered in the early 20th century. It occurs as a large number of subspecies, as a soil inhabiting, gram-positive sporulating bacterium that produces one or more

TABLE 6. EPA Registered Microbial Pesticides and Microbe-related Products.

Microorganism	Year Registered	Pest Controlled
BACTERIA		
Bacillus popilliae & *B. lentimorbus*	1948	Japanese beetle larvae
Bacillus thuringiensis spp. *kurstaki*	1961	Lepidopteran larvae
Agrobacterium radiobacter K84	1979	*Agrobacterium. tumefaciens* (crown gall disease)
B. thuringiensis Berliner	1980	Lepidopteran larvae
B. thuringiensis spp. *israelensis*	1981	Dipteran larvae
B. thuringiensis spp. *tenebrionis*	1988	Coleopteran larvae
B. thuringiensis spp. *kurstaki* EG2348	1989	Lepidopteran larvae
B. thuringiensis spp. *kurstaki* EG2424	1989	Lepidopteran larvae
B. thuringiensis spp. *kurstaki* EG2371	1990	Lepidopteran larvae
B. sphaericus	1991	Dipteran larvae
Pseudomonas fluorescens A506	1992	Ice-crystallizing *Pseudomonas* species
P. fluorescens 1629RS	1992	Ice-crystallizing *Pseudomonas* species
P. syringae 742RS	1992	Ice-crystallizing *Pseudomonas* species
B. subtilis GB03	1992	Damping-off disease
B. thuringiensis spp. *aizawai* GC-91	1992	Lepidopteran larvae
B. thuringiensis spp. *aizawai*	1992	Lepidopteran larvae
Burkholderia cepacia type Wisconsin M36	1992	Damping-off disease & nematodes
Streptomyces griseoviridis K61	1993	Damping-off disease
B. thuringiensis spp. *kurstaki* BMP123	1993	Lepidopteran larvae
B. subtilis MBI 600	1994	Damping-off disease
B. thuringiensis spp. *kurstaki* EG7673	1995	Lepidopteran larvae
P. syringae ESC 10	1995	Post-harvest decay fruit pathogens
P. syringae ESC 11	1995	Post-harvest decay fruit pathogens
B. thuringiensis spp. *kurstaki* M-200	1996	Lepidopteran larvae
B. thuringiensis spp. *kurstaki* EG7841	1996	Lepidopteran larvae
Burkholderia cepacia type Wisc. isolate J82	1996	Damping-off disease & nematodes
B. thuringiensis spp. *kurstaki* EG7673	1996	Coleopteran larvae (Colorado potato beetle)
Bacillus cereus Strain BP01	1997	Plant regulator
B. subtilis spp. amyloliquefacients St.F2B24	2000	PGR & disease control
B. subtilis St. QST713	2000	Several plant diseases
Pseudomonas chloroaphis St. 63-28	2001	Root and stem rots
B. pumilis St. GB34	2003	Soybean soil diseases
B. pumilis St. QST-2808	2003	(Tol. Exempt. only.) Powdery mildew
B. licheniformis SB3086 (Ecoguard)	2003	Broad disease control on turf
Brevibacillus brevis	——	Pending for damping off, gray mold
YEAST		
Candida oleophila I-182	1995	Post-harvest decay fruit pathogens
FUNGI		
Phytophthora palmivora MWV	1981	Citrus stranglervine
Collectotrichum gloeosporioides f. sp. *aeschynomene* ATCC 20358	1982	Northern joint vetch
Trichoderma harzianum ATCC 20476	1989	Tree wound decay
T.polysporum ATCC 20475	1989	Wood rot
Gliocladium virens G-21	1990	Pythium, Rhizoctonia seedling diseases
T. harzianum Rifai KRL-AG2	1990	Damping-off disease
Legenidium giganteum	1991	Mosquito larvae
Metarhizium anisopliae ESF1	1993	Cockroaches & flies
Puccinia canaliculata (Schweinitz) *langerheim* ATCC 40199	1993	Yellow nutsedge
Ampelomyces quisqualis M10	1994	Powdery mildew
Beauvaria bassiana GHA	1995	Grasshoppers, crickets, locusts & whitefly

TABLE 6 *(continued)*

Microorganism	Year Registered	Pest Controlled
Beauvaria bassiana ATCC 74040	1995	Whitefly, boll weevil
Paeciliomyces fumoroseus Apopka strain 97	1997	Insecticide non-food
Gliocladium catenulatum Strain J1446	1998	Seed, stem and root rots
Trichoderma harzianum St. T-39	2000	Root diseases
Coniothyrium minitans St.CON/M/91-08	2001	Sclerotinia in soils
Puccinia thlaspeos St. woad	2002	Dyer's Woad (a weed)
Metrhizium anisopliae St. ESF I	2002	Termites
Metrhizium anisopliae St. F52	2002	Ticks, beetles, diptera and thrips
Beauveria bassiana St. 447	2002	Non- food, fire ant & cockroaches
Pseudozyma flocculosa St. PF-A22 UL	2003	Powdery mildew
PROTOZOA		
Nosema locustae	1980	Grasshoppers
VIRUSES		
Heliothis nucleopolyhedrosis virus (NPV)	1975	Cotton bollworm, budworm
Douglas fir tussock moth NPV	1976	Douglas fir tussock moth
Gypsy moth NPV	1978	Gypsy moth larvae
Beet army worm NPV	1995	Beet armyworm larvae
Corn earworm NPV	1995	Corn earworm, tobacco budworm
Cydia pomonella Granulosis virus	2000	Codling moth larvae
Indian meal moth GV	2001	Indian meal moth
Celery looper NPV	2002	Celery & cabbage loopers, cotton bollworm
NON-VIABLE MICROBIALS		
B. thuringiensis spp. *kurstaki delta-* endotoxin in killed *P. fluorescens*	1991	Lepidopteran larvae
B. thuringiensis subsp. *san diego delta-* endotoxin in killed *P. fluorescens*	1991	Coleopteran larvae
B. thuringiensis Cry1Ac & Cry1c delta-endotoxin in killed *P. fluorescens*	1995	Lepidopteran larvae
B. thuringiensis subs. *kurstaki* Cry1C delta-endotoxin in killed *P. fluorescens*	1996	Lepidopteran larvae
Killed fermentation solids & solubles of *Myrothecium verrucaria*	1996	Nematodes
Agrobacterium radiobacter St. 1026	1999	Crown gall

[1]Modified from: Schneider W (09/04/1998) Microbial Pesticides Registered With the Biological Pesticide Products Division, Environmental Protection Agency. Courtesy, W.L.Biehn, Interregional Research Project No.4, Center for Minor Crop Pest Management, Rutgers University, New Brunswick, New Jersey.

very small parasporal crystals within its sporulating cells. These crystals are composed of large proteins known as delta-endotoxins. These endotoxins act by binding to specific receptor sites on the insect's gut epithelium, leading slowly to degradation of the gut lining and starvation. Several days are required to kill insects that have ingested Bt products. All Bt formulations have as their active ingredients the spores and delta-endotoxin of the indicated variety ("spp." in the scientific name).

Several varieties of Bt were discovered, each with its distinct toxicity characteristics to different insect species. *B. thuringiensis* spp. *kurstaki* was the first and controls most lepidopteran (larvae of moths and butterflies) pests, the caterpillars with high gut pH, e.g. armyworms, cabbage looper, imported cabbage worm, gypsy moth and spruce budworm.

The second was *B. thuringiensis* spp. *israelensis*, used primarily for the control of aquatic insects, e.g. mosquitoes and black flies in their larval forms. It has been exceptionally valuable in mosquito control abatement programs. Next was *B. thuringiensis* spp. *aizawai*, used for control of wax moth infestations in honey comb of honey bees. Following this was *B. thuringiensis* spp. *morrisoni*, effective against a broad spectrum of caterpillars on most crops including the home garden. Next was *B. thuringiensis* spp. *tenebrionis*, developed for Colorado potato beetle on all its hosts, the elm leaf beetle and other beetle larvae on a wide range of shade and ornamental trees. This was the first Bt effective against coleopteran (beetle) larvae. (It later turned out that another Bt product, *B. thuringiensis* spp. *san diego,* was the same variety as *tenebrionis.*)

There are numerous manufacturers of these several varieties of Bt products, in traditional formulations, many of which are registered for use in the home garden and orchard. They are quite safe to use in the presence of children and pets and the treated fruit or vegetables can be eaten with no waiting period after application. A list of the more important caterpillar pests that can be controlled with the delta endotoxins from Bt is shown in Table 7.

TABLE 7. Some of the more important caterpillar pests that succumb from ingesting delta endotoxins of *Bacillus thuringiensis.*

armyworm	eastern tent caterpillar	redbanded leafroller
artichoke plume moth	European corn borer	saltmarsh caterpillar
brown-tail moth	fall webworm	sod webworm
cabbage looper	grape leaffolder	tobacco budworm
California oakworm	grapeberry moth	tobacco hornworm
celery leaf tier	gyspy moth	tomato hornworm
corn earworm	imported cabbageworm	walnut caterpillar
(cotton bollworm)	melonworm	western tussock moth
(tomato fruitworm)	orangestriped oakworm	

Bacillus popillae and *Bacillus lentimorbus*, the milky disease bacteria spores, are closely related to *Bacillus thuringiensis* and were originally developed to control the larval or grub stage of Japanese beetles in lawns and turf. They are also effective against most other white grubs. Only one application is necessary, since the spores multiply in the soil and protect treated areas for years.

Viruses—Several insect viruses were developed, only five of which were ever registered for use by the EPA. Because of cost and lack of efficacy (attributed to environmental instability), none are in production at this writing. Those developed were the *Heliothis* nuclear polyhedrosis virus (NPV), specific against the corn earworm, cotton bollworm and tobacco budworm; the Douglas fir tussock moth virus, specific against that pest; the gypsy moth virus; the *Neodiprion sertifer* virus, specific against pine sawfly larvae; and the codling moth granulosis virus (Decyde®), the most recent to become registered and then discontinued by its manufacturer.

Viruses are quite specific and have modes of action that are not identical throughout. Generally, viruses contain crystalline proteins which are eaten by the larva and begin their activity in the insect gut. The virus units then pass through the gut wall into the blood and there multiply rapidly, taking over complete genetic control of the cells, resulting in the insect's death.

Currently registered viruses include the beet armyworm- and corn earworm NPVs for vegetables, ornamentals and other crops; the Indian meal moth granulosis virus (GV) (Fruitguard V®), for nuts and dried fruits in storage, processing and packaging areas; the celery looper NV for several field and greenhouse crops; *Cydia pomenella* (Galaxy®V4C), for codling moth in deciduous fruits and nuts; and *Mamestra configurata* NPV (Virosoft®) for bertha armyworm in canola.

Fungi—The fungi hold great promise as biorational insecticides. Though no longer produced, Mycar® was a rather good mycoacaricide (fungus-derived miticide). The microorganism was *Hirsutella thompsonii*, a parasitic fungus that infects and kills only the citrus rust mite. When sprayed on plants the spores grow into colonies that attach to the mite. In the presence of ample free moisture the spores germinate and infect the mite, which dies in about three days. The fungus spreads, continuing to propagate itself. However, because the active agent is a fungus, all commercial fungicides, copper chemicals and metal salts, such as zinc and manganese, are detrimental to its success. With its high specificity, *H. thompsonii* would have been totally compatible with other suppression techniques used in the integrated pest management of citrus crops. Those fungi that are currently registered are not yet ready for home garden use.

Protozoa—Protozoa are another form of mibrobial insect control. *Nosema locustae* is a protozoan originally developed in 1980, for the control of locusts and grasshoppers. It is marketed as Nolo Bait®. Depending on the method of application, climatic conditions and grasshopper density, the protozoa kill up to 50% of the insects and sterilize up to 30% of the survivors. Maximum effect occurs over a two- to four-week period. The residual effect of a single treatment continues to control grasshoppers through following generations by transmission through the eggs up to three or four years. This biorationall insecticide is most effective when applied as a bait and is available for use on rangelands as well as for yard and garden.

Nematodes—In the 1990's two commercial nematode products were developed for termite control, Spear® and Saf-T-Shield®. The nematode, *Neoplectana carpocapsae*, is specific for subterranean termites and kills all stages by delivering a pathogenic bacterium, *Xenorhabdus* spp., which is lethal within 48 hours after penetration. Neither product succeeded commercially.

See Appendix G "Pest Control Sources", for the above biorational organisms.

Insect Pheromones

Many insects communicate by releasing minute quantities of highly specific compounds that vaporize readily and are detected by insects of the same species. These delicate molecules are known as *pheromones*, from the Greek *pherein*, "to carry," and *hormon*, "to excite or stimulate".

There are two categories of pheromones, releasers and primers. Releasers are fast-acting and are used by insects for sexual attraction, aggregation (including trail following), dispersion, oviposition and alarm. Primers are slow-acting and cause gradual changes in growth and development, especially in social insects, by regulating caste ratios of the colony (bees/wasps, ants, termites).

Of the several types of pheromones, the sex pheromones presently offer the greatest potential for insect control. For example, a sex pheromone was used in eastern Arizona, in cotton fields. Pheromone traps containing microquantities of the synthetic pink bollworm sex lure, gossyplure, captured a sufficient number of early emerging male moths to prevent mating and reproduction of the first generation of the season. The population was suppressed enough to avoid the use of insecticides for most of the remaining growing season.

There are five general uses for sex pheromones in current insect control programs: (1) male trapping, to reduce the reproductive potential of an insect population; (2) movement studies, to determine how far and where insects move from a given point; (3) population monitoring, to determine when peak emergence or appearance occurs; (4) detection programs, such as around international airports or quarantined areas; and (5) the "confusion or mating disruption" technique. With this technique, the area is saturated with pheromone dispensers such that male insects fail to locate females. The females then either lay no eggs or lay unfertilized eggs that don't hatch.

A list of 108 known synthetic sex pheromones and attractants is given in Table 8 . Several companies manufacture or market pheromone products for use in the home garden/orchard and for agriculture. Among those of interest to the homeowner, but not excluding others, are NoMate® pheromone traps for the codling moth, tufted apple bud moth and tomato pinworm, manufactured by Scentry Biologicals Inc., Hercon® and Disrupt® traps for cockroaches, ants, Japanese beetle, gypsy moth, Nantucket pine tip moth, corn earworm, tobacco budworm, codling moth, peachtree borer and several other vegetable and fruit insect pests, from Hercon Environmental Co.; and Isomate® traps for codling moth in pome fruits, diamondback moth in cole crops, grape berry moth in vineyards, Oriental fruit moth in stone fruits and peachtree borer in orchards, sold by Pacific Biocontrol Inc.

Pheromones have received tremendous publicity regarding their potential for ridding insect pests. However, they are most practically used in survey traps to provide information about population levels, to delineate new infestations, to monitor eradication programs and to warn of new pest introductions. Future insect control programs will more and more rely on the use of pheromones as survey tools and to supress early emerging populations through trapping and confusion.

TABLE 8, Insect sex pheromones and/or attractants, and the species attracted. [a]

Species attracted	*Pheromone common name*	*Species attracted*	*Pheromone common name*
aphid(s)	–	lone star tick	–
alfalfa looper	looperlure	maggot fruit fly	–
almond moth	– [b]	Mediterranean flourmoth	–
ambrosia beetle	sulcatol	Mediterranean fruit fly	trimedlure, siglure
Angoumois grain moth	angoulure	Mediterranean melon fly	cue-lure
apple maggot	–	melon fly	–
artichoke plume moth	–	Mountain pine beetle (Reg 1999)	verbenone
bagworm	–	naval orangeworm (Reg 2000)	–
beet armyworm	Isomate-Baw®	oak leaf roller	–
black carpet beetle	megatomic acid	oblique-banded leafroller	riblure
black cherry fruit fly	–	olive fruit fly	–
black cutworm	–	omniverous leafroller	Checkmate OLR®
boll weevil	grandlure	orange tortrix	–
cabbage looper	looplure	Oriental fruit fly	–
California red scale	–	Oriental fruit moth	orfralure, Checkmate OFM®
carpenterworm	–	Pales weevil	eugenol [c]
cherry fruit fly	–	peachtree borer	–
citrus leafroller	–	peach twig borer	Checkmate PTB®
cockroach (*Periplaneta*)	–	pine bark beetle	–
codling moth	codlelure, Checkmate® CM®	pine beetles	–
confused flour beetle	–	pine tip moth (Nantucket)	–
corn earworm/cotton bollworm	zealure	pink bollworm	gossyplure
dermestid beetle	–	potato tuberworm moth	–
diamondback moth (Reg 2001)	– (Checkmate® DBM-F)	raisin moth	–
Douglas fir beetle	douglure	redbanded leafroller	–
Douglas fir tussock moth	tussolure	red flour beetle	–
Eastern pine shoot borer (Reg 1999)	–	red plum maggot fly	–
Egyptian cotton leafworm	–	San Jose scale	–
elm bark beetle	multilure	smaller tea tortrix	–
European corn borer	nubilure	Southern armyworm	–
European grape vine moth	–	Southern pine beetle	frontalure
European pine shoot moth	– (1999)	soybean looper	looperlure
face fly	–	spidermites	Stirrup M®
fall armyworm	frugilure	spruce budworm	soolure
false codling moth	–	summer fruit tortrix	–
filbert leafroller	–	tentiform leafminer	–
forestry beetles (Reg 1999)	MCH	Texas leafcutting ant	attalure
fruittree leafroller	–	three-lined leafroller	–
fruittree tortrix	–	thrips	–
furniture carpet beetle	–	tiger moth	–
gelechiid moths	–	tobacco budworm	virelure
grape berry moth	vitelure	tobacco hornworm	– [c]
greater peachtree borer	–	tobacco moth	–
greater wax moth	undecanal	tomato fruitworm	–
gypsy moth	disparlure	tomato pinworm	Checkmate TPW®
honey bee	Bee-Here® Bee-Scent® (feeding attractant)	tufted apple budmoth	–
		variegated leafroller	–
house fly	muscalure	walnut husk fly	–
Indian meal moth	–	wasp	citronellal [c]
Japanese beetle	japonilure	Western grapeleaf skeletonizer	–
Khapra beetle	–	Western pine beetle	brevilure
leaf cutter ants	–	Western pine shoot borer	–
leafminer	–	white fly	–
lesser appleworm	–	yellow jacket	–
lesser peachtree borer	–		

[a] All listings may not be available commercially, while some may be sold under several trade names and formulations.

[b] Known only by its chemical name; no common name assigned to pheromone.

[c] Not proven to be the natural pheromone.

See Appendix G "Pest Control Sources", for the above pheromones.

Insect Growth Regulators

Insect growth regulators (IGRs) are a group of synthetic chemicals that alter growth and development in insects. Their effects have been observed in egg, larval and nymphal development, on metamorphosis (changing from one larval stage to another, from larva to pupa, from pupa to adult), on reproduction in both males and females, on behavior and on several forms of diapause (an extended resting stage).

The IGRs disrupt insect growth and development in three ways: As juvenile hormones (JHs), as precocenes and as chitin synthesis inhibitors. JHs include ecdysone (the molting hormone), JH-mimic, JH-analog and are known by their broader synonyms, *juvenoids* and *juvegens*. JHs disrupt immature development and emergence as adults. Precocenes interfere with normal function of glands that produce JH. Also, chitin synthesis inhibitors affect the ability of insects to produce new chitin exoskeletons when they molt.

IGRs are effective when applied in very small quantities and apparently have no undesirable effects on humans and wildlife. They are, however, nonspecific, since they affect not only the target species, but most other arthropods as well. Consequently, when used with precision, IGRs may play an important role in future insect pest management programs.

Only a few IGRs are registered by the EPA. Methoprene, as Altosid®, is used for mosquito larvae control in stagnant and floodwater situations and as Precor® for indoor control of dog and cat fleas. It has other formulations for the pharoah ant, stored peanut and tobacco pests, horn fly control, mushroom growing media to control fungus gnats and for drinking water to control mosquitoes, the latter approved by the World Health Organization.

Hydroprene (GenTrol®, Mator®) is an IGR with JH activity, registered for use against cockroaches, stored grain pests and experimentally against homopteran, lepidopteran and coleopteran pests.

Kinoprene (Enstar II®) is effective against aphids, whiteflies, mealybugs and scales on ornamental plants and vegetable seed crops grown in greenhouses and shade houses. It is specific for Homopteran insects and results in a gradual reduction rather than a quick kill.

Fenoxycarb (Logic®, Award®, Torus®) is a JH-type IGR, which induces ovicidal effects, inhibits metamorphosis to the adult stage, resulting in death of the last larval or pupal stage and it interferes with molting of early instars. Fire ants and all other ants, mosquito larvae, fleas and cockroaches, caterpillars and scale insects have all proved vulnerable to this IGR.

Pyriproxifen (Knack®, Archer®) inhibits molting in a wide array of insects, especially whitefly, citrus scales, fly-breeding sites in poultry and livestock manure and for mosquito control.

Chitin Synthesis Inhibitors

Chitin synthesis inhibitors are a group of insecticides that kill by preventing insects from producing new exoskeletons when molting. The first of these was diflubenzuron.

Diflubenzuron (Dimilin®, Adept®) is not truly an IGR of the juvenoid class, but rather another insecticide with a different mode of action. It acts on the larval stages of most insects by inhibiting or blocking the synthesis of chitin, a vital and almost indestructible part of the hard outer covering of insects, the exoskeleton. Typical effects on the developing insect larva are the rupture of malformed cuticle or death by starvation. It is used against the gypsy moth and most forest caterpillars, boll weevil in cotton, velvet bean caterpillar and cloverworm on soybeans and mushroom flies in mushroom production facilities.

Other chitin synthesis inhibitors with broad ranges of uses are teflubenzuron (Nomolt®), cyromazine (Citation®, Trigard®), triflumuron (Alsystin®), chlorfluazuron (Atabron®) which is not registered by EPA, flucycloxuron (Andalin®), flufenoxuron (Cascade®), hexaflumuron (Consult®, Recruit®) and novaluron (Oscar®, Pedestal®).

IGRs hold intriguing possibilities for future use in practical insect control. They are, however, insect-control chemicals, thus fall within the same legal confines as other synthetic insecticides and must therefore meet the rigorous nontarget and environmental standards set by EPA for all other pesticides.

WEED CONTROL

Herbicides of the future will undoubtedly be developed from the biorational arena rather than from the non-selective chemicals that leave undesirable soil residual carryovers to subsequent crops. Particularly useful among these would be disease organisms, such as fungi and bacteria, that would infect and kill specific weeds without harming the host crop.

Mycoherbicides

A new concept in weed control is the use of disease microorganisms as useful and sometimes self-perpetuating pathogens. The first to be registered in this field is *Phytophthora palmivora* (Devine®), a naturally occurring, highly selective fungal disease of strangler or milkweed vine, a serious pest in citrus groves. Properly applied to the soil beneath citrus, the fungus kills existing milkweed vines, then persists in the vine root debris to carry over to the next generation of germinating vines. This pathogen is selective and does not infect citrus tree roots, fruit, or foliage.

The next myco-herbicide is Collego®, living spores of the fungus *Collectotrichum gloeosporioides* f. sp. *aeschynomene.* This selective, postemergent formulation is applied to rice for the control of coffee weed and to soybeans for the control of Northern jointvetch (curly indigo). These spores are not compatible with fungicides and thus none should be applied within three weeks following application of Collego®. High humidity is required for several hours following application to promote spore germination. Death of the target weeds may require up to five weeks.

A third, *Alternaria cassiae,* a naturally occurring fungal disease of the weed sicklepod (*Cassia obtusifolia*), is being developed. Soybeans and peanuts, both legumes, are particularly plagued by sicklepod, but are immune to this fungus. As with Collego®, dew or mist is required for an extended period for the fungus to germinate and infect the target weeds.

Two new bioherbicides look promising. The fungus *Puccinia canaliculata* (Dr. Biosedge®) is being developed in the South for yellow nutsedge control. Also, X-Po® (Dow AgroSciences) is the bacterium *Xanthomonas campestris*, for the control of annual bluegrass and the golf-green nemesis *Poa annua.*

Application technology is the prime difficulty with this and future myco-herbicides that require substantial surface water on their designated hosts.

PLANT DISEASE CONTROL

The needs in plant disease control are systemics and curatives for the downy mildew diseases, good systemic bactericides and compounds that translocate to new foliage or roots following foliar application. As these are eventually provided, we should begin to lay aside those in current use that suffer from the usual inadequacies.

Mycofungicides

Four fungi are registered for three different forms of fungal disease control. None are yet available commercially. *Gliocladium virens* strain GL21 (SoilGard®) for the control of Pythium and Rhizoctonia seedling diseases of vegetables and field crops.

Trichoderma harzianum strain KRLAG2 (T-22 Planter Box3®) is designed to control *Pythium* seedling diseases, while *T. harzianum* strain ATCC20476 + *T. polysporum* ATCC20475 are aimed at woodrot organisms. *T. harzianum* St. T-39 (Binab-T®) is for several root diseases.

Bacto-Fungicides

The first of these registered with EPA is the bacterium *Pseudomonas fluorescens* Pf-5, marketed as Daggar® G (granules). It is basically a cotton fungicide used for control of *Pythium ultimum* and *Rhizoctonia solani*, fungi that are the causal agents of damping-off of cotton seedlings. The product, now discontinued, was applied in the seed furrow at the time of planting.

Another bacterium *Agrobacterium radiobacter* (strain 84), sold as Galltrol-A®, is registered for control of crown gall infection on deciduous fruits, nuts, vines and ornamentals. It is applied as a preplant dip or spray on cuttings, seedlings and seeds for nonfood and nonbearing plants. Because the formulation is a living bacterial culture, it must be refrigerated, which provides several months' shelf-life.

Bacillus subtilis (Kodiak®) was developed as a seed treatment to prevent seedling diseases of vegetables and some field crops. A somewhat different approach is Milsana®, derived from *Reynoutria sachalinesis*, which is applied to young plants to raise their natural resistance to mildew.

Spot-Less® (*Pseudomonas aureofaciens* St. TX-1) is a bacto-fungicide for turfgrass diseases. BioSave® (*P. syringae*) is for post-harvest fruit rot on apples, pears and citrus in storage and shipment. AtEze® (*P. chlorophis*) is for Rhizoctonia and Pythium root/stem rots in greenhouse vegetables and ornamentals. *P. (Burkholderia) cepacia Wisconsin* St. J82 is registered as a seed treatment to prevent seedling diseases of vegetables and a few field crops.

CHAPTER 5

A good gardener never blames his tools.

American Proverb.

EQUIPMENT

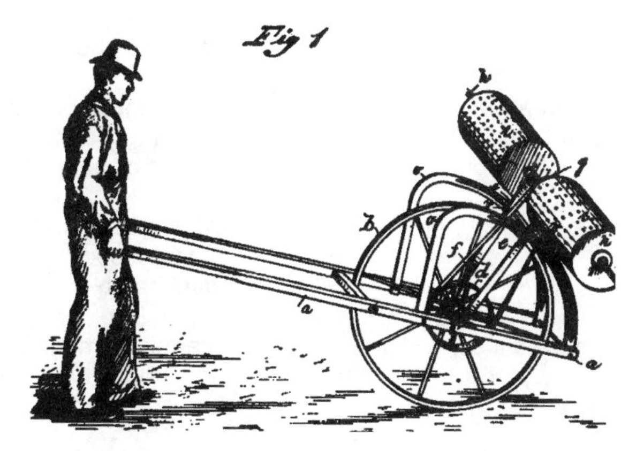

Original 1881 Patent Drawing of a Manual Operated Row-Crop Duster

EQUIPMENT FOR HOME PEST CONTROL

Pesticide application equipment must be used for the purposes and formulations for which it was intended to produce the maximum effect at minimum cost and to avoid hazards to humans, domestic animals, pets, plants and other nontarget organisms. Most application equipment distributes pesticides, whether sprays, dusts, granules, or aerosols, to produce uniform coverage. That describes the scientific aspects of using application equipment. The art depends on the user for good judgment and smooth even movements of the "business end" of the equipment.

SELECTING THE PROPER EQUIPMENT

Much could be written about the selection and proper use of equipment for a particular pest control job. But in the good old American tradition, we "make do" with what we have. Almost any kind of pesticide application could be carried out with any one of the sprayers or dusters described, with of course, a couple of miniature-sized exceptions. Table 9 and Figure 2 contain a list of various home pests and the choices of equipment available to do the job.

In summary, the basic arsenal for the homeowner would be (1) a 2 or 3 gallon compressed air sprayer with 2 interchangeable nozzles, a cone-type and flat fan and (2) a plunger duster. The following figure provides a general guide for selecting the proper equipment for the job. In the left hand column select the outdoor housekeeping task pertinent to your interest. Filled squares at the right indicate types of applicators suggested for that use. Consider all your various jobs to find the type of equipment that will do a wide range of work for you. Applications to lawn grasses can be done with sprayers only. In all other cases your choice can be sprayers or dusters, as you prefer.

TABLE 9. Pest Problems and Equipment to Use.

Pests	Useful Equipment
Inside the Home	
1. Crawling insects and mites (ants, roaches silverfish, clover mites)	Continuous hand sprayer Paint brush Compressed-air sprayer Vacum cleaner hose
2. Fleas	Same as above
3. Ticks	Same as above
4. Pests on houseplants	Same as above
5. Clothes moths and carpet beetles infesting carpets and other fabrics	Continuous hand sprayer Compressed-air sprayer Vacum cleaner hose
6. Spiders	Continuous hand sprayer Compressed-air sprayer Paint brush
7. Flying insects (flies and mosquitoes)	Intermittent hand sprayer Continuous hand sprayer Aerosol bomb
8. Mice	Traps Bait Stations
Outside the Home	
1. Insects and plant diseases	Compressed-air sprayer Knapsack sprayer Hose-end sprayer Power sprayer Plunger duster Rotary duster Knapsack duster Granule spreader
2. Annoying insects	
a. On porch, patio, or terrace	Continuous hand sprayer Compressed-air sprayer Hose-end sprayer Aerosols
b. Refuse disposal areas (garbage cans, compost pile, lawn clippings)	Compressed-air sprayer Hose-end sprayer Plunger duster Rotary duster
3. Weeds (dandelions, plantain, crabgrass)	Compressed-air sprayer with fan nozzle Power sprayer with fan nozzle Killer Kane, pistol sprayer (spot treatment) Granules spreader
4. Rodents and other small mammals	Traps Bait stations

FIGURE 2. Suggested Pesticides Application Equipment Guide for Home Pest Control Jobs.

		Shaker Can	Aerosol	Pistol Sprayer	Contin-uous Sprayer	Hose-End Sprayer	Bucket Pump Sprayer	Slide Pump Sprayer	Compress-ed Air Sprayer	Knapsack Sprayer	Wheel-barrow Sprayer	Power Sprayer	Mist Blower	Plunger Duster	Crank Duster	Drop Spreader	Centrif-ugal Spreader	Root Irrigator
Lawn Weeds	Small Lots			X		X	X	X	X							X	X	
	Large Lots								X	X	X	X				X	X	
Houseplants			X	X	X													
House Interior	Fliers		X															
	Crawlers	X	X	X	X				X					X				
Lawn Insects & Diseases	Small	X				X			X							X	X	
	Large					X			X	X		X				X	X	
Shrubs & Evergreens	24 or less		X	X		X		X	X					X	X			X
	25-50						X	X	X	X	X	X	X	X	X			
	50 or more									X	X	X	X		X			
Gardens Flowers & Vegetables	500 sq ft			X	X	X	X	X	X	X				X	X			
	500-1000							X	X	X	X	X		X	X			
	over 1000								X	X	X	X	X		X			
Trees & Fruit Trees	1-5						X	X	X				X					X
	over 5										X	X	X					X
Outdoor Comfort — Small Lots	Mosquitoes					X		X	X				X		X			
	Flies Wasps		X					X	X					X				
	House Invaders	X		X					X					X	X			
Outdoor Comfort — Large Lots	Mosquitoes							X	X	X		X	X		X			
	Flies							X	X				X					
	House Invaders	X	X	X					X	X	X	X			X			

(Modified from Cornell University's Misc. Bull. 74.)

HAND SPRAYERS

There are basic as well as novelty tools for applying pesticides, but sprayers are the most efficient means and hand sprayers are the most important types of application equipment the homeowner should have. They vary in size and weight from small, hand-operated "flit guns" containing one-half pint to larger, heavier pieces of equipment holding several gallons.

Compressed Air or Tank-Type Sprayer. If you could own but one piece of equipment, it should be this, particularly if you own fruit trees. It is the mainstay of home pest control and can be used by anyone. It will last a lifetime by replacing the various parts that pop, split and crack. It is truly the universal pesticide application unit, as seen in Figure 3. This unit offers a continuous spray pattern expelled by a head of compressed air and has interchangeable or adjustable nozzles. It does require rinsing, draining by inverting and thorough drying after use to preserve its serviceable life.

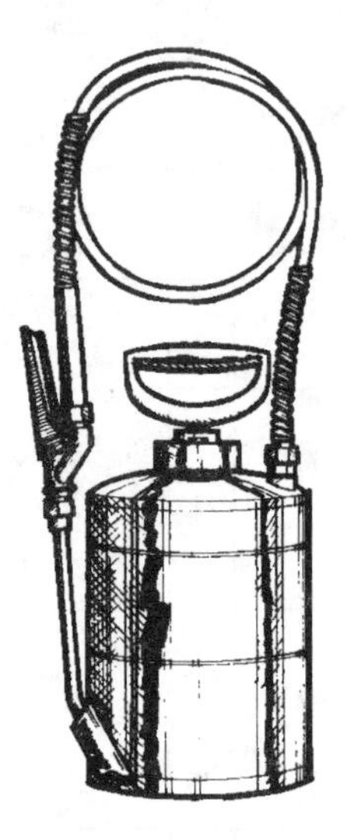

FIGURE 3, Compressed Air Sprayer
(U.S. Public Health Service)

Hand Pump Atomizer. These come in two styles. The first is the old-fashioned flit gun that emits a puff of spray with each plunger action, the intermittent sprayer. The second is a continuous sprayer that forces air into the tank to develop and maintain a constant pressure and deliver a continuous atomized spray. Most continuous hand sprayers have interchangeable or adjustable nozzles and can deliver a fine-, medium—fine-, or coarse-droplet spray. These are designed basically to control flying and crawling insects in the home, particularly flies and mosquitoes. (Fig. 4.).

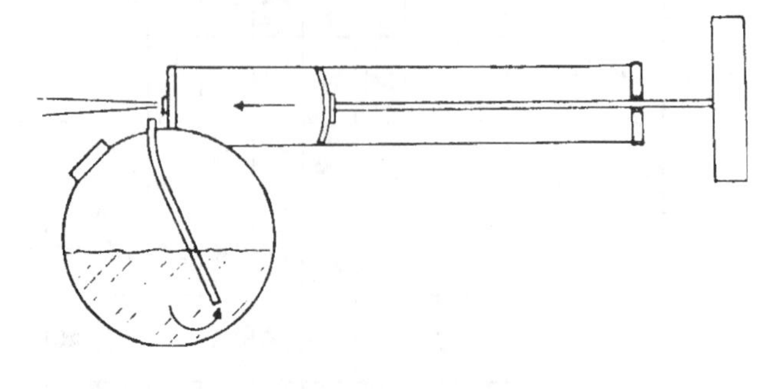

FIGURE 4. Flit Gun Sprayer
(U.S. Public Health Service)

Knapsack Sprayer. These are carried on the back of the operator and have shoulder straps so that weight can be distributed evenly on both shoulders comfortably (Fig. 5). A diaphragm or piston pump and a mechanical agitator are mounted inside the tank and operated by a lever worked with the right hand. The pesticide is under liquid pressure during each stroke of the pump; thus, it produces a surging or pulsed spray. Knapsack sprayers are used chiefly for treatment of gardens, shrubs and small trees. Adjustable nozzles are standard equipment. To extend the life of the pump and tank, rinse, drain and dry after each use. This is not the most useful sprayer for the homeowner.

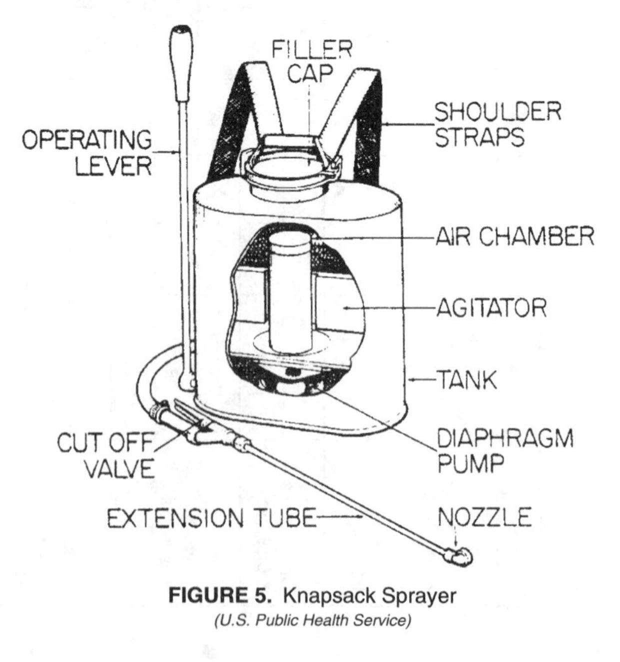

FIGURE 5. Knapsack Sprayer
(U.S. Public Health Service)

Slide Pump or Trombone Sprayer. This bucket-mounted sprayer has a plunger and cylinder moving on each other with a trombonelike action, requiring the use of both hands. It, too, produces a surging or pulsed spray. These can throw a stream of spray 20 to 30 feet vertically and are useful for treating medium sized fruit and shade trees (Fig. 6). This would be the recommended sprayer for home orchards.

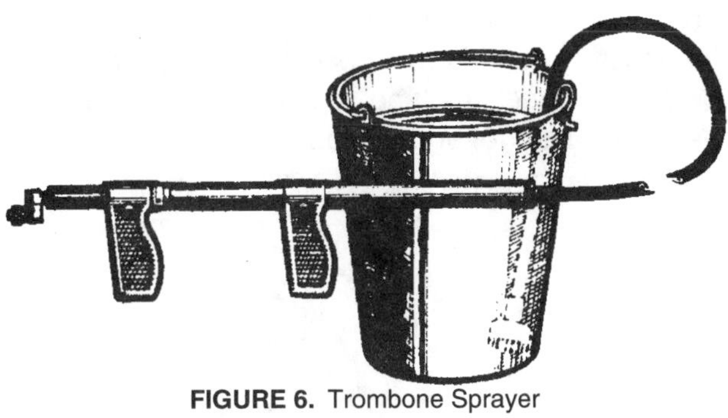

FIGURE 6. Trombone Sprayer
(U.S. Public Health Service)

Garden Hose Sprayer. These are designed to connect to a garden hose and utilize the household water supply and water pressure for dispersing the pesticide. It consists of a jar for holding concentrated spray material, a spray gun attached to the lid and a suction tube from the gun to the bottom of the jar (Fig. 7). The gun siphons the spray concentrate from the jar and mixes it with the water flowing from the garden hose through the gun to produce a dilute spray. Although the primary design of this sprayer is for garden, lawn and shrubbery insects, it is quite useful for controlling fleas, ticks and chiggers in yards and for applying insecticides near the foundations of homes for cockroach, earwig and clover mite control. Its limitation is the length of the garden hose. Be certain the model you purchase contains a device to prevent back-siphoning. This is essential to prevent contamination of the water supply.

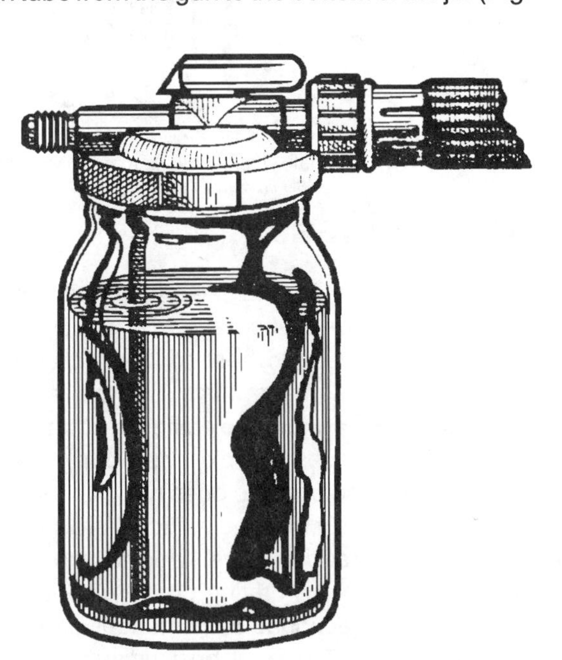

FIGURE 7. Garden Hose Sprayer
(U.S. Public Health Service)

Bucket Pump Sprayer and Wheelbarrow Sprayer. These are essentially the same, having hand pumps that utilize liquid pressure during each stroke of the pump, thus producing surging sprays. The only difference between the two is the mobility of the wheelbarrow and its increased tank capacity. They are useful mainly for spraying shrubs and small fruit and shade trees. The wheelbarrow sprayer can actually be used to treat practically all outdoor situations if a second person can be coerced into pushing the wheelbarrow as you pump and spray. These normally come equipped with interchangeable nozzles (Fig. 8).

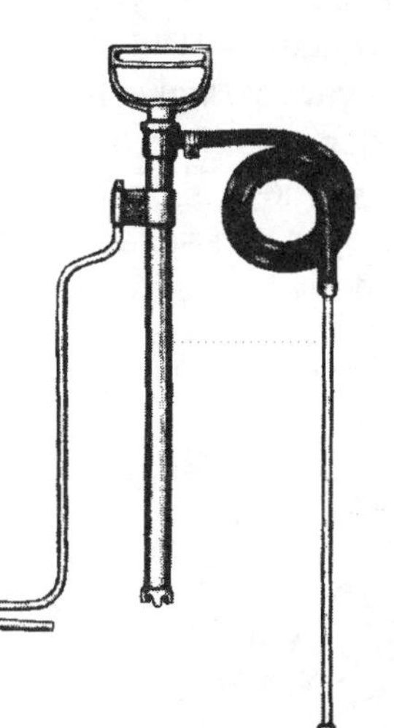

FIGURE 8. Bucket Pump Sprayer
(H.D. Hudson Co.)

Power Sprayers. Power sprayers take just a bit of the sport out of spraying around the home, because of their speed and ease of use. They can be purchased with gasoline or electrically powered pumps (Fig. 9). Personally, I'll take the electric model. (Imagine getting prepared to spray your fruit trees with a dormant spray, first of the year, in mid-March, with no wind and the gasoline engine refuses to start!) Power sprayers are usually piston pumps with a small head of air pressure, which reduces the surging to a minimum. They do require special care, cleaning and preparation for winter storage.

FIGURE 9. A piston Pump Power Sprayer
(H.D. Hudson Co.)

Pistol Sprayer. These are indeed pistol sprayers, for they are held and squeezed like pistols (Fig. 10). Several models are available including the window-cleaner, finger-plunger type. Recently, one of the larger producers of home-use pesticides cleverly included a pistol sprayer and several feet of hose with each gallon of ready-to-use spray.

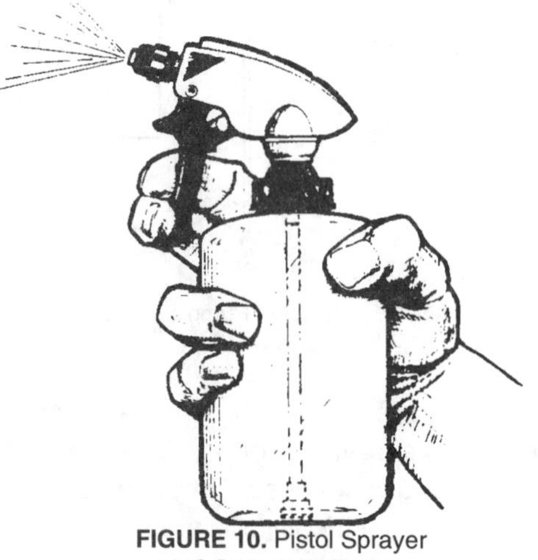

FIGURE 10. Pistol Sprayer
(U.S. Public Health Service)

Aerosols. Pressure cans and aerosol bombs (Fig. 11) containing pesticides are the most convenient of the spray applicators and also the most expensive. Insecticide aerosol cans totaled 199 million sold in 2001 (This accounted for only 6.4% of the total aerosol market for all uses in 2001).* They are useful for making spot treatments to plants, to walls, cracks and crevices both inside and outside the home and as atomizing space sprays for controlling flying insects in confined areas. "Bug bombs" are sometimes made for specific uses. If the aerosol bomb is your choice, be sure to check the label to see if the contents can be safely used for your purposes.

(*Chemical and Engineering News, 2002)

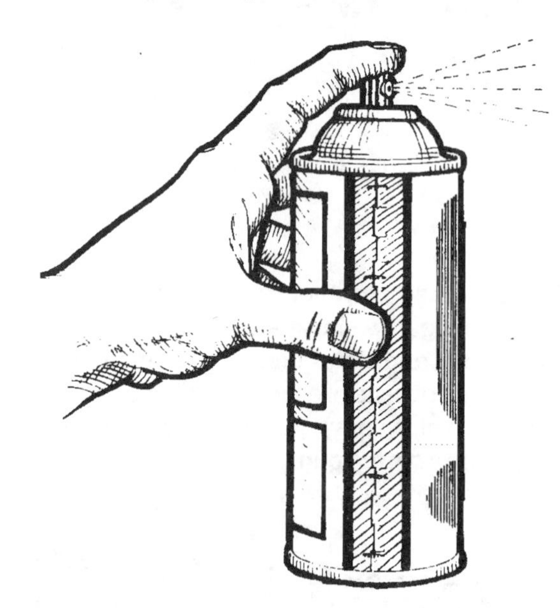

FIGURE 11. Small, Disposable Aerosol Dispenser
(U.S. Public Health Service)

HAND DUSTERS

Hand dusters are easy to use and permit application into dense foliage that sprays will not penetrate. As a rule of thumb, it is best to dust foliage in the mornings when some dew is still on the leaves. This permits a better adherence of the dust that would not stick to the leaves in mid-afternoon. Generally, however, the use of dusts does not give the efficient deposit of sprays, since the particles are smaller and will drift more readily. Give heed to your neighbor when dusting, since drifting dusts are readily seen and may be objectionable. Incidentally, many of the cans of dusts available may be used as self-contained shaker-dusters.

Plunger Duster. This duster is as important to dusts as the compressed air sprayer is to sprays. It consists of an air pump with a reservoir into which the air blast is directed to disperse the pesticide as a fine cloud or as a solid blast (Fig. 12). If the duster is turned so the delivery tube is beneath the dust, very heavy dust patterns will be produced, as needed for ant control. This duster can also be used to apply insecticides to control chiggers, ticks, fleas and clover mites in and around lawns.

FIGURE 12. Hand Plunger Duster
(U.S. Public Health Service)

Knapsack Duster. This is a large bellows duster which is designed to be carried on the back, similar to the knapsack sprayer (Fig. 13). It consists of a large circular tank, the top of which forms the bellows operated by the up-down motion of a side lever. Dust is inserted through a cover on the back of the machine.

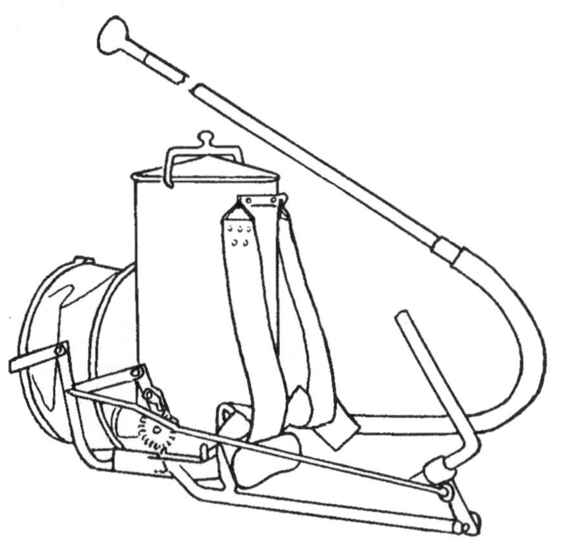

Figure 13. Bellows or Knapsack Duster
(USDA)

It is used for treating flower and vegetable gardens and shrubs. Because it emits puffs of dust rather than a uniform stream, it is not well suited for lawn and turf applications.

Rotary Hand Duster. This unit has a 5 to 10 pound capacity hopper from which dust is fed into a fan case, producing a steady stream of dust when the crank is turned (Fig. 14). Rotary dusters can be adjusted to deliver from 5 to 20 pounds of dust per acre with average walking speeds of 2 to 2 1/2 mph. The advantage these have over knap-sack models is the continuous dust pattern that permits their use on row crops and lawns. Most come equipped with a detachable fan-tip to produce a broad dust band for area treatment or broad plants such as melons and cucumbers.

FIGURE 14. Rotary Hand Duster
(U.S. Public Health Service)

Bulb Duster. This handy little gadget is designed for careful indoor placement of dusts, such as boric acid. A 4-inch rubber bulb is equipped with a screw-cap that holds a small-orifice dust nozzle (Fig. 15). After the bulb is filled with dust and the cap replaced, hand pressure on the bulb dispenses the dust. It is used where careful placement and neatness are required, such as crevices where cockroaches and silverfish hide, or behind baseboards where carpet beetles are found.

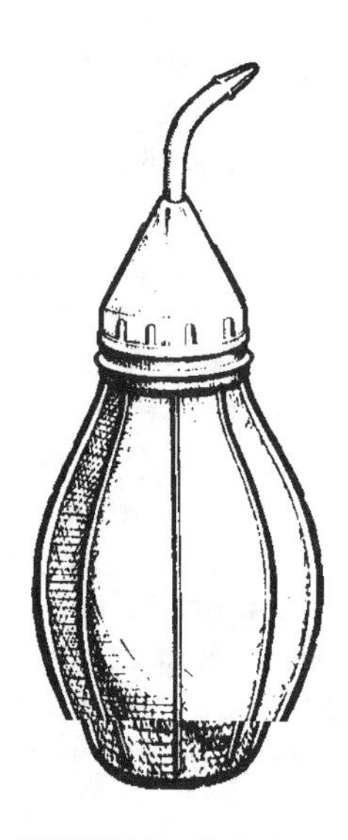

FIGURE 15. Bulb Duster
(U.S. Public Health Service)

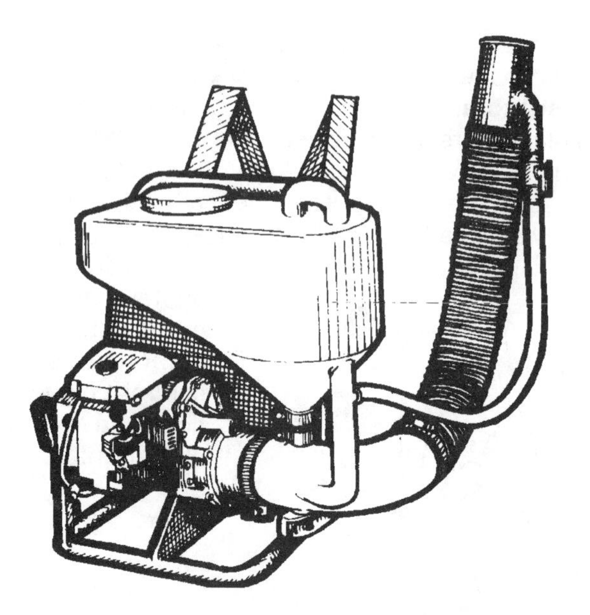

FIGURE 16. Small Portable Power Duster-Mister
(U.S. Public Health Service)

Power Duster. This light-weight unit is designed to be carried on the back of the operator and weighs from 50 to 60 pounds fully loaded, including 10 to 20 pounds of dust. It consists of a hopper, a small gasoline engine which operates a radial fan and an air discharge nozzle into which is fed a metered amount of dust. Some models are mist/dust blowers (Fig.16) and can be used to apply liquids, dusts or granules.

GRANULAR APPLICATORS

Granular applicators have the advantage of serving dual purposes, the application of fertilizers and pesticides. There are two main types used around the home, the two-wheeled fertilizer distributor and the cyclone grass seeder.

Fertilizer Distributor. Little need be said about this common lawn applicator. It has an agitator within the hopper activated by wheel rotation and an adjustable orifice plate to control the rate of flow. Its use is limited to continuous areas such as lawns and flat garden areas not yet planted or thrown up in rows.

Cyclone Seeder. These are available in the larger field-seeding models down to the smaller, plastic 1-quart-capacity units for home lawn use. They produce an amazingly uniform application and cover large areas in a minimum of time.

CHAPTER 6

PROBLEMS — And Solutions

The biggest problem in the world
could have been solved when
it was small.
Witter Bynner, *The way of Life According to Laotzu.*

IDENTIFICATION OF PLANT PROBLEMS: DIAGNOSIS

Determining the causes of poor plant growth and the associated symptoms, such as leafspot and discolored leaves requires years of experience. However, generalizations can be made that will help you become proficient in diagnosing the ills of plants.

They may be injured by animals such as insects and rabbits; by other plants such as fungi and bacteria; by natural causes such as drought and nutritional disorders; and by chemical injury such as phytotoxic symptoms from sprays and air pollution.

Careful observation often reveals the cause. For example, a canker disease may have girdled a twig or limb. The symptoms might include wilted or dead leaves, but one must look carefully along the twig or limb to find the cause. Or an insect may be eating the roots of a plant — the symptoms appear first on the leaves, but until one examines the roots the cause cannot be established.

Two or more causes might produce the same kind of symptom and there are hundreds of causes. With careful observation, knowledge of the plant's history and a general knowledge of possible causes, plant ills can often be diagnosed without the aid of a plant pathologist or entomologist.

Ways in Which Insects Injure Plants

1. Chewing: devouring or notching leaves, eating wood, bark, roots, stems, fruit, seeds, mining in leaves. Symptoms: ragged leaves, holes in wood and bark or in fruit and seed, serpentine mines or blotches, wilted or dead plants, or presence of larvae.
2. Sucking: removing sap and cell contents and injection of toxins into plant. Symptom: usually off-color, misshapen foliage and fruit.
3. Vectoring Diseases: by carrying diseases from plant to plant, e.g. elm bark beetle and Dutch elm disease; various aphids and virus diseases. Symptoms: wilt, dwarf, off-color foliage.
4. Excreting Honeydew: deposits lead to the growth of sooty mold and leaves cannot perform their manufacturing functions, which results in a weakened plant. Symptoms: sooty black leaves, twigs, branches and fruit.
5. Forming Galls: galls may form on leaves, twigs, buds and roots. They disfigure plants, and twig galls often cause serious injury.
6. Scarring by Egglaying: scars formed on stems, twigs, bark, or fruit. Symptoms: scarring, splitting, breaking of stems and twigs, misshapen and sometimes infested fruit.

Ways in Which Diseases Injure Plants

1. Interfere with the supply lines by clogging water-conducting cells. Examples: late blight of tomato and potato. Dutch elm disease. Symptom: wilt.
2. Destroy chlorophyll. Examples and symptoms: blotch, scab, black spots on leaves, brown patch disease of turf.

3. Destroy water- and mineral-collecting tissues. Examples and symptoms: Fusarium wilt, root rot, general stunting of plant.
4. Gall-formation disrupts normal cellular organization. Symptoms: unusual growths on flowers, twigs and roots.
5. Produce flower and seed rots. Examples: fireblight, bacterial rot of potato.

SELECTING THE PROPER PESTICIDE

Ideally, each pest problem should be specifically identified and a particular pesticide chosen for the problem, but this is seldom practical because there are about 10,000 insect pests, 1,500 plant diseases and 600 weeds. For the homeowner or amateur gardener, the purchase of a pesticide for each individual problem is impractical because of cost, lack of storage space and increased danger of misuse and accidental poisoning.

Modern pesticides make pest control relatively simple and only a few are needed around the home. Most of these control a rather wide range of pests, allowing generalizations to be made about control of pest groups such as aphids, mites, beetles, caterpillars, leaf spots, blights, turf diseases. The problem can be even further simplified to protect some groups of plants by use of "multipurpose" or "all purpose" mixtures of pesticides, which contain an insecticide, miticide and fungicide in a single mixture and are effective against insects, mites and diseases rather than against a few insects or a few diseases.

In either case, to select the proper pesticide or pesticide mixture, refer to the proper tables that list the most commonly observed pest problems for that particular problem area. These tables list alphabetically the host, the common name of the pest, the materials and methods of control, with and without chemicals and the approximate time of treatment. The dilution rate for spray chemicals is found in Table 10. Read the label for more specific details and to make certain that the material is safe for the situation and effective for the pest to be controlled.

APPLYING THE PESTICIDE PROPERLY

A basic principle of pest control is the proper application of the pesticide. Generally, two intelligent efforts are involved. The first is to give good uniform coverage of the area to be treated, whether for insect, disease or weed control, or with sprays or dusts. The second effort probably should be first, the proper dilution of the pesticide before application.

Using the proper pesticide dilution is a fairly serious matter. Using too little material will obviously give less than hoped for control of the pest, regardless of its nature. Using too strong a spray mixture could cause plant damage, leave excess residues at time of harvest, or be hazardous to nontarget elements such as pets, sensitive plants and beneficial insects. Using an excessive spray concentration will not kill more pest insects, control the disease longer, or kill more weeds. The recommended dilution has been demonstrated by research to be the most effective for that specific purpose. Heading the list of reasons for not using a heavier concentration than called for is cost. The extra amount of pesticide used is truly wasted. So, read and follow the label directions to obtain the best results.

Table 10 has been designed to aid the home gardener in making one gallon of the proper spray mixtures from wettable powders (WP) or from liquid concentrates (EC). To use Table 10 when mixing a spray of a prescribed percentage of the actual chemical (active ingredient), you need to know only the concentration of the formulation. First find the percentage of actual chemical wanted, then match it with the formulation.

Example: a 1% Baygon® spray is recommended for cockroach control. To make a 1% spray using a 25% emulsifiable concentrate, add 10 tablespoonfuls of the Baygon® concentrate to one gallon of water.

Example: a 0.25% Bravo® spray is recommended for powdery mildew control. To make 0.25% spray using a 15% Bravo® wettable powder, add 7 tablespoonfuls of the Bravo® powder to one gallon of water.

TABLE 10. Pesticide Dilution Table for Home and Garden

(Amount of pesticide formulation for each one gallon of water)

Pesticide Formulation	Percentage of Actual Chemical Wanted								
	.0313%	0.0625%	0.125%	0.25%	0.5%	1.0%	2.0%	3.0%	5.0%
Wettable Powder (WP)									
15% WP	2 1/2 tsp.	5 tsp.	10 tsp.	7 tbsp.	1 cup	2 cups	4 cups	6 cups	10 cups
25% WP	1 1/2 tsp.	3 tsp.	6 tsp.	12 tsp.	8 tbsp.	1 cup	2 cups	3 cups	5 cups
40% WP	1 tsp.	2 tsp.	4 tsp.	8 tsp.	5 tbsp.	10 tbsp.	1 1/4 cups	2 cups	3 1/4 cups
50% WP	3/4 tsp.	1 1/2 tsp.	3 tsp.	6 tsp.	4 tbsp.	8 tbsp.	1 cup	1 1/2 cups	2 1/2 cups
75% WP	1/2 tsp.	1 tsp.	2 tsp.	4 tsp.	8 tsp.	5 tbsp.	10 tbsp.	1 cup	2 cups
Emulsifiable Concentrate (EC)									
10%-12% EC 1 lb. actual/gal.	2 tsp.	4 tsp.	8 tsp.	16 tsp.	10 tbsp.	2/3 pt.	1 1/3 pt.	1 qt.	3 1/4 pt.
15%-20% EC 1.5 lb. actual/gal.	1 1/2 tsp.	3 tsp.	6 tsp.	12 tsp.	7 1/2 tbsp.	1/2 pt.	1 pt.	1 1/2 pt.	2 1/2 pt.
25% EC 2 lb. actual/gal	1 tsp.	2 tsp.	4 tsp.	8 tsp.	5 tbsp.	10 tbsp.	2/3 pt.	1 pt.	1 3/4 pt.
33%-35% EC 3 lb. actual/gal.	3/4 tsp.	1 1/2 tsp.	3 tsp.	6 tsp.	4 tbsp.	8 tbsp.	1/2 pt.	3/4 pt.	1 1/3 pt.
40%-50% EC 4 lb. actual/gal	1/2 tsp.	1 tsp.	2 tsp.	4 tsp.	8 tsp.	5 tbsp.	10 tbsp.	1/2 pt.	4/5 pt.
57% EC 5 lb actual/gal	7/16 tsp.	7/8 tsp.	1 3/4 tsp.	3 1/2 tsp.	7 tsp.	4 1/2 tbsp.	9 tbsp.	14 tbsp.	1 1/2 cups
60%-65% EC 6 lb. actual/gal.	3/8 tsp.	3/4 tsp.	1/2 tbsp.	1 tbsp.	2 tbsp.	4 tbsp.	8 tbsp.	12 tbsp.	1 1/2 cups
70%-75% EC 8 lb. actual/gal	1/4 tsp.	1/2 tsp.	1 tsp.	2 tsp.	4 tsp.	8 tsp.	5 tbsp.	7 1/2 tbsp.	13 tbsp.

gal. = gallon lb. = pound pt. = pint tbsp. = tablespoon tsp. = teaspoon
3 level teaspoonfuls = 1 level tablespoonful 2 tablespoonfuls = 1 fluid ounce 8 fluid ounces or 16 tablespoonfuls = 1 cupful
2 cupfuls = 1 pint 1 quart = 2 pints or 32 fluid ounces 1 gallon = 4 quarts or 128 fluid ounces

INSECT PESTS

Upon his painted wings,
the butterfly Roam'd,
a gay blossom of the sunny sky.
Willis G. Clark

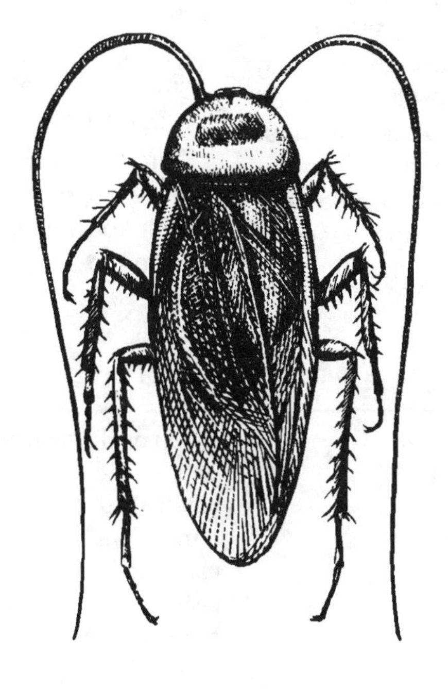

INSECT PESTS

Of the 1 million or more species of insects, only about 1000 are of economic pest status over the entire earth and as you might guess, most of these fall in the agricultural pest class. Less studied and counted are those that are insect pests around and in man's structures, the domestic or urban insect pests. There may be as many as 600 of these but it will probably narrow down to around 200 species that are the normal, everyday type of damaging or simply annoying insects broadly classed as "pests".

But what is an insect pest? We could say that a pest is a pest "in the eyes of the beholder". For what may drive one person up the wall may not even have caught the attention of another. A good example of a pest insect that most people recognize immediately is the earwig, scurrying across the floor. To some, this is a hideous, despicable nuisance. To others they are only amusing, for the latter know that these are only small, harmless guests sharing their home, usually accidentally, but certainly with no malicious intent.

Honeybees are pests to some individuals; especially those who are allergic to bee stings, as well as those who spend long hours digging, transplanting and weeding in the flower garden. The fear of being stung usually becomes exaggerated once stung and all efforts are made to avoid the nectar-foraging friends. Wasps and hornets are feared even more than the lowly bee. Add to this, bumble bees and syrphid flies. Because syrphid flies resemble bees and wasps to the untrained eye, they are frequently mistaken for them and suffer the same classification and as a consequence, fate, as the look-alike pests.

The best way to handle the subject of insects around the home and garden is to break them down into classes: household, vegetable garden, fruit trees, pets, lawn and so on. Diagrams of these insect pests will be used when they are believed useful for gross identification.

VEGETABLE GARDEN

Insects and Mites in the Vegetable Garden

Vegetable gardening is again in vogue. A recent poll of both urban and rural homes indicated slightly more than 50% raised vegetables for their own consumption. The home vegetable garden not only helps reduce the food bill but also provides the gardener with good outdoor exercise and sources of the essential vitamins, nutrients and minerals. To produce vegetables without having them attacked by a wide variety of insects is next to impossible. Thus a control program is essential. This chapter presents important facts to the home gardener about insects and insecticides that will help him control, safely and effectively, the more troublesome pests. Table 11 and Figure 17 contain guides to the types of insect injury commonly found in the garden and their probable cause.

Throughout the nation there are approximately 150 different species of injurious vegetable pests and we never know which ones will be a problem from one season to the next. Nor can we guess how severe

TABLE 11. Guide to Insect Injury in the Garden — Plant Part and Kind of Injury

SEEDS (sprouting)
- Eaten or tunneled (chewing mouthparts)
  - seed corn maggot (beans, corn, melons)

SEEDLINGS
- Stems cut off or girdled (chewing mouthparts)
  - crickets
  - cutworms
  - darkling beetles
  - seed corn maggots (beans, corn, melons)
- Eaten or skeletonized (chewing mouthparts)
  - ants
  - caterpillars
  - armyworms
  - cabbage loopers
  - pillbugs
  - slugs and snails
- Wrinkled or withered (sucking mouthparts)
  - aphids
  - false chinch bugs
  - thrips

STEMS AND VINES
- Chewed or eaten
  - beetles
  - caterpillars
  - crickets
- Tunneling (chewing)
  - cornstalk borer (corn)
  - potato tuberworm (potato)
  - corn borer (corn)
  - squash vine borer (pumpkin, squash)
  - stalk borers (eggplant, potato, tomato)
- Discolored or withered (sucking)
  - aphids
  - plant bugs
  - potato psyllid (potato, tomato)
  - spider mites
  - whiteflies

LEAVES
- Eaten, partly or totally (chewing)
  - ants
  - caterpillars
  - blister beetles (eggplant, pepper, tomato)
  - carrot beetle (larvae and adults)
  - celery leaf tier (celery)
  - crickets
  - cucumber beetles (cucumber, melons)
  - flea beetles
  - grasshoppers
  - hornworms (tomato)
  - June beetles
  - May beetles
  - leaf beetles (beans)
  - Mexican bean beetles (beans)
  - leafrollers
  - potato beetle (eggplant, potato, tomato)
  - slugs and snails
  - squash beetle (squash)
  - tortoise beetles (peppers)
  - vegetable weevil
  - webworms
  - whiteflies
- Tunneled or mined (chewing)
  - potato tuberworm (potato)
  - leaf miners (cucurbits, eggplant, pepper, tomato)
  - tomato pinworm (tomato)
- Discolored, wrinkled, withered or peppered
  - aphids
  - false chinch bug
  - fleahoppers (melons)
  - harlequin bug
  - leafhoppers
  - potato psyllid (potato, tomato)
  - spider mites (melons, beans, tomato)
  - squash bugs
  - thrips (onion)

FRUIT
- Chewed (chewing mouthparts)
  - cucumber beetles (cucumber, melons)
  - grasshoppers
  - tomato fruitworm (tomato)
- Tunneled
  - corn earworm (corn)
  - tomato pinworm
  - pepper weevil (pepper, eggplant)
- Discolored, wrinkled, withered (sucking mouthparts)
  - aphids
  - fleahoppers
  - plant bugs
  - squash bug (cucurbits)
  - stink bugs (tomato)
  - spider mites (tomato)
- Secondary feeders in injuries
  - Sap beetles (corn)

ROOTS
- Chewed or gouged (chewing mouthparts)
  - white grubs
  - several beetle and weevil larvae

ROOTS OR TUBERS
- Tunneled (chewing mouthparts)
  - flea beetle larvae (potato)
  - potato tuberworm (potato)
  - wireworms (potato)

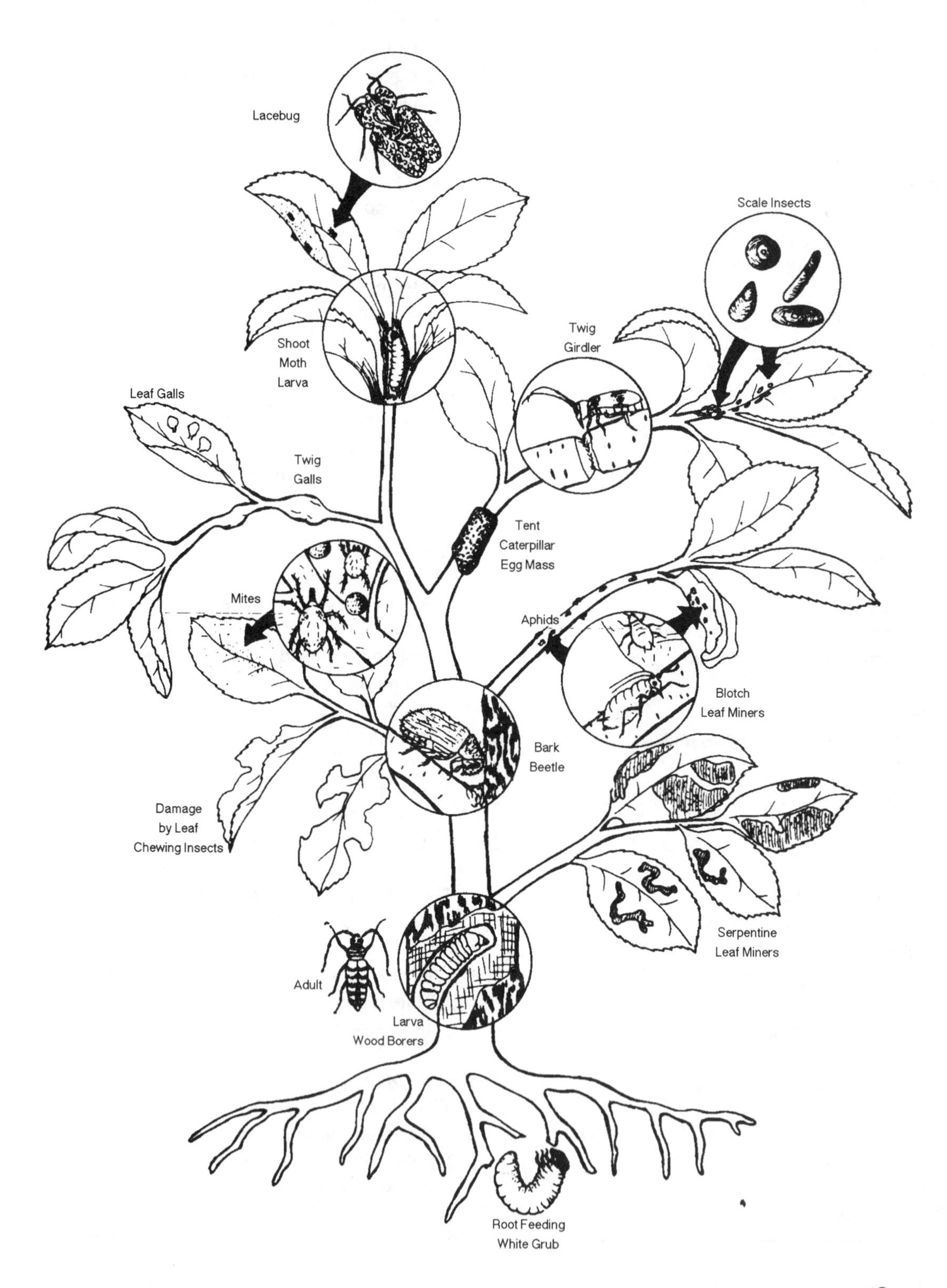

FIGURE 17. Insect Injury to Plants (Courtesy *Cornell University)*

they will be. Some insects, for instance bean beetles, are usually a problem every year. Others become problems only occasionally. When an insect appears in the garden varies with different insects and the weather. Some insects spend the winter in or near the garden and become active as soon as warm days occur. Other insects fly in and out of the garden at random, while still others appear later in the summer. In other words, insects appear in the home vegetable garden from the time it is planted until it is leveled by frost in the fall.

Ants

Everyone recognizes and probably has, ants. There are several species of ants available for practically every garden, and are recognized by their constricted waists. The ones usually seen are the wingless adults of the worker caste. They headquarter in underground nests and are seldom more than minor pests. Their presence is usually their most undesirable trait, particularly when they collect the sweet, sticky honeydew from plants infested with aphids. The more important ant pest is the Argentine ant *(Iridomyrmex humilis).* Being very fond of honeydew, the excretion from plant lice or aphids, it hunts vigorously for it, particularly on plants with moderate to heavy aphid infestations. In their search the ants interfere with the parasitic and predaceous activities of beneficial insects thereby increasing the aphid populations. By controlling the Argentine ant you are probably also assisting in the control of your aphid problem. The leafcutter ants *(Acromyrmex* spp.*)* remove pieces of foliage and carry them in caravans along trails toward underground nests where they are used to support a fungus grown as the basic ant food. Harvester ants *(Pogonomyrmex* spp.*)* harvest and store seeds for food and sometimes remove all the vegetation with a several-foot radius of the nest entrance. Those aggressive little fire ants *(Solenopsis* spp.*)* can inflict painful stings. Their nests are easily recognized by the circular ring of soil around the entrance. There are several species of ants, too numerous to detail and they generally need no chemical control. However, when required, it involves the application of an appropriate insecticide to and around the nest entrance. Pharaoh ants *(Monomorium* spp.*)* are active year round in heated buildings. They are controlled indoors with boric acid baits and the insect growth regulator, methoprene. (See Appendix F)

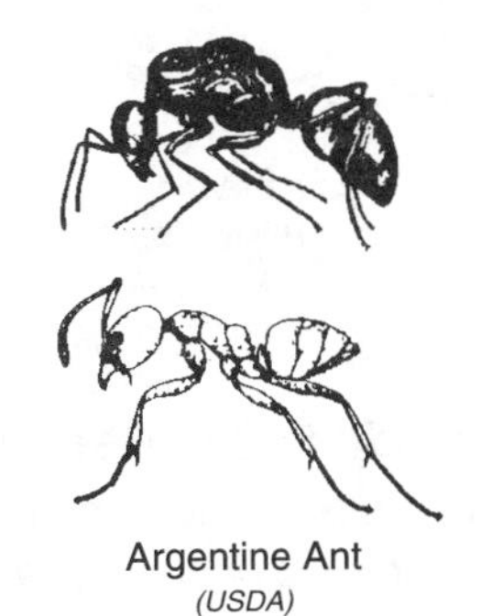

Argentine Ant
(USDA)

Aphids

Frequently referred to as plant lice, aphids are small, slow-moving, soft-bodied insects, winged or wingless when grown. Usually they are found on the young leaves and stem of growing tips. The common garden aphids are the green peach aphid *(Myzus persicae),* which infests several varieties of garden plants, the cotton or melon aphid *(Aphis gossypii),* which is commonly found on melons, squash and cucumbers and the cabbage aphid *(Brevicoryne brassicae),* found on members of crucifers, or the cabbage family. Like ants, there are also many species of aphids found in gardens.

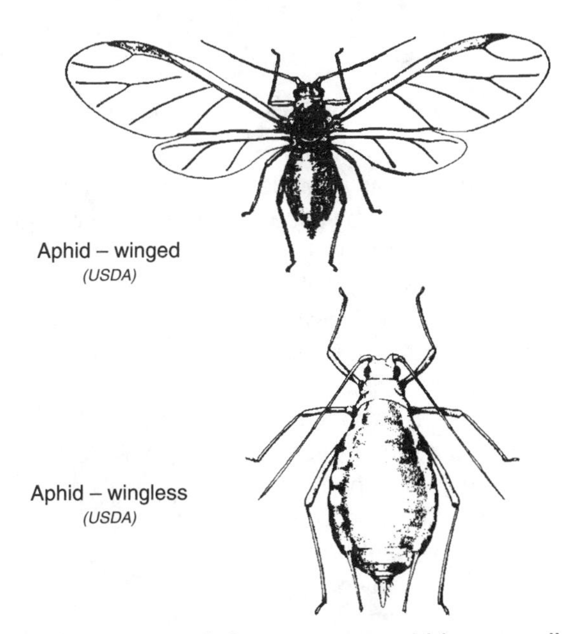

Aphid – winged
(USDA)

Aphid – wingless
(USDA)

In cooler parts of the country, aphids normally produce a fall generation of winged males and females, which mate. The result is an egg, the over-wintering stage. In warmer areas, aphids occur only as females which reproduce by giving birth to living young, with many generations each year. Most aphids are plentiful during the spring months, before their predators and parasites, which have longer life cycles, have gotten off to a good start.

Populations of aphids are frequently held under control by their enemies, which include lacewings, tiny parasitic wasps, lady beetle adults and larvae, syrphid fly larvae and occasionally by disease. The purchase and introduction of lady beetles and preying mantids into the home garden to beef-up those already present is usually of little value. After they've cleaned up the resident aphid populations, they depart in search of richer feeding grounds.

Beetles

Bean Beetles

The Mexican bean beetle *(Epilachna varivestis)* is practically a universal pest of beans, including lima, bush or string and pinto. The adults resemble their

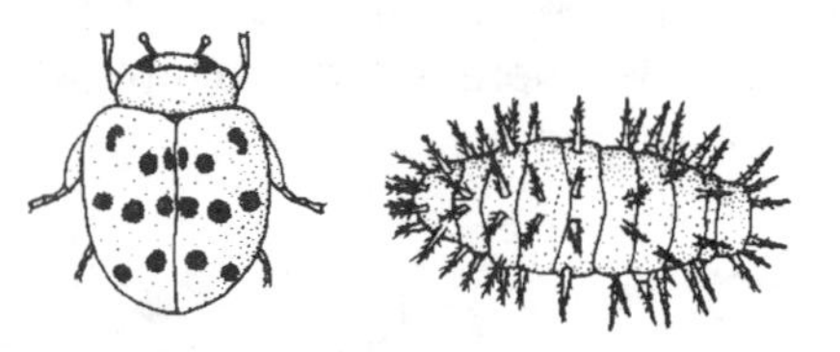

Mexican Bean Beetle – Adult and Larva
(Univ. of Arizona)

close kin, the lady beetles, but are larger, 1/4 to 3/8" long. They are yellow to coppery brown, with 8 black spots on each wing cover. The larvae are oval and about 1/3" long when mature and have 6 rows of long, branched, black-tipped spines along their backs. It can sometimes become a serious pest in the home garden, because both the adults and larvae feed on lower surfaces of bean leaves, leaving only a network of veins. The stripped, skeletonized leaves usually dry up and in severe injury the plants may be killed. Pods and stems are also attacked when heavy infestations occur. The bean leaf beetle can be found all through the United States.

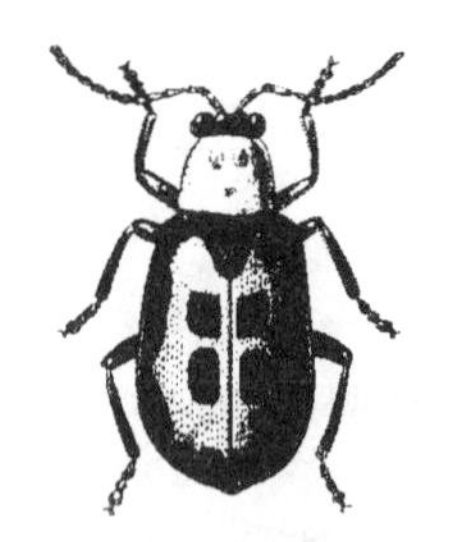

Bean Leaf Beetle
(Union Carbide)

Blister Beetles

There are several kinds of blister beetles, particularly *Epicauta* spp., that eat any of the above-ground parts of tomato, potato, pepper, eggplant and other garden vegetables. They frequently move in small herds and can quickly cause noticeable damage when they appear in large numbers. Blister beetles are identified by their long bodies and slender "shoulders", which are narrower than the head or abdomen. Body colorings vary as do markings and surface textures. Some are quite pretty, and most are from 1/2 to 5/8" long. There is no reason to be concerned about the larvae since they do not feed on foliage.

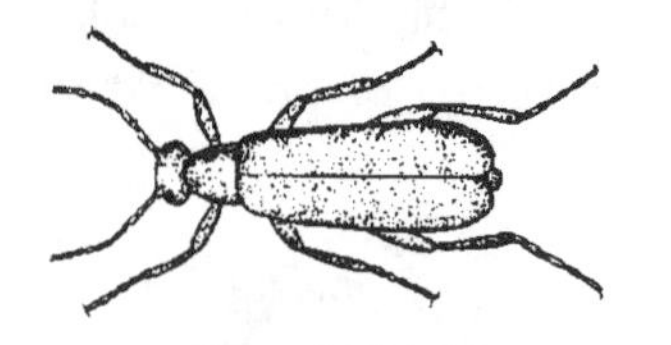

Epicauta fabrici
Ash–Gray Blister Beetle
(U.S. Public Health Service)

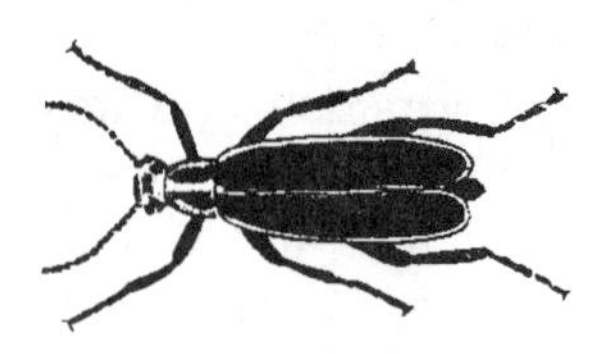

Epicauta pestifera
Margined Blister Beetle
(U.S. Public Health Service)

Colorado Potato Beetles

Both adults and larvae of the Colorado potato beetle *(Leptinotarsa decimlineata)* or just plain potato beetles may be found feeding on potato leaves anywhere in the U.S. They also attack tomato and eggplant and may completely consume infested plants when abundant. The adults are rather rounded, about 3/8" long and 1/4" wide, with the head and thorax black-spotted and the wing covers bearing 10 black and white lengthwise stripes. It can be a serious garden pest in some parts of the nation.

Colorado Potato Beetle – Adult and Larva
(USDA)

Cucumber Beetles

There are several species of cucumber beetles that can be potential problems on several plants, especially the cucurbits. The 12-spotted cucumber beetle *(Diabrotica undecimpunctata tenella)* is related to the western spotted cucumber beetle and to the spotted cucumber beetle known as the corn rootworm of mid-west and eastern states. It is recognized by its black head, yellow prothorax and yellow or greenish-yellow wing covers bearing, naturally, 12 black spots. The western striped cucumber beetle *(Acalymma trivittata)* is about 1/5" long with orange-yellow prothorax and wing covers marked black and yellow lengthwise stripes. The banded cucumber beetle *(Diabrotica balteata)* has a green-yellow prothorax and yellow wing covers marked with 3 transverse green bands. Cucumber beetles feed on leaves, stems and fruit of cucurbits and may produce numerous irregular shot-hole leaf perforations and scars on the rinds of developing fruit. The larvae feed on roots of many common plants and go under the broad identification of root worms. They may also stunt or kill the vines and attack fruit that touch the ground.

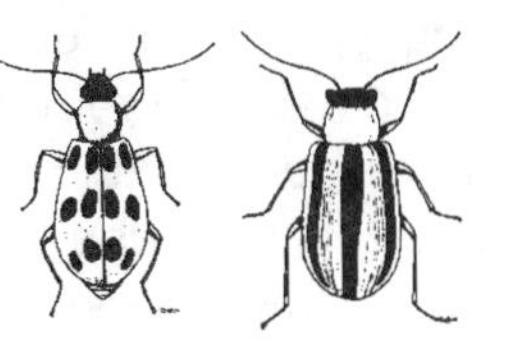
Cucumber Beetles
A. Spotted B. Striped
(USDA)

Darkling Beetles

These are small, dark brown to gray-black beetles, common in western states, that sometimes attack seedlings of practically all garden plants by girdling or cutting off stems at below ground level. Their activities are sometimes blamed on cutworms, since they also are most active at night. When abundant, they may become nuisances by accumulating around outdoor light and occasionally invading homes. The larvae are known as false wireworms and are seldom seen.

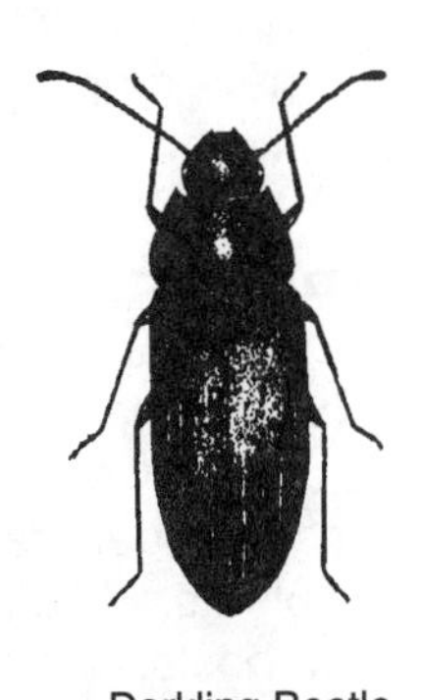
Darkling Beetle
(Univ. of Arizona)

Flea Beetles

Several species of flea beetles attack garden plants in all stages of growth. They range from tiny 1/16" to medium 1/4" long beetles with strong hind legs equipped for jumping, thus the name flea beetles. They come in various colors including black, green, blue-black and yellow. In some species, the head or prothorax may be red, brown or yellow and wing covers may be of a single color or with light and dark stripes. These jumping beetles attack potato, corn, carrot, cauliflower, bean and many other vegetables, eating round or irregular holes in the leaves. This gives the leaves a riddled appearance and some plants, such as corn, may have a scalded appearance resulting from the feeding of adults on the upper leaf surfaces. When flea beetles are numerous, young plants may be killed in 1 or 2 days. The slender white larvae feed on roots and underground stems, or in some species, may feed on lower leaf surfaces or tunnel inside leaves. Some cause potato tubers to become pimply or silvery from small pin holes and burrows just below the surface.

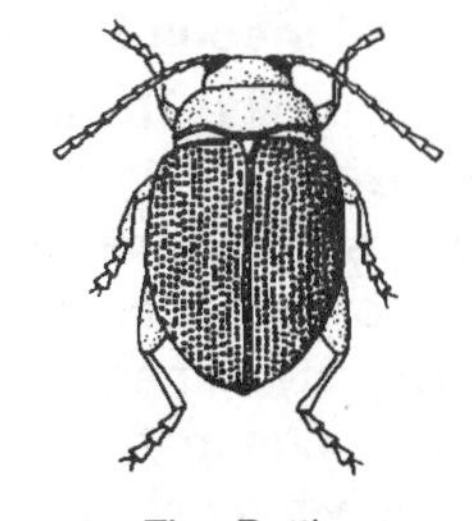
Flea Bettle
(Univ. of Arizona)

May or June Beetles and White Grubs

May beetles of various species, also called June beetles, occasionally feed on foliage of home garden plants. The larvae of May beetles are known as white grubs, are root feeders and are probably more injurious than the adults as garden pests. Adults are from 1/2" to 1" long of various sizes and colors. They become active at dusk and are attracted to lights, flying awkwardly and crashing into objects with their heavy buzzing. The white grubs, or soil-infesting larvae, are whitish with a dark head and 3 pairs of short legs immediately behind the

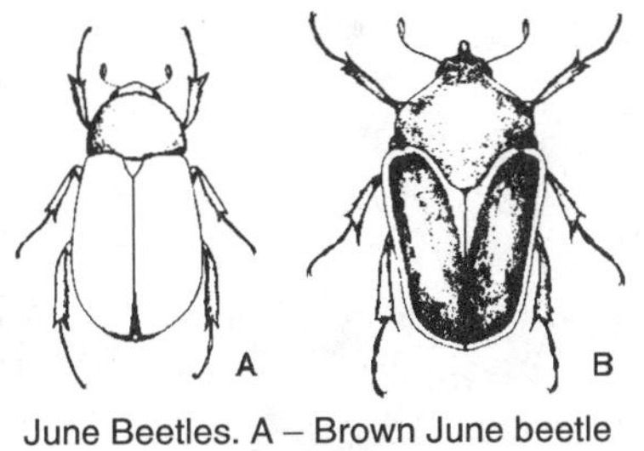

June Beetles. A – Brown June beetle
B – Green June beetle
(Univ. of Arizona)

A White Grub
(Univ. of Arizona)

head. The grubs are always found in their typical "C"-shaped position. Underground damage occurs to practically all varieties of garden plants. The common green June beetle, or fig beetle, *(Cotinis nitida)* is not ordinarily a garden pest, although it may attack over-ripe tree fruit and melons. Their larvae feed on decaying organic matter and not on the living roots of garden plants.

Miscellaneous Beetles

Several other beetles may occasionally become localized or minor pests of home gardens. Sap beetles, of the family Nitiduilidae, may invade maturing ears of sweet corn to feed on fermented kernels previously damaged by the corn earworm. They may also invade maturing melons and tomatoes injured by growth cracks or by earlier infestations of other insects. Adults and larvae of species of tortoise beetles may feed on foliage of peppers and occasionally other vegetable leaves. Wireworms, the larvae of click beetles in the family Elateridae, sometimes tunnel into potato tubers and roots of other plants such as sweet potatoes.

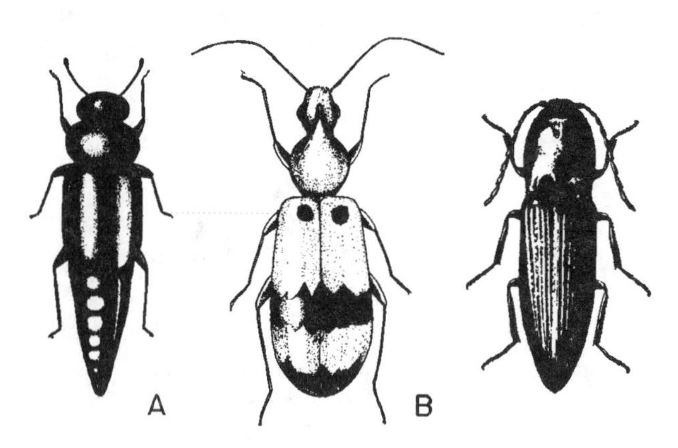

Non-injurious Beetles Found in Gardens. A – Fruit bud beetle; B – Notoxus beetle. *(Univ. of Arizona)*

Click Beetle *(USDA)*

Weevils

There are a large group of beetles whose head tapers into a snout with the mandibles at the tip, the weevils. Although the adults may be destructive pests, the grub-like, legless larvae, often found tunneling within plant tissues, also cause serious injury. Among these are the vegetable weevil, pepper weevil, and several related to the potato stalk borer.

Vegetable weevil larvae *(Listroderes costirostris obliquus)* attack carrots, celery and other vegetables during spring and early summer months. These fleshy, cream-colored grubs feed on foliage and may gouge and tunnel the fleshy roots of carrots and other crops. Adult weevils are brownish-buff, 1/2" long, with a "V" shaped, whitish marking on the wing covers. Adults feed on the foliage of various vegetables during the summer and fall months.

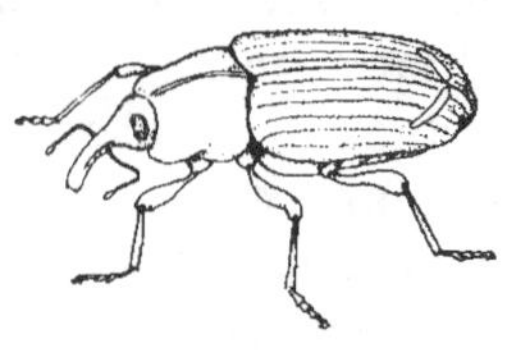

Vegetable Weevil *(Univ. of Arizona)*

The stalk-boring weevils (*Trichobaris* spp.) related to the potato stalk borer, are occasional pests of potato, tomato and eggplant. The graying adults are 1/4" long. The legless larvae are white with darker heads and injure plants by drilling in the stems above and below ground level.

The pepper weevil *(Anthonomus eugenii)* is one of the smaller snout beetles, being about 1/8" long, shiny black and covered with white fuzz. Adults feed on buds, tender pods and leaves of peppers and sometimes eggplants. Eggs are laid within the buds or pods where the larvae hatch and begin feeding. Injured peppers darken and decay, and the smaller peppers may drop in large numbers under heavy infestation.

Caterpillars

Several species of caterpillars commonly attack garden vegetables and may very well be the most serious category of insect pests. Some caterpillars, or worms, are general feeders on many cultivated and wild plants, while others are limited to a single or only a few hosts. Caterpillars are the young or larvae of moths and butterflies, and develop from eggs usually laid on or near the plants they attack. Most garden caterpillars, with a few exceptions, are larvae of dull-colored, night-flying moths. Caterpillars have chewing mouthparts and may damage leaves, buds, fruit, stems or roots. Some may fasten leaves together with silken webs to make protective shelter in which they feed, including the celery leaftier *(Udea rubigalis)*, the omnivorous leafroller *(Platynota stultana)*, and webworms (*Loxostege* spp.).

Cutworms of various species are general feeders on many garden vegetables. Newly transplanted succulent plants are particularly vulnerable, such as tomato and cabbage. Emerging seedlings are cut off at or below the soil surface, and roots and tubers may on occasion be attacked. Cutworms are active during the night and hide during the day beneath clods or in the soil near the plants attacked. A single cutworm may destroy several plants each night, progressing steadily down a row. These voracious larvae are the young of several inconspicuous moths, and when full grown are fleshy, dull in color, with or without markings, up to 2" long, and commonly seen in their distinctive curled position in the soil. To protect transplanted garden seedlings, wrap stems of seedlings with newspaper or foil to prevent damage by cutworms.

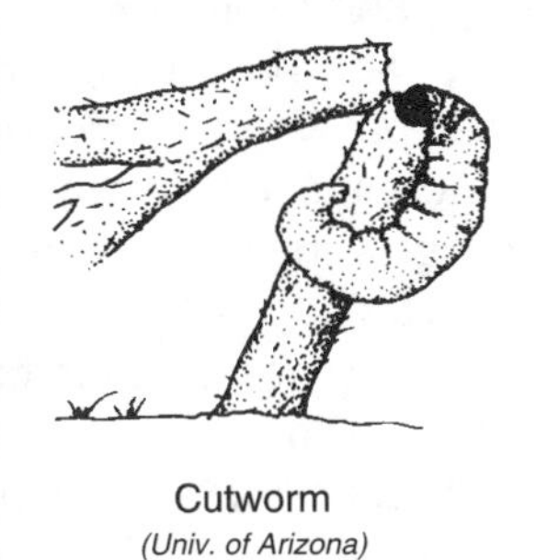

Cutworm
(Univ. of Arizona)

The **cabbage looper** *(Trichoplusia ni)* is perhaps the most common garden caterpillar since they feed on lettuce and several of the crucifers (cabbage family) found in all home gardens. With many generations each year, many kinds of garden plants may be attacked and young plants may be consumed to the ground. Large holes may be eaten in older leaves, however when infestations are light or moderate this damage may be outgrown. The cabbage looper gets its name from its looping movement as it inches forward, and occasionally it is referred to as an "inch worm". Newly hatched loopers are pale green with a black head and are usually found feeding on the undersides of leaves. Older larvae, up to 1-1/2" long, are pale green or tan, with or without white lines along the sides. They may completely devour leaves and consume several times their body weight in plant tissue each day. Small plant-colored fecal pellets are easily identified evidence of their presence.

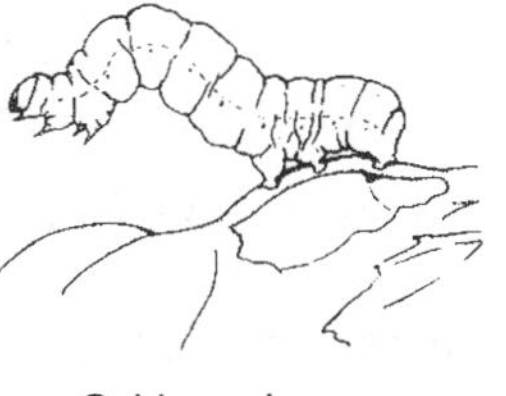

Cabbage Looper
(Univ. of Arizona)

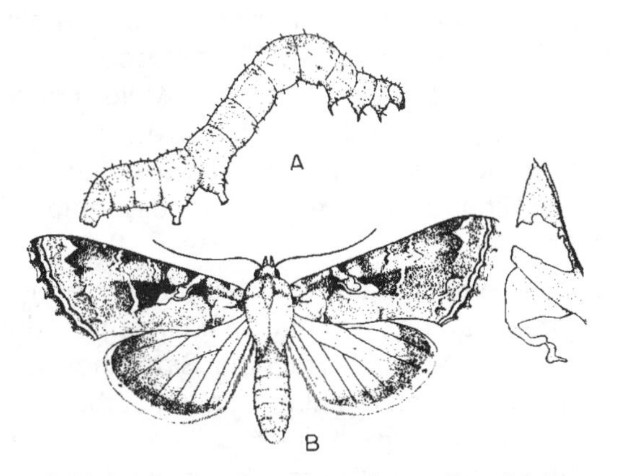

Cabbage Looper. A — Mature Larva; B — Adult
(Univ. of Arizona)

The **corn earworm**, *(Helicoverpa zea)* alias **tomato fruitworm**, alias **cotton bollworm**, prefers developing ears of sweet corn and tomatoes beyond all other food, though it is also a pest of other crops including lettuce. It is the major pest of home garden sweet corn; the principal damage is done when the developing kernels within the ears are eaten. Leaf whorls and developing tassels of younger corn plants may also be attacked. The larvae hatch from eggs laid on fresh corn silks and feed down through the silk channel above the tip of the ear to the kernels where most of the feeding occurs, leaving behind abundant feces or frass. Full-grown larvae are up to 1-1/2" long with brown heads and body coloration varying from green with touches of red or brown to almost black. All stages have a pair of dark lines extending along the back and two other bands along the sides, separated by the usually prominent spiracles, or breathing holes. In its role of a tomato fruitworm, this pest hatches from eggs laid down on tomato leaves. Although foliage may be eaten, the most important feeding damage is to developing tomatoes.

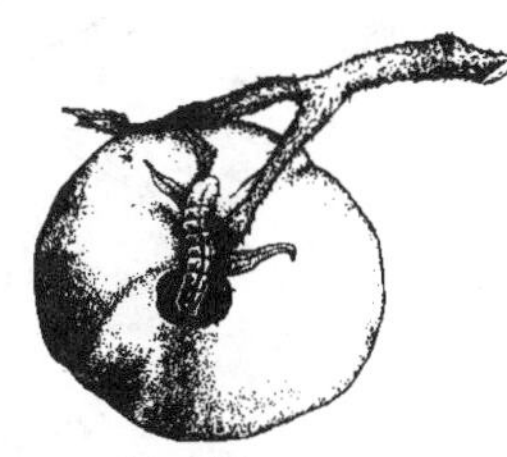

Corn Earworm
in Tomato
(Univ. of Arizona)

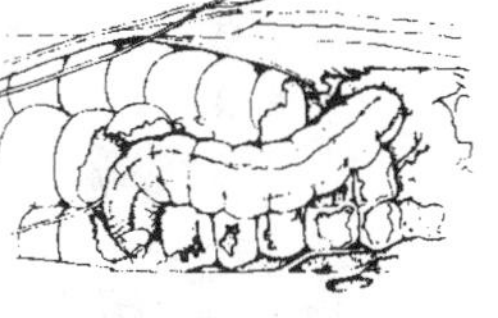

Corn Earworm
in Sweet Corn
(Univ. of Arizona)

Armyworms may on rare occasion invade home vegetable gardens. The beet armyworm *(Spodoptera exigua)* and the yellow striped armyworm *(Spodoptera ornithogalli)* feed on beets and lettuce and some other leafy plants, particularly in the seedling stage. In large numbers they may consume entire plants. Mature beet armyworms are gray-green and up to 1-1/4" long, with an irregular black spot on each side of the thorax above the second pair of legs. Mature yellow striped armyworms are up to 1-3/4" long, generally purplish-brown, velvet-black on the back and have two outstanding yellow stripes on each side. Younger larvae of these two are variable in color but similar in appearance and feeding

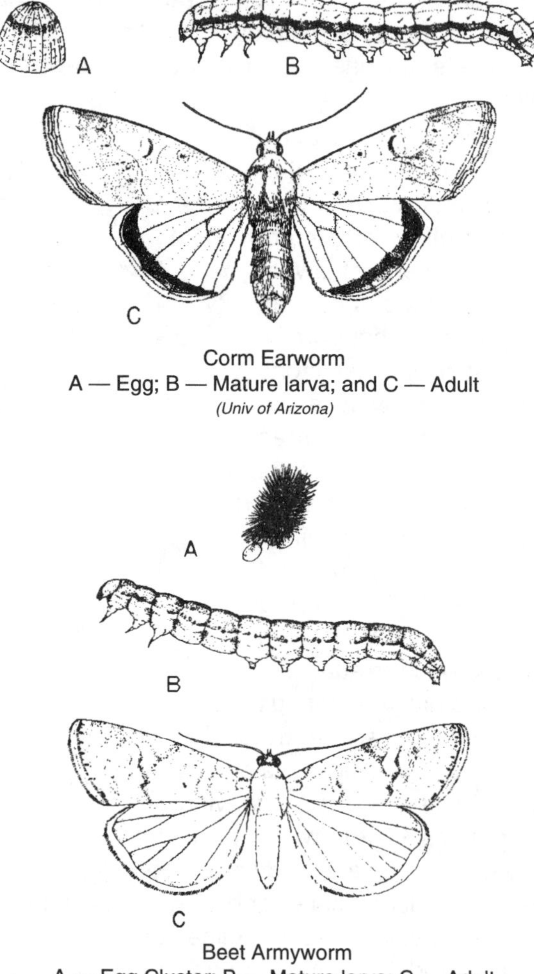

Corm Earworm
A — Egg; B — Mature larva; and C — Adult
(Univ of Arizona)

Beet Armyworm
A — Egg Cluster; B — Mature larva; C — Adult.
(Univ. of Arizona)

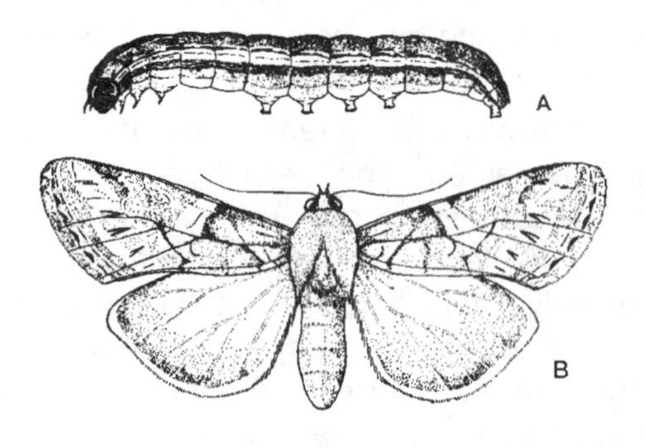

Yellow Striped Armyworm. A—Mature larva; B—Adult
(Univ. of Arizona)

habits. They can be distinguished from cabbage loopers by their 4, instead of 2, pairs of abdominal legs and their absence of looping movements. Fall armyworm larvae *(Spodoptera frugiperda)* feed inside the leaf whorls of developing sweet corn plants before tassels emerge and they may weaken or kill the growing tip. This damage is usually not observed until the ragged, shredded-looking leaves grow out. Fall armyworms resemble corn earworms in size and shape. When mature they have shiny, brown bodies with black spots. Their black head has a very distinguishing, white marking resembling an inverted "Y".

Tomato Hornworm
(Union Carbide)

Tobacco Hornworm
(Univ. of Arizona)

Tomato and tobacco hornworms are occasional foliage pests on tomatoes. Hornworms are large, sphinx moth larvae, much larger than any other garden caterpillar. They derive their names from the spine-like horn at the end of the abdomen, which appears menacing to the untrained. The tomato hornworm *(Manduca quinquemaculata)* has 8 forward-pointing, V-shaped, white markings on each side, each enclosing a spiracle or breathing hole. Its horn is black and uncurved. The tobacco hornworm (*Manduca sexta*) has 7 diagonal white markings, extending upward and to the rear on each side of the body. Its horn is red and curved. Neither require chemical control in that they are readily removed by hand-picking.

Woolly worms are the saltmarsh caterpillar *(Estigmene acrea),* and may become a pest during the late summer and fall months. Since they usually migrate from commercial plantings, they are a problem only to gardeners adjacent to farming operations. The larvae develop in farm fields and then move out in hordes, feeding on nearly all kinds

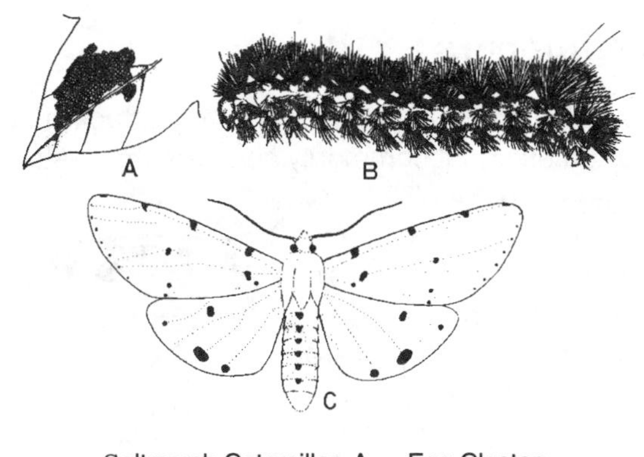

Saltmarsh Caterpillar. A — Egg Cluster; B — Mature larva; C — Adult.
(Univ. of Arizona)

of vegetation in their path. Lettuce and other leafy vegetables are particularly vulnerable. Full-grown caterpillars are up to 2" long, buff-colored, and covered with dense dark hairs, thus wooly worms.

Lesser cornstalk borers *(Elasmopalpus lignosellus)* kill or weaken seedling corn plants by entering at or below the ground line and tunneling inside often destroying the growing tip. Beans, turnips, peanuts and other plants may also be attacked, and one larva is capable of injuring or killing several plants. The larvae are usually found outside the infested plants resting in silk tubes in the soil adjacent to the entrance holes. They move rapidly when disturbed, and reach 3/4" in length when mature. Johnsongrass is a favored host and should be removed from the home garden area.

Lesser Cornstalk Borer
(Univ. of Arizona)

Potato tuberworms *(Phthorimaea operculella)* bore into potato tubers while growing and also in storage. They enter near the soil surface and make tunnels which become brown and corklike, as well as mining in growing potato foliage and stems. Other potato-related plants such as tomato and eggplant may also be attacked. There may be 2 or 3 generations per year, and the mature larvae are about 1/2" long, pink or white, and a red-purple band down the back. It sometimes develops in volunteer plants and in old tubers left in the ground or in storage.

European Corn Borer
(Union Carbide)

Corn borers *(Diatraea spp.)* include the European and southwestern and are primarily pests of field corn, but sweet corn is also attacked. The young caterpillars feed on leaf surfaces, producing translucent or skeletonized areas. They may also feed, several together, in leaf whorls of younger corn plants and may destroy the growing tip, causing a condition known as dead heart. Early in their development the larvae bore into the stalks, tunneling both upward and downward. Tunnels may extend from bases of the ears to the tap roots. Internal girdling, especially near the ground, may cause the stalks to break over. Mature larvae are white, about 1-1/4" long, and may be "peppered" with conspicuous dark spots, depending on the season of the year.

The squash vine borer *(Melittia calabaza)* is one of the more aggravating of the caterpillar pests. It tunnels in vines of squash, pumpkins, melons, and cucumbers and sometimes the fruit are attacked. Coarse yellow borings are pushed out through holes made in the tunneled areas, and the vines may suddenly wilt and die, with no warning to the green thumb owner. Mature borers are white with brown heads and up to 1-1/4" long. The simplest control method is to slit each infested vine area with a blade and remove borers individually. Good preventive methods have not been worked out.

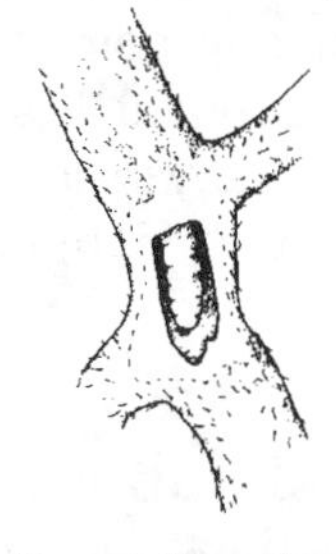

Squash Vine Borer and Injury
(Univ. of Arizona)

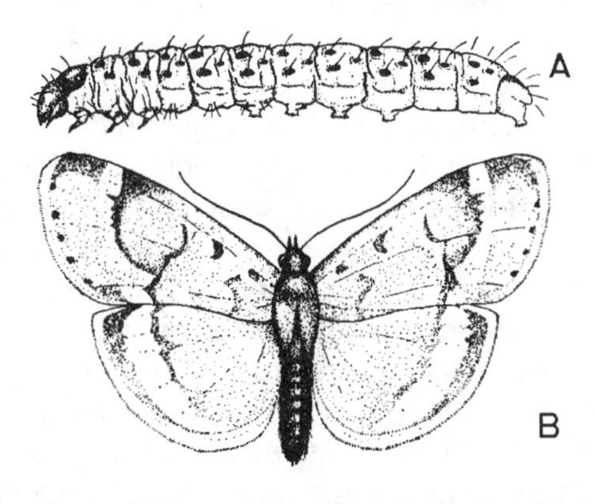

Garden Webworm. A — Mature larva; B — Adult
(Univ. of Arizona)

Garden webworms may be occasional pests in home gardens, and would be classed as general feeders. They may attack foliage of beets, beans, cabbage, cucurbits, potatoes and other related vegetables. They web leaves together to construct shelters within which they feed until the enclosed areas are skeletonized, thus the name webworms. Younger plants may be killed. Several species occur across the United States.

Crickets

Field crickets, (*Gryllus assimilis*) can become pests feeding on the tender seedlings of most vegetables, during late summer and early fall. They are attracted to young lettuce plants, but also nibble on cole crops, beets, tomatoes and most other vegetables. This cricket may eat almost any part of any garden plant when abundant. It is nocturnal, feeding by night and hiding by day in soil cracks or under other cover, so can cause considerable injury before being discovered.

Field Cricket
(Univ. of Arizona)

Grasshoppers

Several species of grasshoppers are occasional pests of home gardens and may feed on the foliage and tender buds of the common vegetables including lettuce, cole crops, potatoes, beets and carrots. One large grasshopper, feeding on only one plant, may consume a surprising amount of foliage before being discovered. There are two common routes that grasshoppers enter gardens: Some develop from eggs laid in soil in or near gardens, and others may migrate by flying in from distant vegetation. In small gardens, on cool mornings when they are relatively inactive, grasshoppers may be caught and eliminated by hand. When chemical control is required treat infested areas while grasshoppers are still small. Biological control is available in the form of baits containing the protozoan, *Nosema locustae*, a disease organism of grasshoppers and crickets. It is sold as Nolo Bait® and other proprietary names. (See Appendix G.)

Grasshopper
(USDA)

Leafhoppers

Several species of leafhoppers may attack garden vegetables. Most commonly seen are the V-shaped hoppers, up to 1/8" long, green to yellow in color and which are active fliers when disturbed. The nymphs resemble the adults, but are smaller, lack wings and consequently cannot fly. Leafhoppers, adults and nymphs, have the habit of walking sidewise, in cautious retreat. They have sucking mouthparts, and damage to foliage varies from unimportant to fatal, depending on the species, numbers and weather conditions. Vegetables commonly attacked by one or more species of leafhoppers are cantaloupes and other cucurbits, tomato, potato, pepper, eggplant, lettuce, beans, beet, chard and radish. Damage to leaves is caused by their sucking sap, removing the green chlorophyll and causing pale, circular spots or specks to appear. The potato leafhopper *(Empoasca fabae*) interferes with fluid movement inside leaves by plugging the phloem tubes, resulting in browning of leaf edges and later the entire leaf. Some species cause stunting and downward curling of leaves. Beet leafhoppers (*Circulifer tenellus*) sometimes transmit curly top virus, which infects various garden vegetables and may be particularly serious to tomatoes. Protection of tomato plants can be achieved by draping them with insect netting (Reemay®) in the presence of dense populations.

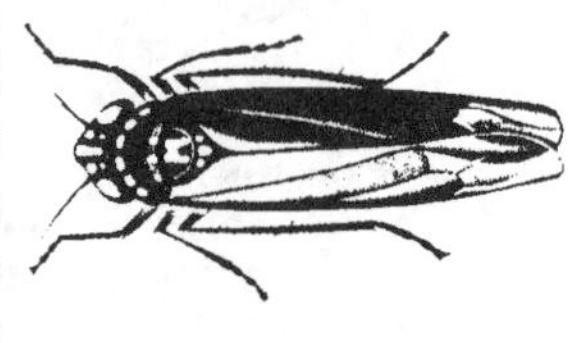
Potato Leafhopper
(Union Carbide)

Leafminers

There are several flies and moths whose larvae tunnel between leaf surfaces of vegetation, known as leaf miners. One common group is the serpentine leafminers *(Liriomyza* spp.), tiny larvae of small flies. Adults are black, about 1/16" long, with yellow markings on the face and back. Eggs are laid in leaf tissues, which hatch into pale larvae that mine gradually enlarged tunnels in random directions within the leaves, leaving dead tissue trails. Several larvae may mine the same leaf, resulting in yellowing and

Leaf Miner Injury
(Univ. of Arizona)

dropping of the leaf. Damage, however, is usually of a minor and sporadic nature, and occurs more commonly on seedling leaves than on larger leaves. Certainly their trails are not as noticeable on mature leaves as on recently emerged seedlings. Plants susceptible to attack are melons and lettuce. Other vegetables that may be infested by these general feeders are beans, eggplant, pepper and tomato. Generally, leaf miners are kept under control by several species of small parasitic wasps, and dead parasitized miners can occasionally be found in leaf tunnels.

Plant Bugs

Various species of plant bugs are sometimes seen in home gardens, including chinch bugs, false chinch bugs, lygus bugs, tarnished plant bugs, flea-hoppers, squash bugs and stink bugs. These all belong to the Order Hemiptera, which are true bugs. They are usually of minor importance, although sporadic infestations may cause serious injury.

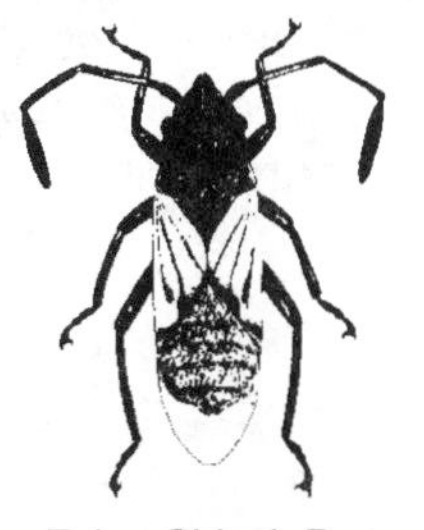

False Chinch Bug
(Univ. of Arizona)

Chinch Bug
(Union Carbide

Chinch bugs and false chinch bugs normally develop on weeds and grasses until they become unpalatable or die then migrate over the ground in swarms to other sources of green food. Vegetable and flower gardens, lawns and young fruit trees are among the common targets. When numbers are heavy most garden vegetables may be seriously weakened or killed. Prevention centers on the control of weeds in neglected areas and on the borders of home gardens.

Fleahoppers may be found on foliage and tender fruit of melons in late spring and early summer. Fleahoppers are 1/8" to 1/4" long, green, black, yellow or sometimes mixed in color and usually jump great distances when disturbed. Fleahoppers are also found on lettuce, carrot, chard and other garden plants. They are not usually considered pests in the home vegetable garden, but more as very active "guests".

The **harlequin cabbage bug** *(Murgantia histrionica),* also known as the calico bug from its red, black, and white coloration, attacks cabbage and other crucifers including cauliflower, broccoli, turnip and radish. Occasionally it will feed on squash, sweet corn and garden beans.

Harlequin Bug
(Univ. of Arizona)

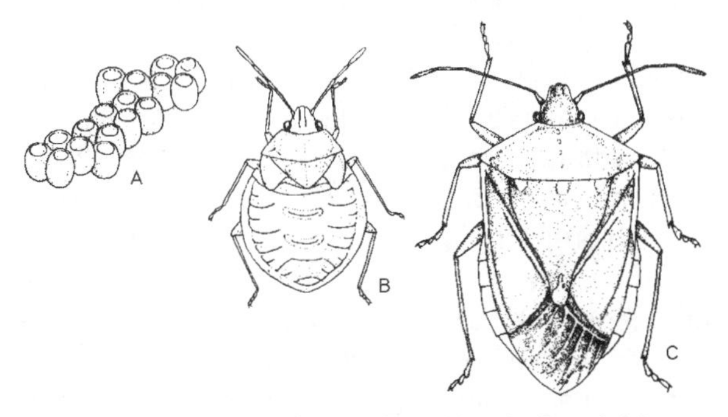

Stink Bugs. A — Egg Cluster; B — Nymph; C — Adult.
(Univ of Arizona)

Stink bugs are probably the largest plant bugs found in the garden. They are recognized by their shield-shaped body and the awful protective odors emitted when molested. They are occasionally minor pests in gardens.

Lygus bugs are minor pests of western home garden vegetables and are most commonly seen when alfalfa, their favorite host, is grown close by. They are mainly pests of buds, flowers and seeds and on many crops and weeds. They have on occasion caused minor injury to beans, cabbage and other crucifers, celery, corn and potato. Their feeding with piercing-sucking mouthparts through tender bean pods may cause the young beans within to become misshapen, with discolored areas around the points of penetration. Similar discolorations surround the punctures of other plants attacked. Lygus bugs vary in color as much as almost any other insect. They may be yellowish green, brown, or reddish brown. The antennae are whiplike and about 2/3 the length of the body. Control measures are seldom needed.

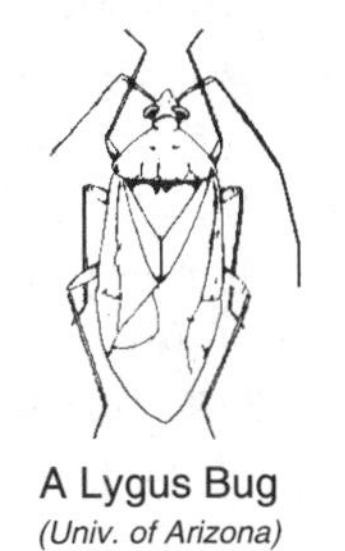

A Lygus Bug
(Univ. of Arizona)

Squash bugs *(Anasa tristis)* are generally found throughout the U.S. They are pests particularly of squash, pumpkin, cantaloupes, cucumbers and watermelons. Both adults and nymphs feed on plant juices. The nymphs hatch from brown eggs laid in clusters on the lower sides of leaves and feed together in colonies. The salivary juices of squash bugs are toxic to leaves and cause large discolored areas to develop, causing leaves to wilt, curl and turn brown. The fruit may be attacked after the vines have been killed. The greater part of squash bug damage occurs in mid-and late summer when nymphs are abundant. Adults are flat-backed, and about 5/8" long. They are blackish brown on top and mottled yellow beneath. Both nymphs and adults emit a disagreeable odor. The most direct control is to collect and destroy adults and egg clusters by hand. In late season bugs may be trapped and destroyed under pieces of board or burlap where they seek overnight shelter.

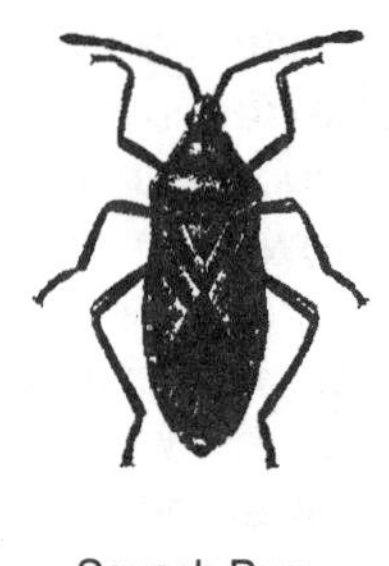

Squash Bug
(Union Carbide)

The **squash mirid** *(Pycnoderes quadrimaculatus)* is one of the smaller plant bugs related to lygus bugs and black fleahoppers. The adults are 1/8" long, black mottled with gray and white and yellow legs. It is quite active and flies readily, being much smaller than the adult squash bug. It feeds on the foliage of squash and other cucurbits, lettuce, beans, and several weeds, usually in late summer and fall. Adults and nymphs leave feeding punctures in the lower leaf surface, which causes the upper surfaces to become gray. Attacked plants may be destroyed under heavy infestations.

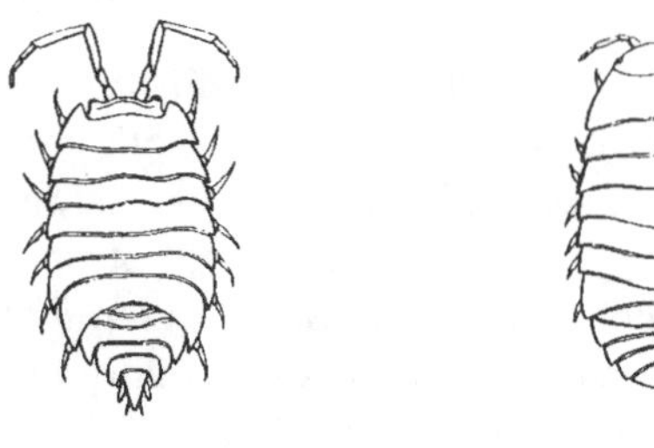

Sow Bug Pill Bug
(U.S. Public Health Service)

Sowbugs and Pillbugs

Sowbugs and pillbugs are not insects, but belong to the crustaceans, somewhat related. They grow to about 1/2" in length, are gray, with 7 pairs of short legs, and a hard, shell-like covering. When disturbed pillbugs frequently roll into a ball resembling an armadillo, thus the name. Sowbugs do not. Unlike insects they breathe by gills and live only in moist environments. Though they normally feed on decaying vegetable material, they may also attack seedlings, new roots and tender stems of growing plants. They are nocturnal and hide by day in damp environments such as under boards, flower pots, rocks and in mulches and decaying plant material. Ventilation and dryness will help reduce their numbers, and control measures are seldom if ever needed.

Spider Mites

Several species of spider mites in the genus *Tetranychus* are among the plant feeding mites frequently found in varying numbers on garden vegetables. They remove plant juices from the leaves through piercing-sucking mouthparts, similar to certain insects, though they are not even distantly related to insects.

These tiny spider mites are almost invisible to the naked eye. Their color varies from a reddish yellow to light green, usually with some dark markings. The young resemble adults, newly hatched nymphs have only 6 legs while older nymphs and adults have eight. Spider mites live on the undersides of leaves, sometimes covering them with dusty webbing. Their feeding results in bronzed leaves, with yellow or brown discolorations. Heavily attacked leaves may become dry and brittle and drop from the plant. Mites are particularly a problem on cucurbits, beans and tomatoes.

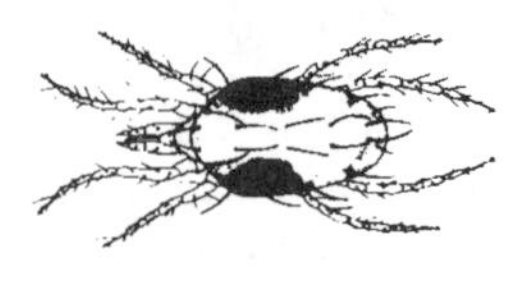

A Spider Mite
(Univ. of Arizona)

On tomatoes can occasionally be found the ultra-small tomato russet mite, which is only 1/125th of an inch in length, and virtually invisible to the naked eye. It is cream colored, elongated and has but two pairs of legs. It attacks only tomato and feeds on the upper surfaces of leaves, causing the usual bronzing. It is rarely found in home gardens and may not be recognized because of its small size. Injury normally appears first on the stalks at ground level and moves upward. In severe cases the fruit may be attacked.

Psyllids

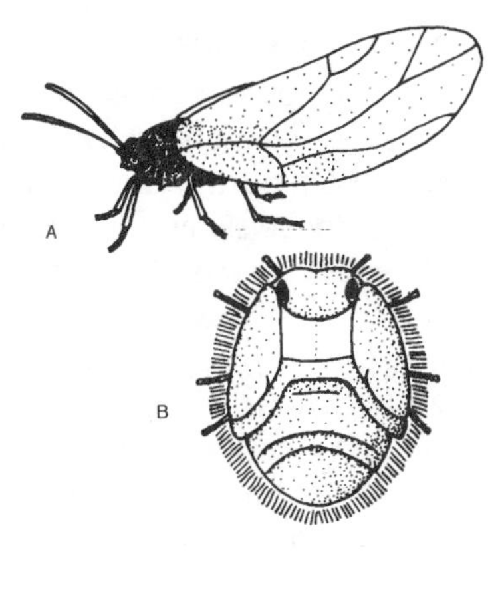

Potato Psyllid. A — Adult; B — nymph.
(Univ. of Arizona)

In the West, the potato (or tomato) psyllid *(Paratrioza cockerelli)* removes the plant juices from potato, tomato, pepper, eggplant and related weeds of the family Solanaceae. The adults are small, about 1/8" long, with dark gray-brown bodies. They are marked above with parallel whitish stripes on the thorax and an inverted white "Y" on the rear of the abdomen. Their wings are held rooflike and are longer than their bodies. When disturbed they jump easily with their strong hind legs, suggesting their rather close relationship to the leaf- or plant-hoppers. Sometimes psyllids are referred to as jumping plant lice. The young (nymphs) are flat, greenish-yellow and resemble their other cousins, the scales. The older nymphs have bright red eyes. Psyllid nymphs are normally found attached to the under sides of leaves, feeding on plant juices through their piercing-sucking mouthparts. Their injury, however, is the result not of removing the plant juices, but rather from injecting their salivary fluid into the leaves while feeding during the nymphal stages, resulting in "psyllid yellows" on potato and tomato. The first symptoms of psyllid yellows are a slight yellowing of the growing tips, the midribs and leaf edges and curling at the base of leaves. Progressively the entire plant becomes yellow to purple and growth ceases. Heavy infestations of nymphs early in the growing season invariably result in reduced yields, both in size and number of tubers.

Seedcorn Maggot

Seed and small seedlings of corn, beans, melons and other large-seeded vegetables may be attacked by the seedcorn maggot *(Hylemya platura).* Their feeding may weaken or destroy cotyledons and growing tips. The full grown maggots, which are the larvae of a gray fly, are 1/4" long, and resemble all other fly-maggots—white, legless and tapered toward the headless front end which bears two destructive feeding hooks. Injury occurs in the spring as seeds begin to emerge, in damp, heavy soils heavy in organic matter especially when cool weather has slowed emergence. (Beware, organic gardeners!) Little damage occurs under ideal growing conditions, as in late summer or early fall gardens.

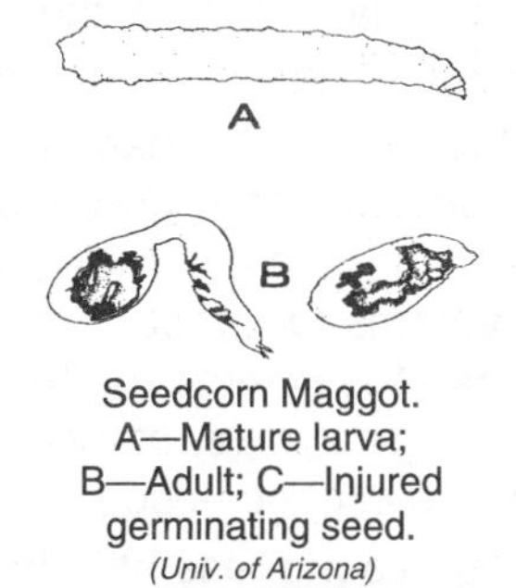

Seedcorn Maggot. A—Mature larva; B—Adult; C—Injured germinating seed.
(Univ. of Arizona)

Slugs and Snails

Snails and slugs belong to the mollusks, which include the shellfish oysters, clams, mussels, barnacles and cuttlefishes. Slugs and snails are the malilcious dry-land equivalent to the marine barnacles, in their damage to plants. They come in colors of grey, orange and black. Snails have distinct shells and slugs appear not to have shells but actually have a very small rudimentary shell plate on the upper side near the head. Slugs and snails both feed on young and succulent plants, especially in moist,

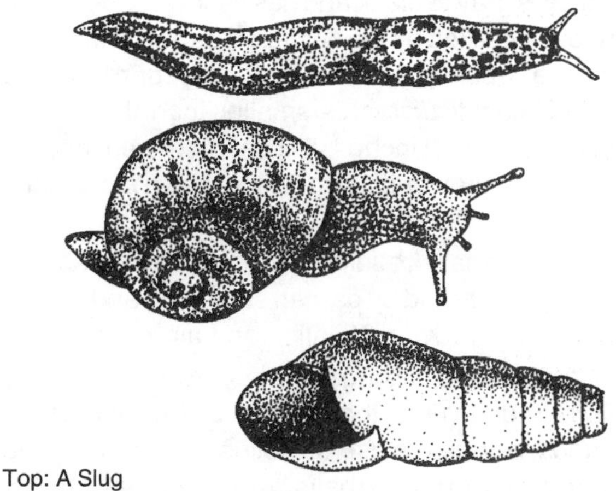

Top: A Slug
Center: Brown Garden Snail
Bottom: Decollate Snail (shell)
(Univ. of Arizona)

well-watered situations. They injure plants by rasping irregular designs in leaves, flowers and stems, normally at night and during damp weather. In contrast, caterpillars feed on plants all the time, leaving large amounts of droppings instead of a slime trail. Often the silvery, slimy trails of snails and slugs are seen on walks, stepping stones, soil, grass, and foliage before their damage is found. Some crawl on houses or damage painted siding.

Slugs are more damaging than snails and both are often present in greenhouses throughout the year and in many home gardens and border flower plantings for most of the summer. Eggs are found almost any time of the year and most species overwinter in this stage. A few species overwinter as adults in concealed places.

The young resemble the adults and begin feeding as soon as they hatch. The life span is usually less than a year.

Two common garden snails are the brown garden snail or European brown snail *(Helix aspersa),* which has a roughly globular spiral shell up to an inch or more in diameter, and the decollate snail *(Rumina decollata),* which has a cone-shaped spiral shell about 1-1/2 inches long and 1/2 inch wide at the open end.

Three species of slugs may be found in home yards and gardens. The spotted garden slug *(Limax maximus)* is the largest and may reach 4 inches in length. It is yellowish-brown to gray and has large black spots which may appear as black bands in the adults. The tawny garden slug *(Limax flavus)* is somewhat smaller, reaching 3 to 3-1/2 inches in length. It is gray to black and has yellow spots. Unlike the clear slime left by the spotted slug, the tawny slug leaves a trail of yellow. The gray garden slug *(Agriolimax reticulatus)* is smaller than the others, measuring up to 2 inches. It is flesh-colored, heavily spotted with gray or black and it leaves a clear slime trail.

Slugs and snails require moist, shaded areas and can be discouraged by garden sanitation and foliage removal to improve ventilation and air movement.

Control by hand picking provides limited success, is tedious, but avoids the use of chemicals. Other methods include the use of molluscicides (snail and slug poisons) as poison baits, trapping under boards, and using beer pan traps. In the poison bait category are two compounds: Metaldehyde and carbaryl, sold as ready-to-use baits. Dusts or liquid formulations are recommended for greenhouse slug problems, while baits are more satisfactory for home garden and flower problems. Baits are more successful when distributed in infested areas just after a rain or after watering when these pests are most active.

Non-chemical control can be achieved by laying boards of any size one inch thick on moist garden soils. The next morning the underside should be heavily laden with these slimy culprets which can be removed and destroyed. (See Diatomaceous Earth, Appendix F.)

Beer pan traps are the most interesting of all control methods. To make these fascinating traps, place small cans in the ground with the lip at soil level, spaced at 3- to 10-foot intervals. Then fill to about one-half with beer. Empty the traps and refill twice weekly.

Thrips

One thrips or ten thrips, the name is both singular and plural. These are slender, tiny, active insects about 1/20" long. Their colors vary from light yellow to dark black, depending on the species. They can readily be found in the blossoms of most garden flowers by thumping the blossom over a sheet of light-colored paper. Their wings are narrow and fringed, but may be absent. Thrips are somewhat unusual in that they possess rasping-sucking mouthparts. Most species are plant feeders, but intermingled among them may be predatory species which feed on mites and other insects. The plant feeders rasp surfaces of young leaves and growing tips to expose the plant juices within, causing the leaves to become distorted. Older tissues become blotched appearing silvery or leathery in affected areas. Two common thrips are the onion thrips *(Thrips tabaci)* and the western flower thrips *(Frankliniella occidentalis).* The latter may be found in both vegetable and flower gardens on practically all species. The onion thrips, when abundant, cause onion leaf blades to become spotted and wither at the tips, giving the plants an aged, grayish appearance. The flower thrips cause occasional damage, particularly in the slow-growing

Onion Thrips
(Univ. of Arizona)

period of early spring and late fall, by feeding on the tender growing tips of developing garden plants. Because adults are strong fliers, there is really no way to avoid their emigration.

Whiteflies

Whiteflies first become noticeable as swarms of tiny 1/16th inch, white, mothlike insects that fly up from plants when they are disturbed. These are the adults. The nymphs are very different, tiny and scale-like, flat and fringed with white waxy filaments. They have sucking mouthparts, are confined to the undersides of leaves, and secrete sticky honeydew.

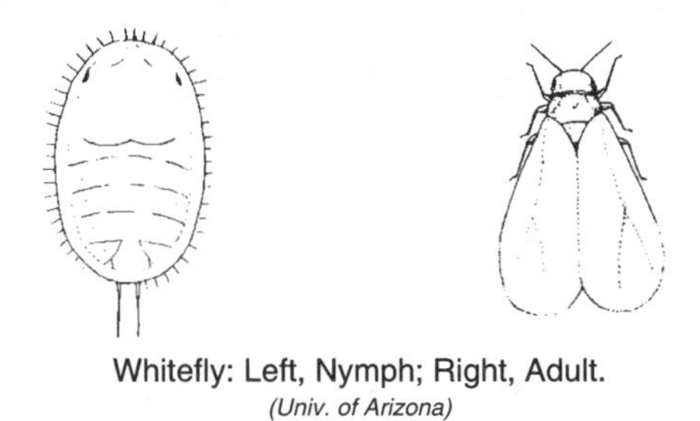

Whitefly: Left, Nymph; Right, Adult.
(Univ. of Arizona)

Whiteflies are very common pests in greenhouses and are much more noticeable because they fly around the bench plants at eye level. In the garden they are usually a pest on tomatoes, though they may be found on other bushy plants such as beans. Whiteflies are most frequent early in the season, but may develop later also, moving from weed hosts in the area. Damage is usually slight to otherwise healthy plants, so insecticidal control is seldom suggested.

Applying Insecticides Properly in Vegetable Gardens

Table 12 shows generally which of the natural or organic controls is preferable for insect groupings. In Table 13, Vegetable Insect Control, insecticidal controls are emphasized only because there simply are not enough non-chemical methods available to match the efficacy of currently available insecticides. In the application of insecticides to garden plants, it is extremely important to observe the DAYS-WAITING-TIME to harvest. This is the waiting period required by federal law between the last application of a pesticide to a food crop and its harvest. These waiting periods are expressed as the number of days from the last application of the insecticide until the vegetable can be harvested. The number of days that you must wait before harvest varies depending upon the insecticide and the crop; therefore, you must look on the insecticide label for this information. If you observe the days-to-harvest waiting time, there will be no reason to fear that your vegetable will contain harmful insecticide residues as your family sits down to enjoy the home-grown evening repast.

Occasionally it is convenient to have a multi-purpose spray mix that will control insects, mites and diseases in one fell swoop. For the vegetable garden mixture, add 2 tablespoons of carbaryl (50% WP), 4 tablespoon of malathion (25% WP) and 2 tablespoons of maneb (80% WP) to one gallon of water. Do not prepare ahead of time, but only as needed.

In every case, it is very important to read and observe all directions and precautionary statements on the label. Also in the application of all pesticides, cover plants thoroughly with the suggested materials for maximum protection and benefit.

Because Table 13 carries mostly insecticidal recommendations, the reader will assume that this is where the emphasis is placed. It is not. Instead, it should be strongly emphasized that an insecticide, or any other pesticide for that matter, is to be used only when needed. They are not to be used as a routine gardening procedure or scheduled, but only when and as needed. Any other system is environmentally unsound and counter to good judgement.

TABLE 12. General Selection Chart For Pest Control in the Garden and on Ornamentals Using Natural Controls.

	Bacillus thuringiensis	Dusting Sulfur*	Hand Removal	Lime Sulfur	Milky Spore Disease	Petroleum Oils	Pyrethrins	Rotenone (cubé)	Ryania	Sabadilla	Insecticidal Soap
Sucking											
Aphids						x	x	x	x	x	x
Leafhoppers							x	x	x		
Mealybugs						x					x
Scales			x	x		x					x
Spider mites		x		x		x			x		x
Spittlebugs								x			
Thrips						x	x	x	x	x	x
Whiteflies						x	x	x			x
Chewing											
Beetles							x	x	x		
Japanese beetles			x		x (grubs)				x		
Weevils			x				x				
Caterpillars	x		x				x	x			
Grasshoppers			x							x	
Snails & slugs			x								
Burrowing											
Codling moths			x						x	x	
Leaf miners			x					x			
Corn earworms			x						x		
Borers			x								
Soil Insects											
Cutworms			x								
Grubs			x		x (lawns)						
Lawn moths			x				x	x			
Nuisance											
Ants			x				x				x

* Do not use sulfur on vegetables to be canned.

TABLE 13. Insect Control Suggestions for the Vegetable Garden, With and Without Chemicals[1]

Plant	Pest	Chemical Control	Non-Chemical Control
Asparagus	Asparagus beetle	Carbaryl or Malathion. Treat spears and vines when beetles appear and repeat as needed. Do not harvest until 1 day after treatment,nor repeat applications within 3 days.	Remove beetles by hand Remove plant debris in winter to prevent adults from overwintering.
Beans	Aphids	Acephate or Malathion. Apply to foliage when aphids appear. Repeat as needed. Wait 1 day before harvest.	Knock off with hard stream of water. Insecticidal soap spray.
	Bean beetles Flea beetles	Carbaryl or Malathion. Apply to underside of foliage at first sign of leaf-feeding and repeat as needed. Wait 1 day before harvest.	Remove beetles by hand.
	Spider mites	Dicofol or Ethion. Apply at first sign of off-color stippling of foliage. Repeat in 2 weeks if needed. Wait 2 days before harvest for Ethion.	Insecticidal soap spray or sulfur dust.
	Potato leaf-hopper	Carbaryl or Malathion. Treat when leafhoppers appear and repeat as needed. Wait 1 day before harvest.	Insecticidal soap spray.
	Seedcorn maggot	Plant only insecticide-treated seed.	Avoid planting in soil containing too much humus. Don't overwater.
	Leafhoppers, Spider mites, Bean beetles and Thrips	Disulfoton, 1% granules, will control these pests if applied near seed at planting time. CAUTION: Use only as directed on label. Do not apply within 60 days of harvest.	
Beets	Flea beetles	Carbaryl or Acephate. Treat at first sign of small circular holes in leaves and repeat as needed. Wait 3 days before harvest, and 14 days if tops are to be eaten.	Rotenone dust or spray.
	Leafminer and Aphids	Malathion or Acephate. Spray when mines appear and at 7 day intervals as needed. For aphids treat on first appearance and as needed. Wait 7 days before harvest.	Wash aphids off with hard stream of water.

[1] Refer to Appendix B for trade or proprietary names of recommended insecticides.

Plant	Pest	Chemical Control	Non-Chemical Control
Cabbage Cauliflower Collards Broccoli Brussels sprouts	Aphids	Malathion. Spray on appearance and weekly as needed. Wait 7 days before harvest.	Remove with hard stream of water, or insecticidal soap.
	Cabbage worms	Carbaryl or Spinosad. Apply when worms are very small and repeat every 10 days until harvest. Wait 3 days before harvest for Carbaryl.	*B. thuringiensis.* Repeat weekly as needed. Begin treatment when worms are small, or rotenone.
	Flea beetles	Carbaryl. Treat when tiny holes are found in foliage and repeat as needed. Wait 3 for Carbaryl before harvest.	
	Cabbage maggot	Acephate. Use as a transplant water treatment. CAUTION: follow label directions exactly. Apply 1 cupful of the mixture into each transplanting hole.	
	Aphids, Flea beetles, Leafhoppers, Spider mites, and Thrips	Disulfoton, 1% granules, will control these pests if applied as a band on each side of the seed furrow at planting time, or as a side dressing after plants become established, or mix in with the soil in transplant. CAUTION: Use only as directed on label.	Insecticidal soap with pyrethrins.
Cantaloupes (see watermelons)			
Carrots	Six-spotted leafhopper	Carbaryl or Malathion. Leafhopper transmits carrot yellows disease. Apply when leafhoppers are first seen and repeat as needed. Wait 0 days for Carbaryl and 7 days for Malathion before harvest.	Rotenone dust or spray.
Corn, sweet	Corn earworm Corn borer	Carbaryl or Spinosad. Spray or dust foliage and silks. Apply when tassels begin to emerge and at 2-3 day intervals through silking. There is no waiting period when used as directed.	Examine silks daily and remove newly laid white eggs. Inject 1/2 medicine dropperful of mineral oil into silk channel as silks start to dry. Follow planting dates prescribed for your area. Ryania dust on silks for earworm.
	Seedcorn maggot	Purchase insecticide-treated seed.	Avoid planting in soil containing too much humus; don't overwater.

Plant	Pest	Chemical Control	Non-Chemical Control
Corn, Sweet (Continued)	Flea beetles	Carbaryl or Acephate. Flea beetles transmit Stewart's disease in the North Central U.S. Apply when plants emerge and repeat 2 or 3 times at 5-day intervals. Early applications are necessary if infection is to be avoided.	Rotenone or ryania dust.
	Sap beetles	Carbaryl or Acephate. Apply to damaged ears of corn as described under corn earworm.	Beetles feed on overripe and cracked fruit including corn ears. Harvest corn when ripe and remove all fallen or rotten fruit from vicinity of garden.
Cucumbers	Aphids	Malathion or Propoxur. Spray as needed when vines are dry. Wait 1 day before harvest for Malathion.	Remove with hard stream of water. Insecticidal soap spray.
	Cucumber beetles	Carbaryl or Propoxur. Apply when seedlings emerge and at 5-day intervals. Repeat after a rain.	Rotenone or ryania dust in early morning.
	Spider mites	Dicofol or Ethion. Apply when mites appear and as needed. Wait 0 days for Ethion and 3 days for dicofol before harvest.	Abamectin or insecticidal soap spray to undersides of leaves.
	Squash vine borer	Carbaryl or Spinosad. Begin a 3-treatment series, at 7 day intervals, when first borer entrance signs are found. Wait 1 day before harvest.	Break growing tips to ensourage branching. Cover main stems with soil or mulch to prevent egg laying by moths. Destroy crop residue after harvest.
Eggplant	Aphids	Malathion or Acephate. Apply when aphids appear and repeat as needed. Wait 3 days for Malathion or Acephate before harvest.	Remove with hard stream of water or insecticidal soap spray.
	Flea beetles	Carbaryl. Treat at first sign of tiny holes in leaves and repeat as needed. 0 days waiting period.	Rotenone spray.
	Potato beetle See Potatoes		

Plant	Pest	Chemical Control	Non-Chemical Control
Eggplant (Continued)			
	Spider mites	Malathion or Dicofol. Apply when mites appear and repeat as needed.	Insecticidal soap spray to undersides of leaves, or sulfur dust.
Lettuce	Leafhoppers	Carbaryl or Ethion. Spray or dust when they appear and weekly as needed. Wait 14 days on leaf and 3 days on head lettuce for Carbaryl and Ethion before harvest.	D-Limonene spray in early morning.
	Cabbage loopers	Malathion. Wait 7 days for head and 14 for leaf before harvest.	*B. thuringiensis.* Repeat weekly as needed. Begin when worms are small.
	Aphids	Malathion or Acephate. Treat when aphids appear. Wait 14 days on leaf and 7 days on head lettuce before harvest.	Carefully remove with hard stream of water. Light insecticidal soap spray.
Muskmelons	Aphids	Acephate or Malathion. Aphids may transmit a mosaic disease. Treat when aphids appear and repeat as needed. Wait 1 day before harvest.	Remove with hard stream of water. Insecticidal soap spray, or rotenone.
	Cucumber beetles	Follow procedures as prescribed under Cucumbers	
	Spider mites	Follow procedures as prescribed under Cucumbers.	Abamectin or insecticidal soap spray.
Mustard		Follow procedures prescribed under Cabbage, etc.	
Okra	Aphids	Malathion or Acephate. Wait 1 day before harvest.	Insecticidal soap spray.
Onions	Onion maggot	Carbaryl or Acephate. Spray very dilute mixture in the row at planting time.	Avoid intensive use of humus or overwatering.
	Thrips	Malathion. Treat when thrips appear and at 5-10 day intervals as needed. Wait 3 days for harvest.	Insecticidal soap spray, cinnamon oil or neem.

Plant	Pest	Chemical Control	Non-Chemical Control
Peas	Aphids	Malathion or Acephate. Apply when aphids first appear and repeat as needed. Wait 3 days before harvest.	Remove with hard stream of water. Insecticidal soap spray, or cinnamon oil.
Peppers	Aphids	Malathion or Acephate.. Apply when aphids first appear and repeat as needed. Wait 3 days before harvest.	Remove with hard stream of water, or apply insecticidal soap spray.
	Flea beetles	Carbaryl or Acephate. Spray or dust as needed.	
	Hornworms	Carbaryl. Spray or dust as needed.	Remove larvae by hand, or use rotenone spray.
	Pepper weevil	Carbaryl or Malathion. Treat as needed when pods begin to set.	
	European corn borer	Carbaryl or Ryania. In North Central states apply as preventive treatment every 3 days after blossoms appear and fruit forms. Damage occurs in late July and August.	Plant resistant varieties, or apply ryania.
Potatoes	Aphids	Malathion or Acephate. Apply to foliage when aphids appear and repeat as needed.	Remove with hard stream of water, or apply insecticidal soap spray.
	Flea beetles	Carbaryl or Rotenone. Treat when beetles appear and weekly as needed.	Rotenone spray.
	Leafhoppers	Carbaryl. Treat when they are first seen and repeat as needed.	D-Limonene, spray in early morning.
	Potato beetle	Carbaryl or Rotenone. Spray or dust on appearance of larvae (slugs) and weekly as needed.	Remove larvae by hand.
	Soil insects (cutworms, white grubs, wireworm)	Imidacloprid or Disulfoton granules. Apply to soil as a preplanting treatment to spaded ground and work into 3-5", then plant potatoes. Lasts only one season.	Don't plant potatoes in same ground two successive years. Rotate crops.

Plant	Pest	Chemical Control	Non-Chemical Control
Pumpkins and Squash	Aphids	Malathion or Acephate. Apply when aphids first appear and repeat as needed. Wait 3 days for Malathion on pumpkin and 1 for squash before harvest. Wait 3 days for Acephate on squash.	Remove with hard stream of water. Insecticidal soap spray, or rotenone.
	Cucumber beetles	Carbaryl or Malathion. Follow procedures as prescribed under Cucumbers. Wait 1 day for materials except 3 days for Malathion on pumpkin before harvest.	Rotenone or ryania dust in early morning.
	Squash bugs	Carbaryl or Acephate. Treat when bugs appear and repeat as needed.	Hand-pick adults and brown egg-masses from plants. Trap adults under boards laid beneath plants. Sabadilla is also very effective.
	Squash vine borer	Malathion or Acephate. Apply 5 times at 7-day intervals when first borer entrance signs are found. Wait 1 day for Malathion on squash and 3 days for pumpkin before harvest.	See Cucumbers Butternut squash is resistant to vine borer. Destroy residue after harvest.
	Spider mites	Pyrethrins or Dicofol. Treat as needed.	Dust sulfur undersides of leaves.
Radishes	Aphids	Malathion. Treat when aphids first appear and repeat as needed. Wait 7 days for Malathion.	Carefully remove with hard stream of water, or apply insecticidal soap spray, neem or rotenone.
	Flea beetles	Carbaryl. Treat when small holes are seen in leaves and repeat as needed. Wait 3 days before harvest.	
	Radish maggot	Carbaryl. Spray or dust soil at 7 day intervals after planting. Wait 3 days before harvest.	Plant at weekly intervals. Some plants will avoid being damaged.

Plant	Pest	Chemical Control	Non-Chemical Control
Spinach	Leaf miner	Malathion or Acephate. Spray when mines appear at 7-day intervals as needed. Wait 7 days before harvest.	There is no non-chemical control.
Tomatoes	Aphids	Malathion. Treat when aphids first appear and repeat as needed. Wait 1 day before harvest.	Usually do not require control, however insecticidal soap spray is very effective.
	Cutworms	Carbaryl or Acephate. Apply when plants are first set out and repeat twice at weekly intervals. Carbaryl bait works well.	Search beneath clods for cutworms. Look for plant parts in or near hiding place. Place a paper collar in the soil 1" deep and 2" high, 1" from plant. This is a barrier that cutworms will not climb.
	Flea beetles	Carbaryl. Treat after plants are set and weekly for two weeks.	
	Blister beetles	Carbaryl. Treatment is needed later in season when first seen on plants. Repeat as needed. Wait 1 day before harvest.	"Herd" out of garden with small limb or broom to weedy margin where they originated, and/or rotenone dust.
	Hornworms and Fruitworms	Carbaryl or Acephate. Spray or dust 3 to 4 times at 10-day intervals. Begin when first fruits are small.	Hand-pick worms when nibbled foilage is noticed Apply *B.thuringiensis.*
	Whitefly	Malathion or Acephate. Treat when whiteflies can be shaken from foliage, at 15-day intervals for 3 applications and repeat as needed. Wait 1 day before harvest.	Inspect undersides of leaves at place of purchase. Do not buy if tiny, oblong, white, motionless insects are found. Insecticidal soap spray, or rotenone.
	Sap beetles	Carbaryl or Malathion. Treat when beetles appear and repeat as needed	Buy crack-resistant varieties. Harvest fruit when ripe and remove all fallen or rotten fruit from vicinity of garden.
	Aphids, Flea beetles, Leafhoppers, Leaf miners, and Mites	Disulfoton or Imidacloprid granules. Apply at planting time only. Mix in with the soil in the transplant hole before setting out plant. CAUTION: use only as directed on label.	Rotenone, insecticidal soap, or neem sprays as needed.

Plant	Pest	Chemical Control	Non-Chemical Control
Turnips	Aphids	Malathion or Acephate. Treat when aphids are seen and repeat as needed. Wait 3 days before harvest.	Remove with hard stream of water, or apply rotenone.
	Flea beetles	Carbaryl or Malathion. Treat when small holes are first seen in leaves and repeat as needed. Wait 3 days for Carbaryl (14 days if tops are eaten) and 7 days for Malathion before harvest.	Insecticidal soap spray.
	Turnip maggot	Carbaryl. Spray soil at 7-day intervals after planting. Wait 10 days before harvest.	
Watermelons Cantaloups Muskmelons	Aphids	Malathion or Acephate. Treat when aphids appear and repeat as needed.	Remove with hard stream of water.
	Cucumber beetles	Follow procedures as prescribed under Cucumbers.	
	Spider mites	Follow procedures as prescribed under Cucumbers.	

HOME ORCHARD

APPLE, PLUM, CHERRY, PEACH AND GRAPE

What greater pleasure can man derive from his domestic garden activities than offer a friend fresh fruit, preserves and jellies, or a glass of wine, with the casual statement, ". . . from this year's crop." There is many a slip from the cup to the lip in the production of the wine, the preserves, or that bowl of fresh fruit, since diseases and insects are also fond of your fares, from the time of blossom to the rich colors of ripening.

In the case of plum, cherry and peach pest control, you can improve your overall pest control program by keeping trees pruned and the areas around the trees and nearby fence rows free of brush and weeds. In the fall or winter collect and bury peach and plum mummies (rotted and dried fruit) to reduce brown rot and the survival of certain insects. If brown rot is a problem, when the fruit rots before it ripens, spray with captan at the full bloom stage. Prune and burn all infected branches if black knot is prevalent on plums. Spray sour cherries right after harvest with dodine or thiophenate-methyl to control cherry leaf spot, which causes defoliation.

Grapes must be pruned every year. Do not use 2,4-D anywhere near grapevines, for it will damage foliage and reduce production more than you can imagine. Symptoms of this damage are elongated terminal growth, downward cupping of old leaves and fanshaped growth of new leaves.

As for apples, you have been spoiled by those beauties brought from the grocer. It's extremely difficult to grow fruit as high in quality as that produced commercially, but you can produce healthy apple trees and quality fruit through tree and orchard sanitation, pruning and training trees properly and fertilizing properly. Of the five fruits listed, apples are the most difficult to carry to perfection.

The universal multi-purpose spray mix including disease control that can be used on fruit trees and grape vines is as follows: to one gallon of water add 2 tablespoons of carbaryl (50% WP), 3 tablespoons of malathion (25% WP) and 2 tablespoons of captan (50% WP). Do not prepare ahead of time, but only when needed.

In all cases, it is very important to read and observe all directions and precautionary statements on the label. In the application of pesticides cover plants thoroughly with the suggested materials for maximum protection and benefit. (Table 14).

Pheromone (sex lure) and brightly colored bait traps are available for many tree fruit insects. Become acquainted with the use of pheromones by reading Chapter 4, *Biorationals* and the suppliers of pheromone and food attractant traps in Appendix G, *Pest Control Sources.*

Apple Maggot

Apple Maggot — Adult
(Union Carbide)

Of the orchard fruits attacked by the apple maggot, apples are the most seriously damaged. Other hosts are hawthorn, plum, pear and cherries. The adult is a small, dark brown fly, with light and dark markings on the body and wings. Damage is caused by the larvae, which are white tapered maggots slightly smaller than those of house flies. The adults emerge from the overwintering puparia beginning in mid-June and continue for about a month. A week or more after emergence the females begin laying their tiny white eggs beneath the skin of the young fruit. Susceptible varieties of apple are the early maturing, such as Delicious, Cortland and Wealthy, those that are sweet. The eggs hatch in a few days and the maggots mine the fruit with brown tunnels, which may cause premature fruit drop. The larvae will mature in as little as 2 weeks in early varieties and as much as 3 months in hard winter apples. On maturing the maggots drop to the ground, enter the soil and become puparia. Usually there is only one generation per year.

Cultural control consists of collecting all infested fallen apples and the elimination of hawthorn in the immediate vicinity. The fruit of ornamental flowering crabapple are a prime source of developing populations and should be sprayed or have all fruit collected periodically. Chemical control depends on using insecticide sprays to kill the adults before they lay their eggs. These can be included in the third through fifth or sixth cover sprays.

The bright yellow, sticky panel, bait traps that are commercially available, should be hung in trees to capture the female flies before they begin laying their eggs. Traps should be hung in mid-June, one-third of the way into the tree, on the south side, with foliage and twigs removed from at least 1 foot around the traps. One to three traps per tree will provide good control of the newly emerged flies. Maggot fly emergence continues through mid-September, so traps should be cleaned weekly and replaced every two weeks through harvest.

Cherry Fruitworm

This is the larva of a very small gray moth related to the Oriental fruit moth. Its original or native host was probably wild cherry, but it is also known to attack blueberry. Mature larvae hibernate in galleries in the bark and pupate in the spring. The moths emerge from early June through mid-July and lay small, flat eggs on the fruit. After one week, hatching takes place and the white larvae with black heads bore into the green cherries and feed around the pit. Their development is completed in about 3 weeks and they leave the cherries to begin hibernation. There is usually only one generation each year.

Pheromone traps are now available to capture male cherry fruitworm moths as they emerge and begin their search for females. One trap should be hung in each tree, beginning in early June. Follow the manufacturer's instructions regarding location, cleaning of the traps and replacement of the sex lure.

Cherry Fruit Flies

These are the larvae of the cherry fruit fly, the black cherry fruit fly and the western cherry fruit fly. These maggots all feed in the flesh of cherries around the pit, often causing deformed fruit. Both sweet and sour cherries are attacked. The adults are about one half the size of the housefly and have dark bands across their wings. Their bodies are dark with yellow markings.

Typically winter is passed as puparia in the soil. The adults emerge in late spring and lay eggs in the fruit. The maggots complete their development in the cherries usually at harvest time and drop to the ground to change to puparia. There is only one generation each season.

Cultural control consists of destroying infested fruit and cultivation to destroy the puparia in the soil. Insecticidal control is directed at the adults just as they emerge and before they begin egg laying, requiring 2 to 4 applications.

Pheromone and yellow sticky traps are available for cherry fruit flies and should be hung one per tree beginning about one week before the first adults typically appear, which should be in late spring. Season-long trapping is not necessary since there is but one generation per year.

Codling Moth

The codling moth occurs wherever apples are grown and is generally considered one of its most important pests. Pear, quince, English walnut and sometimes other fruit are also injured. Damage is caused by the worm or larval stage which tunnels into fruit, usually to the core.

The moth is less than one inch in wingspread, with the front wings a gray brown, crossed with lines of light gray and bronzed areas near the tips. The larva is white, occasionally slightly pink, with a brown head. They survive the winter as fully developed larvae, hibernating in cocoons on or near apple trees. In the spring the larvae pupate and emerge as active moths the same time that apples are in bloom

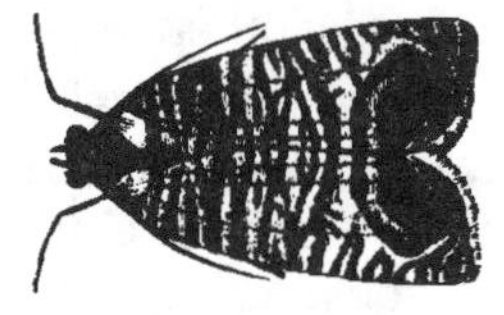

Codling Moth
(Union Carbide)

and live 2 to 3 weeks. Eggs are white, about the size of a pin head and are deposited on leaves, twigs and fruit. The eggs hatch in a few days and the small larvae bore into the fruit, often through the calyx. After 2 to 3 weeks the larvae spin cocoons and pupate, to emerge in 2 weeks as the next generation of moths. In the southern apple-growing areas, there may be three and a partial fourth generations.

Weather and several parasitic wasps play important roles in the numbers of codling moths that appear each year, but the most practical control is achieved with a spray schedule. This will usually combine a fungicide with an insecticide. Proper timing of sprays is essential and moth activity can be determined by using pheromone traps baited with the synthetic sex lure codlelure. The first spray is referred to as the petal-fall and is applied when nearly all petals, have fallen and before the calyx closes. It is important to follow this with several cover sprays and a second-generation spray applied about 10 weeks after petal-fall, usually late July or early August for most fruit-growing areas.

Pheromone traps for the codling moth are best used to help determine the timing of insecticide applications. However, in the home orchard setting, mass trapping and disorientation may work quite well with the placement of one trap in every tree. Insecticide application should be made 14 days after the first male moth is trapped, about petal-fall stage, or immediately after the fifth male is caught. This spray will kill hatched larvae before they penetrate the fruit.

European Red Mite

This spider mite is one of the most important fruit tree pests in the northern United States and adjacent areas of Canada. It attacks apple, pear, peach, plum,

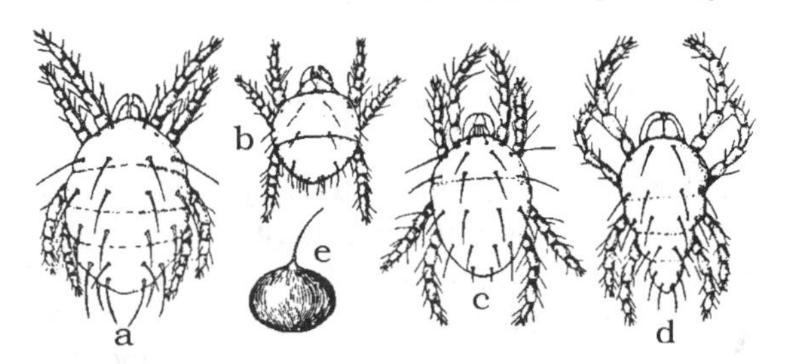

The European Red Mite, *Panonychus ulmi* (Koch): a, adult female; b, first instar; c, deutonymph; d, adult male; e, egg.
(Newcomer & Yothers, USDA)

prune and may be found on other deciduous trees and shrubs. Its damage results from the removal of plant juices with its piercing-sucking mouthparts, causing bronzed and off-colored foliage. Under heavy infestations, defoliations and undersized poorly colored fruit are produced.

The mites are rust-colored, though colors range from dark green to bright red. Winter is passed as the egg, usually on branches and twigs of the host trees. Hatching takes place during the pink bud stage and damage may be seen shortly after young foliage appears. There may be 6 or 8 overlapping generations per season.

Considerable natural control by predators occurs in the home orchard, however control with sprays is usually necessary. Good control can be had by killing the overwintering eggs using a delayed dormant (1/2 inch green tip stage) application of a superior-type oil emulsion. The eggs become increasingly susceptible to control as hatching time approaches. If mite control is required in mid-summer use a miticide least damaging to the predators, such as dicofol, where resistance is not a problem.

Eyespotted Bud Moth

Also referred to as bud moth, this pest is distributed in all principal apple producing areas, but has been a serious pest only in the northeastern and northwestern states. The moth is a bit smaller than the codling moth, is dark brown with a light-colored band. The mature larva has legs, is 1/2 inch long, brown with a light stripe down its back and has a black head.

There is only one generation a year, beginning when the adults appear in midsummer and begin egg laying on either side of leaves. After several days the eggs hatch and the larvae begin feeding on leaves, where their damage is most important. Occasional damage occurs to fruit that are in contact with infested leaves. Early in the fall the immature larvae hibernate in protected portions of the tree. In the spring they begin their feeding again as the buds swell and open, thus the name, bud moth. In early summer they mature, pupate in silk chambers usually in crumpled leaves and emerge as adults beginning in mid-June. The life cycle is started again with egg laying. Very low winter temperatures (—21°F) and numerous parasites and predators play important roles in its natural control. Chemical control is for the larvae feeding in the buds and is applied at petal fall and the first cover spray.

Pheromone traps are available for the bud moth males. Traps should be hung in apple trees in mid-summer, before the first adults are known to appear. Because there is but one generation per year, trapping is not necessary beyond August 1.

Leafrollers

There are three leaf-rolling caterpillars in this group: the fruittree leafroller, obliquebanded leafroller and redbanded leafroller. Their life cycles vary considerably and each will be discussed briefly.

The fruittree leafroller attacks apples and all other orchard fruits and may require special control measures because larval feeding on blossom buds may prevent fruit setting. Damage to fruit results where leaves are attached to fruit with silk, the larvae feeding within. Foliage injury is not as important. The insect overwinters in the egg stage and hatching begins when buds open. The larvae feed on buds, blossoms, leaves and fruit, reaching maturity in June. They pupate inside rolled leaves and emerge as moths in about 2 weeks. Eggs are layed shortly after and the adults die. There is only one generation each year.

The obliquebanded leafroller is of lesser importance. The caterpillars attack foliage in the spring, and fruit and foliage in the summer and fall. They feed on many different hosts including greenhouse crops and roses. After hatching the young act as leafminers, then feed inside rolled leaves tied with silk. They overwinter as immature larvae in silk cases and renew their feeding with the appearance of young leaves. The moths emerge in June and begin egg laying shortly afterward. There may be two generations per year, depending on latitude.

Redbanded Leafroller
(Union Carbide)

The redbanded leafroller is most abundant in the Cumberland-Shenandoah apple region and attacks apple, cherry, plum, peach, grape, other fruit, vegetable crops, ornamentals and weeds. The caterpillars feed on rolled or folded leaves held together with silk. Fruit are blemished when they are attached to infested leaves with silk. This pest overwinters in the pupal stage in protected cases on the ground. Moths emerge in April and lay their eggs in clusters on the tree bark. The eggs hatch near petal-fall and the larvae feed, develop the pupate and emerge as adults in July. The cycle is repeated and there may be 3 to 4 overlapping generations, again depending on latitude.

There are several parasites and predators of the eggs and larvae of all three leafrollers which serve to some extent as natural controls. Chemical control is for the larvae and is included in the petal-fall and first through fifth cover sprays.

The redbanded leafroller pheromone provides a good demonstration of how effective and powerful pheromones are. Males are usually attracted within 5 minutes of exposing a cap to air currents on a warm afternoon in late April. Traps should be set out in early April. Change the cap and trap every four weeks. Mass trapping and disorientation are effective in the home orchard, but not in large commercial plantings.

Oriental Fruit Moth

Peach and quince are the most commonly attacked hosts, however apple, pear, plum and other fruit are occasionally attacked, particularly if they are near infested peach trees. Injury to trees occurs when the larvae tunnel into the tips of growing twigs, which prevents normal tree development. After attack new

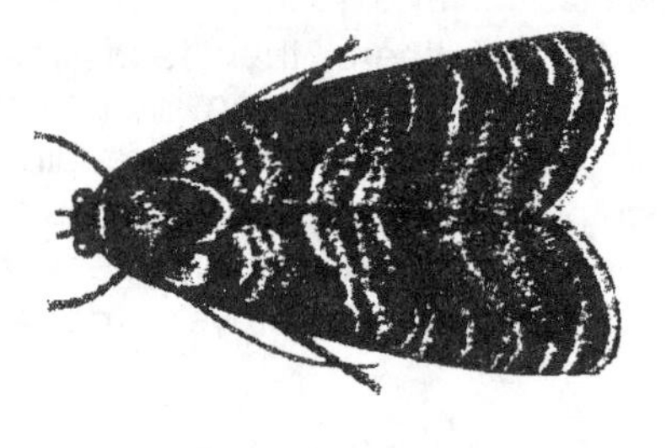

Oriental Fruit Moth
(Union Carbide)

lateral shoots develop just below the damage, giving the tree a brushy appearance. Larvae also attack the fruit at the end of the growing season when twigs harden.

This pest is closely related to the codling moth and both the larval and adult stages resemble each other. The larvae hibernate in cocoons on the tree, on the ground, or on near-by objects. Pupation takes place in the spring and moths emerge at peach tree blooming. The eggs are laid on twigs and foliage and later on the fruit. The larvae mature in two weeks when they leave the twigs to spin a cocoon and pupate. After 10 days the adults emerge and the cycle begins again, requiring about one month. There may be 4 to 5 generations a year.

There are several parasites and predators that attack eggs and larvae, thus exerting some level of natural control. However, chemical control is usually necessary, and is achieved by including the appropriate insecticide in the second, third and remaining cover sprays.

The pheromone for this insect is quite effective in capturing males. Use caps and traps similar to the pattern used for the codling moth, however, replace traps before each new generation of adults appears. Mass trapping and mating disruption will not work quite as well in the home orchard as for the codling moth. The traps will indicate when to apply insecticides to achieve maximum adult control.

Peachtree Borer

The peachtree borer attacks not only peach, but plum, prune, cherry, almond, apricot and nectarine. The injury is caused by the larvae boring beneath the bark near ground level, sometimes girdling the trunk. It occurs all over the United States where peaches are grown. The adult is truly beautiful. It is a clearwing moth with blue, yellow and orange markings. The moths fly by day and resemble wasps. The larvae overwinter in their tunnels at the bases of trees. They complete their development and leave their tunnels to pupate in the soil near the tree base. Pupae and adults can be found most of the growing season. Eggs are laid usually on tree trunks, requiring 8-9 days to hatch. The young bore into the trunk causing

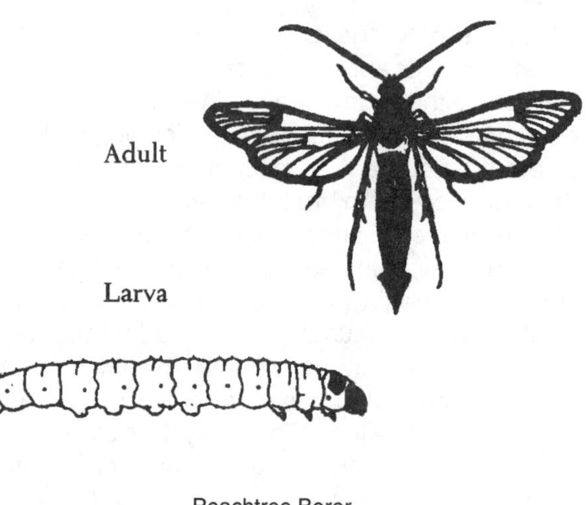

Peachtree Borer
(USDA)

gum to seep from the holes along with their borings. There is one generation per year.

Natural control is not complete and must be aided by chemical control. This consists of fall treatment of the soil around each tree trunk with paradichlorobenzene which acts as a fumigant. Spraying the tree trunks with an insecticide to kill adults, eggs and newly hatched larvae before they enter the bark is also effective. Pheromone, or sex attractant, baited traps will help determine the best timing for sprays.

Pheromone traps are best used to indicate when trunk sprays should be applied for adult control to prevent egg laying. Recent research in the South, however, indicates that these traps can be used in the home orchard for mating disruption as a non-chemical control.

Peach Twig Borer

As implied by its name, the peach twig borer is primarily a pest of peach, but it also attacks plum, apricot and almond. Its injury is to the twigs and fruit similar to the damage of the oriental fruit moth. The immature hibernating larvae emerge in the spring when growth of the twigs begins and bore into twigs and buds. Their tunneling stops growth or kills the shoot. The larvae leave the twigs when development is completed and pupate on branches or trunk. Some 2 weeks later the adults emerge and lay eggs on leaves and fruit. As the season progresses each new generation tends to feed more and more on the fruit. There may be up to 4 generations per year.

Peach Twig Borer
(Union Carbide)

The twig borer is not a problem in trees receiving dormant sprays of lime sulfur or oil emulsion each year. Otherwise, insecticidal control is achieved beginning with the petal fall and first through third cover spray.

Pheromone traps need to be in place by late petal fall and should be maintained up to harvest time, since there may be as many as 4 generations during the growing season.

Pear Psylla

Psyllids, or jumping plant lice, resemble tiny cicadas that hop or fly when disturbed. The nymphs may be misidentified as woolly aphids since they produce large amounts of the white, cotton-like covering on their ends. The pear psylla is found in the Pacific Northwest and Eastern U.S. It is reddish brown with green or red markings, has transparent wings, and lays yellow eggs at the base of buds, on the upper leaf surfaces, or on twigs. There are three to five generations per year and the adults hibernate in ground trash.

The pear psylla is specific to pear and quince, though look-alike species attack other plants. The nymphs and adults feed on fruit and foliage. The honeydew they excrete supports the growth of sooty mold. Their feeding results in leaves developing brown spots, scarred fruit and buds that fail to develop. Dormant oil sprays in the spring are the most effective control.

Plum Curculio

The plum curculio is a hard-bodied snout weevil that attacks stone fruits, and is broadly distributed east of the Rockies. It also attacks apple, pear, quince and related fruits. Injury to fruit is extensive and begins in the spring from feeding of the adults, followed by punctures in the fruit from females laying their eggs, then feeding of larvae within the fruit and last from fall feeding by adults.

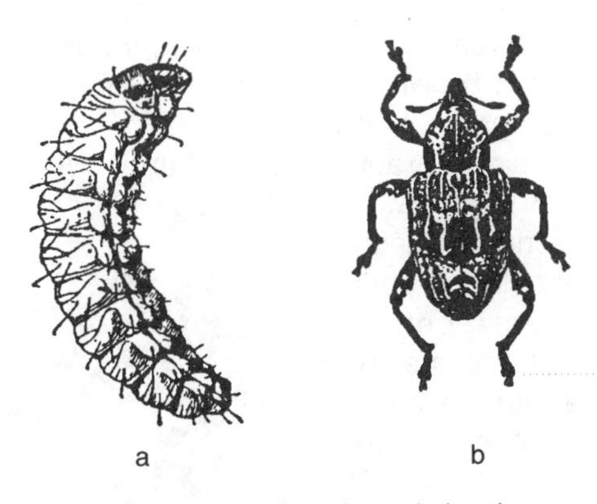

Plum curculio; a, larva; b, beetle.
(USDA)

The adults are brown with gray markings and have 4 small bumps on their backs. The larvae are curved, legless and white with a brown head. The plum curculio overwinters as the adult in protected places near the fruit trees and appears in the spring about the time trees begin to leaf. The adults feed for a month or more during which time they lay their white eggs in the holes where they have fed. These holes are cresent-shaped and remain with the fruit until harvest marring them and lowering their quality. Eggs hatch in 4-5 days and the grubs or larvae feed for 2-3 weeks when they mature. The larvae then bore their way out of the fruit and drop to the ground to pupate in the soil. The new generation of beetles emerge in a month and usually feed on the fruit before going into hibernation to await spring.

Many of the hibernating beetles die during the winter, while birds and other predators take their toll. There are several species of wasps that attack the eggs or larvae and a fungus is known to infect both the larvae and weevils. An old method of control consists of jarring the weevils from the tree onto a sheet in the early morning and destroying them. Destruction of infested, fallen fruit and cultivation beneath the trees to kill the pupae is quite helpful. Chemical control is for the adults and is usually applied at the petal-fall and first and second sprays.

Scale Insects

The two most important scale pests of orchard fruits are the San Jose and the oystershell scales. The San Jose scale attacks most cultivated fruits and a large number of ornamental shrubs and trees. The osage orange is often a reservoir for reinfestation. Tree health declines and fruit are blemished by the nymphs and adults removing plant juices from any part of the plant but particularly the wood. Heavy infestations may kill trees.

The mature female scale is circular, the size of a pinhead, dark brown with a yellow center. Young scales are lighter, becoming dark with age. The males are smaller and oval in shape. Winter is passed in the immature scale stage beneath the scale coverings attached to the trees. In the spring, development continues and winged males appear and mate with the females under the edge of their scale coverings. The female produces young or "crawlers" which move from under her covering to sites on the tree. The young are bright yellow, soon settle down to insert mouthparts into the bark, leaves or fruit to feed, and lose their legs and antennae during the first molt. Their growth continues and is completed in about 6 weeks. Two or more generations may occur each season.

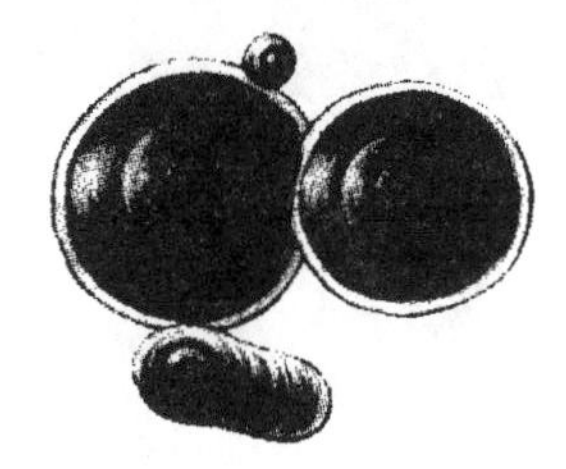

San Jose Scale
(Union Carbide)

Spread within the tree is by crawlers, and from tree to tree on the feet of birds and on other insects. Some perhaps are carried by wind. Several parasitic wasps and predacious lady beetles and mites exercise a fairly high degree of natural control. Dormant sprays of lime-sulfur or superior-type oil emulsions are more effective than mid-summer efforts with the broad-spectrum insecticides.

The oystershell scale is a pest of apple, pear, occasionally other fruit trees and many shade and ornamental plants. Lilac and ash trees are particularly susceptible. The scales are brown to gray and considerably larger than the San Jose scale. Winter is passed as tiny white eggs beneath the scale of the female. In the spring the eggs hatch and the young crawlers migrate over the fruit tree to settle down and remove plant juices with their piercing mouthparts. The remainder of the process and cycle is identical to the San Jose scale, as are the controls.

The pheromone for San Jose scale is one of the better trapping methods for males, since the females do not fly. Because this scale attacks most cultivated fruit trees, traps can be used anywhere they are suspected of infesting.

Woolly Apple Aphid

The aphid occurs in most of the apple-producing areas of the world. Not only is it a serious pest of apple, it attacks, elm, mountain ash and species of hawthorn. It feeds on plant juices from the usual above-ground plant parts and roots, making it a complex pest. The most serious injury is the below-ground, gall-forming feeding. These aphids are recognized by the woolly covering at the rear end of their blue-black bodies.

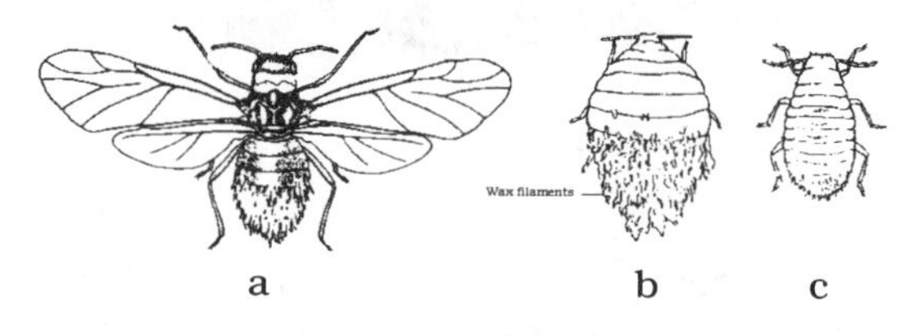

Eriosoma lanigerum, the Woolly Apple Aphid.
a, Winged female; b, apterous female; c, male.
(From Baker, 1915)

The alternate host is the elm tree and they move back and forth depending on the season, partially explaining their somewhat complex life cycle. Syrphid fly larvae, lady beetles and a chalcid wasp all are important in keeping the aphid checked. Dormant sprays listed in the table of spray schedules are quite effective in controlling this pest.

Grape Berry Moth

The berry moth, which occurs east of the Rockies, is the only common insect that does heavy damage to grape berries. The small brown moths begin to emerge from overwintering pupae when grape foliage has unfolded, usually the early part of June. The small scalelike eggs are laid on grape stems, blossom clusters or the small berries. The larvae feed on blossoms and small berries, usually leaving a silken thread as they move, resulting in webbed clusters.

Grape Berry Moth Larva
(Union Carbide)

One larva may practically destroy an entire cluster. After 3 to 4 weeks feeding they each cut a semicircular slit in the leaf, fold it over and tie it with silk, thus forming a cocoon where they pupate.

Second-generation moths emerge in 10 to 14 days and begin to lay their eggs on the berries. It is this second generation of larvae that cause the greatest damage, usually in late August, by eating their way into the berries and feeding on the pulp and seeds.

There are 2 generations per year and the mature larvae of the second generation pupate inside the cocoons made in the leaves, which fall to the ground in the fall. Cultural control is quite valuable, especially cultivation or plowing under and burying the cocoons containing the overwintering pupae, thus preventing adult emergence the following spring. Chemical control is necessary for backyard vineyards with a berry moth problem, involving at least 3 and possibly 4 applications. The first is at postbloom, the second 7 to 10 days later and the third around the first of August. Controlling the berry moth with insecticides will also control the grape flea beetle.

Pheromone traps should be installed in vineyards near the first of June, when the grape foliage begins to unfold. In home vineyards mass trapping and mating disruption can be used by increasing the frequency of traps. This is not effective for commercial plantings.

Grape Flea Beetle

The grape flea beetle occurs only in the eastern United States, from the Mississippi valley eastward. It damages plum, apple, pear, quince, beech and elm. Neglected vineyards, however are its favorite host. Their damage is caused by the adults eating the buds and unfolding leaves of grape and the larvae skeletonizing the leaves.

The adults are about 1/4 to 1/6 inch long, robust, with a metallic blue-green color. They emerge from hibernation and begin egg laying in the bark and on the leaves and buds. The brown larvae marked with black spots hatch from the eggs in a few days, feed on the undersides of the leaves for 3 to 4 weeks, then drop to the ground to pupate in the soil. They emerge as adults in 10 to 14 days, feed the rest of the summer and go into hibernation in the fall.

Flea Beetle
(Union Carbide)

There is only one generation each season. Control of the grape berry moth also controls the grape flea beetle.

Grape Leafhoppers

Wherever grapes are grown in the United States and Canada, several species of leafhoppers can be found sucking plant juices from the lower leaf surfaces. The foliage becomes blotched with small white spots and under intense damage will turn yellow or brown, even to defoliation. The result of this is a reduction in the quantity and quality of the grapes. Leafhoppers may also attack apple, plum, cherry, currant, gooseberry, blackberry and raspberry.

There may be 8 or more species involved in the heading of leafhoppers. They are all small, usually no longer than 1/8 inch, pale yellow, with red, yellow, or black markings on the wings.

Leafhoppers pass the winter as adults in protected areas under plant debris. As the spring days warm they become active and feed on any green plant before grape leaves develop. The eggs are laid in the leaf tissue, usually from the underside. Hatching occurs within 2 weeks and the tiny nymphs move about and feed on the lower leaf surfaces, molting 5 times before becoming winged adults. This development requires 3 to 5 weeks and there are 2 to 3 generations each season.

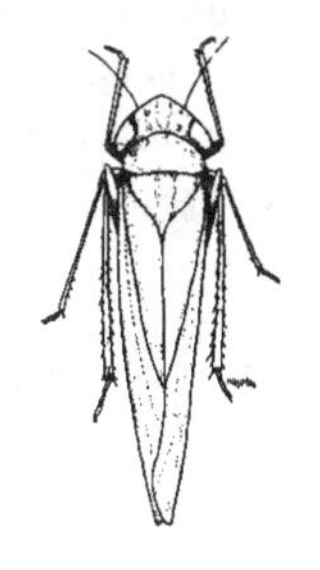

Leafhopper — Adult
(Univ. of Arizona)

Partial natural control of leafhoppers results from several parasites and predators and on fungus. In dense populations, however, chemical control is necessary and is included in prebloom and postbloom sprays and later, when needed. The heavy use of insecticides on grapes may result in mite problems.

Grape Leaf Skeletonizer

These small yellow caterpillars are a pest all over the United States, but usually only on vines grown in home gardens. They attract attention because of their feeding only on the upper surfaces of the leaves, sometimes side by side in rows across the entire leaf. When fully developed the caterpillars are a little less than 1/2 inch long. The adult is a small, black, narrow-winged moth. Winter is passed as pupae in cocoons on leaves in ground trash. The adults appear late in spring and lay their bright yellow eggs in clusters on the lower leaf surfaces. There are 2 and sometimes 3 generations a year. Vines receiving regular sprays for other pests are not usually infested with skeletonizers.

Rose Chafer

Actually the rose chafer is a general feeder and attacks not only grapes and roses but many tree fruits, raspberry, blackberry, strawberry and several garden flowers. The chafer occurs generally east of the Rockies and is more abundant in areas having light sandy soil. The beetles feed on leaves, flowers and fruit, while the larvae, which resemble small white grubs, feed in the soil on roots of various weeds and grasses. The larvae pass the winter in the soil and pupate near the surface in May. The tan, long-legged beetles, 1/2 inch long, appear in late May and June. The adults may cause serious damage in the years when they reach abundance.

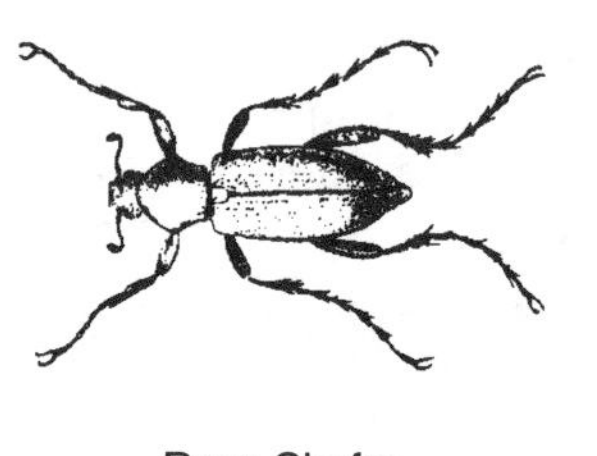

Rose Chafer
(USDA)

Feeding, mating and egg-laying occur into early July. The eggs are deposited one at a time below the soil surface near the larval host plant. After hatching, the grubs feed until cold weather, becoming nearly full grown. There is one generation each year. The petal-fall spray for berry moth on grape also controls rose chafers, however, control is seldom needed on grape or other fruits.

CITRUS

If you are one of the rare "northerners" who manage to produce citrus in your backyard by the skillful use of winter protection in the form of tents, plastic tarps and temporary or permanent greenhouses, then you probably have few if any of the pest control problems plaguing your southern citrus growing neighbors. However, for those gardeners who raise citrus as easily as they do bermudagrass, they may indeed have pest problems (Table 15).

As a general practice, citrus trees should receive a moderate pruning every year, removing especially those limbs infested with scales. The trees should be fertilized twice, once in the spring and again in the fall. In the spring, fallen and shriveled fruit still on the tree should be removed and destroyed, since they invite secondary pests such as flies, wasps and bees and ground beetles. At all times, weeds and grass should be removed or maintained at lawn mower height, since they too harbor pests, including gnats, flies and mosquitoes.

Citrus grown strictly for ornamental purposes, such as sour orange requires little or no pest control

TABLE 14. Pest Control Spray Schedules for Apple, Peach, Plum, Cherry and Grape in the Home Orchard.[1]

Schedule (When to Spray)	Pesticides to use	Pest to Control
Apple (Monitor insect pests with pheromone traps)		
Dormant — before any green starts to show	Dormant oil or lime sulfur	Aphids, mites and scales
Prebloom — leaves 1/4 inch long until flowers start to open (3 sprays, 7 days apart)	Fungicides: Captan, Thiophenate-methyl or Dodine.	Apple scab
Bloom — at peak of blossom	Fungicide: same as prebloom. Do not use insecticides during bloom.	Apple scab
Petal Fall — When 3/4 or more of blossom petals have dropped.	Fungicide: same as prebloom	Apple scab
	Insecticide: Ryania, Phosmet, or Malathion + Propoxur	Aphids, codling moth, leafroller, plum curculio

Schedule (When to Spray)	Pesticides to use	Pest to Control
First Cover - 7 to 10 days after petal fall	Fungicide: same as prebloom	Apple scab
	Insecticide: same as petal fall	Codling moth, oystershell scale, plum curculio
Additional Covers - apply regularly at 10 to 14 day intervals	Fungicide: same as prebloom	Apple scab
	Insecticide: Ryania, Permethrin, Phosmet, Carbaryl or Malathion +Propoxur	Aphids, codling moth, leafroller, apple maggots, fruit moth
Post-Harvest - before leaf drop	Fungicide: Thiophanate-methyl + Captan	Apple scab (over-wintering stage)
Peach, Plum and Cherry (Monitor insect pests with pheromone traps)		
Dormant — before growth in spring	Lime sulfur	Aphids, mites, peach leaf curl, black knot
Petal Fall — when 90% of blossoms have fallen	Fungicide: Thiophanate-methyl + Captan or Iprodione plus	Brown rot
	Insecticide: Carbaryl, Permethrin, Propoxur or Phosmet	Plum curculio, eyespotted bud moth, prune moth, leaf rollers,
First Cover — 10 days after petal fall	Fungicide: Captan or Thiophanate-methyl plus Insecticide: Carbaryl or Rotenone	Brown rot, leaf spot Plum curculio, cherry fruitworm
Additional Covers — continue to spray at 10 to 14 day intervals until fruit begins to ripen	Fungicide: Captan or Thiophanate-methyl plus Insecticide: Carbaryl, Phosmet or Rotenone	Brown rot, leaf spot Cherry fruit fly,leaf-roller, fruit moth
Grape (Monitor insect pests with pheromone traps)		
When new shoots are 6 to 8 inches long	Fungicide: Captan or Thiophanate-methyl	Black rot
Prebloom — just before blossoms open	Fungicide: same as above	Black rot
	Insecticide: Carbaryl, Rotenone, Phosmet, or Malathion + Propoxur	Leafhoppers, rose chafer, flea beetles
Post-bloom — just after blossoms have fallen	Fungicide: Captan and fixed copper + hydrated lime	Downy mildew
	Insecticide: Rotenone, Phosmet, Propoxur or Permethin	Berry moth, rose chafer, leafhoppers, skeletonizers
When berries begin to touch in the cluster or are about pea-size	Insecticide: same as above	Berry moth, leafhoppers, skeletonizers
	Fungicide: same as above	Downy mildew

[1] Refer to Appendixes B & D for trade or proprietary names of recommended insecticides and fungicides.

attention. Scale insects, however, can eventually produce partial defoliation and should be given attention when present. Please, pick up your fallen oranges. They ruin the desired effect.

Whether your citrus are grown for gustatory or visual purposes, in the use of pesticides, cover trees thoroughly with the suggested materials for maximum benefit and read and observe all direction and precautionary statements on pesticide containers.

Citrus Thrips

Thrips are found on all varieties of citrus in all citrus-growing areas. Their damage is caused by both the nymphs and adults by their rasping the surfaces of leaves and fruit. Growth is stunted and the leaves and fruit take on a leathery silver appearance. Buds and blossoms are sometimes killed under dense populations and new leaves may be gnarled and dwarfed.

Thrips overwinter as eggs deposited in leaves and stems, which hatch in early March. The young are pale and wingless with red eyes. As they mature they grow larger and become orange in color. The adults are only about 1/10 inch in length, and have 4 narrow, fringed wings. The life cycle from egg to adult requires from 2 to 4 weeks, depending on the temperature, and there may be as many as 12 generations a year.

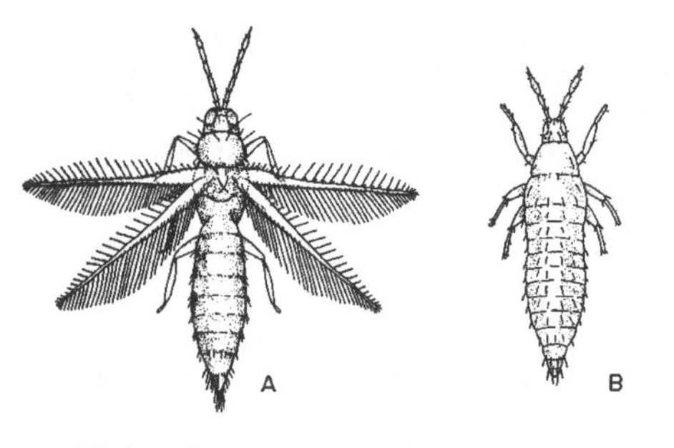

Thrips. A — Adult; B — Mature nymph
(Univ. of Arizona)

There are other species which may be found on citrus, but the home gardener cannot distinguish between them. They are similar in life cycle and are more often found in blossoms where their feeding is considered unimportant.

Thrips are controlled to some extent by predator thrips, mites, spiders and lady beetles. Insecticides are usually not necessary in that damage is primarily only of cosmetic or appearance value and does not actually alter the quality or quantity of fruit. Only rarely will chemical control be of value. (See also Thrips under Garden Pests).

Orangedog

The orangedog caterpillar is a master of disguise in a citrus tree, resembling more a large bird dropping than it does a worm. These unusual larvae will pupate and develop into beautiful swallowtail butterflies. They overwinter as pupae attached to the tree or more commonly in ground trash. In the spring, the adults emerge, mate, and the females lay their eggs singly and randomly throughout the plant during the months of May through August. There may be up to 3 generations a year.

California Orangedog
(Union Carbide)

Surprisingly, the orangedog is usually a pest only in the home yard and never a problem in commercial groves. Even in the backyard it is seldom a problem, since the larvae normally eat only foliage. Because they occur in small numbers they can be easily removed by hand.

Scale Insects

There are several scales that infest citrus, some occurring in one area and not others, while others may be more or less universal. Among these are California red scale, yellow scale, purple scale, Florida red scale, citrus snow scale, chaff scale, black scale, citricola scale, cottonycushion scale, brown soft scale and others. They are classified as to their scale or covering into armored and unarmored scales, but in general their life cycles are similar as are their controls.

The scales overwinter either as eggs or young beneath the female scale, or the female gives birth to living young in the spring. The young are "crawlers" and move about the plant until they settle down and insert their piercing mouthparts to remove plant juices. They lose their legs and antenna and begin developing the scale, or grow larger without a scale depending on their classification. There are usually

California Red Scale
(Union Carbide)

several generations a year and natural controls by predators such as syrphid fly larvae and lady beetles sometimes are adequate to hold their numbers to unimportant levels. Chemical control is usually best achieved through the use of summer oil emulsions or dormant oils, though several of the organophosphate insecticides can be used in emergency conditions. (See also Scale Insects under deciduous fruits).

STRAWBERRIES

Strawberries, at one time available only to the affluent, can be produced by any interested person in a garden, barrel planter, window box, or even pots. Generally, there are few pests that specifically attack strawberries, but several seem to thrive on them simply because nothing else is available at the time.

To keep the protection of strawberries simple, insect and disease control have been combined into a spray schedule (Table 16). Chemical weed control is not suggested for home gardeners or amateurs. However, you should keep the plant rows narrow and free of weeds, for they not only inhibit growth and are unsightly, but they harbor pests. Heavy mulches are the best weed control.

TABLE 15. Insect Control Suggestions for Citrus With and Without Chemicals (Includes grapefruit, kumquat, lemon, lime, mandarin, orange, tangerine and tangeloe)[1]

Pest	Chemical Control	Non-Chemical Control
Orange dog	Malathion. Apply when worms appear.	Remove larvae by hand or spray with *B.thuringiensis.*
Scales	Malathion or Phosmet. Treat when crawlers appear in spring and as needed. Summer oil emulsion (2%) or in fall.	Allow adequate time for biological control by vedalia and other lady beetles, before applying insecticides.
Spider mites	Propargite or Oxythioquinox. Apply when mites become numerous and as needed. Sulfur dust is also effective and the least disruptive to non-pests.	Mites may be hosed off with strong stream of water, or use insecticidal soap solution.
Whitefly	Malathion or Dimethoate. Apply as spray when young begin to increase on leaf undersides.	Insecticidal soap, highly refined oils or neem.
Thrips	Dimethoate or Malathion. Apply when 75% of blossoms have fallen, to protect foraging honeybees. Sulfur dust is also effective and the least disruptive with regard to non-pests.	Ryania, cinnamon oil, or neem.
Aphids	Malathion or Dimethoate. Apply when aphids appear.	Remove with strong stream of water. Insecticidal soap, or neem.
Leafhoppers	Malathion or Acephate. Apply only if damaging leafhopper populations appear	A light spray of whitewash, though unattractive, is repellent to leafhoppers.
Leafroller	Carbaryl, Acephate or Permethrin. Apply when leaves are closed.	Larvae can be removed by hand or controlled with *B. thuringiensis.*

[1] Refer to Appendixes B and D for trade or proprietary names of recommended insecticides and fungicides.

As the best safeguard against disease, choose plants grown from virus-free stock and select those known to be disease-resistant. When using sprays, mix all materials together and apply at one time and apply enough to wet the foliage thoroughly. When using dusts, apply the material under leaves, but avoid overuse. Do not use dusts after berries begin to ripen.

If you intend to plant strawberries in areas that were previously in sod or weeds, you should control resident soil insects by first treating the soil with disyston 1% granules and stirring the soil thoroughly. Follow the label instructions. Combine the insecticide with the upper 4-6 inches of soil immediately after application.

As fruit is produced, it is best to remove ripe and overripe or damaged berries from the garden, because they attract sap beetles, birds, wasps and hornets, bees, sowbugs or pill bugs and a host of other hungry pests. Don't however, throw away such damaged fruit. Instead, wash them, cut out the damaged area and make strawberry preserves for next winter's hot biscuits.

Root Weevils

There are several species known as root weevils which, in addition to strawberries, attack raspberries, loganberries, blueberries, grapes, nursery stock and flower garden plants. Damage is caused by the larvae feeding on the roots and the adults feeding on the leaves. The adults are light brown to black snout beetles with rows of small pits on their backs. The larvae are white legless grubs, with pink tinge and brown heads. Most overwinter as nearly grown larvae in the soil among the roots of the host plants. In the spring they pupate and emerge as adults in June. Egg laying begins in about 2 weeks, in the crown of the plants. After hatching the larvae burrow into the soil and begin feeding on the roots.

Cultural control consists mainly of making new plantings in uninfested soil. Control with insecticides is directed at the adult stage since the larvae are protected beneath the soil.

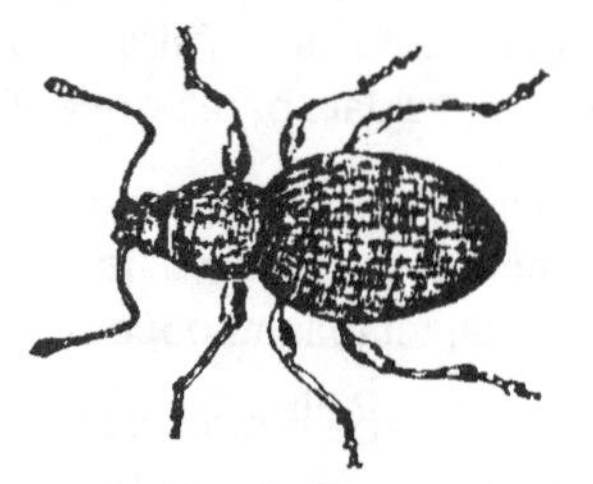

Strawberry Root Weevil
(USDA)

Spittlebugs

Spittlebugs are pests of perennial plants as strawberries, nursery stock and legume forage crops. They injure plants both as nymphs and adults by removing plant juices with their piercing-sucking mouthparts. Some of the effects are stunting of growth, shortening of internodes, dwarfing, rosetting and a general loss of vitality. The adults resemble robust leafhoppers, with many varied color patterns. Eggs are deposited in rows between sheaths and stems of plants near the soil surface. The yellow nymphs become green as they reach full development. Nymphs form the spittle mass by mixing air with the excretion from their alimentary canal, thus giving it the appearance of spittle.

Control with chemicals is best when applied before the adults begin egg laying. In small strawberry patches the nymphs can be removed by hand as easily as the plants can be sprayed.

Strawberry Crown Borer

Crown borer damage is caused by the larvae boring into the crown of the strawberry plant and feeding on the interior, sometimes so completely that growth is stopped or the plant killed. Some injury results from the feeding cavities made by adults in which they lay their eggs, as well as small holes they eat in the leaves.

The dark brown snout beetles cannot fly and hibernate in or near the host plant in the winter. They become active when strawberries bloom and begin laying eggs and continue into August. The small white, legless grubs feed in the crowns, pupate and emerge as adults before summer ends. After feeding for a time they enter hibernation in protected debris. One generation develops per year.

Because the adults do not fly, they must crawl to their hosts. Consequently, it is best to avoid planting strawberries near infested areas, as well as purchasing borer-free plants when establishing new plantings. Chemical control is directed both toward the adults and larvae exposed in the crown.

Strawberry Leafroller

Damage caused by this leafroller, or leaf folder, results from the caterpillars feeding within the folded, rolled, or webbed leaves causing them to turn brown and die. Adults are rusty brown moths with light yellow markings, while the larvae are pale green to brown and up to 1/2 inch in length when fully developed. It overwinters as larvae or pupae in folded leaves or in other nearby shelters. Adults emerge in April or May and lay their eggs on foliage. The eggs hatch in a week and the larvae complete their feeding, pupate and emerge as adults in 40 to 50 days. There are 2 generations each season.

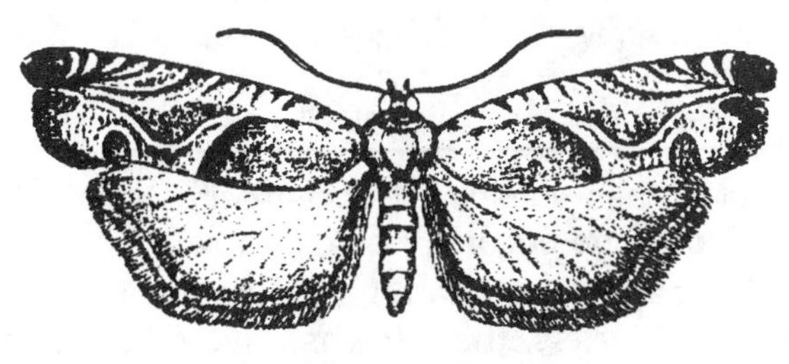

Strawberry Leafroller Moth
(J. B. Smith)

Several parasites act to suppress the leafroller, though it may be necessary to use insecticides under heavy population pressure.

Strawberry Weevil

This small, brown snout beetle with the black patches on its wings is a native of North America and is a pest not only of strawberry, wild and domestic, but also of dewberry, brambles and redbud. The larvae are small white curved grubs. The hibernating weevils emerge in early spring and lay eggs in their feeding punctures in strawberry blossom buds. They then nibble away part of the bud stem, causing it to wilt, fall over and usually drop to the ground. The larvae complete their development in these buds, pupate and emerge as adults. After a short feeding period they enter hibernation, thus making it a 1-generation per year pest. Only staminate flowers provide proper nutrition and developmental conditions making the pistillate varieties of strawberries relatively immune from attack. If insecticide applications are made in the spring at the first sign of weevil activity, there should be very little damage.

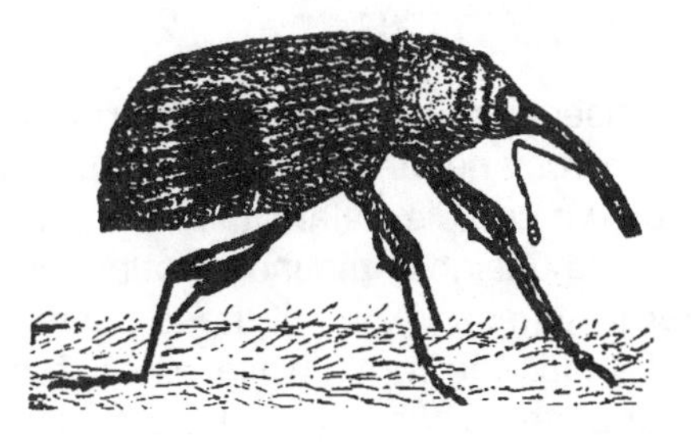

Strawberry Weevil, *Anthonomus signatus.*
(Baerg, Ark. Agr. Exp. Sta.)

TABLE 16. Pest Control Spray Schedule for Strawberries[1]

Schedule (When to spray)	Pesticides to Use	Pest to Control
Prebloom – just as first blossom buds appear in spring	Insecticide: Dicofol + Malathion	Aphids, spittlebugs, plant bugs, spider mites, leafroller, weevils
	Fungicide: Basic Copper, Captan or Thiophanate-methyl	Leaf scorch, leaf spot, leather rot
Every 10 days following first spray until blossoms appear	Insecticide: Same as above	Aphids, spider mites, plant bugs, leafroller

[1] Refer to Appendixes B and D for trade or proprietary names of recommended insecticides and fungicides.

Schedule (When to spray)	Pesticides to Use	Pest to Control
Every 10 days following first spray (Continued)		
	Fungicide: Same as above.	Gray mold, leaf spot
Every 10 days starting in early bloom until 3 days before first picking	Fungicide: Captan or Thiophanate-methyl **Do not apply insecticides during bloom.**	Berry mold and rot, leaf spot
After bloom and on Sept. 1 and Oct. 1	Fungicide: Any listed in Prebloom spray Insecticide: Malathion + Permethrin or Rotenone	Fungus diseases Aphids, leaf roller

LAWN AND TURF

Several insects, mites and related arthropod pests damage lawns and other turf. They cause the grass to turn brown and die, retard its growth and spread, or they build unsightly mounds that may eventually smother the grass. Some of these pests infest the soil and attack the roots, while others feed on the blades and stems, while yet others suck juice from the plants.

Other insects and insect-like pests inhabit lawns but do no damage. They may merely annoy by their appearance, or they may attack man and his pets.

Both of these categories can be controlled with insecticides and a few can be managed with non-chemical methods. The suggestions in Table 17 are applicable not only to home lawns, but also to athletic fields, golf courses, parks, cemeteries and to the areas along roadsides.

Not every pest, not every type of turf and not every control method are listed. However, a general scope is presented, representing most of the common lawn situations. The solutions were derived from the latest research and recommendations of state agricultural experiment stations. If your problem isn't answered, or you want more detailed information, contact your Cooperative Extension Office (County Agent). (See Appendix H)

Billbugs

Injury to turf and lawns by billbugs, or snout beetles, is caused by larvae eating the roots and crowns of grasses and by the adults feeding on stems and foliage. Their easily identified feeding punctures in grasses show up as a series of transverse holes of the same size and shape in the leaf. They result from a single puncture through the leaf in the bud stage before it has unfolded. Punctures in stems cause more damage but are less noticeable.

Billbugs are likely to be more numerous in well watered grass than lawns given haphazard treatment. The adults vary in color from light olive-yellow to brown and black and they have the characteristic elongated mouthparts or snout. The larvae are white, short, thick-bodied, curved, legless grubs. There is only one generation a year. The overwintered adults are produced in late summer and may be active and feed for a while, or remain in the pupal cells until spring. In the spring the adults appear and feed almost continuously, laying eggs in feeding punctures near the base of the grass stems. The larvae develop quickly feeding in crowns and larger roots. In midsummer pupation occurs in the soil or in feeding cavities near the base of plants. In

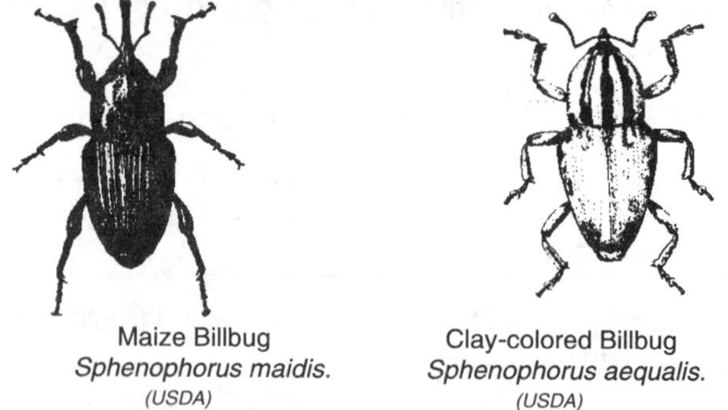

Maize Billbug *Sphenophorus maidis.* (USDA)   Clay-colored Billbug *Sphenophorus aequalis.* (USDA)

10 to 14 days the adults emerge completing the one-year cycle.

Control by natural enemies is ineffective and requires the use of one or more applications of insecticide when adults first appear in the spring.

Chiggers

Sometimes called red bugs, chiggers are actually the larval, parasitic stage of a small mite. They are about

1/150 inch long and bright red. Young chiggers attach to the skin of humans, domestic animals, wild animals including reptiles, poultry and birds. More commonly than not they will enter hair follicles or pores and insert their mouthparts to suck up cellular fluids. They become engorged in about 4 days then drop off to change to nymphs.

Unlike the larvae, nymphs and adults feed on insect eggs, small insects and other small creatures found near woody decaying material.

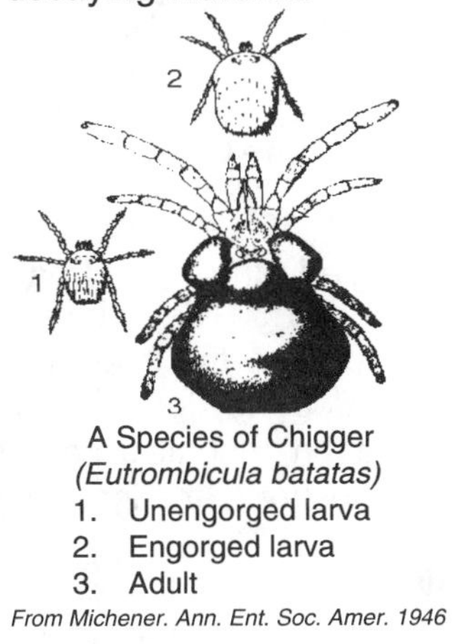

A Species of Chigger
(Eutrombicula batatas)
1. Unengorged larva
2. Engorged larva
3. Adult

From Michener. Ann. Ent. Soc. Amer. 1946

Once the lawn is established and the area becomes "civilized", the chigger problem gradually disappears. Insecticide dusts or sprays will aid immensely in resolving the problem. For personal protection use one of the two repellents, Deet or dusting sulfur or oil of citronella, on the skin, socks and trousers. Repeat applications several times if exposure is continuous. Sulfur dust applied to the lawn at 7-10 day intervals is quite helpful in holding down populations.

Fiery Skipper

Skippers are the small, butterfly-like insects that have short, rapid, darting flights, usually at dusk and not the long, sustained flights of the moths and other butterflies. They are small, yellowish-brown butterflies. Larvae of the fiery skipper feed on the leaves of common lawn grasses, but attack bentgrass most severely. Early infestation is indicated by isolated round bare spots, 1 to 2 inches in diameter. The spots may become numerous enough to destroy most of the grass on the lawn. The fiery skipper is occasionally a pest of lawns in California.

Frit Flies

The frit flies, grass stem maggots and eye gnats are all closely related and may be found in the same general environments. Frit flies are true flies and the immature stages are maggots. The larvae feed in the lower portion of grass stems of healthy, well watered lawns that are left rank. Close mowing and usually one treatment of some short-lived insecticide will dispel the problem of hovering flies.

Japanese Beetles

The Japanese beetle is destructive in both the larval and adult stages. The larvae feed in the soil, devouring the roots of a large number of plants and are especially injurious to turf in lawns, parks, golf courses and pastures and are sometimes a pest in nurseries and gardens. The adults feed on foliage, flowers and fruits during their period of activity, which is only on warm sunny days. More than 275 kinds of plants, including fruit and shade trees, ornamental shrubs, flowers, small fruit and garden crops are often damaged.

The beetles are 1/2 inch in length, shiny metallic green with coppery brown wing covers, making it one of the prettier beetles. The larvae resemble all of the other white grubs, being C-shaped and having 6 legs.

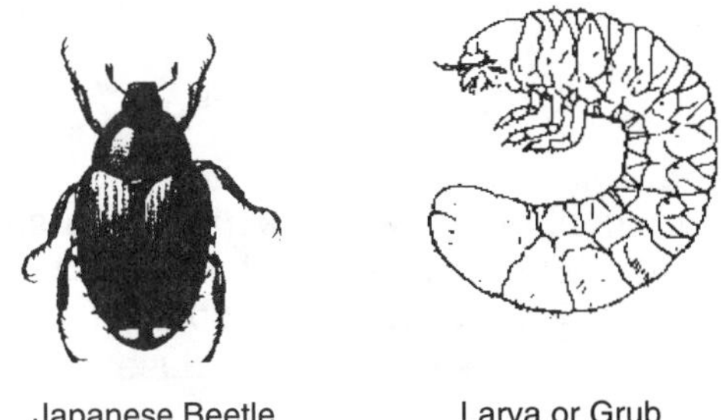

Japanese Beetle *(Union Carbide)*   Larva or Grub *(USDA)*

It overwinters in the soil in the half-grown larval stage. In the spring they crawl near the surface and feed. In late May and June they pupate and adults appear in late June and are active through September. Eggs are laid in the soil at 1-inch depth, with hatching 2 weeks later. The young feed until cold weather sets in then retreat below the frost line. There is only 1 generation per year.

Larvae are controlled by incorporating milky disease spores in the soil of the bacteria, *Bacillus popilliae.* Moles, skunks and birds also aid in the reduction of the grubs. The only effective method of adult control is with the use of insecticides which is less than satisfactory. Adult beetles are readily trapped in commercial traps from suppliers listed in Appendix G.

Leaf Bugs

Leaf bugs are those tiny, 1/10 to 1/4 inch long true bugs that are easily seen crawling through a lawn. Most everyone who sits down in a grassy, weedy spot in early summer has seen these green, black, or red, often flecked, spotted, or striped insects. Both the nymphs and adults cause damage to grass by removing the plant juices from tender grass leaves. Their injury is recognized at first by small yellow dots on the leaves where their first scattered punctures are made. Later, with increased activity, leaves will yellow and die, leading to small areas and eventually to spots as much as a yard wide that die, appearing as if left unwatered. Prevalent among these leaf bugs are Lygus bugs, tarnished plant bugs and apple redbugs.

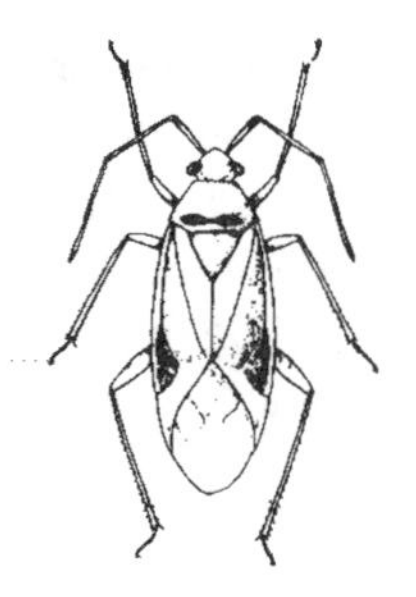

Tarnished Plant Bug
(Univ. of Arizona)

Cultural control consists of keeping the area free of weeds and mowing regularly. Chemical control is the only effective route when dead spots are observed and plant bugs identified as the cause.

Lucerne Moth

The larvae of this insect prefer clover and other legumes that may be found in your lawn, but they also infest grass. The adult is a grayish-brown moth and has 2 pairs of dark spots on each forewing. The larvae resemble the sod webworms, but are slightly larger. Their damage occurs mostly in late summer.

Mole Crickets

Mole crickets are voracious feeders and damage is known to occur to garden vegetables and strawberries as well as grasses. Their damage is the result of underground feeding by the nymphs and adults at or near the soil surface, where they chew roots, tubers, underground stems and most fruit that touch the ground. Of the four common species, the southern mole cricket is the most abundant. They all resemble each other except for size. Adults are about 1-1/2 inches in length, brown, tiny eyes and short mole-like legs equipped for digging. The winter is passed as nymphs or adults in the soil. In spring the females lay their eggs in cells constructed in the soil, several to a cell. After hatching the young develop rapidly, with most of them becoming adults by early fall.

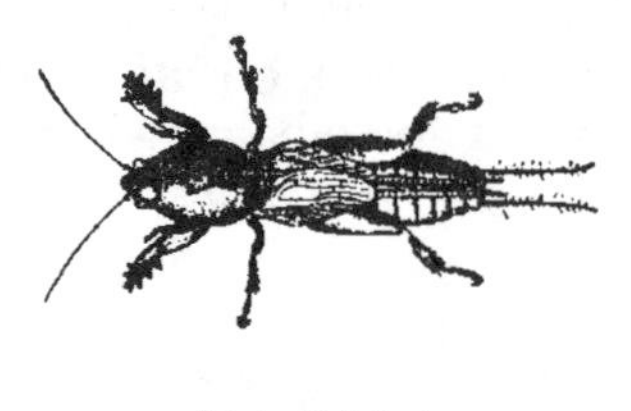

Mole Cricket
(USDA)

There is very little in the way of natural controls. Chemical control can be aided by watering lawns heavily prior to applying the insecticide. When the soil becomes flooded they leave their burrows and swim about in an effort to locate dry quarters and may increase their exposure to the insecticide.

Sod Webworms

Sod webworms, or grass moths, are not only pests of lawns, but also of golf courses and occasionally cultivated crops planted in soil which was previously grassy sod. The yellow-to-white larvae spin threads as they move about and feed, which webs soil and leaves together near the surface, often forming tubes in which they live. They feed at the soil surface and slightly below. The light brown moths are less than an inch in length.

There are several species, but their life cycles are all similar. They overwinter as larvae in silk cases covered with soil near ground level. They continue feeding in the spring and pupate close to their feeding tunnels. The adults appear in early to mid-summer and lay their eggs on grass at dusk. The eggs hatch in 5 to 7 days and the larvae begin the cycle anew. One to three generations may occur each year, depending on the latitude and species.

Sod Webworm
(Union Carbide)

Natural control depends on insect and vertebrate predators, especially ants and birds, several insect parasites, as well as the use of *Bacillus thuringiensis*. Chemical control is quite simple and always effective.

Cutworms and Armyworms

Cutworms and armyworms are the larvae of black or brown moths that may be seen at dusk flitting over lawns. The black and variegated cutworms are the most common late spring and early summer worms found in turf. The bronze cutworm is a late fall and early spring pest, while armyworms appear in mid-summer. An infestation can be revealed by flushing an area of about 1-2 square yards with soapy water, a soap flush. If worms are present they quickly crawl to the surface and are easily seen. Light traps and pheromone traps can be used to monitor moth activity. Cutworms are generally more numerous in well-watered areas. The chemical control threshold is any cumulative population of the above that exceeds one worm per square yard. If possible, mow, rake and water before making an insecticide application. Do not water for at least 3 days after a spray application.

Wireworms

Wireworms are the shiny, slender yellow or brown larvae found at all times of the year in most any type of soil. The adults are known as click beetles, so named because of their snapping into the air when turned upside down. Wireworms feed on many food crops as well as grass and turf lawns. They injure plants by eating seeds in the soil, by cutting off small underground stems and roots and by boring in the larger stems, roots and tubers.

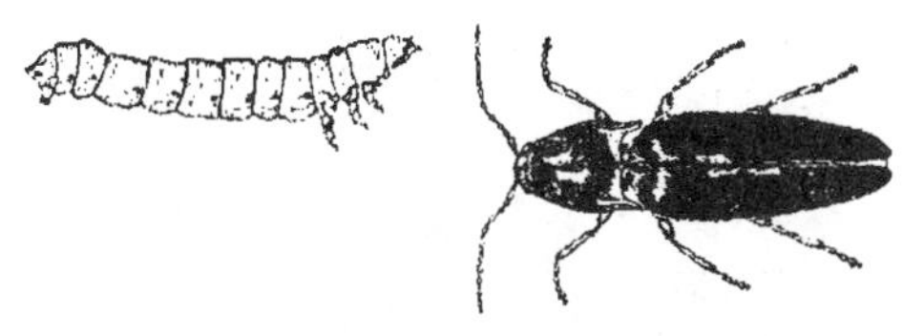

Wireworm Larva (top) and Adult (Bottom)
(USDA)

Wireworms usually overwinter in the soil as larvae and complete their life cycles in one year, while others may take more than one year. The eggs are laid one at a time beneath the soil surface during the spring or early summer by the female click beetles. After hatching, the young begin moving through the soil in search of food. All sizes of the larvae may be found feeding on the roots of grass during the spring and summer.

Cultural control is not very effective, since the most direct method is to deprive them of their food source. Either flooding or drying of the soil during the summer

TABLE 17. Insect Control Suggestions for Lawn and Turf With and Without Chemicals [1]

Pests	Chemical Control	Non-Chemical Control
Ants	Acephate, Malathion, Permethrin or Carbaryl. Apply commercial ant baits to nest as drench and lightly along trails.	Hot, soapy water will act as a deterrent for several days. Repeat as needed.
Aphids (green bugs)	Acephate, Spray, 4 gal./1000 ft ^{2} of lawn area. Don't mow for 24 hours after treatment.	
Armyworms	Acephate, Carbaryl, Permethrin or Trichlorfon. Apply at first sign of feeding damage.	Mow closely and roll lawn at dusk with heaviest roller manageable.
Billbug	Acephate, Carbaryl or Propoxur. Adults can be controlled by treating in April-May and larvae with May-June application.	Beneficial nematodes, *Heterorhabditis bacteriophora.*
Chinch bugs	Carbaryl, Bifenthrin or Cyfluthrin. Water lawn before treatment. Apply preferably as granular formulation in early June with second application 2-3 weeks later. If second generation occurs in August repeat applications.	Plant resistant varieties. Avoid drought condition of lawn.

[1] Refer to Appendix B for trade or proprietary names of recommended insecticides.

Pests	Chemical Control	Non-Chemical Control
Clover mite	Dicofol, Ethion or Acephate. Apply treatment at first sign of an infestation.	Dusting sulfur, light application.
Cutworms	Acephate, Bifenthrin or Cyfluthrin. Apply at first sign of feeding damage.	Mow closely and roll lawn at dusk with heaviest roller manageable.
Earthworms	Not really a pest. No chemical control.	Probably indicates over-watering. Rake or roll worm casting mounds into surrounding lawn.
Earwigs	Chemical control not recommended since they do no damage. If enough of a pest Bifenthrin or Cyfluthrin are effective.	Probable indication of over-watering. Trap in flower beds with boards, folded newspaper or dark plastic and destroy.
Fiery skipper	Carbaryl, Acephate or Bifenthrin. Apply when bentgrass lawns show isolated, round 1/2" diameter bare spots.	
Fleas and Chiggers	Acephate, Bifenthrin, Cyfluthrin or Carbaryl. Spray all areas thoroughly, especially pet sleeping areas. Repeat in 10 days.	Fleas — open up bedding and expose for long periods to sun. Chiggers — keep lawn mowed to minimum height. Make light application of dusting sulfur.
Frit Fly	Acephate or Bifenthrin. Apply to lawns on first appearance of hovering flies.	
Grasshoppers	No chemical control recommended.	Mow lawn frequently enough to qualify as well-kept.
Japanese beetle grubs & other white grubs	Imidacloprid, Carbaryl or Trichlorfon. Apply uniformly to lawn when grubs are small, early spring.	Apply milky disease spore granules and water into soil any time when ground not frozen.
Leaf bugs	Acephate or Malathion. Apply when first dead spots are observed in lawn.	
Leafhoppers	Carbaryl, Acephate or Bifenthrin. Apply when leafhoppers are numerous, but before grass shows yellowing.	
Lucerne moth	Carbaryl or Cyfluthrin. Usually a pest on clovers, this caterpillar can be controlled with one application.	Remove clovers from lawn by hand as soon as they appear.
Millipedes	Carbaryl, Acephate or Bifenthrin. Treat when millipedes become abundant.	
Meadow spittlebug	No chemical control recommended.	Regular mowing.
Mole crickets	Carbaryl, Cyfluthrin or Bifenthrin. Apply when pests are present and use adequate water to thoroughly wet grass.	

Pests	Chemical Control	Non-Chemical Control
Scales	Disulfoton 1.5% G. Apply granules and water into the soil as near the time of crawler stage as can be determined.	
Sod webworm	Carbaryl, Bifenthrin, Acephate or Cyfluthrin. Spray area for both generations, June and August.	Beneficial nematodes, *Steinernema carpocapsae*
Thrips	Not a lawn pest but may annoy humans. Bifenthrin or Cyfluthrin spray.	Keep down nearby rank grass, flowers and weeds.
Ticks	Acephate, Bifenthrin, Cyfluthrin or Carbaryl. Spray entire lawn giving special attention to dog runs and sleeping areas. Fit dogs with flea/tick collars.	Check dogs and remove all ticks every 3 days during tick season.
Wild bees	Malathion, Acephate or Bifenthrin. Spray individual burrows as they appear.	Discourage by filling burrows with sand or coffee grounds and soap or detergent water.
Wireworms	Imidacloprid or Disulfoton. Apply as granules in early spring to control these larvae of various click beetles.	

will kill many wireworms, but this too is impractical. This leaves us only with chemical control which is best achieved with the use of the granular formulations of several effective insecticides. After application a thorough watering of the lawn improves the effectiveness of the insecticide.

TREES, SHRUBS AND WOODY ORNAMENTALS

Millions of dollars are spent annually by homeowners to beautify their homes and surroundings through the artistic and utilitarian planting of trees and shrubs. This beautification is not only pleasing to the neighbor and the homeowner, but it also increases the value of the property.

This increasing interest in and growing demand for more ornamental plants have been the incentives for nurserymen to experiment and grow an even greater selection of plants, including many exotics. The results are that never before have we had such variety of plant materials and accessories from which to choose.

Usually, however, each of these species is the favorite food of one or more species of insect or mite. Most of the time natural enemies and other factors keep these pests at levels low enough that plants are not damaged. However, a season seldom passes that conditions are not favorable for the multiplication of at least one pest and control measures must be applied to prevent serious injury to the plant.

We can never anticipate just which pests will become a problem, nor how severe the infestation will become. But there will always be potentially injurious pests in the area from the first warm spring days until frost and freezing in the fall and winter.

Tables 18 and 19 have been prepared with the intent that this information will help you to identify the pest and know when and how to safely and effectively control the more troublesome pests of ornamental plants. As a result you can derive more enjoyment from your landscaping efforts.

Before closing, let me leave with you a universal insect and mite control spray mixture recipe that can be used on practically all ornamental trees, shrubs and flowers. It is a multi-purpose spray mix, including disease control. To one gallon of water, add 2 tablespoons of carbaryl (50% WP), 1 tablespoon of dicofol (18% WP), 4 tablespoons of malathion (25% WP) and 1 tablespoon of maneb (65% WP). Do not prepare ahead of time, but only when needed.

In every case, it is very important to read and observe all directions and precautionary statements on the label. And in the application of all pesticides cover plants thoroughly with the suggested materials for maximum protection and benefit.

TABLE 18. Insect and Mite Control Suggestions for Trees, Shrubs and Woody Ornamentals, With and Without Chemicals.[1]

Plant	Pest	Chemical Control	Non-Chemical Control
Andromeda (Pieris)	Lace bug	Carbaryl, Acephate or Malathion. Treat as needed, underside of leaves.	Insecticidal soap spray
	Spider mites	Dicofol or Ethion. Repeated applications may be necessary.	Insecticidal soap spray or neem.
Alder	Leafminer	Acephate. When first mines appear.	
	Woolly aphid	Malathion, Acephate or Bifenthrin. Treat when aphids appear and as needed.	Insecticidal soap or neem.
Arborvitae	Bagworms	Carbaryl, Spinosad, Acephate or Malathion. Treat when bagworms are small, usually early to mid-June.	*B. thuringiensis* or neem. Remove bags by hand in the fall and destroy.
	Leafminer	Malathion, Bifenthrin or Cyfluthrin. Spray twice at 10-day intervals in June.	
	Scales	Dormant oil, Acephate, Malathion or Bifenthrin. Use oil as dormant treatment in spring or apply others against crawlers from late June to early July. May have to repeat in early September.	Insecticidal soap spray or neem.
	Spider mites	Acephate or Ethion. Apply after new growth begins and in fall.	A light dusting with sulfur; insecticidal soap spray.
Ash	Aphids	Malathion, Propoxur or Acephate. Treat when aphids appear and repeat when needed.	Remove with hard stream of water on smaller trees and lower limbs of large trees.
	Borer	Acephate. Apply to trunk and large branches in late May or early June, again in August-September.	Use pheromone traps 3 weeks earlier than usual treatment time.
	Ash flower gall mite	Dormant oil, Carbaryl or Ethion. Use oil as dormant treatment; others when first blossoms begin to form.	Insecticidal soap or neem.
	Fall webworm	Carbaryl or Acephate. Apply when webs appear.	*B. thuringiensis*.
	May beetles	Carbaryl or Bifenthrin. Treat when adults are seen and as needed.	Remove by hand and collect beetles at nearby brightlights at night.

[1] Refer to Appendix B for trade or proprietary names of recommended insecticides. Many of the pests mentioned in this table are illustrated on pages 134-138, Figure 18.

Plant	Pest	Chemical Control	Non-Chemical Control
Ash (Continued)			
	Scale	Dormant oil + Ethion as dormant spray. Acephate. Treat against crawlers in May-June. Repeat in 10 days.	
	Leafhoppers	Carbaryl or Cyfluthrin. Treat when first seen, repeat as needed.	Insecticidal soap.
	Leafroller	Carbaryl or Acephate. Apply on first appearance of leaves rolled together.	*B. thuringiensis.*
	Plant bugs	Carbaryl or Malathion. Apply when nymphs first appear and as needed.	Insecticidal soap or neem.
Azalea	Bark scale	Malathion, Acephate or Carbaryl. Treat for crawlers in June-July.	
	Leafminer	Malathion, Acephate or Bifenthrin. Make 2 applications, mid-June and mid-July.	
	Spider mites	Dicofol, Oxythioquinox or Ethion. Treat when mites appear and repeat as needed.	Remove with hard stream of water. Insect-icidal soap spray.
	Whitefly	Bifenthrin, Cyfluthrin or Acephate. Treat when flies appear and repeat at 5-day intervals. Repeat as needed.	Insecticidal soap spray leaf undersides or neem.
	Lace bug	Malathion, Bifenthrin, Carbaryl or Acephate. Treat when first bugs are seen, repeat as needed.	
	Borer	Acephate. Treat trunk and large branches in May, repeat twice at 3-week intervals.	Use pheromone traps 3 weeks before usual treatment time.
Bald-Cypress	Bagworms	Carbaryl, Spinosad, Acephate or Malathion. Treat when bags are small, usually in June.	*B. thuringiensis.*
Barberry	Aphids	Malathion, Acephate or Cyfluthrin. Treat when aphids become numerous.	Remove with hard stream of water.
	Scale	Carbaryl or Acephate. Treat when crawlers appear.	Insecticidal soap spray.
	Webworms	Carbaryl, Bifenthrin or Acephate. Treat when larvae appear, usually.	*B. thuringiensis.*
Birch	Aphids	Malathion, Cyfluthrin or Acephate. Treat when winged aphids appear and as needed.	Remove with hard stream of water on smaller trees.
	Bagworms	Carbaryl, Spinosad, Malathion or Acephate. Apply when bags are small, usually June.	*B. thuringiensis.*

Plant	Pest	Chemical Control	Non-Chemical Control
Birch (Continued)			
	Fall webworm	Carbaryl or Acephate. Treat when webs appear.	*B. thuringiensis.*
	Japanese beetles	Carbaryl or Phosmet. Apply when beetles or riddled foliage are seen and as needed.	Place traps on periphery of yard (5 per acre).
	Scale	Malathion, Acephate or Carbaryl. Apply for crawlers, usually May, and repeat in 10 days.	Insecticidal soap spray.
	Borer	Acephate. Difficult to do, but entire tree should be treated twice, May-June.	Wrap trunks with heavy fabric.
	Leafminer	Carbaryl, Phosmet, Malathion or Acephate. Apply when adults are present, usually mid-May and again in late June.	There is no non-chemical contol.
	Leafhopper	Carbaryl, Acephate or Bifenthrin. Treat when leafhoppers appear and repeat as needed.	Insecticidal soap spray, or rotenone.
Bittersweet	Scale	Malathion, Carbaryl or Acephate. Apply in May-June when crawlers are present, repeat three times at 10-day intervals.	Insecticidal soap spray, or rotenone.
Box elder	Box elder bug	Carbaryl or Acephate. Spray bugs in early summer.	Trap under boards or folded paper and destroy.
Boxwood	Leafminer	Malathion, Bifenthrin, Carbaryl or Acephate. Apply when first mines appear, April-May.	See Birch.
	Psyllid	Acephate or Carbaryl. Treat when young appear, usually May, repeat as needed.	Remove with hard stream of water.
	Scale	Ethion + Oil as dormant spray. Acephate or Bifenthrin. Treat when crawlers appear, usually June.	Insecticidal soap spray.
	Spider mites	Dicofol, Oxythioquinox or Acephate. Treat when mites appear and again in 10 days. Repeat if needed.	Remove with hard stream of water, or insecticidal soap spray.
Buckthorn	Bagworms	Carbaryl, Bifenthrin, Trichlorfon or Acephate. Apply when bags are small, usually June.	*B. thuringiensis.*

Plant	Pest	Chemical Control	Non-Chemical Control
Buckthorn (Continued)			
	Japanese beetle	Carbaryl, Cyfluthrin or Imidacloprid. Apply when beetles or riddled foliage is seen and as needed.	Rotenone or neem.
Catalpa	Catalpa sphinx	Carbaryl or Acephate. Treat at first sign of small larvae.	Shake larvae off and use for fish bait.
Cercis (see Red Bud)			
Chestnut	Leafhoppers	Malathion, Bifenthrin or Carbaryl. Treat when hoppers appear and as needed.	
Clematis	Blister beetle	Acephate or Carbaryl. Spray or dust beetles as needed.	Remove by hand. Wear gloves and long-sleeve shirt.
	Borer	Acephate. Spray 3 times at 10-day intervals, beginning usually in May.	Wrap trunk with heavy tree wrap.
Cotoneaster	Aphids	Malathion or Acephate. Treat when aphids appear and as needed.	Remove with hard stream of water. Wait for predator control.
	Lace bug	Malathion, Carbaryl or Acephate. Treat when young are seen and repeat as needed to protect young foliage.	Insecticidal soap.
	Leafhoppers	Acephate, Carbaryl, or Malathion. Treat on first appearance and as needed.	Insecticidal soap or rotenone.
	Pear slug	Malathion or Cyfluthrin. Treat foliage 2 weeks after petal-fall and again in 2 weeks.	Remove by hand where possible.
	San Jose scale	Horticultural oil or Acephate. Treat against crawlers in late June, repeat twice at 10-day intervals.	Import vedalia beetles, their natural enemies.
	Webworm	Trichlorfon, Cyfluthrin or Carbaryl. Treat when webs first appear.	*B. thuringiensis* or neem.
	Spider mites	Dicofol, Ethion or Oxythioquinox. Treat twice at 10-day intervals and as needed.	Remove with hard stream of water; insecticidal soap spray.
Crab-Apple	Aphids	Malathion. Treat as needed.	Knock off with strong stream of water.
	Borers	Acephate. Spray trunk twice at 10-day intervals beginning usually in June.	Wrap trunk with heavy tree wrap.

Plant	Pest	Chemical Control	Non-Chemical Control
Crabapple (Continued)			
	Leafhoppers	Carbaryl, Cyfluthrin or Malathion. Treat when leafhoppers appear and as needed.	Insecticidal soap or rotenone.
	Scale	Acephate or Malathion. Treat when crawlers appear, usually late May.	Insecticidal soap spray.
Deutzia	Aphids	Malathion, Acephate or Bifenthrin. Apply when aphids appear and as needed.	Remove with hard stream of water. Insecticidal soap spray.
	Leafminer	Cyfluthrin. Treat when first mines appear and as needed.	
Dogwood	Borer	Acephate. Treat trunk and bases of low branches in May and repeat 3 times at 3- week intervals.	Wrap trunk with heavy tree wrap.
	Leafhopper	Carbaryl, Malathion or Cyfluthrin. Treat when leafhoppers appear and as needed to protect new foliage.	
	Scale	Malathion, Acephate or Carbaryl. Apply against crawlers in late May and repeat in 10 days.	Remove with hard stream of water or insecticidal soap.
Dutchman's Pipe	Mealybug	Malathion or Imidacloprid. Spray 3 times at 14-day intervals begining mid-May.	Insecticidal soap spray.
	Scale	Horticultural oil or Acephate. Apply against crawlers in early June and repeat in 10 days.	Insecticidal soap spray.
Douglas Fir	Gall aphid	Horticultural oil, Cyfluthrin or Carbaryl. Apply in early spring before budbreak or in October.	Insecticidal soap spray.
	Bagworms	Carbaryl, Bifenthrin, Trichlerfon or Acephate. Apply when bags are small, usually June.	*B. thuringiensis* or rotenone. Remove by hand in fall.
Elm	Woolly aphid	Bifenthrn, Acephate or Malathion. Treat in spring when leaves are expanding.	Remove with hard stream of water on smaller trees.
	Bark beetles	Acephate or Carbaryl. Call your local County Agent or U.S. Forest Service Office.	
	Elm leaf beetle	Carbaryl, Phosmet, Cyfluthrin or Acephate. Apply when first larvae appear, usually when leaves are fully opened, and again in July	*B. thuringiensis* or rotenone.

Plant	Pest	Chemical Control	Non-Chemical Control
Elm (Continued)			
	Scale	Ethion + Oil as dormant spray. Carbaryl or Acephate. Apply against crawlers in June-July.	Insecticidal soap spray or rotenone.
	Fall webworm	Carbaryl, Trichlorfon or Cyfluthrin. Treat when webs appear.	*B. thuringiensis* or rotenone.
	Leafhoppers	Carbaryl, Malathion or Acephate. Treat when hoppers appear and as needed to protect new foliage.	
	Japanese beetle	Carbaryl, Cyfluthrin or Phosmet. Apply when beetles or riddled foliage is seen and as needed.	Japanese beetle traps, 5 per acre.
Euonymus	Aphids	Malathion or Acephate. Treat when aphids appear and as needed.	Knock off with strong stream of water.
	Scale	Ethion + Oil dormant spray. Cyfluthrin, Carbaryl or Acephate. Use against crawlers in May-June and repeat twice at monthly intervals.	Insecticidal soap spray.
	Bagworms	Carbaryl, Bifenthrin, Acephate or Malathion. Treat when bags are small, usually mid-June.	*B. thuringiensis* or rotenone. Remove by hand in fall.
	Leafhoppers	Carbaryl, Malathion or Acephate Apply when hoppers are seen and as needed.	Insecticidal soap or rotenone.
	Spider mites	Oxythioquinox, Dicofol or Ethion. Treat when mites appear and again in 10 days or as needed.	Remove with hard stream of water.
Fir	Bagworms	Bifenthrin, Acephate or Malathion. Apply when bags are small, usually late June.	*B. thuringiensis* or rotenone. Remove by hand in fall.
	Balsam twig aphid	Acephate or Malathion. Treat when aphids are first noticed, usually early May.	Knock off with heavy stream of water.
	Pales weevil	Acephate or Cyfluthrin. Treat young trees in May and again in July.	Pull and burn stumps before July.
	Pine needle scale	Bifenthrin, Malathion or Acephate. Apply against crawlers.	
	Spider mites	Oxythioquinox, Dicofol or Ethion. Apply twice, 10 days apart, and as needed.	Remove with heavy stream of water.
Firethorn (Pyracantha)	Aphids	Malathion, Acephate or Cyfluthrin. Apply when aphids appear and as needed.	Remove with hard stream of water.

Plant	Pest	Chemical Control	Non-Chemical Control
Firethorn (Continued)			
	Lace bug	Carbaryl, Malathion or Acephate. Apply when bugs first appear and as needed to protect new foliage.	
	Spider mites	Dicofol or Acephate. Apply twice, two weeks apart.	Remove with hard stream of water.
Flowering Fruit Trees (Ornamental apricot, cherry, crabapple, peach, pear, plum, quince)			
	Aphids	Malathion, Acephate or Cyfluthrin. Treat as needed.	Knock off with heavy stream of water.
	Borers	Acephate. Apply to trunks in May-June and repeat 4 times at 3-week intervals.	Wrap trunks during this period with tight trunk wrap
	Fall webworm	Carbaryl, Cyfluthrin or Acephate. Treat when webs appear.	*B. thuringiensis* or rotenone.
	Leafhoppers	Carbaryl, Malathion, Bifenthrin or Acephate. Treat when leafhoppers appear and as needed.	
	Peach tree borer	Acephate. Make first application to trunks June-July, then two more spaced 3 weeks apart. Or use paradichlorobenzene crystals around base of trunk in the fall.	See Borers.
	Lesser peach tree borer	Acephate. Follow instructions for peach tree borer only beginning one month earlier.	See Borers.
	Pear slug and psylla	Ethion + dormant oil. Apply during dormant season only. For psylla, use Carbaryl or Acephate 2 weeks after petal fall and repeat in 2 weeks.	Remove slugs by hand where possible, or rotenone.
	Japanese beetle	Carbaryl or Malathion. Apply when beetles or riddled foliage is observed and as needed.	Install Japanese beetle traps between trees, 5 per acre.
	Bagworms	Carbaryl, Bienthrin or Acephate. Apply when bags are small, usually early to mid-June.	*B. thuringiensis* or rotenone. Remove by hand in the fall.
	Scales	Dormant Oil, Acephate or Carbaryl. Use oil only in dormant stage. Apply others when crawlers appear.	Insecticidal soap spray.
	Spider mites	Dicofol, Tetradifon or Ethion. Apply when mites appear and again in 10 days, and as needed.	Knock off with hard stream of water.

Plant	Pest	Chemical Control	Non-Chemical Control
Flowering Fruit Trees (Continued)			
	Spring cankerworm	Phosmet, Acephate, Malathion or Carbaryl. Apply when worms appear. early to mid-May.	*B. thuringiensis* or rotenone.
	Tent caterpillars	Carbaryl, Cyfluthrin or Acephate. Eggs usually hatch when buds break in spring. Treat then or when webs appear.	*B. thuringiensis* or rotenone. Remove by hand.
	Woolly aphid	Malathion, Acephate, Bifenthrin or Carbaryl. Treat when bluish-white threads from aphids appear, May-June.	Hard stream of water or insecticidal soap spray.
Forsythia	Spider mites	Oxythioquinox, Dicofol or Acephate. Begin applications when mites appear and repeat in 10 days, and as needed.	Knock off with hard stream of water.
Golden Raintree	Leafhoppers	Carbaryl, Acephate or Cyfluthrin. Apply when leafhoppers appear and as needed.	
Hackberry	Lace bugs	Malathion, Carbaryl or Acephate. Treat when first small nymphs appear, usually in May.	Rotenone or neem.
	Nipple gall psyllid	Acephate or Carbaryl. Apply in April-May.	Insecticidal soap spray.
	Scale	Dormant Oil, Carbaryl or Acephate, Use oil only as dormant application and the others against crawlers in May.	Insecticidal soap spray.
Hawthorn	Aphids	Acephate, Malathion or Bifenthrin. Treat when aphids appear and as needed.	Knock off with strong stream of water, or Insecticidal soap spray.
	Bagworms	Bifenthrin, Carbaryl or Acephate. Treat when bags are small, usually June.	*B. thuringiensis* or rotenone. Remove by hand in fall.
	Fall webworm	Carbaryl, Cyfluthrin or Acephate. Treat when webs appear.	*B. thuringiensis* or rotenone.
	European red mite	Dicofol, Ethion or Oxythioquinox. Apply just prior to beginning of growth.	Remove with hard stream of water.
	Japanese beetle	Carbaryl or Phosmet. Apply when beetles or riddled foliage appear and as needed.	Japanese beetle traps, 5 per acre.
	Tent caterpillars	Malathion, Bifenthrin or Acephate. Apply at first sign of leaf feeding while larvae are small.	*B. thuringiensis* or rotenone.

Plant	Pest	Chemical Control	Non-Chemical Control
Hawthorn (Continued)			
	Lace bugs	Acephate, Carbaryl or Malathion. Treat when first small nymphs appear, usually in May.	Rotenone or neem.
	Leafhoppers	Carbaryl, Acephate or Cyfluthrin. Treat when hoppers appear and as needed.	
	Leafminer	Acephate. Apply when leaves are fully expanded or at sign of browning early to mid-May.	Remove mined leaves when first noticed.
	Scurfy scale	Dormant Oil, Malathion or Carbaryl. Use oil only as dormant treatment, and other materials against crawlers in May.	Insecticidal soap or rotenone.
	Terrapin scale	Dormant Oil, Carbaryl or Acephate, Use oil only as dormant application and others when crawlers appear in June.	Insecticidal soap spray.
	Oystershell scale	Dormant Oil, Carbaryl or Acephate. Use oil only as dormant application and others against crawlers in May and again in 10 days.	Insecticidal soap spray.
	Pear slug	Malathion or Cyfluthrin. Make foliage application 2 weeks after petal fall and repeat in 2 weeks.	Remove by hand where possible.
Hemlock	Bagworms	Bifenthrin, Carbaryl or Trichlorfon. Treat when bags are small, usually in June.	*B. thuringiensis* or rotenone.
	Looper	Acephate or Bifenthrin. Apply when worms appear, July-August.	Rotenone or ryania.
	Scale	Dormant Oil, Carbaryl or Acephate. Use oil only as dormant application, and others when crawlers appear, usually July.	Insecticidal soap spray.
	Strawberry root weevil	Acephate. Apply to foliage and soil around infested plants in June.	
	Spider mites	Dicofol, Ethion or Acephate. Treat when mites appear, and again in 10 days or as needed.	Knock off with hard stream of water or soap spray.
Holly	Southern red mite	Dicofol, Oxythioquinox or Acephate. Treat when mites appear and again in 10 days or as needed.	Remove with hard stream of water.
	Leafminer	Imidacloprid or Acephate. Apply non-systemic material to foliage in May for adult control. Apply systemic in June to control larvae in mines.	Remove mined leaves when first noticed.

Plant	Pest	Chemical Control	Non-Chemical Control
Honey Locust	Bagworms	Carbaryl, Acephate or Bifenthrin. Treat when bags are small, usually in June.	*B. thuringiensis* or rotenone. Remove by hand in fall.
	Locust borer	Acephate. Spray trunks thoroughly in early September.	Wrap trunks with heavy wrap.
	Spider mites	Dicofol, Acephate or Ethion. Treat when mites appear and again in 10 days and as needed.	Insecticidal soap spray.
	Plant bugs	Carbaryl or Bifenthrin. Apply when bugs appear and as needed.	Rotenone or neem.
	Pod gall midge	Carbaryl. Apply to growing tips in spring and again at 10-day intervals until controlled.	
	Honey locust scale	Dormant Oil, Acephate or Carbaryl. Use oil only in dormant season, and others when crawlers appear.	Insecticidal soap spray.
	Leafhoppers	Malathion, Acephate or Cyfluthrin. Apply when leafhoppers appear and as needed.	
	Mimosa webworm	Carbaryl, Acephate or Trichlorfon. Apply at first signs of foliage browning from first generation in July and second in August.	*B. thuringiensis* or rotenone. Remove by hand.
	Oystershell scale	Dormant Oil, Malathion or Acephate. Use oil only in dormant season, and other materials when crawlers appear in May-June.	Insecticidal soap spray.
Honeysuckle	Aphids	Acephate, Malathion or Bifenthrin. Apply when aphids appear and as needed.	Knock off with strong stream of water or soap spray.
	Leafminer	Acephate or Cyfluthrin. Treat foliage at first signs of mines, May-June.	
	Spider mites	Dicofol, Oxythioquinox or Ethion. Apply when mites appear and as needed.	Remove with hard stream of water or soap spray.
Hornbeam	Bagworms	Carbaryl, Malathion or Acephate. Treat when bags are small, usually June.	*B. thuringiensis* or rotenone. Remove by hand in fall.
	Leafhoppers	Acephate, Carbaryl or Bifenthrin. Apply when leafhoppers appear and as needed.	
Inkberry	Leafminer	Carbaryl or Acephate. Apply in April and again in 10 days.	

Plant	Pest	Chemical Control	Non-Chemical Control
Ivy (Boston and English)	Scale	Malathion or Acephate. Spray at 2-week intervals during June-July.	Insecticidal soap spray, or rotenone.
	Spider mites	Dicofol, Oxythioquinox or Ethion. Apply when mites appear and repeat as needed.	Remove with hard stream of water or soap spray.
	Leafhoppers	Acephate, Bifenthrin or Acephate. Apply when leafhoppers appear.	Rotenone or neem.
	Japanese beetle	Acephate or Malathion. Treat when beetles first appear, usually mid- to late June and as needed.	Remove by hand, and Japanese beetle traps.
Juniper	Bagworms	Carbaryl, Malathion or Acephate. Apply when bags are small, usually June.	*B. thuringiensis* or rotenone. Remove by hand in fall.
	Midges	Carbaryl to foliage in mid-May. Disyston. Drench soil around plants in April and treat foliage in May. Repeat foliage treatment at 40 day intervals if needed.	
	Scale	Dormant Oil, Acephate or Malathion. Use oil only in dormant season and other materials beginning in May and repeat at 10 day intervals until controlled.	Insecticidal soap spray.
	Webworm	Acephate, Naled or Carbaryl. Apply in April-May and again in September if needed.	Remove by hand.
	Spruce spider mite	Dicofol, Oxythioquinox or Disyston soil drench. Apply when mites appear in April and as necessary.	Remove with hard stream of water or soap spray.
	Tip dwarf mite	Oxythioquinox, Dicofol or Ethion. Apply May-June as needed.	Make light application of sulfur dust.
Larch	Bagworms	Malathion, Carbaryl or Acephate. Apply when bags are small, usually early to mid-June, and again in 10 days if needed.	*B. thuringiensis* or rotenone. Remove by hand in fall.
	Woolly aphid	Acephate, Malathion or Bifenthrin. Apply in early May when aphids appear.	
Laurel	Borer	Acephate. Spray main stem and branches in mid-May and again in June.	Wrap trunk with heavy fabric.
	Lace bug	Carbaryl, Acephate or Malathion. Make two applications, May and July, to undersides of leaves.	

Plant	Pest	Chemical Control	Non-Chemical Control
Laurel (Continued)			
	Taxus weevil	Acephate or Malathion. Spray foliage and soil surface in late June.	
	Whitefly	Malathion or Acephate. Make two applications, June and July.	Insecticidal soap spray.
Lilac	Borer	Acephate. Treat stems in May-June.	
	Leafminer	Acephate or Bifenthrin. Apply at first indication of mining.	
	Japanese beetle	Carbaryl or Acephate. Make 2 applications, July and August.	Remove by hand, and traps.
	Oystershell scale	Ethion + Dormant Oil, Malathion or Acephate. Use Ethion and oil only as dormant spray and other materials against crawlers in May-June.	Insecticidal soap spray.
Linden	Aphids	Bifenthrin, Acephate or Malathion. Apply when aphids appear.	Remove with hard stream of water or soap spray.
	Cankerworms	Phosmet, Acephate or Carbaryl. Treat at first sign of infestation, usually May.	*B. thuringiensis* spores, or rotenone.
	Fall webworm	Cyfluthrin, Acephate or Carbaryl. Apply when webs appear, usually in June.	*B. thuringiensis* spores, or rotenone.
	Japanese beetles	Carbaryl, Acephate or Malathion. Apply when beetles appear in June and as needed.	Beetle traps, 5 per acre.
	Leaf beetles	Carbaryl or Acephate. Spray at first sign of beetles or feeding injury, June-July.	
	Scurfy scale	Dormant Oil, Acephate or Carbaryl. Use oil only as dormant treatment, and others against crawlers in May.	
	Bagworms	Carbaryl, Bifenthrin or Acephate. Apply when bags are small in June.	*B. thuringiensis* or rotenone. Remove by hand in fall.
	Cottony maple scale	Dormant Oil, Carbaryl or Acephate. Apply oil only in dormant season and other materials in late June-July, concentrating on leaf undersides.	
Locust	Pod gall	Acephate. Begin in May, and repeat at 2-week intervals.	

Plant	Pest	Chemical Control	Non-Chemical Control
Locust (Continued)			
	Borers	Carbaryl or Acephate. Treat trunks in early September.	Wrap trunk with heavy fabric.
	Leafminer	Naled, Bifenthrin or Acephate. Apply as foliage is developing and again in July.	
	Spider mites	Dicofol, Oxythioquinox or Ethion. Apply when mites appear and as needed.	Remove with hard stream of water or soap spray.
London Plane	Lace bug	Carbaryl or Malathion. Spray in May and again in June.	
	Plum borer	Acephate or Carbaryl. Apply 3-6 times to trunk and limbs, beginning in June.	Wrap trunk with heavy fabric.
	Scale	Dormant Oil, Cyfluthrin or Acephate. Apply oil only in dormant season and other materials against crawlers in June.	Insecticidal soap spray, or rotenone.
Magnolia	Scale	Ethion + Dormant Oil, Malathion or Acephate. Use oil only as dormant treatment and other materials against crawlers in August-September and as needed.	
	Yellow poplar weevil	Carbaryl or Acephate. Treat when adults appear, June-July.	
	Leafminer	Naled or Acephate. Apply when mines first appear and as needed.	
Mahonia	Barberry aphid	Malathion, Bifenthrin or Acephate. Treat when aphids appear.	Knock off with strong stream of water.
	Scale	Carbaryl, Malathion or Acephate. Apply when crawlers appear.	Insecticidal soap spray.
	Webworm	Trichlorfon, Carbaryl or Acephate. Apply when worms appear.	*B. thuringiensis* or rotenone.
Maple	Aphids	Malathion, Bifenthrin or Acephate. Treat when aphids appear and as needed.	Remove with hard stream of water.
	Bagworms	Carbaryl, Bifenthrin or Acephate. Treat when bags appear, usually June.	*B. thuringiensis* or rotenone. Remove by hand in fall.
	Borers	Acephate or Carbaryl. Apply to trunk and lower branches in May-June-July.	Wrap trunk with heavy fabric.

Plant	Pest	Chemical Control	Non-Chemical Control
Maple (Continued)			
	Cottony scale	Dormant Oil, Acephate or Carbaryl. Use oil in dormant season only or one of the other materials in July and repeat in 10 days.	
	Eriophyid mite	Ethion, Dicofol or Oxythioquinox. Apply in early spring when buds start to open or when red or yellow patches appear on the undersides of leaves.	Insecticidal soap spray.
	Leafhoppers	Carbaryl or Acephate. Apply when Leafhoppers appear and every 3 weeks as needed.	
	Bladder gall mite	Carbaryl or Acephate. Apply when leaves are 1/4 expanded and again in 10 days, with special attention to leaf undersides.	
	Oystershell scale	Acephate, Carbaryl or Ethion + Oil. Apply oil or Ethion and oil late in dormant season or others in May and repeat in 10 days.	
	Spider mites	Malathion, Oxythioquinox or Ethion. Treat when mites appear and again in 2 weeks. Repeat as needed.	Insecticidal soap spray.
	Webworm	Carbaryl, Acephate or Bifenthrin. Treat when webs appear.	*B. thuringiensis* or rotenone.
	Japanese beetle	Carbaryl, Acephate or Phosmet. Apply when beetles or riddled foliage appear and as needed.	Japanese beetle traps, 5 per acre.
	Cankerworm	Acephate, Phosmet or Carbaryl. Treat at first sign of infestation, usually May.	*B. thuringiensis* or rotenone.
Mimosa	Webworm	Carbaryl, Bifenthrin or Acephate. Apply late June-July and in August if foliage begins to brown.	*B. thuringiensis* or rotenone. Remove by hand.
Mountain Ash	Aphids and Woolly aphids	Bifenthrin, Malathion or Acephate. Treat in early May and as needed.	Remove with hard stream of water on smaller trees.
	European red mite	Dicofol or Ethion. Treat in early June and again 10 days later.	See Aphids. Insecticidal soap spray.
	Japanese beetle	Carbaryl, Malathion or Acephate. Apply when beetles or riddled foliage appear and as needed.	Japanese bettle traps, 5 per acre.

Plant	Pest	Chemical Control	Non-Chemical Control
Mountain Ash (continued)			
	Sawfly	Carbaryl or Acephate. Apply 2 weeks after petal fall and again in 2 weeks.	Remove by hand in smaller trees or rotenone.
	Lace bug	Malathion, Carbaryl or Acephate. Apply in May and repeat in 10 days.	
Mountain Laurel	Leafminer	Malathion, Bifenthrin or Carbaryl. Apply when mines appear and as needed to protect new growth.	
	Scale	Malathion, Acephate or Carbaryl. Apply when crawlers appear in June and again in 10 days.	
	Lace bug	Malathion, Acephate or Carbaryl. Apply in early June.	
	Borer	Acephate. Apply to trunk and large branches in May and twice again at 3-week intervals.	Wrap trunk with heavy fabric.
Oak	Aphids	Bifenthrin, Malathion or Acephate. Apply when aphids appear and as needed.	Remove with hard stream of water.
	Bagworms	Carbaryl, Bifenthrin or Acephate. Treat when bags are small, usually June.	*B. thuringiensis* or rotenone. Remove by hand in fall.
	Borers	Acephate. Spray trunks in May, June, July, and August.	Wrap trunk with heavy fabric.
	Webworms	Acephate, Carbaryl or Bifenthrin. Apply when webs appear, usually June.	*B. thuringiensis* or rotenone.
	Galls	No chemical control. Galls rarely harm trees.	Prune and destroy stem and twig galls while still green to reduce further infestations.
	Golden scale	Dormant Oil, Carbaryl or Acephate. Use oil only as dormant spray in spring, and other materials against crawlers beginning in May and repeat twice 10 days apart.	
	Japanese beetle	Carbaryl or Phosmet. Apply when beetles or riddled foliage appear and as needed.	Remove by hand at night on small trees, & traps 5 per acre.
	Leafminers	Carbaryl, Cyfluthrin or Acephate. Apply against adults, when leaves are 1/2 expanded, covering upper leaf surfaces.	

Plant	Pest	Chemical Control	Non-Chemical Control
Oak (Continued)			
	Leafhoppers	Carbaryl or Acephate. Apply when leafhoppers appear and as needed.	
	May beetles	Carbaryl or Acephate. Treat when leaves are being eaten, usually June. (Beetles are night feeders).	Remove by hand at night.
	Lace bug	Carbaryl, Acephate or Malathion. Apply when bugs appear, usually May and as needed.	
	Spider mites	Oxythioquinox or Ethion. Apply when mites appear and as needed.	
	Pin oak sawfly	Carbaryl or Acephate. Apply when larval feeding is seen.	
	Skeletonizers	Carbaryl or Acephate. Treat when riddled leaves are seen, usually June and again in August.	
	Cankerworms	Malathion, Acephate or Carbaryl. Apply when worms are small, usually May.	*B. thuringiensis* or rotenone.
	Tent caterpillars	Acephate, Bifenthrin or Carbaryl. Apply when webs appear, usually April.	*B. thuringiensis* or rotenone.
	Twig pruner	Chemical control usually ineffective.	Rake and destroy fallen twigs by last of April.
Pachysandra	Euonymus and Oystershell scales	Carbaryl, Cyfluthrin or Acephate. Apply when crawlers appear, May-June and again in 10 days.	Insecticidal soap spray.
	Spider mites	Dicofol, Acephate or Ethion. Apply when mites appear and as needed.	Remove with hard stream of water or soap spray.
Pieris (Andromeda)	Lace bug	Carbaryl, Acephate or Malathion. Apply in May, and again in 10 days.	
	Spider mites	Dicofol, Ethion or Oxythioquinox. Apply when mites appear and again in 10 days or as needed.	Remove with hard stream of water or soap spray.
Pine	Aphids	Bifenthrin, Acephate or Malathion. Apply once in May and possibly again in August if aphids persist.	Remove with hard stream of water on smaller trees.
	Bagworms	Carbaryl, Acephate or Bifenthrin. Treat when bags are small usually June.	*B. thuringiensis* or rotenone. Remove by hand in fall.

Plant	Pest	Chemical Control	Non-Chemical Control
Pine (Continued)			
	European shoot moth	Acephate, Malathion or Carbaryl. Apply to terminal growth in April and again in June.	Prune off infested terminals in May
	Nantucket tip moth	Bifenthrin, Acephate or Trichlorfon. Begin treatment in early spring when growth starts and as needed.	Hang pheremone traps.
	Northern pine and Pales weevil	Acephate or Carbaryl. Spray seedlings and young twigs in April, May, July and September.	Pull and destroy stumps by June.
	Pine bark aphid	Dormant Oil, Acephate or Phosmet. Use oil as dormant treatment or other materials when aphids appear, usually May.	Remove with hard stream of water.
	Needle scale	Ethion + Oil, Acephate or Cyfluthrin. Apply Ethion + oil as dormant treatment or others when crawlers appear in May, July and August.	
	Tortoise scale	Dormant Oil, Bifenthrin or Acephate. Apply oil as dormant treatment in spring or other materials when crawlers appear in June.	
	Webworm	Carbaryl, Acephate or Trichlorfon. Apply when larvae appear, usually in July and August.	*B. thuringiensis* or rotenone.
	Sawflies	Carbaryl, Cyfluthrin or Acephate. Apply when larvae appear and begin feeding on needles, usually early May.	Remove by hand on smaller trees or rotenone.
	Spruce spider mites	Acephate, Dicofol or Oxythioquinox. Apply when mites appear, again in 10 days and as needed.	
	White pine weevil	Acephate. Apply to leaders when weevils appear, usually April.	
	Zimmerman pine moth	Acephate, Bifenthrin or Trichlorfon. Apply in early April or September to control larvae.	Hang pheromone traps.
	Eriophyid mite	Ethion, Dicofol or Trichlorfon. Apply when mite activity begins, usually May.	
	Root collar weevil	Acephate. Treat 3 times, May August and September.	

Plant	Pest	Chemical Control	Non-Chemical Control
Poplar	Oystershell scale	Dormant Oil, Carbaryl or Acephate. Apply oil before growth begins, or other materials against crawlers in June as needed.	
	Tentmaker	Carbaryl or Acephate. Apply when first feeding or webbing appears.	*B. thuringiensis* or rotenone.
Privet	Rust mite	Dicofol, Oxythioquinox or Ethion. Apply when mites appear and as needed.	Remove with hard stream of water or insecticidal soap spray.
	Thrips	Carbaryl, Acephate or Bifenthrin. Apply when thrips become active. Beat growing tips on dark paper to count.	Control seldom needed.
	White peach scale	Carbaryl, Acephate or Cyfluthrin. Apply when crawlers appear, usually May-June.	Insecticidal soap spray.
Pyracantha (see Firethorn)			
Red Bud (Cercis)	Leafhoppers	Malathion, Carbaryl or Acephate. Apply when leafhoppers appear and as needed.	
Rhododendron	Azalea bark scale	Malathion, Acephate or Carbaryl. Apply when crawlers appear in June.	
	Black vine weevil	Acephate or Bifenthrin. Treat plants and soil around plants.	
	Borer	Acephate. Apply to trunk and branches in May and again twice at 3-week intervals.	Wrap trunks with heavy fabric.
	Lace bug	Malathion, Acephate or Carbaryl. Treat in May when bugs appear and as needed to protect new growth.	
Roses	Aphids	Malathion or Carbaryl. Apply when aphids appear and as needed.	Knock off with strong stream of water.
	Chafer	Carbaryl or Acephate. Spray or dust foliage when beetles or damage appear and as needed.	Protect with gauze netting or rotenone.
	Japanese beetle	Carbaryl or Cyfluthrin. Spray when first beetles appear and at weekly intervals until beetle season is over.	Protect with gauze netting, or rotenone, neem or beetle traps.
	Spider mites	Dicofol, Oxythioquinox or Ethion. Apply when mites appear and as needed.	Knock off with strong stream of water, or insecticidal soap.

Plant	Pest	Chemical Control	Non-Chemical Control
Roses (Continued)			
	Leafhoppers	Acephate or Malathion. Treat 3 times at 3-4 week intervals beginning May-June.	
Spirea	Aphids	Malathion, Acephate or Carbaryl. Apply when aphids appear and as needed.	Knock off with strong stream of water.
	Leaftier	Phosmet or Acephate. Apply when leaves are observed folded or tied.	Remove tied leaves by hand or rotenone.
Spruce	Aphids	Acephate, Malathion or Bifenthrin. Apply when aphids appear and as needed.	Remove with hard stream of water.
	Bagworms	Carbaryl, Acephate or Bifenthrin. Treat when bags are small in June.	*B. thuringiensis* or rotenone. Remove by hand in fall.
	Balsam twig aphid	Malathion, Acephate or Bifenthrin. Apply when aphids appear in April-May.	Remove with hard stream of water.
	Gall aphids	Malathion or Acephate. Apply just before buds break in the spring or after galls open in late summer or fall.	
	Needle scale	Cyfluthrin, Malathion or Acephate. Apply against crawlers in May and again in July and August.	
	Sawflies	Carbaryl or Acephate. Apply when larvae appear in early spring.	Remove by hand where possible.
	Bud scale	Carbaryl or Acephate. Apply June-July when crawlers appear.	
	Needle miner	Carbaryl or Acephate. Treat foliage in mid- late June.	Insecticidal soap spray.
	Spider mites	Dicofol, Ethion or Oxythioquinox. Apply when mites appear in April and a second spray 10 days later and as needed.	Remove with hard stream of water or soap spray.
	Pine weevil	Oxydemeton-methyl or Bifenthrin. Treat leaders in spring when beetles appear in mid-April.	
Sweet gum	Bagworms	Carbaryl, Malathion or Acephate. Apply when bags appear in June.	*B. thuringiensis* or rotenone. Remove by hand in fall.

Plant	Pest	Chemical Control	Non-Chemical Control
Sweet gum (continued)			
	Fall webworm	Trichlorfon, Acephate or Carbaryl. Apply when worms and webs first appear.	*B. thuringiensis* or rotenone.
	Leafminer	Acephate or Bifenthrin. Apply when first mines appear and as needed.	
	Scale	Dormant Oil, Acephate or Carbaryl. Apply oil only as dormant treatment in spring and others against crawlers on leaves in June or September when young scales appear on twigs and buds.	Soap spray on small trees.
Sycamore	Aphids	Malathion, Acephate or Cyfluthrin. Apply when aphids appear and as needed.	Remove with hard stream of water.
	Bagworms	Carbaryl, Malathion, Trichlorfon or Acephate. Apply when bags appear in June.	*B. thuringiensis* or rotenone. Remove by hand in fall.
	Fall webworm	Acephate, Naled or Carbaryl. Apply when first webs or worms are seen in June.	*B. thuringiensis* or rotenone.
	Japanese beetle	Carbaryl or Acephate. Apply when beetles or riddled foliage appear and as needed.	Japanese beetle traps, 5 per acre.
	Leaf folder	Carbaryl or Acephate. Apply when leaves are folded together.	Remove folded leaves by hand or rotenone.
	Leafhoppers	Malathion, Acephate or Bifenthrin. Apply when Leafhoppers appear and as needed.	
	Lace bug	Malathion, Acephate or Carbaryl. Make 2 applications 10 days apart beginning in May.	
	Terrapin scale	Carbaryl or Acephate. Apply when crawlers appear on leaves in June.	
Taxus (see Yew)			
Tulip Tree	Aphids	Malathion, Acephate or Bifenthrin. Apply when aphids appear and as needed.	Remove with hard stream of water.
	Leafminer	Acephate or Phosmet. Apply when first mines appear and as needed.	

Plant	Pest	Chemical Control	Non-Chemical Control
Tulip Tree (continued)			
	Scale	Dormant Oil, Carbaryl or Acephate. Apply oil only as dormant treatment and others when crawlers appear in August.	
	Yellow poplar weevil	Carbaryl. Apply for adult control in June-July.	
Viburnum	Aphids	Malathion, Acephate or Bifenthrin. Apply when aphids appear and as needed.	Remove with hard stream of water.
	Spider mites	Dicofol, Oxythioquinox or Acephate. Apply when mites appear in spring and again in 10 days and as needed.	See Aphids.
Walnut (If walnuts are to be eaten check insecticide labels for days-waiting-time from last application to harvest.)	Aphids	Malathion, Oxythioquinox or Acephate. Apply when aphids appear and as needed.	Remove with hard stream of water on smaller trees.
	Fall webworm	Carbaryl or Acephate. Apply when webs first appear, mid-June thru July.	*B. thuringiensis* or rotenone.
	Leafhoppers	Malathion, Acephate or Carbaryl. Apply when leafhoppers appear and as needed.	
	Gall mite	Carbaryl, Ethion or Oxythioquinox. Apply when leaves are half open or when mites appear on leaf petioles.	
	Walnut caterpillar	Acephate or Carbaryl. Apply when caterpillars appear, usually in June.	*B. thuringiensis* or rotenone.
Willow	Aphids	Malathion, Acephate or Oxydemeton-methyl. Apply when aphids appear and as needed.	Remove with hard stream of water.
	Borers	Acephate. Treat trunk thoroughly monthly from May thru August.	Wrap trunk with heavy fabric.
	Fall webworm	Naled, Acephate or Carbaryl. Apply when webs appear, usually June.	*B. thuringiensis* or rotenone.
	Leaf beetles	Carbaryl, Malathion or Acephate. Apply on appearance of leaf feeding and as needed.	Remove with hard stream of water.

Plant	Pest	Chemical Control	Non-Chemical Control
Willow (continued)			
	Spider mites	Dicofol, Ethion or Oxydemeton-methyl. Apply when mites appear, again in 10 days and as needed.	Remove with hard stream of water.
	Oystershell scale	Dormant Oil, Acephate or Carbaryl. Apply oil only as dormant treatment in spring or other materials against crawlers in May-June.	Insecticidal soap spray.
	Sawflies	Carbaryl or Acephate. Apply when larvae appear but before heavy leaf feeding.	Remove by hand on small trees.
	Tent caterpillars	Malathion, Bifenthrin or Acephate. Apply when worms or webs first appear, usually April.	*B. thuringiensis.* Remove by hand on small trees.
Wisteria	Leafhoppers	Acephate, Malathion or Carbaryl. Apply when leafhoppers appear and as needed.	Insecticidal soap spray, or neem.
Yew (Taxus)	Black vine weevil	Acephate, Bifenthrin or Phosmet. Spray foliage and soil around infested plants in June.	
	Fletcher scale	Dormant Oil, Carbaryl or Acephate. Apply oil only as dormant treatment in spring, other materials against crawlers in June-July.	
	Mealybug	Dormant Oil, Malathion or Acephate. Apply oil only as dormant treatment in spring or other materials against over wintering nymphs in May and again in June and July.	In small numbers remove with hard streams of water.
	Scales	Carbaryl, Acephate or Malathion. Treat when crawlers are first seen and repeat in 3 weeks.	Insecticidal soap spray.
Yucca	Aphids	Acephate, Cyfluthrin or Malathion. Treat when aphids appear and repeat as needed.	Remove with hard stream of water.

FIGURE 18. Some Insect Pests of Trees, Shrubs, Woody, Annual and Perennial Ornamentals.

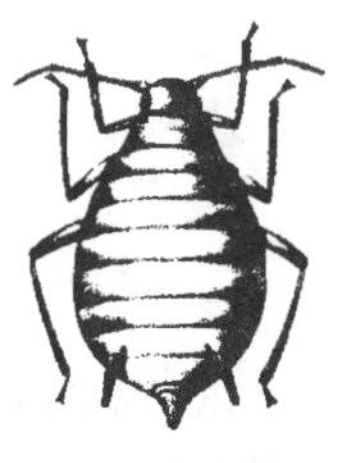

Aphid
(Union Carbide)

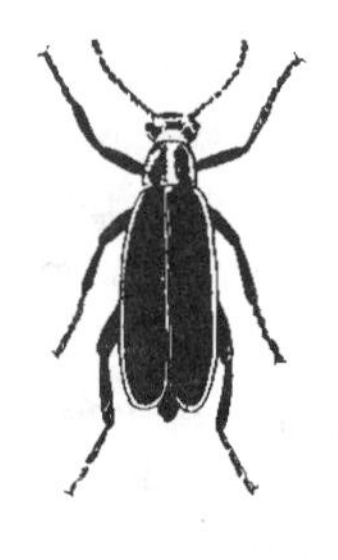

Margined Blister Beetle
(U.S. Public Health Service)

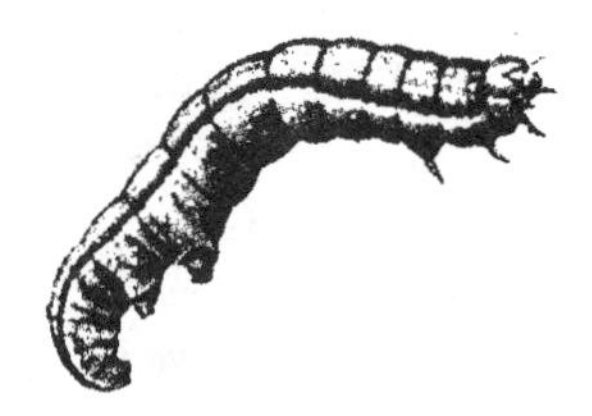

Cankerworm
(Union Carbide)

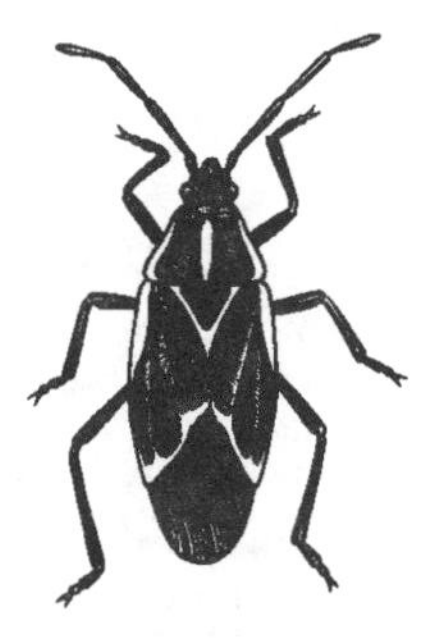

Boxelder Bug
(Univ. of Arizona)

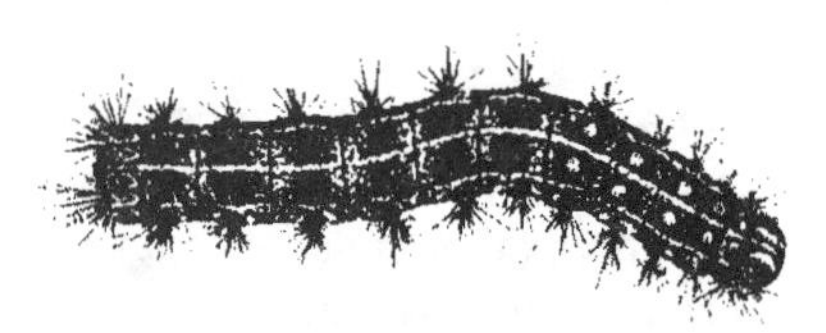

Gypsy Moth Larva
(Union Carbide)

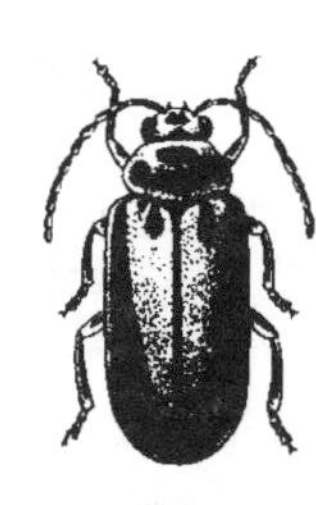

Elm Leaf Beetle
(Union Carbide)

Flea Beetle
(Union Carbide)

Horntail
(USDA)

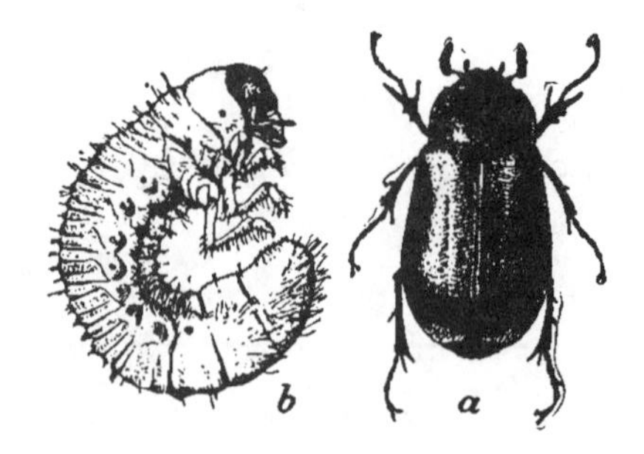

A May Beetle
A, adult; B, fully grown larva
(USDA)

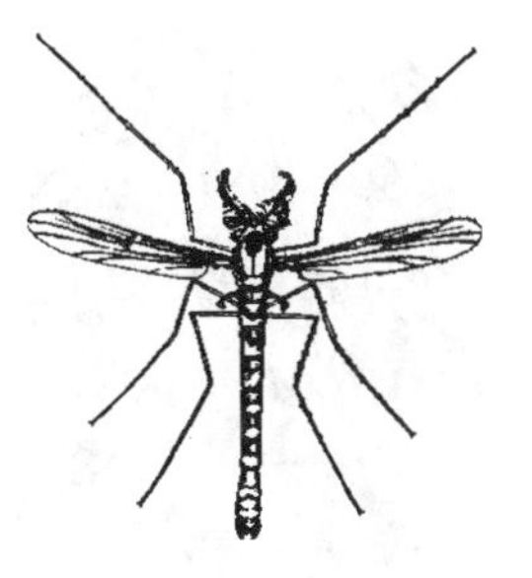

A Midge
(USDA)

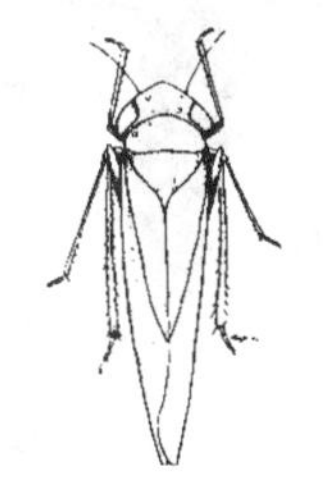

Leafhopper Adult
(Univ. of Arizona)

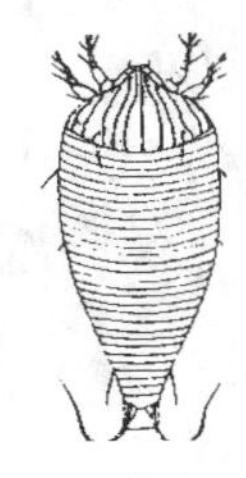

Eriophyid Gall Mite
(Enlarged 50 X)
(USDA)

Birch Leafminer Adult
(Union Carbide)

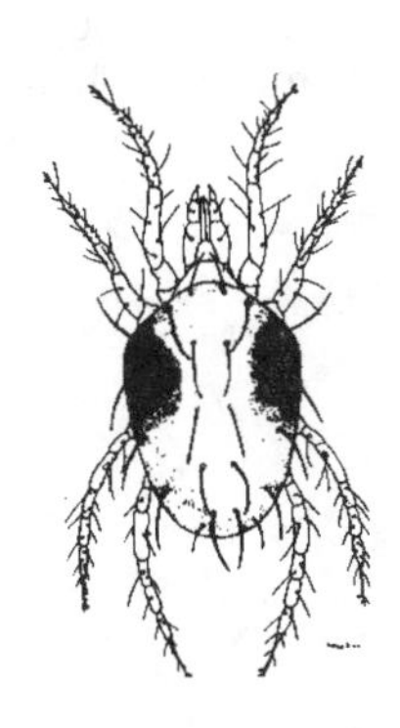

A Spider Mite
(Univ. of Arizona)

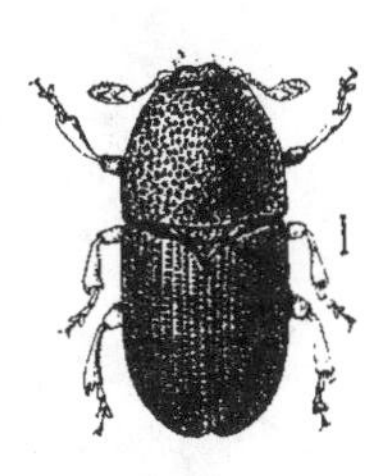

Shot-hole Borer Adult
(USDA)

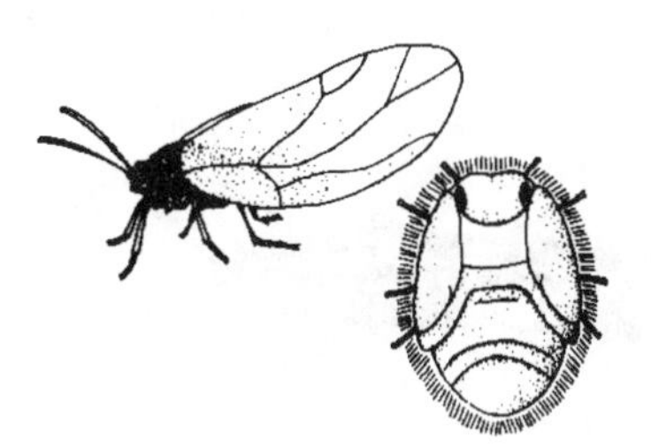

Psyllid
Above: adult
Right: nymph
(Univ. of Arizona)

Peach Twig Borer
(Union Carbide)

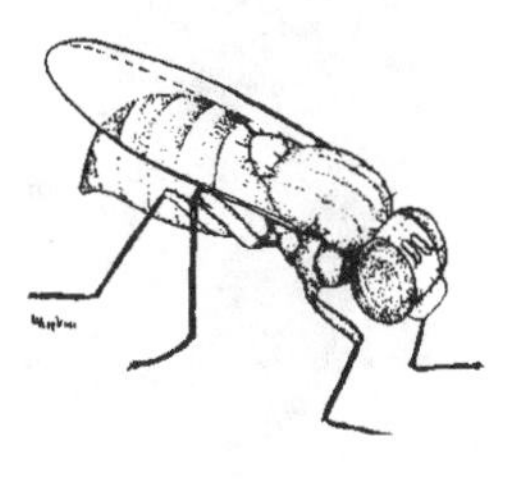

Leafminer Fly Adult
(Univ. of Arizona)

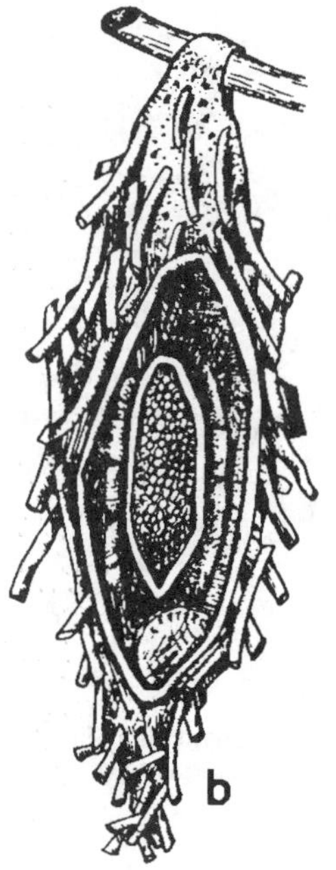

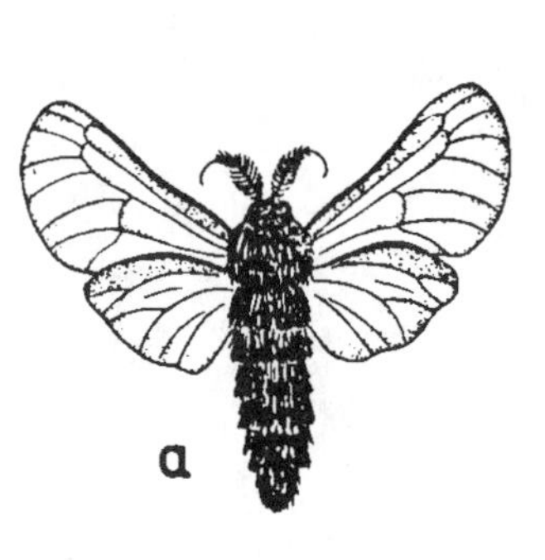

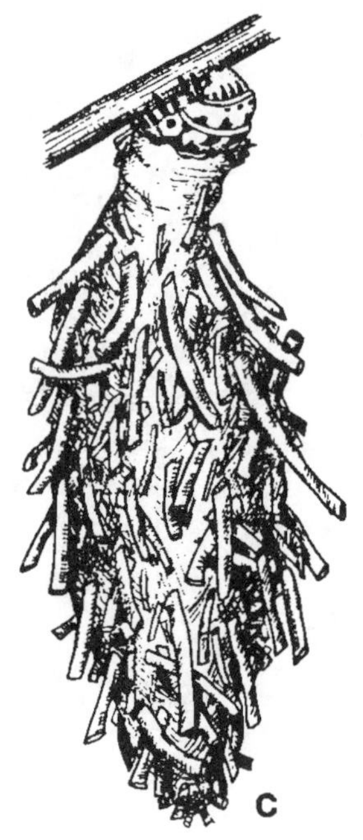

A typical bagworm; (A) adult male moth; (B) larval case in winter, cut open to show pupal case and eggs, and (C) full grown larva moving within its case. *(USDA)*

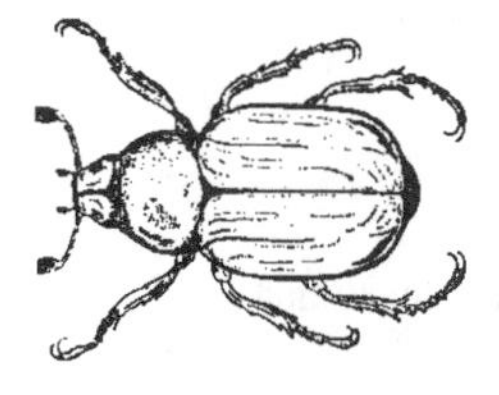

Garden Chafer
(Union Carbide)

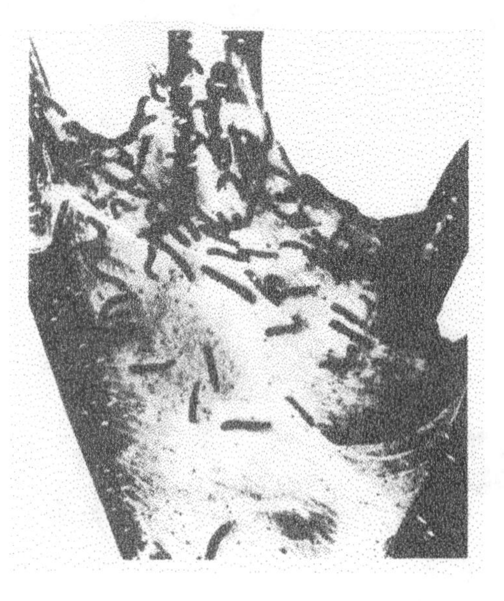

Nests of the Eastern Tent Caterpillar; note the layers of silk.
(USDA)

Eastern Tent Caterpillar
(Union Carbide)

A Thrips
(Univ. of Arizona)

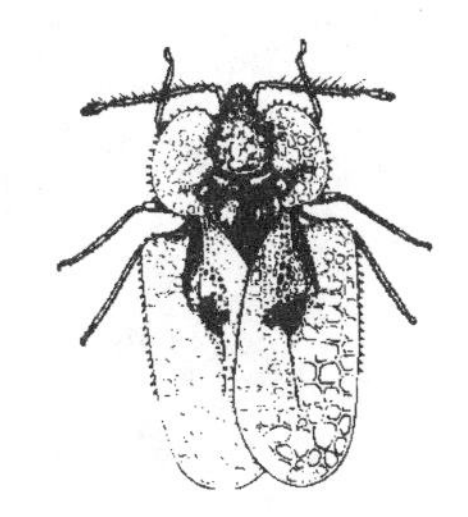

Azalea Lacebug
(USDA)

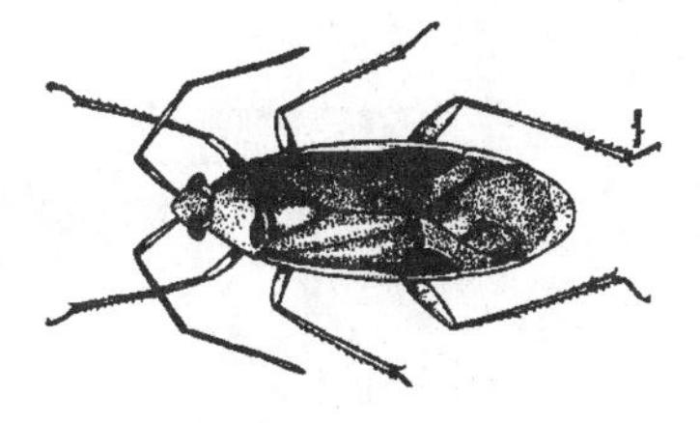

Plant Bug
(Univ. of Arizona)

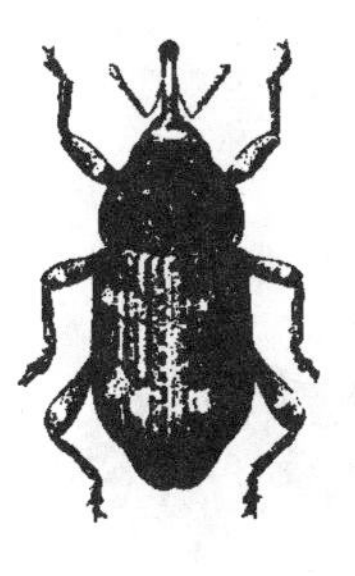

White Pine Weevil
(USDA)

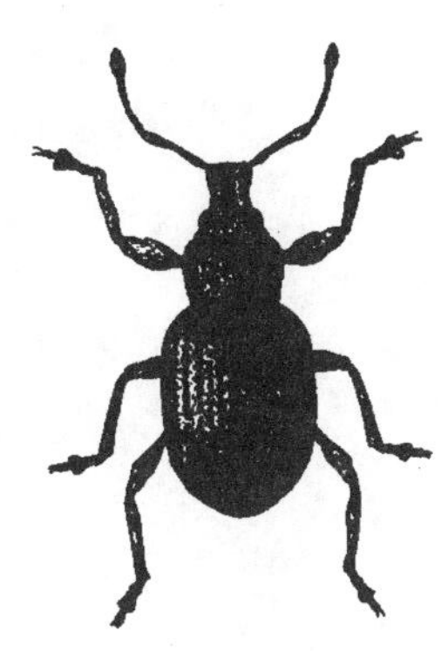

Black Vine Weevil
(USDA)

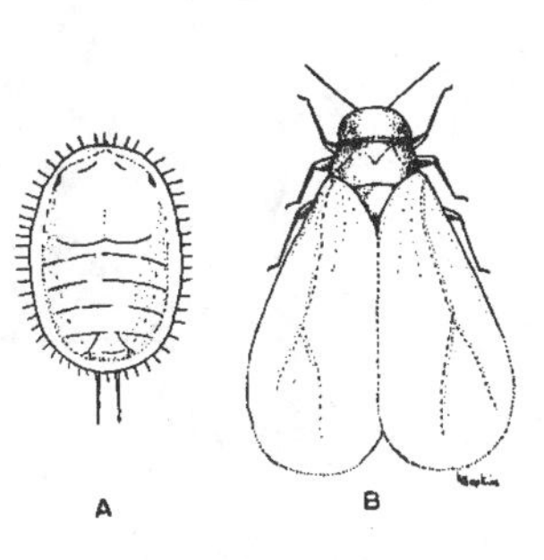

Whitefly. A — Nymph; B — Adult.
(Univ. of Arizona)

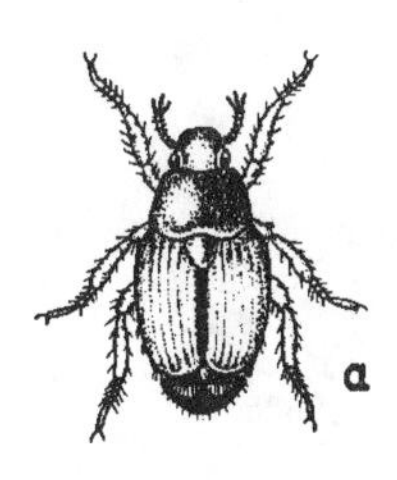

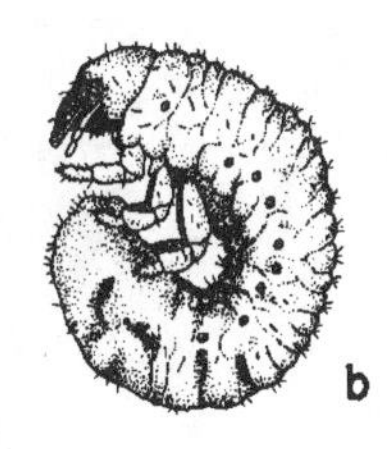

Japanese Beetle *(Popillia japonica)*
(A) Adult, (B) Larva (grub).
(USDA)

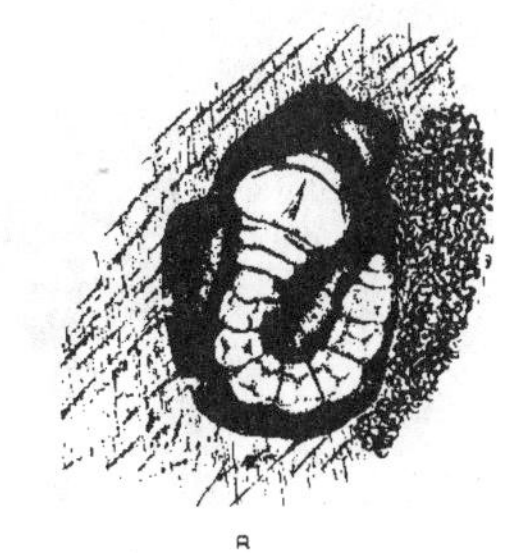

Flatheaded Appletree Borer
(Chrysobothris femorata)
A. Adult, B. Larva.
(USDA)

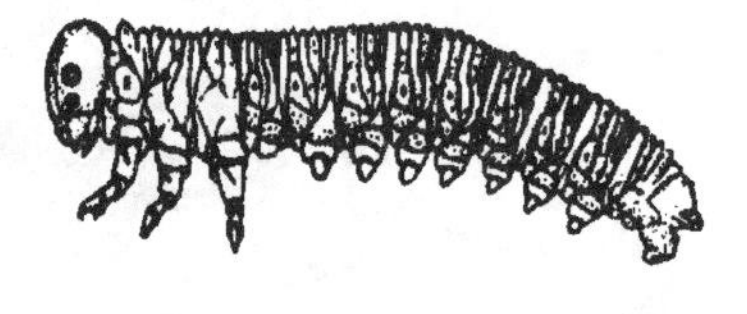

Sawfly Larva, *Allántus cintus*
(USDA)

Pear Sawfly, *Caliroa cerasi*
(USDA)

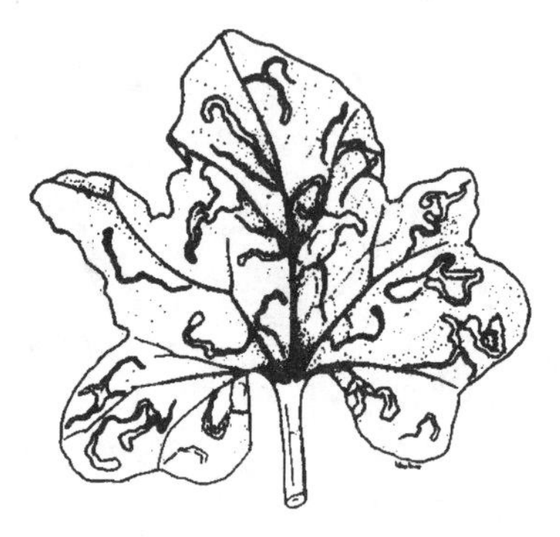

Leafminer Injury
(Univ. of Arizona)

California Orange Dog
(Union Carbide)

California Red Scale
(Union Carbide)

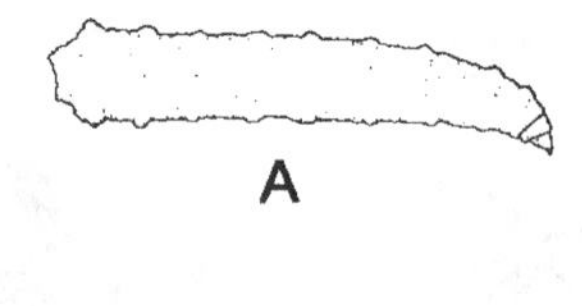

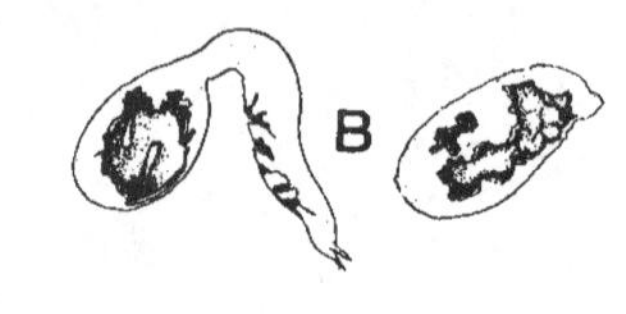

Seedcorn Maggot
A — Mature larva; B — Seed damage
(Univ. of Arizona)

TABLE 19. Insect and Mite Control Suggestions for Ornamental Annuals and Perennials With and Without Chemicals.[1]

Plant	Pest	Pesticides to Use	Non-Chemical Control
Ageratum	Whitefly	Malathion, Acephate or Bifenthrin. Spray 2 or more times at 5-day intervals.	Insecticidal soap spray.
	Spider mites	Dicofol or Ethion. Spray as needed.	Remove with hard stream of water.
Chrysanthemum	Aphids	Bifenthrin, Acephate or Malathion. Spray as necessary. Disulfoton 1% G.	Remove with hard stream of water or soap spray.
	Beetles	Carbaryl or Acephate. Spray or dust as needed.	Remove by hand, or rotenone.
	Caterpillars (including borers)	Carbaryl, Acephate or Malathion. Spray or dust as necessary.	Remove and burn large-stemmed weeds nearby, or rotenone.
	Plant bugs	Acephate, Malathion or Carbaryl. Spray or dust as needed. Disulfoton 1% G.	Insecticidal soap, or rotenone.
Cockscomb	Spider mites	Dicofol, Ethion or Propargite. Spray as needed.	Remove with hard stream of water.
Columbine	Aphids	Malathion, Acephate or Malathion. Spray as necessary. Disulfoton 1% G.	Insecticidal soap spray.
	Leafminer	Acephate or Naled. Spray 3 times at 5-7 day intervals, beginning when leaves are about half developed.	
Citrus	Scales	Malathion, Acephate or Bifenthrin. Apply as spray or with paint brush.	Insecticidal soap spray.
	Whitefly	Malathion or Cyfluthrin. Apply as spray.	Insecticidal soap spray.
	Orange dog	Chemical control not normally required.	Remove by hand or apply *B. thuringiensis*.
	Spider mites	Dicofol, Propargite or Oxythioquinox. Apply when mites appear and as needed.	Remove with hard stream of water. Insecticidal soap spray.
Dahlia	Aphids	Acephate or Malathion. Treat as needed.	Insecticidal soap spray.
	Beetles and borers	Malathion and Bifenthrin. Mix and spray on 14-day schedule.	
	Leafhoppers	Acephate or Carbaryl. Treat when first leaf damage is noticed and repeat as needed.	Insecticidal soap spray.

[1] Refer to Appendix B for trade or proprietary names of recommended insecticides. Many of the pests mentioned in this table are illustrated on pages 134-138, Figure 18.

Plant	Pest	Pesticides to Use	Non-Chemical Control
Dahlia (Continued)			
	Spider mites	Dicofol, Ethion or Oxythioquinox. Spray as needed.	Remove with hard stream of water.
	Thrips	Acephate or Malathion. Spray when flowers begin to open.	Control usually unnecessary.
Day Lilies	Thrips	Malathion and Bifenthrin. Mix and spray 2 or 3 times at 10-day intervals, beginning just after flower buds form.	Destroy all plant residue in fall.
Delphinium (see Larkspur)			
Forget-Me-Not	Aphids	Acephate or Malathion. Spray or dust as needed.	Insecticidal soap spray.
	Flea beetles	Carbaryl or Acephate. Spray or dust as needed.	Rotenone spray.
Hollyhock	Beetles	Carbaryl or Malathion. Spray as needed.	Remove by hand, or rotenone.
	Leafhoppers	Acephate or Malathion. Spray as needed.	
	Spider mites	Dicofol or Malathion. Treat as needed.	Insecticidal soap spray.
Hydrangea	Aphids	Malation or Acephate. Spray as needed.	Insecticidal soap spray.
	Leaftier (caterpillar)	Carbaryl and Malathion. Mix and spray as necessary.	Remove rolled leaves by hand.
	Spider mites	Naled, Dicofol or Malathion. Treat as needed.	See Aphids.
Impatiens	Aphids	Malathion. Spray or dust as needed.	Insecticidal soap spray.
	Spider mites	Naled, Dicofol or Malathion. Treat as needed.	See Aphids.
Iris	Aphids	Acephate or Malation. Spray as needed.	Insecticidal soap spray.
	Borers	Acephate. Spray twice at 2-week intervals, beginning when leaves are 3/4 inches long.	Mulch heavily with bark chips.
Larkspur	Aphids	Acephate or Malathion. Spray as needed.	Insecticidal soap spray.
	Borers	Malathion and Bifenthrin. Mix and spray 3 times, beginning at early growth and at 2-week intervals thereafter.	Collect and burn large-stemmed weeds. Burn all larkspur and delphinium after fall frost.
	Spider mites	Naled, Dicofol or Malathion, Treat as needed.	See Aphids.

Plant	Pest	Pesticides to Use	Non-Chemical Control
Lilies	Aphids	Acephate or Malathion. Spray or dust as needed.	Insecticidal soap spray.
Lupine	Aphids	Acephate or Malathion. As needed.	Insecticidal soap spray.
	Whitefly	Malathion or Cyfluthrin. As needed.	Insecticidal soap spray.
Marigold	Spider mites	Naled, Dicofol or Malathion. Treat as needed.	Insecticidal soap spray.
Nasturtium	Aphids	Malathion or Acephate. Spray or dust as needed.	Remove with hard stream of water. Insecticidal soap spray.
	Spider mites	Naled, Dicofol or Malathion. Treat as needed.	See Aphids.
	Plant bugs	Malathion and Bifenthrin. Mix and spray as needed.	
Pansy	Aphids	Malathion or Acephate. Spray as needed.	Insecticidal soap spray.
	Mealybugs	Acephate or Malathion. Spray in spring or as needed.	Insecticidal soap spray.
	Spider mites	Naled, Dicofol or Malathion. Frequent treatment may be needed.	Insecticidal soap spray.
Peony	Ants	Malathion or Carbaryl. Spray as needed.	Ants are nuisance but not pest. Attracted to sweet nectars.
	Scales	Acephate or Malathion. Spray when insects are first seen.	Destroy all plant residue in fall.
	Thrips	Malathion or Bifenthrin. Spray lightly with fine mist daily through bloom.	Destroy all plant residue in fall.
	Japanese beetle and rose chafer	Carbaryl or Acephate. Spray when insects are first seen and repeat weekly or as necessary.	Protect plants with gauze netting.
Phlox	Plant bugs	Malathion and Bifenthrin. Mix and spray as needed.	Nymphs can be controlled with insecticidal soap spray.
	Spider mites	Naled, Dicofol or Malathion. Spray as needed.	Insecticidal soap spray.
Poppy	Aphids	Malathion or Acephate. Spray or dust as needed.	Insecticidal soap spray.
	Leafhoppers	Malathion and Bifenthrin. Mix and spray at 5-day intervals or as needed.	Nymphs can be controlled with insecticidal soap spray.
Shasta Daisy	Aphids	Malathion or Acephate. Treat as needed.	Remove with hard stream of water. Insecticidal soap spray.
	Beetles	Carbaryl or Acephate. Spray or dust as needed.	Remove by hand. as needed.

Plant	Pest	Pesticides to Use	Non-Chemical Control
Snapdragon	Aphids	Acephate or Malathion. Spray as needed.	Insecticidal soap spray.
	Beetles	Carbaryl. Spray or dust as needed.	Remove by hand.
	Spider mites	Dicofol or Propargite. Treat as needed.	See Aphids.
Sweet Alyssum	Leafhoppers	Malathion or Carbaryl. Spray as needed.	Insecticidal soap spray.
Sweet Pea	Aphids	Malathion or Acephate. Spray as needed.	Insecticidal soap spray.
	Leafminer	Acephate or Naled. Spray or dust weekly.	Remove and destroy infested leaves.
	Seedcorn maggot	Disyston granules placed with seed at planting.	
Verbena	Aphids	Malathion or Acephate. Spray as needed.	Insecticidal soap spray.
	Plant bugs	Malathion and Bifenthrin. Mix and spray as needed. Or apply Naled.	See aphids.
	Leafminer	Acephate or Naled. Spray at first sign of mines.	
	Spider mites	Naled, Dicofol or Malathion. Treat as needed.	See Aphids.
Zinnia	Japanese beetle	Carbaryl or Malathion. Mix and treat as needed.	Protect with gauze netting, or rotenone.
	Spider mites	Naled, Dicofol or Malathion. Spray as needed.	Insecticidal soap spray.

HOUSEHOLD — UNWANTED HOUSE GUESTS

Controlling household pests is important throughout the year. There are several species of cockroaches which are obnoxious, odorous and leave trails of fecal droppings on foods and house furnishings during their nocturnal frolicking. Termites (heaven forbid!) move into the structure from the ground and attack wood, books and papers. Ants can contaminate food and cause homeowners great annoyance. Clothes moths and carpet beetles often damage or destroy woolen clothing, furs, rugs and that record-holding trophy head-mount in the den. Fleas attack pets and then their masters. House dust mites, carpet beetle cast skins, cockroaches and unknown others produce allergies. Flies and mosquitoes are tremendous nuisances outdoors and occasionally indoors.

If the householder can determine the identity of the pest and the extent of the infestation, has the proper insecticide or non-chemical control and is confident of his ability to solve the problem, he won't need a pest control professional to do the job. When he follows the instructions properly, he will derive immense pride in a job well done and especially in the savings realized.

Before we start describing those pesky indoor insects and their habits and life cycles, you must be given the wisdom of the ages about two amazing tools. The first is the ***caulking gun,*** the other is the ***vacuum cleaner***.

The caulking gun, with appropriate caulking compound, is used to exclude insect pests from your dwelling. It is the safest and surest of all tools, including the array of insecticides at your disposal. Cracks, holes, pipe entry openings, wall separations and every small port-of-entry inside and outside the house should receive a careful caulking. Don't forget the foundation and those gaps between the siding and foundation. Every conceivable pest is then prevented from entering your castle—ants, boxelder bugs, centipedes, chinch bugs, clover mites, cockroaches—the entire list. Then you have only to control those that managed to make your home their home.

Of the several caulking materials readily available, the latex base, siliconized acrylic latex and silicone are recommended for sealing and excluding invading insect pests. The old oil base caulk tends to have a short lifespan and lacks flexibility, while butyl rubber caulk is difficult to apply, resulting in a stringy appearance, thus neither of these is recommended.

The vacuum cleaner with a hose extension is a marvel in removing reservoirs of usually unseen and unwanted guests, such as pantry pests, e.g. cereal beetles, grain moths and weevils. Carpet beetles and their larvae and house dust mites that accumulate in corners and where carpeting adjoins the kick board or wall molding, are easily removed using the narrow edging attachment to your vacuum. Ants, clover mites, tiny ticks crawling up the walls, cockroach nymphs and a host of others can be sucked into the disposable vacuum bag and destroyed. It is important to dispose of that bag in a way that doesn't allow the pests to reenter the home. As soon as vacuuming is completed, remove the disposable bag from the vacuum *out-of-doors* and immediately place it in a plastic bag that can be sealed or closed with a rubber band. After your former guests are sealed in the plastic bag, it can then be safely tossed in the trash.

Now, let's learn about those indoor pests!

Ants

Ants of various sizes and species may be found in houses. They have constricted "waists," distinguishing them from termites, which have straighter bodies without such narrow constrictions. Those most commonly seen are wingless, nonreproducing adults of the worker caste.

Ant
(U.S. Public Health Service)

They feed on a variety of substances including sweets, starches, fats, dead insects and other animal matter and are particularly fond of man's foods.

Ants build outdoor nests in soil or other protected areas. Some species, particularly carpenter ants, nest in damp or unsound house timbers. Fire ants, which occur in the South, can inflict painful stings, build nests in gardens and under turf, with mounds of excavated soil extending above the grass level. Most ant species commonly travel in search of food along "trails" which are easily recognized when individuals are numerous.

Ants normally will not enter homes where food is not accessible. Control of ants in homes should begin with the removal of crumbs and other attractive material from floors and areas where food is prepared, stored or consumed. Food should be stored in ant-proof containers.

Spot treatments of an approved residual insecticide may be applied to ant trails and other traveled areas for temporary relief. Such applications are not a substitute for locating and eliminating the nesting site. Approved ant baits, preferably in sealed containers placed in secluded locations beyond the reach of small children or pets, may be a helpful supplementary control measure.

Ants may be controlled by using an approved insecticidal spray applied to nesting sites, which may be near foundations or at some distance from invaded houses. Additional protection can be provided by an insecticidal spray applied to the lower foundation walls and to the soil area surrounding the house. Special attention should be given to door and window sills and other areas of ant entrance. Carpenter ants and other house-nesting ants may be controlled by applying an approved insecticide to the nesting site. Damaged areas may require structural repair. (See Boric Acid Baits in Appendix F.)

Aphids

Aphids or plant lice, feed on plant juices and are occasional pests of house plants or annoying temporary invaders on cut flowers. They may be winged or wingless, of differing sizes and colors. Feeding may weaken or disfigure house plants by causing leaves to curl or to become covered with honeydew. Potted plants should be inspected for aphids and treated outside, if necessary,

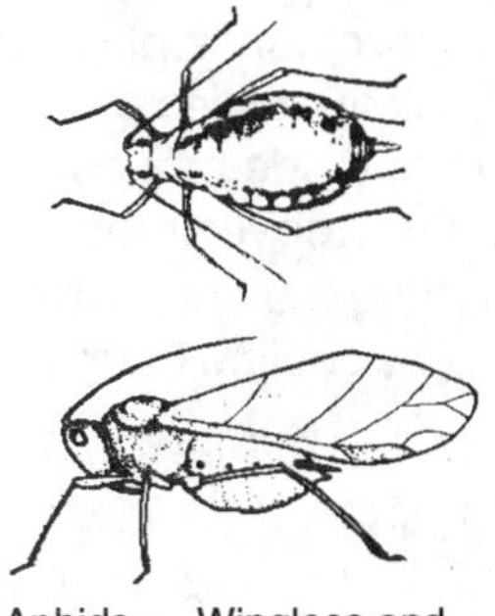

Aphids — Wingless and Winged
(USDA)

before being brought indoors. Light infestations may be controlled by washing plants in lukewarm water or removing the aphids with tweezers or a cotton swab dipped in rubbing alcohol. On cut flowers, aphids are at most a temporary nuisance best avoided by not bringing infested blooms indoors or by removal by hand as described.

Assassin Bugs — see Kissing Bugs

Bark Beetles

There are several species of small (1/4 inch) beetles, black or dark brown and oval-shaped, commonly called bark beetles. Almost never are they generated within the home, but rather hitchhike into the home by means of firewood or are occasionally attracted to lights in the spring and early summer. They should be of no concern. Removing the firewood usually removes the source of the problem.

Bark Beetle
(U.S. Public Health Service)

Bean Weevils

Stored dry beans and peas in loose or uncovered containers may become infested with these short-snouted weevils, which can breed continuously in homes. Beans are destroyed by the internal feeding of the larvae with several individuals often found in a single bean. The adults, which emerge through holes bored in the seed coat are 1/8 inch long and mottled gray and do not feed on beans.

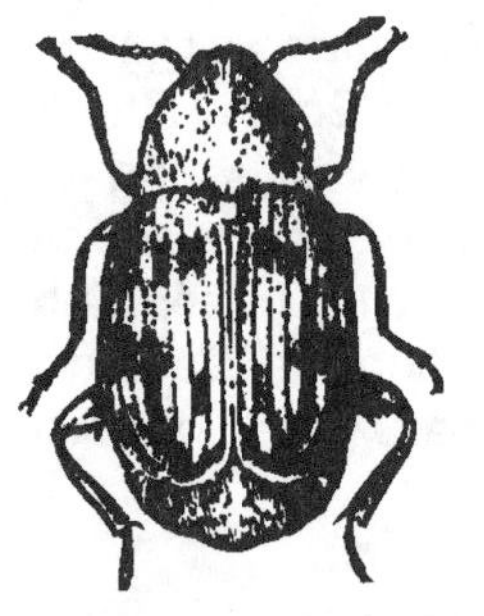

Bean Weevil
(Univ, of Arizona)

Storing beans in tight metal or glass containers and avoidance of spillage may prevent infestations. Small quantities of heavily infested beans should be promptly destroyed when salvage is not practical. Light weevil infestations may be controlled by heating the beans in an oven for 30 minutes at 130° For by cooling them for 4 days in a freezer at 0° F. or below.

Bed Bugs

Adult bed bugs are flat, wingless, reddish-brown insects, 3/16" long and 1/8" wide. The nymphs and adults feed on human blood, although pets and other animals may be attacked. Bed bugs usually hide during the day in floor cracks, behind base-boards, in cracks in wooden beds, inside mattresses and upholstered furniture and in other similar situations. Infested homes have a sweetish, indescribable odor. Bed bugs are mainly night feeders, but during the day they may emerge from hiding to attack persons using infested furniture such as padded chairs or stools. Bed bugs may be widely transported through infested luggage, clothing, bedding and furniture. They are commonly associated with, but not restricted to, homes with poor sanitation.

Bed Bug
(Univ. of Arizona)

When bedbugs have infested an occupied house they can be removed only by use of an insecticide. Prevention through good housekeeping and proper living conditions, is preferred.

Bees, Africanized ("Killer") Bees, Carpenter Bees and Leaf-Cutter Bees

Swarms of honey bees may enter cracks or other openings in walls of houses to establish new colonies. Under favorable conditions these colonies may exist for years, causing continued annoyance. Weak or poorly located colonies may die and leave an agglomeration of wax, honey, dead bees and other debris attractive to new swarms and to secondary invaders such as cockroaches, carpet beetles and wax moths. An objectionable odor will probably develop.

Control of honey bees in houses involves transferring or killing the bees, removal of the combs, honey and debris from the nest area and sealing the wall to prevent further invasions of bees or other insects. With the cooperation of an interested beekeeper, who has the required skill and equipment, the bees may be transferred to a hive and taken away. A colony may be killed outright by blowing an insecticidal dust into the nest. This will also kill newly emerging young bees and new insect invaders from the outside.

Honey Bee
(U.S. Public Health Service)

Africanized or *"killer"* bees, are a much more aggressive honey bee than the domesticated European bees raised for honey and used for crop pollination in the U.S. To the novice they are indistinguishable from the European bees. Only experts can differentiate them. Africanized bees entered Texas from Mexico in 1989 and Arizona from Mexico in 1993. Its exact range of distribution is not known. Because they are wild bees and colonize in relatively unprotected locations, they likely will be confined to the warmer Southern states, since low winter temperatures kill wild bees in colder areas of the U.S. Africanized honey bee venom is no more poisonous than their European counterparts. However, they are more defensive or aggressive if provoked, resulting in 10-15 times more stings than from the European variety. "Killer bees" is a name given the Africanized bee by media "hype" in the 1960s after several persons died from massive attacks in Brazil and Argentina in the late 1950s. If Africanized bees are suspected to have colonized indoors or outdoors, a professional beekeeper or pest control operator should be employed to handle the case. The consequences of doing otherwise can be quite serious!

Carpenter bees may weaken exposed, unpainted or weathered boards or timbers of softer woods by boring nesting tunnels. These large bluish-black (female) or tan (male) insects resemble large bumble bees but have smooth rather than hairy abdomens. The tunnels are approximately 1/2" in diameter when first made and may extend either vertically or horizontally. The wood is not eaten. The tunnels are used only as nesting sites and if not disturbed, may be reused, widened and lengthened by further generations over a period of years. Carpenter bee tunnels are not usually found in wood with well-painted surfaces. Control includes use of an insecticide to destroy the developing bees within the tunnels, followed by repair, sealing and painting the damaged wood.

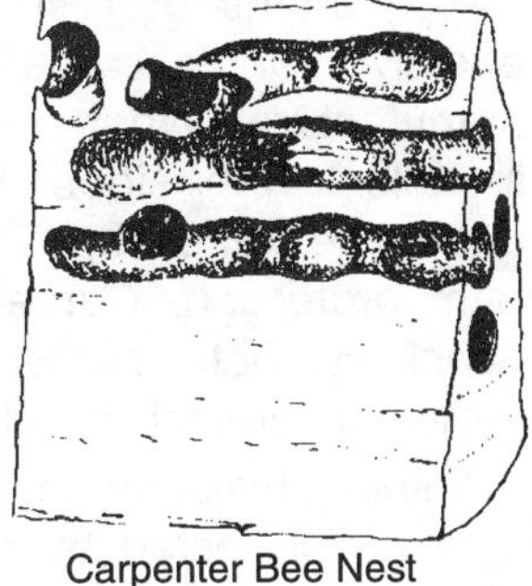

Carpenter Bee Nest
(U.S. Public Health Service)

Leaf-cutter bees are a nuisance because they cut C-shaped discs from rose leaves, bougainvillea, cottonwood and other juicy leaves. These discs are to provision their nests in the ground where they raise their young. They are the best of pollinators and should not be killed under any circumstances. Don't use insecticides. Protection of prize roses and other plants can be obtained by covering the plants with a light mesh netting or gauze to exclude them.

Bee stings may be relieved by application of ice packs to the affected areas. Persons allergic to bee stings should remain away from areas where bees are active. Such persons should consult a physician regarding medication available on prescription for emergency treatment of stings and wear a medical alert bracelet.

Booklice or Barklice

Booklice, sometimes called barklice, are psocids, tiny (1/16 inch) soft-bodied insects which closely resemble small, bleached out aphids. They may occasionally become pests by feeding on paper, starch, grain and other substances in damp places. Sometimes they are found in the mulch of potted plants, around the bases of trees and stone walls where lichens are available as food. One species of booklice is commonly found in libraries and museums and in deserted bee hives and wasp nests. They are usually of no importance and are readily controlled with any of the commonly used household insecticides.

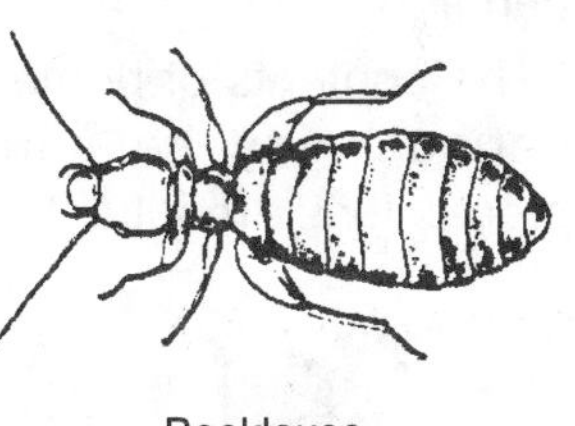

Booklouse
(U.S. Public Health Service)

Boxelder Bugs

These red and black plant-feeding insects may invade houses where the boxelder or ash-leaf maple, is grown as a shade tree. Adults have three length-wise red lines on the shield or pronotum, behind the head and other conspicuous red lines on the front wings. These insects feed mainly on seed-bearing or female, boxelder trees. In the fall the adults may enter homes in search of shelter. Although they do not feed they become nuisances by their presence, by their foul odor when crushed and by their fecal stains on curtains, furniture, clothing and similar objects.

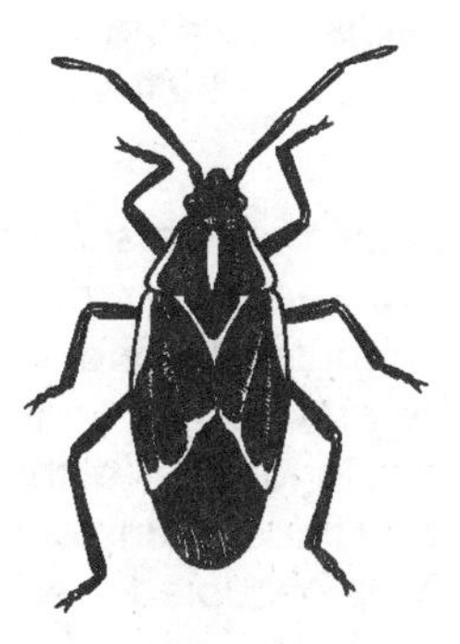

Boxelder Bug
(Univ. of Arizona)

The best control of boxelder bugs is elimination of seedbearing (female) boxelder trees. Staminate (male) trees are not attacked and should be chosen for planting. Insecticides may be applied to control outdoor infestations and to provide barriers surrounding homes. Control indoors can be achieved with a vacuum cleaner.

Carpenter Bees — see Bees

Carpet Beetles

Carpet beetle larvae (also known as Dermestids, Hide or Larder beetles) feed on a variety of articles of animal origin and on dried foods high in protein. These include woolen rugs, blankets, clothing, upholstery, piano felts, feathers, furs, hides, bristles, powdered milk, cereals, spices, dog food and bird seed. Irregular holes are eaten in rugs and stored woolens which may become localized patches of damage.

The contents of homes unoccupied for extended periods may be seriously damaged unless precautions are taken.

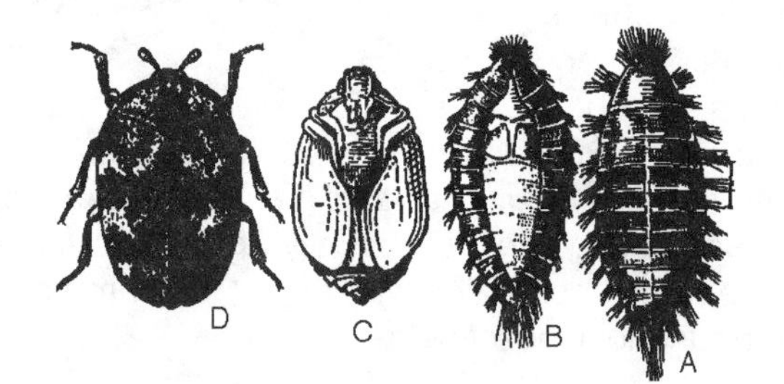

Carpet Beetle or Buffalo Moth. (A) larva; (B) pupa in larval skin; (C) pupa from below; (D) adult.
(USDA)

Larvae may be short and robust or wedge shaped and more slender, depending on the species. They are about 1/4" long, with circular rows of stiff dark hairs around the body segments and usually with tufts of longer hair at the rear of the body. They prefer to feed in dark, undisturbed situations rather than in actively occupied areas exposed to light.

The adults are about 1/8" long, mottled gray, brown and white and are attracted to windows and other lighted areas. Adults of most species feed outdoors on pollen and nectar.

Carpet beetles are best controlled by protective care of articles subject to attack and by eliminating surplus or discarded materials which might attract new infestations. Hides, furs, feathers and woolens no longer of value should be destroyed. Lint and hair should be regularly removed by vacuuming cracks and crevices behind baseboards, moldings and similar locations. Storage closets, indoors and out, should be regularly aired and inspected for susceptible material, including bodies of dead insects or mice. Rugs in continuous use and subject to regular vacuum cleaning, are seldom damaged by carpet beetles provided heavy pieces of furniture are occasionally moved for examination of the vulnerable compacted areas beneath. Storage of cereals, spices and dog foods in open containers for long periods should be avoided. Small amounts should be promptly used or discarded.

Woolens and blankets should be cleaned before summer storage and placed in tightly sealed containers. For storage over extended periods moth (PDB) crystals should be scattered on paper between layers of articles to be protected. In sealed containers one application per year is adequate. Plastic products, such as buttons, should be kept from direct contact with PDB crystals to prevent their disintegration by "melting." Cedar chests or cedar-lined closets give but limited protection beyond the tightness of their construction. Insects in infested materials may be killed by exposure to temperatures of 120° or higher, for at least 2 hours. In summer these temperatures are reached in uninsulated attics or during outdoor exposure to the open sun. When woolens are aired outdoors they should be thoroughly brushed, especially in seams, folds and pockets to remove lint, eggs and larvae.

Applications of residual insecticides are very effective. Some carpeting is mothproofed during manufacture. When new untreated carpeting is installed both sides of the underpad and the underside of the carpeting may be sprayed for long-term protection. Carpet beetles do not normally attack synthetic carpets but do feed on lint and proteinaceous foods imbedded in such carpets.

Clothing, blankets and other susceptible materials may be protected by hanging them outside and spraying them with a stainless formulation of an approved household insecticide.

Centipedes

These "100 legged" relatives of insects have one pair of legs on each body segment. Most species live outdoors and only incidentally invade homes although the house centipede, 1-1/2 inches long when full-grown, may regularly live indoors.

They are usually considered nuisances because of their presence but may also be regarded as beneficial because they feed on insects, spiders and other small animals. They overpower their prey with

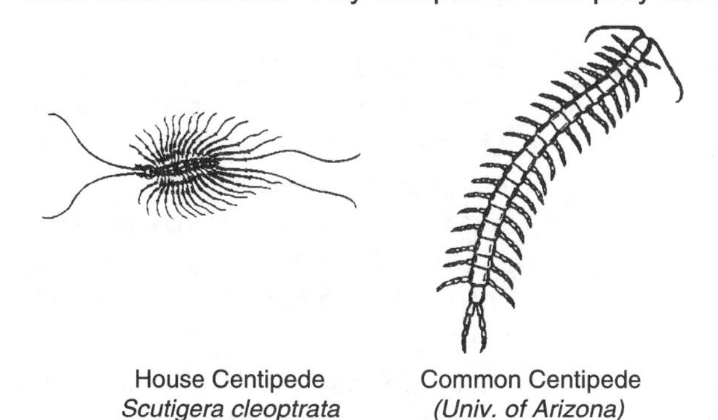

House Centipede
Scutigera cleoptrata
(U.S. Public Health Service)

Common Centipede
(Univ. of Arizona)

prominent jaws and a relatively mild venom. Humans are rarely attacked, but if bitten a temporary swelling, less painful than a bee sting, may result from the injected venom. The effects of such bites may be more emotional than physical. Deep puncture wounds from larger species should be examined by a physician.

Centipedes are active at night. In homes they are most commonly found in places of high humidity, such as basements, bathrooms and damp, dark storage areas. Outdoors they are usually found under stones, damp boards and other debris. Invasions of homes may be reduced by eliminating these sources of shelter, especially from near foundations.

Chinch Bugs

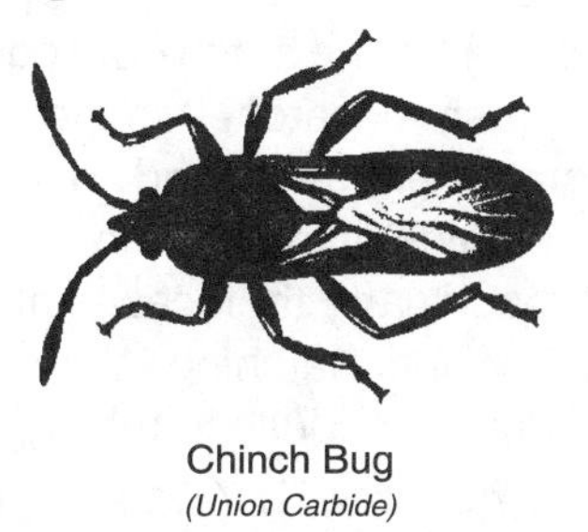

Chinch Bug
(Union Carbide)

Chinch bugs are pests of grain crops and grasses and are a serious pest of lawns in Florida. These small black bugs with clear wings may invade homes in the fall as cold weather begins, particularly near agricultural areas. The insects give off a vile odor when crushed and should be removed with a vacuum. Their fall intrusion is rather brief and they are no more than a nuisance. (See Plant Bugs under Vegetable Garden Insect Pests)

Click Beetles

Click beetles are the adults of wireworms, brown, hard-shelled larvae that feed on growing plants. They are attracted to night lights and may be especially troublesome in early spring and late fall as they emerge from nearby cropland. They vary in size and color, being solid, striped or spotted and are identified by their clicking action when held between the thumb and forefinger. The click or snapping action is their rapid, spring-loaded movement of the abdomen hinging at the thorax. This is a defensive behavior and usually surprises the holder causing it to be released. They cause no harm within the home and can be avoided by using yellow bug-lights outside and by maintaining good, tight window and door screen. (See Wireworms under Lawn and Turf Insect Pests)

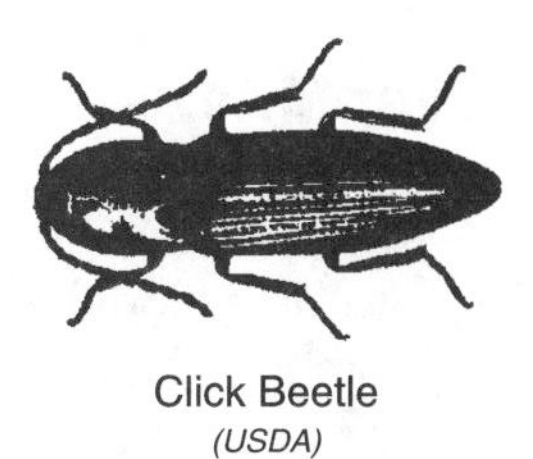

Click Beetle
(USDA)

Clothes Moths

Clothes moths and carpet beetles are similar in that the damage of each is caused by the larvae, the products attacked are almost the same and the same preventative and remedial control methods are used.

Two species of clothes moths are widely distributed in the United States. Larvae of the webbing clothes moths feed within silken tubes or webs spun over the infested material. Survival is much lower on clean wool than on wool soiled with food, beverages or body stains. Infestations are thus centered in such areas. Larvae of the casebearing clothes moth have similar food habits but feed within a portable case rather than under a web.

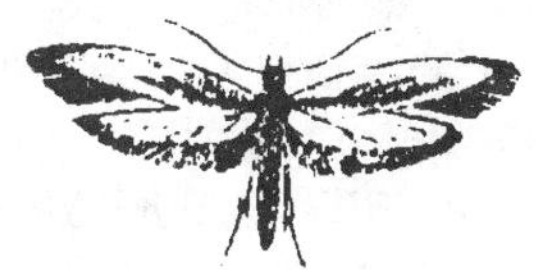

Webbing Clothes Moth
Tineola bisselliella
(U.S. Public Health Service)

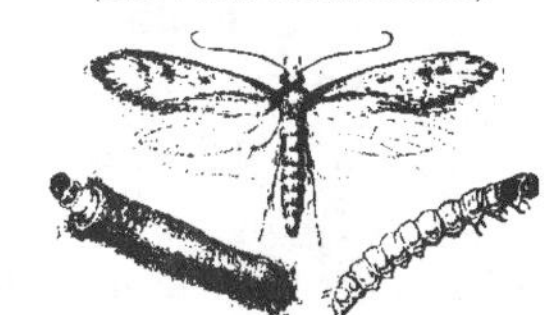

Case Making Clothes Moth
Tinea pellionella
(U.S. Public Health Service)

Infestations of clothes moths may be prevented or controlled by the procedures suggested for carpet beetles.

The following list of home remedies was proved useless by U.S. Dept. of Agriculture entomologists in the mid-1940's: allspice, borax, formaldehyde, lead-oxide, sodium bicarbonate, tobacco, black pepper, red cedar leaves, hellebore, lime, salt, lavender-flowers, cayenne pepper, eucalyptus leaves, lead-carbonate, quassia chips and sulfur.

Any sucess achieved from using any of the above would come from the fact that moths were absent or woolens so thoroughly cleaned and well stored that moths did not find them.

Clover Mites

These bright red mites invade homes in search of warmth or shelter, particularly in the early spring. They are objectionable by their presence, often in large numbers and because of the reddish-brown stains made on floors, carpets and other surfaces or fabrics when they are inadvertently crushed.

Clover mites are about the size of a small pinhead. The front legs are longer than the others. They may become abundant in lush lawns and other vegetation near houses, especially when soils are rich in organic matter. After emerging from winter shelter they may become household pests when vegetation, including winter lawns, is maintained close to foundation walls. On warm, early spring days, they may crawl up the walls of houses, particularly white walls and enter through doors or other openings.

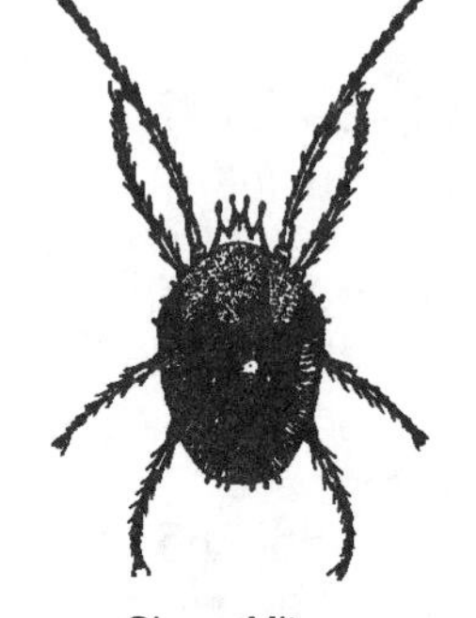

Clover Mite
(USDA)

They are controlled by maintaining a grass-free strip of loose soil 2 feet wide next to foundation walls. This area and adjacent doorways and window sills may be sprayed with a residual insecticide. Larger shrubs may be safely grown in this area although common weeds and annual flowers may attract clover mites.

Indoor infestations may be removed by a vacuum cleaner, using care to avoid surface staining from the bodies of crushed mites. Recommended spray treatments may also be needed.

Cluster Flies or Attic Flies

Cluster flies (*Pollenia rudis*) are common in the colder climates. The adults frequently collect in attics and other protected places in late summer and fall to hibernate. Occasionally their numbers are enormous, such that they become a pest. The flies are larger and darker colored than houseflies and noticeably sluggish in habit. The larvae are parasites of earthworms.

Cockroaches or Waterbugs

Cockroaches have lived in or near human dwellings for thousands of years and are among the most familiar and objectionable of household pests. They are oval, flattened insects with a prominent shield extending forward over the head. Most species are active at night and hide in darkened areas, often in cracks, during the day.

Roaches and their egg cases may be brought into homes in boxes, cartons, grocery and produce bags and in household goods.

Light infestations are common in homes although heavy, continuing infestations are usually associated with sloppy housekeeping. These pests may carry on their bodies disease organisms which can be transferred from garbage, filth and decaying organic matter to food, dishes and kitchen utensils. Further contamination results from spots of regurgitated food and fecal matter on household articles. Articles with starch or glue sizing such as paper, fabrics and book bindings, may be attacked. Cockroaches may seek shelter or food in closets, dressers, desks, electric clocks and even TV sets. Contaminated articles, particularly foods have a distinct roach odor.

Several species of cockroaches are major pests while others are only minor pests. These pest species may be distinguished by their size, appearance and habits (Figure 19.) Young hatch from eggs deposited in groups within small satchel-shape cases. These egg cases are often deposited in nearby debris although the females of the German cockroach and the field cockroach carry their cases attached to the rear of the body until hatching. Young roaches or nymphs are similar to adults in feeding habits.

The **German cockroach** *(Blatella germanica)* is probably the most abundant and prolific pest species. The adults are pale yellowish-brown, about 5/8 inch long, with two blackish-brown lengthwise stripes on the pronotum. Both sexes have wings but rarely fly. With ample food and moisture one female is capable of producing over a million descendants in a year. German roaches are usually found in or near kitchens and are general feeders, although fermented foods are particularly preferred.

FIGURE 19.

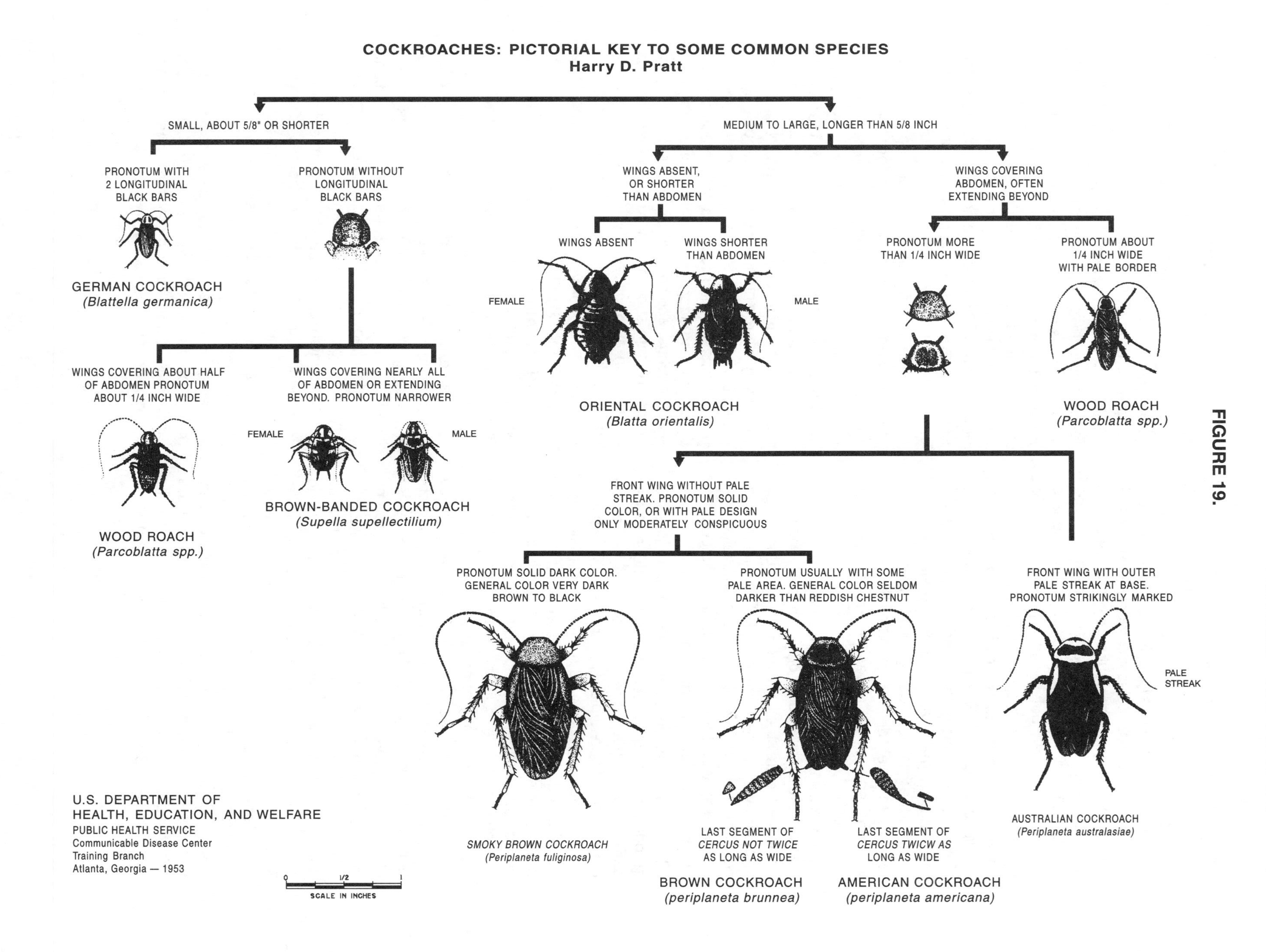
COCKROACHES: PICTORIAL KEY TO SOME COMMON SPECIES
Harry D. Pratt
SMALL, ABOUT 5/8" OR SHORTER
MEDIUM TO LARGE, LONGER THAN 5/8 INCH
PRONOTUM WITH 2 LONGITUDINAL BLACK BARS
PRONOTUM WITHOUT LONGITUDINAL BLACK BARS
GERMAN COCKROACH
(Blattella germanica)
WINGS COVERING ABOUT HALF OF ABDOMEN PRONOTUM ABOUT 1/4 INCH WIDE
WINGS COVERING NEARLY ALL OF ABDOMEN OR EXTENDING BEYOND. PRONOTUM NARROWER
WOOD ROACH
(Parcoblatta spp.)
FEMALE
MALE
BROWN-BANDED COCKROACH
(Supella supellectilium)
WINGS ABSENT, OR SHORTER THAN ABDOMEN
WINGS COVERING ABDOMEN, OFTEN EXTENDING BEYOND
WINGS ABSENT
WINGS SHORTER THAN ABDOMEN
FEMALE
MALE
ORIENTAL COCKROACH
(Blatta orientalis)
PRONOTUM MORE THAN 1/4 INCH WIDE
PRONOTUM ABOUT 1/4 INCH WIDE WITH PALE BORDER
WOOD ROACH
(Parcoblatta spp.)
FRONT WING WITHOUT PALE STREAK. PRONOTUM SOLID COLOR, OR WITH PALE DESIGN ONLY MODERATELY CONSPICUOUS
FRONT WING WITH OUTER PALE STREAK AT BASE. PRONOTUM STRIKINGLY MARKED
PRONOTUM SOLID DARK COLOR. GENERAL COLOR VERY DARK BROWN TO BLACK
PRONOTUM USUALLY WITH SOME PALE AREA. GENERAL COLOR SELDOM DARKER THAN REDDISH CHESTNUT
PALE STREAK
SMOKY BROWN COCKROACH
(Periplaneta fuliginosa)
LAST SEGMENT OF CERCUS NOT TWICE AS LONG AS WIDE
LAST SEGMENT OF CERCUS TWICW AS LONG AS WIDE
BROWN COCKROACH
(periplaneta brunnea)
AMERICAN COCKROACH
(periplaneta americana)
AUSTRALIAN COCKROACH
(Periplaneta australasiae)
U.S. DEPARTMENT OF HEALTH, EDUCATION, AND WELFARE
PUBLIC HEALTH SERVICE
Communicable Disease Center
Training Branch
Atlanta, Georgia — 1953
0
1/2
1
SCALE IN INCHES

The American cockroach *(Periplaneta americana)* is the largest and one of the commonest pest species in the southern states. It is reddish-brown, 1-1/2 to 2 inches long, with darker blotches on a yellowish pronotum. It is commonly found in dark, moist areas near bathtubs, clothes hampers, basements and sewers. It may enter homes through plumbing traps. The American cockroach feeds on many substances but seems to prefer decaying organic matter.

The brown-banded cockroach *(Supella longipalpa)* is slightly smaller than the German cockroach and has a pale band across the base of the wings. It may have one or two similar bands towards the wing tips. It is a general feeder and tends to be more widely distributed over a home, including warmer and drier areas, than the German cockroach.

The **Oriental cockroach** *(Blatta orientalis)* is shiny black or brown and about one inch long. The wings of the female are small, rudimentary and functionless and hardly more than small pads. The males' wings cover only about 75% of the abdomen. Neither sex flies and they prefer dark, damp areas indoors and warm humid situations outdoors.

The **Turkestan cockroach** *(Blatta lateralis)* was introduced into the U.S. in about 1980 and has spread rapidly throughout the southern half of the country. Adults are typically an inch long. The males are slender with long, tan wings. Females are broader with short, nonfunctional wings and dark brown to black in color, with light markings on the shield and similar markings on the short wings. Nymphs are wingless, light tan in front and black on the rear. Any quantity of organic matter, such as a mulch pile can house very large populations. They are mainly an outdoor insect, seen usually at night but almost never during the day. They seldom enter dwellings, but can be easily controlled with the same techniques used for other roaches.

The **Surinam cockroach** *(Pycnoscelus surinamensis)* is normally associated with humid tropical areas such as Florida and Brownsville, Texas. They are about one inch long, have yellowish-brown wings that extend beyond the abdomen and a black shield or pronotum with the front margin cream colored. Surprisingly, only females have been found in the U.S. The nymphs are shiny reddish brown and pear-shaped. It typically lives outdoors in loose soil, under stones or burrowing. This roach is not considered a household pest. However, it has become a problem in greenhouses, sometimes damaging roots of plants. The species cannot survive winter nighttime temperatures of 24° F or less.

The **desert cockroach** *(Arenivaga genitalis)* is found only in the dry, Southwestern U.S. It looks like a cockroach, is about three-quarters of an inch long and light tan in color. The males fly readily while the females are wingless. These cockroaches spend most of their lives burrowing in soft soil, feeding on organic debris. They are commonly found in pack-rat nests, where soil is loosened and contains abundant organic matter. They seldom are found in dwellings, but because they are nocturnal outdoor residents, may on occasion be attracted to indoor lights.

The **Asian cockroach** *(Blattella asahinai)* is fairly new to North America, having first been identified in 1985. It is very similar in size and color to the German cockroach and is frequently confused with our old friend. The Asian roach is a strong and accomplished flier, living in large numbers outdoors, but is equally content to inhabit homes where it becomes a serious domestic pest. Asian roaches are attracted to the German roach aggregation pheromone and recent research has shown that Asian males may on occasion mate successfully with German females, resulting in a hybrid variety. Currently, the Asian cockroach distribution includes the East and Southern coastal areas. Their control should be identical to that of the German cockroach.

The **Nicaraguan cockroach** *(Ischnoptera begrothi)* is probably our most recent roach invader, first identified in 1990. It is native to Central America, from Panama to Nicaragua. It is well established in Florida and Louisiana, preferring tropical environments. Both males and females fly readily and are attracted to lights. There is no indication that it prefers or is able to breed under indoor conditions, since it is a known outdoors type. Large populations have been observed in lawns and plantings around houses which are adjacent to prime habitats. Treatment of the perimenter of homes, especially near doors and other entry ports and shrubbery and beds near the house with a residual spray or bait provides control.

Other cockroaches that may be encountered as minor pests are the very small **Caribbean cockroach** *(Cariblatta mimina)*; the common **wood cockroaches** *(Parcoblatta pennsylvanica* and *P.*

virginica), which are large, flying, outdoor types that resemble the American cockroach; and the **smokybrown cockroach** *(Periplaneta fuliginosa)*, also an outdoor type that resembles its cousin, the American cockroach, but is slightly smaller. Again, the secret to controlling any type of cockroach is to find and destroy its primary habitat. This may mean reorganizing a woodpile, raking up decomposing leaves, grass cuttings and ground trash, or selecting trash and garbage cans with tight-fitting lids. Because most outdoor cockroaches are attracted to lights, using yellow bug lights at entrances and closing curtains and blinds at night will reduce your home's attractiveness and perhaps provide a bit more freedom from pests.

Roach control requires good housekeeping and thorough application of insecticides when required. There is no control unless it is complete. It is important that insecticide residues remain on treated surfaces for continued control action. Applications to warm areas, such as heater rooms, may need to be made more often than elsewhere because of more rapid disappearance of residues at warmer temperatures. In difficult situations, such as multiple dwellings, a community effort may be required. Several roach baits containing insecticides are quite attractive to roaches and very effective. (See Table 20) Boric acid as a powder or tablets is also very effective. Ultrasonic pest control devices are totally ineffective for any kind of pest control including cockroaches. Some very effective remedies can be made at home and are described in Appendix F.

Crickets

Our chirping friends are general feeders on most kinds of outdoor vegetation. When abundant they may enter homes for shelter or food, particularly when vegetation becomes dry or scarce, or at the approach of winter. They are attracted to lights outside of buildings where they may become nuisance pests when abundant.

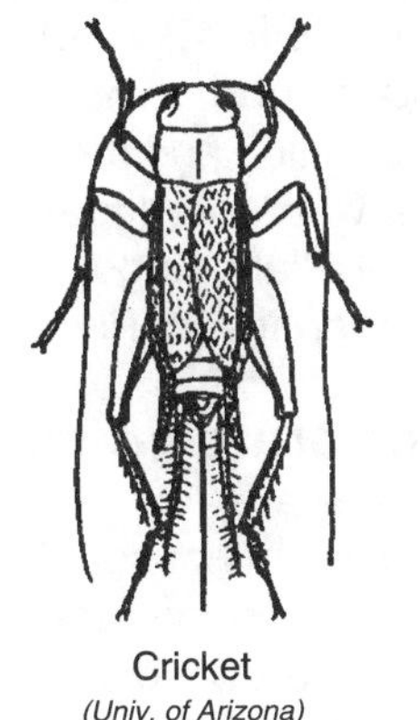

Cricket
(Univ. of Arizona)

Inside homes they may feed on clothing and fabrics spotted or stained by spilled food or per-spiration. The resulting feeding holes have frayed edges.

Two species may become house pests, field and house crickets. Adults are 3/4 to 1 inch long, brown to gray and with thread-like antennae longer than the body. Females have a long, needle-like ovipositor at the rear of the body. The chirping call of crickets, which may be an annoyance if crickets are numerous, is produced only by the males.

Crickets may be excluded from homes by tightly-fitting doors, windows and screens. Foundation perimeter treatments with insecticide, particularly near doors and under lights, may be needed when crickets are abundant. At such times the use of yellow, non-attractive light bulbs is helpful. Residual sprays may be applied to baseboards and other surfaces in indoor areas where crickets are found. Roach sticky-traps work miracles, as do some of the homemade remedies described in Appendix F.

Darkling Beetles

In the West and Southwest these outdoor insects may become nuisance pests by invading homes in search of darkness or shelter. Adult beetles are 1/4" long and dark brown to grayish black. They may be found in agricultural areas and in flower beds or leafy trash near foundations of houses. These beetles feed on living or decaying vegetable matter and may girdle stems of living plants, especially seed-lings, near the soil level.

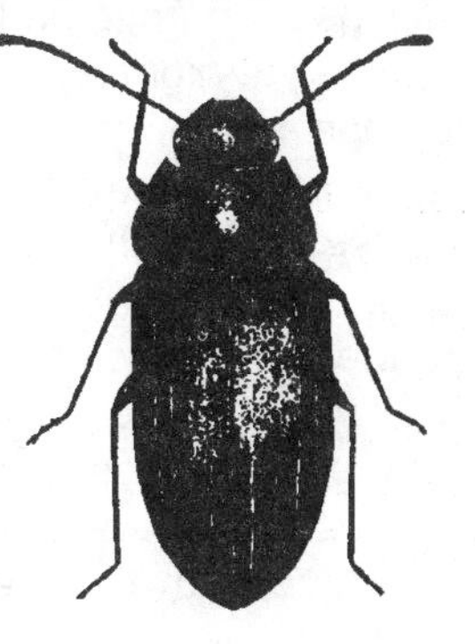

Darkling Beetle
(Univ. of Arizona)

When infested crops or plantings become dry these beetles may invade houses, often to escape the summer heat. In summer, they may be found in large numbers under outdoor lights, especially near agri-cultural areas or extensive home plantings. The larvae, known as false wireworms, live below ground and are seldom seen.

Invasions of darkling beetles may be reduced by tight-fitting doors, windows and screens. Dead leaves and rubbish should be removed from outside areas near foundation walls. An insecticide barrier surrounding the house, as used against other invading insects, may also be helpful.

Indoor infestations may be controlled with a fly swatter, vacuum cleaner, hand picking, or by surface applications of a residual-type insecticide to door sills, baseboards and dark corners where beetles may hide.Dermestids — **see Carpet Beetles**

Drain Flies or Moth Flies

Drain or sewage flies, also known as moth flies are usually less than 1/8 inch long, with scaly wings held roof-like over the body. They may develop inside homes in the organic matter found in sink drains, plumbing traps, or garbage. They may also invade homes from nearby sewage plants or other areas with moist, decaying organic matter. These nuisance pests can be controlled by eliminating breeding places through good housekeeping, including regular treatment of sink drains and traps with a drain cleaner or flushing the overflow drains with scalding water.

Drain or Moth Flies
(USDA)

Earwigs

Earwigs are harmless nuisance invaders often found in homes at night and are easily recognized by the "forceps" at the rear of the body. Wings are short or lacking and movement is usually by crawling. Several species occur although two species, the striped, or riparian earwig (3/4" long) and the ring-legged earwig (5/8" long) are most common in houses. They emit a substance with an offensive odor when disturbed.

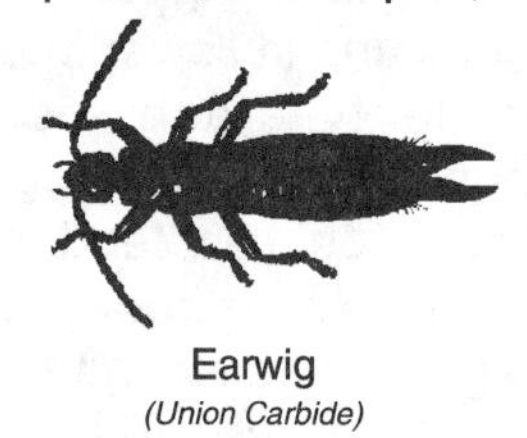

Earwig
(Union Carbide)

Earwigs develop outdoors, often in flower beds and lawns near bases of houses. The striped or riparian earwig is predaceous on insects or earthworms. The ring-legged earwig feeds on decaying plant or animal matter.

Earwigs are best controlled outdoors by eliminating plant debris from areas close to foundations of houses and by insecticidal treatment of breeding areas.

Elm Leaf Beetles

This is a major foliage pest of elm trees. It occasionally becomes a household nuisance when adult beetles seek winter shelter in homes or when larvae seek protected areas for pupation in the spring and later in the season.

The buff-colored adult beetles are 1/4" long, with an olive green stripe along the outer edge of each wing cover. They commonly spend the winter in protected outdoor locations, but in colder areas, may also be found indoors, behind curtains, under carpets, between books and in similar locations. Mature larvae are 1/2" long, dull yellow and with two stripes along the back. They may seek protected areas for pupation in homes or porches.

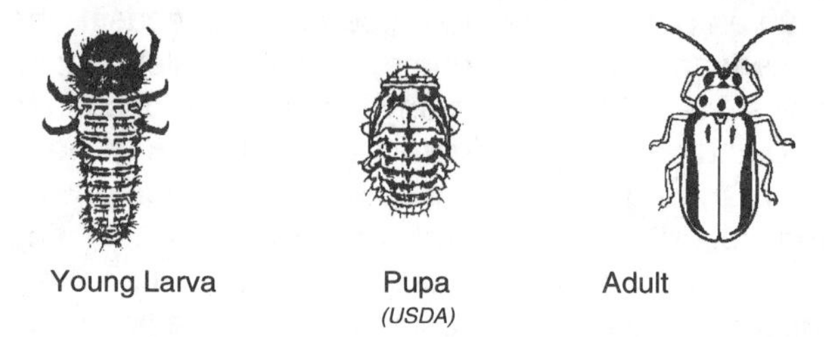

Young Larva — Pupa — Adult
(USDA)

Basically, their control involves treatment of outdoor infestations in elm trees. Invasions of homes may be reduced by tight-fitting doors, windows and screens. Indoor infestations may be removed by vacuum cleaning or use of a recommended insecticide.

False Chinch Bugs

These are pests in the West and Southwest. Adults are less than 1/4" long, with grayish bodies and paler outer wings. These plant-feeding insects may invade homes in mid- to late spring, particularly near agricultural areas or fallow lands. These invasions are sporadic, of short duration and usually of minor importance. They usually occur after winters with higher than average rainfall, when there is an abundant early spring growth of annual weeds and desert vegetation. Migration toward homes may occur after these plants mature and wither with the approach of warmer, drier weather.

Invasions of homes may be discouraged by tightly fitting doors, windows and screens, by keeping foundation areas free of weeds and annual vegetation and providing an insecticidal barrier near the wall as suggested for other invading insects. (See Plant Bugs and illustration under Vegetable Garden Insect Pests)

Firebrats and Silverfish

These are two closely related, widely distributed species of household insects of similar appearance, habits and importance. The firebrat prefers warm, dry areas while silverfish are moisture-loving.

Adults are 1/2" to 5/8" long, mottled gray, flattened, boat-shaped, covered with scales, without wings and with 3 long "tails" at the rear of the body. It hides during the day, runs swiftly and is most active at night. It commonly falls into bathtubs, from which it

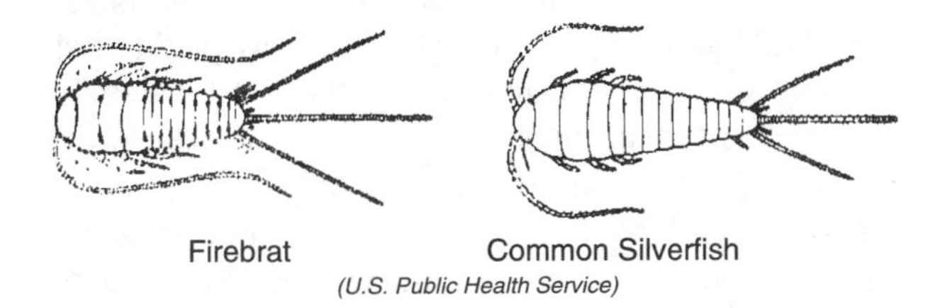

Firebrat Common Silverfish
(U.S. Public Health Service)

cannot escape. It eats and contaminates starchy foods, fabrics and paper and may attack book bindings, leaving irregular roughened areas.

Infestations may be reduced by good housekeeping to remove starchy food scraps, elimination of hiding places, sealing cracks and other structural repairs and by use of insecticides. Sprays should be applied to cracks and surface areas where they are observed and to floors or shelves under books and cartons in moist areas, especially garages and basements.

Fleas

Fleas found in homes are mainly pests of cats and dogs although humans may also be attacked. The wingless blood-feeding adults are brown to black, with laterally flattened bodies less than 1/8" long and with hind legs adapted for jumping. The worm-like larvae are not blood feeders but feed on organic refuse in animal bedding, on floors and in outdoor runways used by pets. Fleas may complete their development in the absence of animals. Persons entering infested homes unoccupied during vacation periods or between tenants may be attacked, particularly on the legs, by hungry, newly developed adult fleas that can survive for long periods without feeding and lose no time in attacking new arrivals. Bedding used by infested pets should be destroyed or thoroughly cleaned to kill the developing fleas. Floors, carpets and upholstery in infested areas should be cleaned to remove all debris and animal refuse. Outdoor animal runs should also be kept free of debris and manure which might harbor developing fleas. After cleaning, the previously infested areas should be protected with an approved insecticide. Infested animals should be treated by one of the methods listed in the table on pet pest control, in addition to the use of a flea collar.

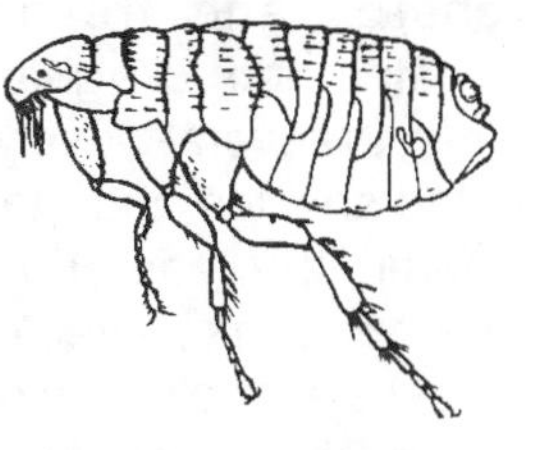

Flea
(Univ. of Arizona)

A new approach to flea control in dogs and cats is the once-a-month flea control tablet, available only by veterinary prescription. It is, lufenuron (Program®), a systemic insect growth regulator manufactured by Syngenta. When a female flea bites a treated animal, the flea ingests this ingredient that prevents her eggs from developing. This breaks the flea life cycle so that infestations are virtually eliminated. It is reported to be safe, even for pregnant and nursing females, breeding males and for puppies and kittens as young as six weeks. It does not kill adult fleas, but when a female flea bites a treated animal, she will be unable to reproduce.

Flies

Several species of flies may be annoyances in and around homes, especially during the summer.

Female **"filth" flies** of various species may deposit eggs in manure piles, dead animals, meat, garbage and other decaying plant and animal matter. During warm weather flies quickly pass through egg, larval (maggot) and pupal stages (8-24 days, depending on the species.) These dull, grey-black, or shiny, blue or green adult flies disperse freely and may enter homes in search of food or shelter. Their frequent movements between "filth" and our food and our persons may spread disease.

Of these flies, the **common house fly** needs little introduction. It is 1/4 inch long, with gray to black markings. It is probably the most persistent nuisance pest and is an efficient disease carrier because of its ability to breed in a variety of substances. It has been implicated in the spread of typhoid, dysentery, tuberculosis and other diseases.

The **little house fly** is a smaller, closely related species. It seldom alights on man or his food and is less likely to contaminate food than the common house fly. This species is characterized by its habit of flying to and fro in the middle of a room or in shaded entryways.

Musca domestica
House Fly
(USDA)

Stable flies resemble house flies very closely, but bite with a sharp, blood-sucking proboscis that projects forward when the fly is at rest. Larval breeding sources include manure, especially when mixed with straw and lawn clippings.

The **false stable fly** is slightly larger and darker gray than the house fly. It may enter homes and deposit eggs in well-decayed vegetable matter. It is not a persistent nuisance. Flesh flies are the largest of the gray-colored flies. They have a checkerboard pattern on the abdomen with the tip being red in many species. These flies breed in dead animals and excrement. Occasionally they alight on man or his food.

Fruit flies or **vinegar flies** *(Drosophila),* are small, yellowish-brown and commonly associated with decaying fruit, vegetable or garbage cans. Eggs hatch in 24 hours and a life cycle takes from 8 to 11 days. These flies are most abundant around larval breeding sources but they may also enter the home as hitchhikers with fruit.

Blow flies are shiny, metallic blue, green, black or copper colored flies. They lay their eggs in exposed meat, garbage or pet droppings. The loud droning buzz created by flight is an annoyance in most confined situations. These flies may be particularly abundant during the spring and fall.

Phaenicia sericata
Green Blow Fly
(U.S. Public Health Service)

The first step in the control of all these flies is sanitation, the elimination of breeding places. Regular removal of garbage, livestock and pet manure and all other decaying plant and animal matter is essential for successful fly abatement. Garbage cans should have tight-fitting lids. The use of plastic garbage or trash bags is very successful.

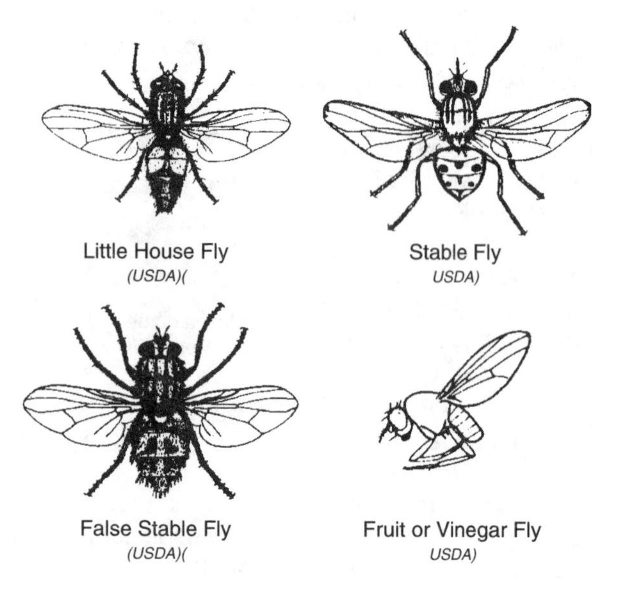

Little House Fly
(USDA)(

Stable Fly
USDA)

False Stable Fly
(USDA)(

Fruit or Vinegar Fly
USDA)

Homes should have tight-fitting screens in good repair.

Fly swatters are still one of the best methods for control of the occasional flies that get into the tightest house. Keep one handy in the kitchen. Adhesive fly paper strips can help when hung discretely in selected areas.

In areas of high fly density, insecticides may be needed both inside and outdoors. Certain aerosol bombs may be used indoors as space sprays. Residual surface sprays are useful for killing flies outside the home. Insecticide-impregnated No-Pest® Strips are amazingly effective indoors and in well-protected areas. The use of "Big Stinky®" and similar fly traps outdoors can also be of value.

Areas to be treated include porches, walls, garbage cans, carports and other surfaces where flies may rest. Bait formulations may be effective but should be kept out of reach of children and pets. Community action is usually necessary to achieve fly control in areas where breeding sources are extensive, such as trash dumps, animal feeding facilities and pet areas.

Gnats

Gnats can ruin any outside activity. These annoying patio pests are tiny flies, less than 1/8 inch long, which are particularly attracted to the moist regions of the eyes, nose and mouth. They are most active during the warmer months, especially near fruit orchards and freshly cultivated land. Eye gnats belong to several closely-related species and may transmit the bacterium causing pink eye of domestic animals as well as man. The strong-flying adults are active during daylight hours. The larvae live in the soil and feed on decaying matter. There may be several generations a year.

Gnat
(U.S. Public Health Services)

Personal protection involves applications of a repellent to exposed body areas. It is difficult to control infestations on individual home properties since it is usually an area rather than an individual problem. More satisfactory control results may be produced from soil treatments to control the larvae on an area-wide basis. A community effort, much as for mosquitoes, may be required. The insecticide-impregnated No-Pest® strips are very effective against gnats indoors.

Ground Beetles — **See Chapter 1, "Predators"**

Honey Bees — See Bees

Hornets — See Wasps

Kissing Bugs

Kissing bugs, which are blood-feeders, are also known as assassin bugs, conenose bugs and Walapai tigers. They commonly live in outdoor nests of rodents, such as pack rats and are more frequently reported from rural and desert locations than from urban areas. The adults are from 3/4" to 1-1/4" long and resemble large squash bugs, but with the head more slender than the rest of the body. Several species of varying color patterns, usually dark, are found in Southwestern desert areas.

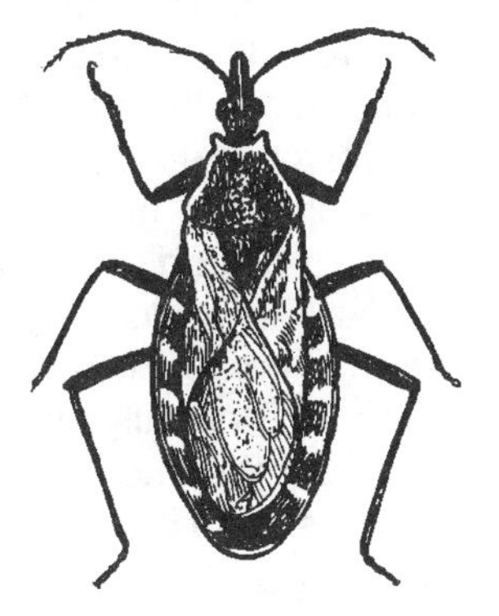

Kissing Bug
(niv. of Arizona)

Adult kissing bugs mature during the spring months, are most active at night and are attracted to the lights of homes. Upon entry they may remove blood from sleeping persons without waking them, although painful swellings and occasional allergic reactions may later develop. During the day they may conceal themselves in bedding. Older homes and outbuildings may harbor pack rat nests in open crawl spaces beneath wooden floors and provide a source of kissing bugs which may attack persons living in the area.

Preventive control includes elimination of rodent nests under buildings and in nearby desert areas. Pack rat nests have characteristic mounds of debris, often including remnants of cholla cactus. Invasions of homes may be reduced by using yellow bug lights outside and by tightly fitting doors, windows and screens, sealing cracks in wooden floors and walls and screening ventilation ducts.

Treatment of bites of kissing bugs may require the aid of a physician. Application of ice packs to the affected area may give partial relief.

Larder Beetle — see Pantry Pests also Carpet Beetles

Lice

Three species of lice feed on man and both males and females require blood meals. The head louse and crab louse are more commonly reported than the body louse. Lice are usually associated with crowded, unsanitary living conditions, including frequent contacts between persons. Recent trends toward greater cultural permissiveness and longer hair have led to a more uniform distribution of lice among all socio-economic groups.

The head louse or cootie is bluish-gray to whitish, wingless, up to 1/8" long, usually found among the hairs of the scalp. The eggs, or nits, are attached to hairs close to the skin. Head lice may be spread through shared objects such as hats, hair brushes, combs and towels. Infestations may be innocently obtained from upholstered furniture, bedding and very commonly from contacts between children at school.

The crab louse, or crab, is a short, broad, thick-legged insect about 1/5" long found in the crotch and other body areas with pubic hair. The eggs are also attached to hairs. Crab lice are spread mainly by personal contact but may also be received from bedding or from toilet seats (rarely) recently used by infested persons.

The body louse resembles the head louse but is found mainly in seams of clothing, worn close to the body. The eggs are attached to clothing. In other countries, but not recently in the United States, it has been a vector of typhus and other diseases, particularly in times of war or disaster.

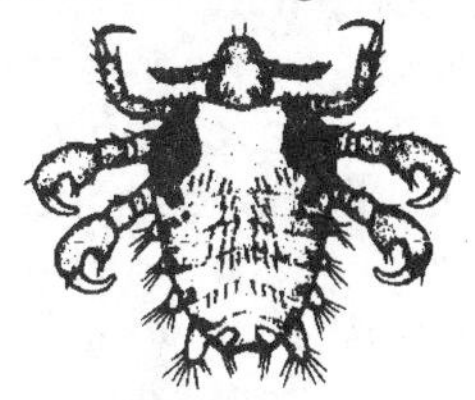

Crab Louse
(Univ. of Arizona)

Louse control involves sanitation and personal hygiene based on a knowledge of lice and their methods of transfer. Infested persons should be promptly treated to destroy both the lice and their developing eggs. Several suitable louse ointments, lotions and shampoos are available and most require a physician's prescription.

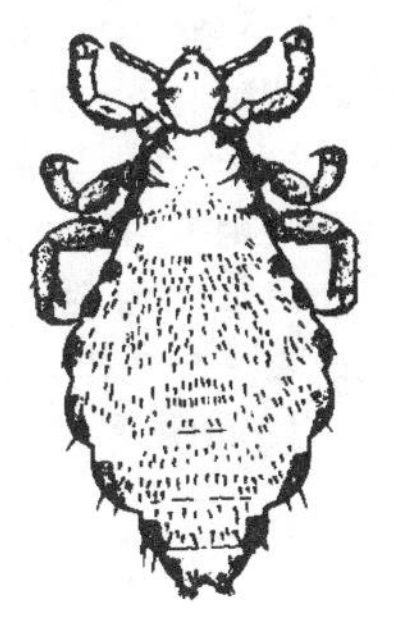

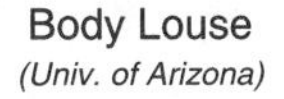

Body Louse
(Univ. of Arizona)

When isolated from human contact nymphal lice usually survive not more than a day and adults not more than 5 to 6 days. Eggs may remain alive until hatching, from 8 to 10 days after being laid. Lice are eliminated when clothing is stored for 3 weeks.

Lice in clothing or bedding are destroyed by laundering or dry cleaning. Immersion in hot water, 125°F or higher kills adults in 5 minutes and eggs in 10 minutes. Thorough exposure of infested materials to the heat of the midday summer sun is also effective.

Light-Attracted Insects

Many insect species are naturally attracted to lights and may become nuisances both outside and inside homes, particularly in the warmer months. Smaller flying insects, including tiny beetles, embiids, (web spinners), gnats, leafhoppers and midges, may pass through screens and collect in lighting fixtures. Embiids are 1/2 inch long, dark and slender and, with some of the smaller beetles, are probably the largest insects able to pass through screens. Leafhoppers are tiny, wedge-shaped and usually green or yellowish. Gnats and midges are flies. These and larger insects such as moths, larger beetles, crickets and flies may become abundant at outdoor entrance and patio lights. Some of these insects may be attracted for considerable distances.

Exclusion of these insects from houses may be difficult during their periods of greatest abundance. Invasions may be reduced by tight-fitting doors, windows and screens, use of light-proof window drapes and avoiding unnecessary use of lights inside or outside. Space sprays may be of temporary value. Dead insects in and beneath light fixtures should be removed to prevent secondary invasions of pests such as carpet beetles.

Insects are attracted to lights to varying degrees—some not at all. Despite exceptions, most insects are attracted in greater number to white incandescent, fluorescent, or blue mercury-vapor-type lamps than to yellow lamps. The use of yellow bulbs or "bug lights" for entrance or patio lighting is therefore helpful in suppressing light-attracted insects. White lights located at distances from human activity can be used to lure insects away from the party.

Long-Horned Beetles

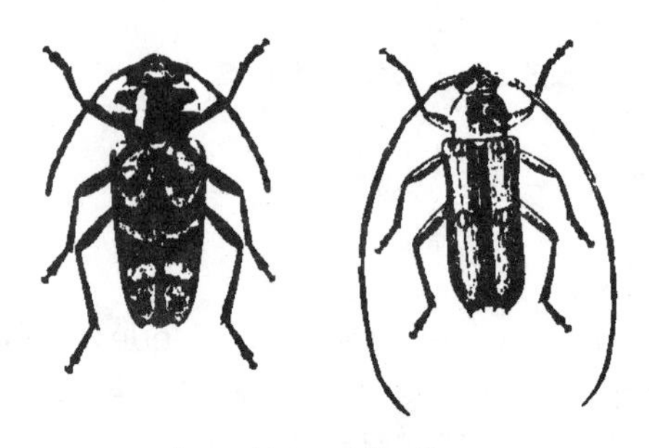

Long Horned Beetles
(Ohio State Univ.)

There are several long-horned beetles that may hitchhike their way into homes. They receive their name from their long antennae, longer than their bodies, held to the side. The larvae are wood borers and are found in dead wood. When fire wood is brought into the home, adults sometimes emerge from the logs and are attracted to windows seeking an escape. They do not generate from house timbers unless the home is less than a year old. Long-horned beetles are normally not destructive, but may gnaw their way through curtains in their attempt to reach light and should be removed when found. They can bite!

Millipedes

The "1000-legged" near-relatives to insects are millipedes. They have many body segments and often 30 or more pairs of legs, grouped 2 pairs to the segment. Species vary in length from less than an inch to 2 or more inches. Most millipedes are scavengers, feeding on decaying plant matter, although some species attack roots of living plants. They tend to avoid the light and may form a coil when at rest or disturbed.

Millipede
(Union Carbide)

They occasionally find their way into homes as casual invaders although they do not feed or cause damage, except for possible stains when they are inadvertently crushed. These invasions may be discouraged by tight fitting doors and screens and by the removal of decaying vegetation, leaf litter and similar materials from foundation areas of houses.

Mites

There are several kinds of mites that may attack man. These tiny relatives of insects have sucking mouthparts and may attack persons outdoors, as in the case of the mites known as chiggers, or may enter homes with birds and other pets, rats and mice, grains, cheese, dried fruits and infested clothing. Species of very tiny mites found in house dust may cause allergic reactions. Mites cause such human ailments as grocer's itch, mange, or "7-year itch." The clover mite is a plant-feeding outdoor pest that may invade homes but does not attack humans.

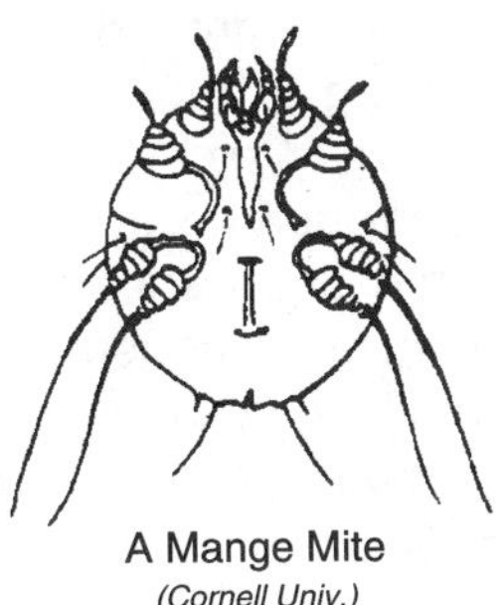
A Mange Mite
(Cornell Univ.)

Tropical rat mites are a common problem when rats occupy attics. When a rat dies, large numbers of mites leave it and find their way into the living space of the home. Some people get a severe itch from the bites of these mites and suffer for years before finding the cause. (Refer to the section on rat control for means of controlling rats and keeping them out of buildings.) Bird mites can also be a problem, coming from bird nests, such as those built under the eaves of houses.

Some mites can be detected with a hand lens, while others require microscopic examination of skin scraping or even dust from vacuum cleaner bags. Mite irritations can be confused with *entomophobia* (a fear of insects) in certain people. Mite diagnosis is often difficult because specimens must be collected and identified by trained specialists before treatment can be made. Some of the other more prevalent and annoying mites are the *Northern fowl mite*, which may come from poultry, pigeon and starling nests or dead carcasses; the *chicken mite*, which feeds on poultry, sparrows, canaries and other birds and may also come from their abandoned nests or dead carcasses; the *follicle mite* is microscopic in size and lives in the hair follicles or sebaceous glands of most humans; the *itch* or *scabies mite,* which tunnels into the skin, especially hands and wrists and is transmitted usually by direct contact with an infested person; the *straw itch mite,* which causes epidemics of dermatitis during harvesting and post-harvesting operations in straw, hay or certain grains; *grain* and *mold mites,* which are found in a wide variety of stored products and can cause mild dermatitis known as "grocer's itch". For treatment of dermatitis, scabies, or other skin disorders thought to be caused by any of the above, contact your physician.

House dust mites receive considerable attention in the media because of their potential to cause allergic reactions in susceptible persons. Allergic reactions are the result of inhaling the house dust allergen produced by these mites, causing bronchial asthma, sneezing and coughing in some individuals. There are two species that commonly occur in dwellings, the European (*Dermatophagoides pteronyssinus*) and the American house dust mite (*D. farinae*). Both are found on human skin, stored food and animal skins and throughout the home associated with house dust and lint. They are microscopic in size, almost invisible to the naked eye. Limited control can be achieved by frequent vacuuming, shampooing carpets, laundering bed clothing and drycleaning non-washable clothing, draperies, throw-rugs and furniture covers. Chemical treatment consists of sprayable or dust forms of amorphous silica gel, or sprayable, pyrethrins, resmethrin, or other pyrethroids.

Treatment varies with the species of mite and in some cases may require the services of a physician. Control involves removal of mite food sources through good housekeeping or sanitation and for some species, the use of a suitable repellent on the person. Infestations on pets should be controlled as a part of the treatment process.

Also see "Chiggers" "Clover Mites" and "Spider Mites".

Mosquitoes, Midges and Black Flies

Several species of mosquitoes may become nuisances in and about homes. Adults are generally 1/4 inch long, with two wings, long slender legs and a needle-like proboscis. Only females suck blood. Their bites may be painful at first, followed later by swelling and itching. Feeding typically occurs from dusk to dawn.

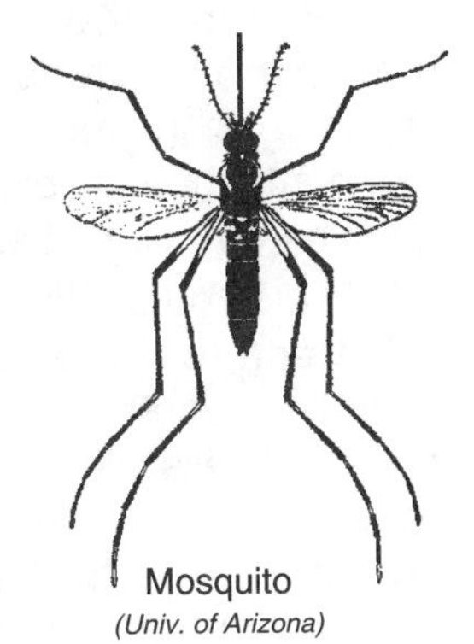
Mosquito
(Univ. of Arizona)

Mosquitoes are dependent on water for egg laying and development of their young. Eggs are deposited singly or in rafts on moist soil and on the surface of standing water in pools, ditches and discarded containers such as tin cans or tire casings. Eggs hatch into larvae or wigglers that feed on organic matter in the water. In a few days, larvae stop feeding and transform into pupae or tumblers that later become adult mosquitoes. The complete cycle from egg to adult may take as little as 9 days during the summer.

The elimination of breeding sources is the first and best step toward controlling mosquitoes around the home. Low areas or ditches where standing water may collect should be eliminated. Remove or cover open containers that may hold water and permit breeding.

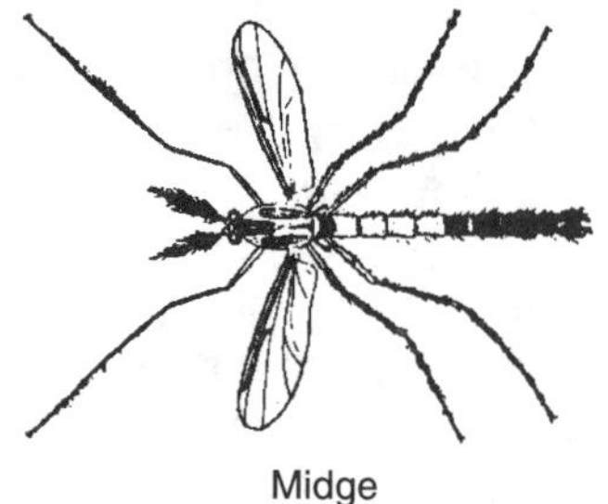

Midge
(U.S. Public Health Service)

Several species of small midges may be attracted in large number to lights. The males have typical plume-like antennae. Midges resemble mosquitoes, but their proboscis is short and not adapted for sucking blood. Midges are widely distributed and may be extremely abundant in areas near standing water. When attracted to lights they may create a major annoyance.

Black flies are small, humpbacked, gray to green flies that bite most warm-blooded animals. They range from 1/25 to 1/5 inch long with clear wings and stiff horn-like antennae. They make very little sound as they swarm about the head and other parts of the body. Bites cause itching or swelling that may persist for several days.

Black flies are usually a problem only in mountainous areas near swift running water. Immature forms are found attached to objects such as gravel, rocks, or plants in flowing streams.

Black Fly
(U.S. Public Health Service)

Protective clothing including boots, gloves and head veils, along with repellents, provide the best protection from these biting flies. Area-wide control of adults and larvae is difficult and usually not practical. Residual or space sprays may give temporary relief in confined areas such as patios.

The repellents Deet (Off®, etc) and Avon's "Skin-So-Soft" applied to exposed skin areas, give temporary protection from attacks of mosquitoes, black flies, stable flies and ticks.

Palm Flower Caterpillar

This dark-cream-colored, cutworm-like caterpillar is about 1 inch long and usually feeds on the blossoms of fan palms. When fully developed, the caterpillars drop to the ground and search for protection while changing to moths. At this time they may invade homes and excavate oval, inch-long holes in carpets to form pupal cells. Palm trees in largely paved patios near houses should be watched for this insect. Invasions of houses may also be discouraged by tight-fitting doors and screens and by a foundation pesticide barrier similar to that used against other pests.

Pantry and Stored Product Pests

Pantry pests (Figure 20) are those that can live and develop in the food and other products of low moisture content stored in home cupboards. Materials attacked include whole grains and seeds, flour and cereal products, dried milk, dried dog and other pet food, dried fruit, dried or cured meats, baked goods, candy, nuts, chocolate and cocoa spices, chili powder and even drugs.

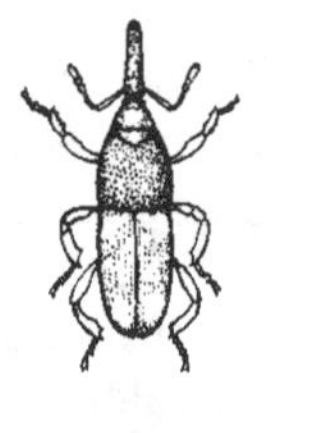

Rice Weevil

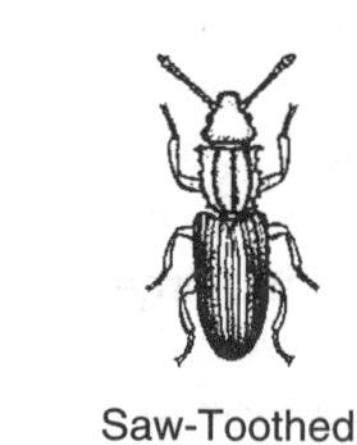

Saw-Toothed Grain Beetle

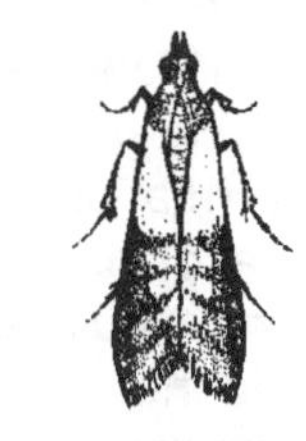

Indian Meal Moth

(Univ. of Arizona)

The rice weevil is dark brown, 1/8 inch long, with a prominent snout. It feeds on whole grains and hard cereal products such as macaroni.

More general feeders include flour beetles and saw-toothed grain beetles, each slender brown and 1/8 inch long and the drug store beetle, slightly smaller,but with fairly long antennae. The sawtoothed grain beetle has prominent rows of saw-like projections on each side of the upper surface in the area behind the head. Dried fruits and nuts may be attacked by these general feeders and by the

FIGURE 20.

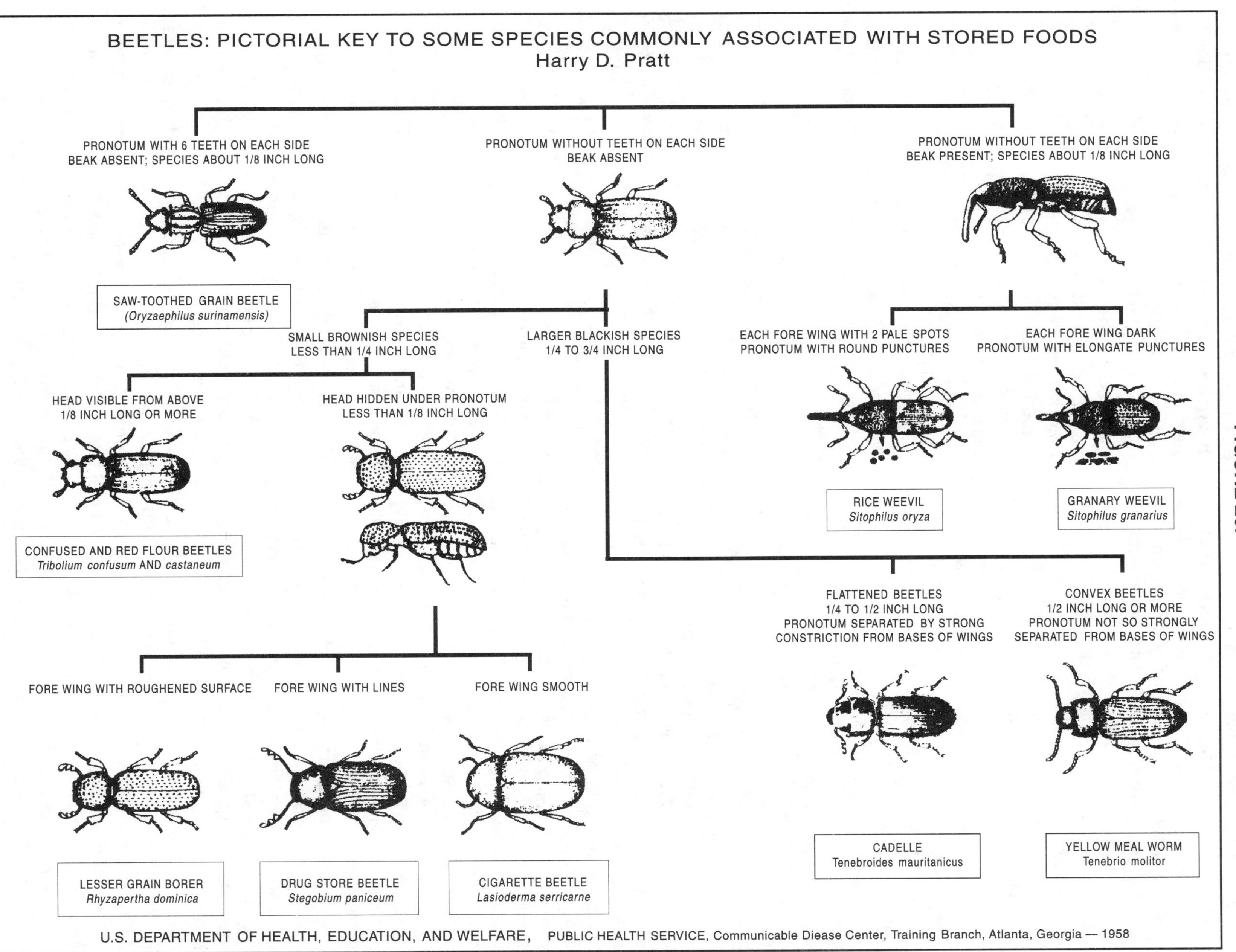
BEETLES: PICTORIAL KEY TO SOME SPECIES COMMONLY ASSOCIATED WITH STORED FOODS
Harry D. Pratt
PRONOTUM WITH 6 TEETH ON EACH SIDE
BEAK ABSENT; SPECIES ABOUT 1/8 INCH LONG
PRONOTUM WITHOUT TEETH ON EACH SIDE
BEAK ABSENT
PRONOTUM WITHOUT TEETH ON EACH SIDE
BEAK PRESENT; SPECIES ABOUT 1/8 INCH LONG
SAW-TOOTHED GRAIN BEETLE
(Oryzaephilus surinamensis)
SMALL BROWNISH SPECIES
LESS THAN 1/4 INCH LONG
LARGER BLACKISH SPECIES
1/4 TO 3/4 INCH LONG
EACH FORE WING WITH 2 PALE SPOTS
PRONOTUM WITH ROUND PUNCTURES
EACH FORE WING DARK
PRONOTUM WITH ELONGATE PUNCTURES
HEAD VISIBLE FROM ABOVE
1/8 INCH LONG OR MORE
HEAD HIDDEN UNDER PRONOTUM
LESS THAN 1/8 INCH LONG
RICE WEEVIL
Sitophilus oryza
GRANARY WEEVIL
Sitophilus granarius
CONFUSED AND RED FLOUR BEETLES
Tribolium confusum AND castaneum
FLATTENED BEETLES
1/4 TO 1/2 INCH LONG
PRONOTUM SEPARATED BY STRONG
CONSTRICTION FROM BASES OF WINGS
CONVEX BEETLES
1/2 INCH LONG OR MORE
PRONOTUM NOT SO STRONGLY
SEPARATED FROM BASES OF WINGS
FORE WING WITH ROUGHENED SURFACE
FORE WING WITH LINES
FORE WING SMOOTH
LESSER GRAIN BORER
Rhyzapertha dominica
DRUG STORE BEETLE
Stegobium paniceum
CIGARETTE BEETLE
Lasioderma serricarne
CADELLE
Tenebroides mauritanicus
YELLOW MEAL WORM
Tenebrio molitor
U.S. DEPARTMENT OF HEALTH, EDUCATION, AND WELFARE, PUBLIC HEALTH SERVICE, Communicable Diease Center, Training Branch, Atlanta, Georgia — 1958

dried fruit beetle, recognized by its dark, oval body, 1/8 inch long, with 2 large amber-brown areas at the tips of the front wing covers. Larvae of Indian meal moths are general feeders on many stored foods and may be recognized by the conspicuous webbing spun over and through infested materials. Adult moths are 1/2 inch long with wings gray at the base and copper-brown at the tip. Foods stored in cupboards may also be attacked by such temporary invaders as ants, roaches and firebrats. Also see "Bean Weevils.

Control of pantry pests requires elimination of existing infestations and prevention of new outbreaks by good housekeeping, frequent inspections applications of insecticides.

Shelves should be regularly and thoroughly cleaned to eliminate spilled food particles on exposed surfaces and from cracks and corners. Use of a vacuum cleaner with suitable accessories may be helpful. Small quantities of materials in open containers should be promptly used or destroyed, particularly if the house is to be vacated for the summer or for extended vacation periods. Usable but infested products may be sterilized by heating in an oven for 30 minutes at 130°F or by exposure for a longer period outdoors in the heat of the noonday sun or by placing the products in the freezer at 0°F for 4 days. New food purchases should be inspected for pests before being stored. Uninfested or heat-sterilized dry foods should be stored in glass or metal containers with tight-fitting insect-proof lids. Coffee cans with tight-fitting plastic lids make good canisters. Clean surfaces of cupboards may be coated with an approved residual insecticide. The pesticide deposit should not be removed by washing although treated shelf surfaces may be covered with clean shelf paper.

For larger, emergency reserves of grain and other dry, but perishable products, fumigation with carbon dioxide is an exceptionally good and safe method of controlling stored grain insects. Heavy plastic or metal cans, 5-gallon capacity, make good storage containers. As the containers are being filled with grain, spread about 2 ounces of crushed dry ice (solid carbon dioxide) on 3 to 4 inches of grain in the bottom of the container, then add the remaining grain until the container is filled. For larger quantities, use 6 ounces of dry ice per 100 pounds of grain.

The vaporizing dry ice produces carbon dioxide fumes that are heavier than air, which displace the existing air in the container. **Allow from 30 minutes to an hour for complete vaporization before placing the lid on tightly. Premature sealing can cause bulging of the container. If plastic bags are used in the cans as liners, do not seal until vaporization is complete.** Carbon dioxide in closed containers destroys most adults and larvae, however, some eggs and pupae may escape. Dry ice can produce serious burns and should be handled with caution.

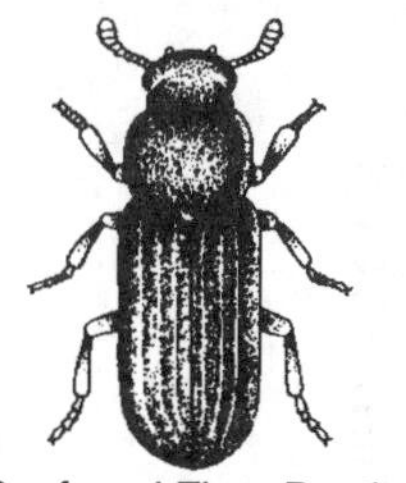

Confused Flour Beetle
(U.S. Public Health Service)

Deep freezing these larger containers of grain for 3 to 4 days will destroy all stages of all stored grain pests.

Powder-Post Beetles

Several small, wood-boring beetles are commonly called "powder-post" beetles because of the powdered frass discharged from larval tunnels. Species in homes are usually of a single family and are more accurately call "lyctid beetles." Injury is caused by larval feeding on the starch of recently-seasoned sap wood. Tunnels the diameter of a pencil lead are made in wooden objects, usually of hardwood, such as floors, furniture, gun stocks and tool handles. Similar larvae have been found in imported bamboo baskets and furniture. "Powder" is produced in the infested articles until larvae are full grown. Adults are reddish brown to black, 1/8" to 1/5" long are attracted to lights. Powder-post beetle larvae of a number of species and families are found outdoors in dead limbs or branches of native woody plants.

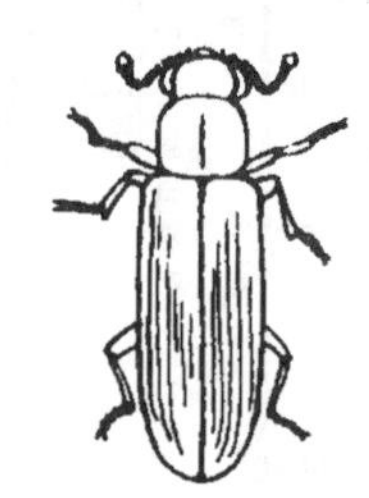

Powder Post Beetle
(U.S. Public Health Service)

Attacks of powder-post beetles may be prevented by protecting surfaces of wooden objects with wax, varnish or paint, to prevent egg laying in the tiny, exposed pores forming the "grain" of the wood. Active infestations may be destroyed by "cooking" the larvae inside small, infested articles in the heat of the noonday summer sun or in moderate heat from other sources. Use of fumigants or residual-type insecticides may be required in some cases. Newly purchased articles suspected of being infested should be carefully inspected outdoors and suitably treated, if necessary, before being brought inside.

Adults and larvae of other kinds of wood-boring insects may be inadvertently carried into homes in fireplace wood. Frass, usually coarser than that from powder-post beetle larvae, may be discharged from tunnels in such wood. Adults may be attracted to light and are capable of cutting through the fabric of closed drapes to reach windows. Fireplace wood should be kept outdoors until the time of actual use.

Roaches — see Cockroaches and Waterbugs

Scorpions

These menacing creatures have stout bodies elongated in front, with a large pair of pincers and 4 pairs of legs, followed by a slender, segmented tail-like abdomen with a stinger at the tip. Scorpions are relatives of spiders, ticks and mites and are most often seen in the warmer parts of the country, particularly the South and Southwest. Scorpions do not usually attack man unless directly or accidently provoked. All may produce painful stings and the stings of two species may be fatal, particularly to small children and older persons.

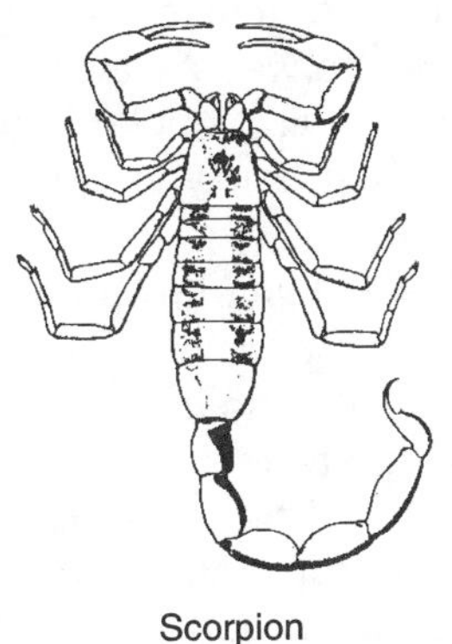

Scorpion
(Univ. of Arizona)

Scorpions are normally found outdoors and search for food at night. During the day they may be found under the bark of trees or in moist areas under boards or debris. They may invade homes in search of moisture and hide during the day in bathrooms, closets, garments, shoes or bedding.

Sanitation is the first step in scorpion control. Loose boards, wood piles, rocks and debris should be eliminated from areas about homes, particularly near foundation walls. Insecticides may also be used in such areas. This will also reduce populations of insects fed upon by scorpions. Scorpions may be trapped under moist burlap and later destroyed. Care should be used in handling boards or other objects under which scorpions may be hiding.

All scorpion stings should be promptly treated by a physician. Immediate action is particularly important when small children are involved. First aid: promptly cool the affected area with ice within a cloth bag and take the victim to a physician. PROMPT ATTENTION BY A PHYSICIAN IS IMPORTANT. The offending scorpion should be shown to the physician for identification, since proper treatment may depend on the species involved.

Silverfish—see Firebrats

Solpugids

Solpugids or sun spiders, are ferocious-looking but harmless relatives of spiders and are strictly confined to the desert Southwest. They are usually yellowish-brown, with bodies an inch or more in length, huge jaws, a pair of long, leg-like palps and 4 pairs of legs. They have no poison glands and prey on insects and other small animals. Most species are active only at night. They are usually seen in patios and other lighted outdoor locations. They are not usually found indoors, except possibly in adobe buildings with dirt floors.

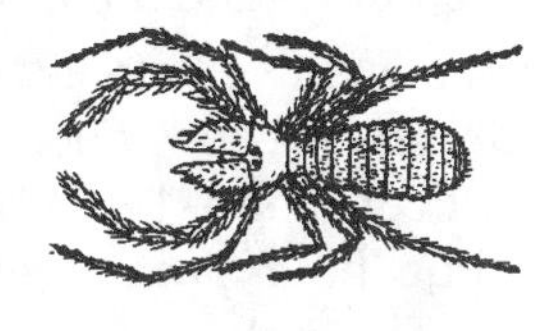

Solpugid or Sun Spider
(U.S. Public Health Service)

Sowbugs and Pillbugs

These are not insects but rather crustacea and are about 1/2" long, usually grayish, with 7 pairs of short legs and a hard, shell-like body covering. They breathe by gills and can live only in moist areas. Sowbugs have 2 short, tail-like rear appendages. Pillbugs can roll themselves into a tight ball for protection. They feed on decaying matter but may also attack living plants. They are commonly found outside homes in damp locations such as under stones, boards, flower pots and plant debris. They hide during the day and are active at night. They may occasionally enter homes but are not pests except by their presence.

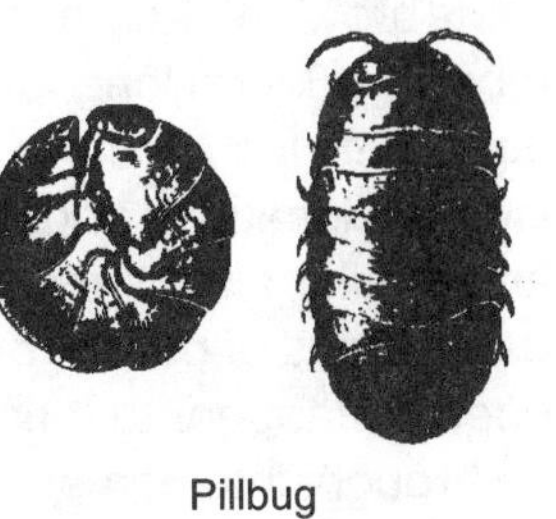

Pillbug
(U.S. Public Health Service)

Control measures are as for other house-invading pests, including tight-fitting doors, foundation cleanup, elimination of damp places of shelter, caulking cracks and seams and insecticide barriers. (See Sowbugs under Vegetable Garden Insect Pests)

Spiders

Spiders are only distantly related to insects. Unlike insects, they have 4 pairs of legs but lack wings and antennae. A huge number of species of varying size and color are found throughout the U.S. and are usually regarded as pests merely because of their presence. Most spider species are really beneficial, since they feed on flies and other insects. The black widow is the only seriously poisonous spider commonly found in homes. The brown recluse spider, found in Eastern states and the Arizona brown spider are two closely related species of very similar appearance and are both very poisonous.

The adult female black widow has a rounded, glossy-black body about 1/2 inch long and an overall length of up to 1-1/4 inches. It is identified by a red to orange hour-glass-shaped marking on the under side. The male is much smaller, with white and pale brown markings and lacks venom. Insects and other spiders are aggressively attacked by the female black widow but humans are bitten only when she is accidentally touched or threatened. Black widows construct loose, irregular webs in dark protected corners of homes, carports, outdoor storage areas under outside chairs, benches or privies. Bites are made by paired jaws with internal ducts from which a venom is discharged. A sharp pain, like that of a needle puncture, is first produced, followed by a variety of other discomforts, depending on location, as the venom penetrates the nervous system. Prompt treatment by a physician is recommended.

The brown recluse and Arizona brown spiders have a body 1/3 inch long and an overall span, including legs, of an inch or more. Their normal habitats are in isolated locations under pieces of wood, dead cacti and similar situations but they may also move into dark places of nearby buildings. The Arizona spider does not appear to thrive in irrigated areas but may be brought into homes on wood or cactus skeletons from the desert. The Arizona brown spider and the related local species more frequently reported from California, where it is known as the "brown spider," are capable of inflicting painful bites on humans. The effects of these bites are reported to be less severe than those of the brown recluse spider found in more easterly states. Persons bitten by spiders should consult a physician for relief. Treatment will be aided if the offending spider is preserved for positive identification.

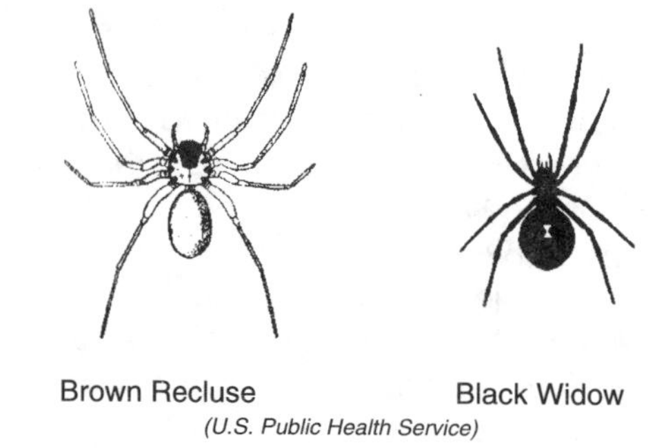

Brown Recluse Black Widow
(U.S. Public Health Service)

Among the non-injurious spiders commonly found in or near homes and objectionable only by their presence, are patch spiders, giant crab spiders and daddy-long-legs. Patch spiders spin silken "patches" about 1/2 inch in diameter, particularly in slight depressions of plastered walls and in mortared joints of brick walls. Giant crab spider measure as much as 2 inches across the legs and prey on insects but make no webs. Daddy-long-legs are recognized by their small, globular bodies 1/8 to 1/4 inch long and their extremely long legs, extending for an inch or more.

Tarantulas are probably the most bizarre of the spiders, principally because of their size and a few monster movies which have starred our furry friends. The tarantula is quite harmless, does not produce venom and is not easily aroused. The males are usually seen in the late summer and fall, on the march, in search of females, which are somewhat more secluded and seldom seen. Tarantulas should never require control. Rather they should be permitted to move on in their nuptial pursuits.

Tarantula
(USDA)

Black widows and other spiders may be controlled by good housekeeping indoors and outdoors, including tight-fitting doors and screens, elimination of trash and debris, removal of webs by sweeping including the crushing of the spiders and egg sacs and by application of insecticides to infested areas. Night inspections with a flashlight of suspected outdoor hiding areas will reveal most black widows that hide during the day.

Springtails

Springtails (Collembola) are small, wingless insects with a jumping mechanism at the rear underside of the body. Species usually seen about homes are bluish gray and scarcely 1/16 inch long. They live in moist areas rich in organic matter such as the soil in flower pots or in newly-seeded lawn areas well covered with composted manure. They are usually but a temporary nuisance and are not found in hot, dry environments.

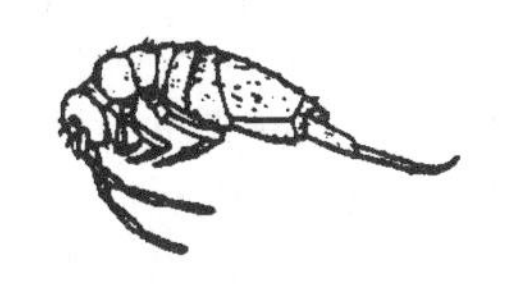

Springtails
(Ohio State Univ.)

Stinging Caterpillars

Hairy caterpillars, capable of causing skin irritations when touched or handled, may be found in patios, especially in late summer and early fall. These leaf feeders have hollow, brittle hairs or spines containing a toxic venom which, when injected into the skin, may produce reddened welts and blisters, accompanied by itching and burning sensations. Although nettling or stinging hairs are found in caterpillars of many species, those most often reported are the puss caterpillar and the larvae of the western grape leaf skeletonizer.

The puss caterpillar is about 1 inch long, covered with a dense mat of soft, buff to gray hairs. It has a prominent hairy ridge or crest extending along the top of the body and ending in a slender tail. It is most often reported from communities where it feeds on oaks, mulberries, fruit trees and shrubs. Curious school children are frequently stung by these caterpillars. Severe attacks have been accompanied by pain, swelling and nausea. Discomfort may last several days.

SOME OF THE MORE COMMON

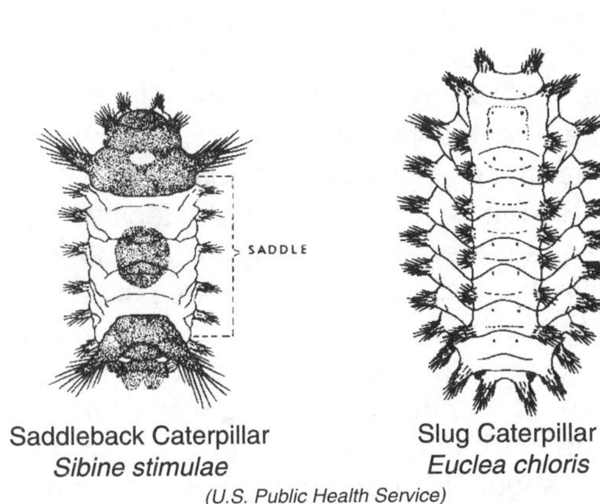

Saddleback Caterpillar *Sibine stimulae* — Slug Caterpillar *Euclea chloris*
(U.S. Public Health Service)

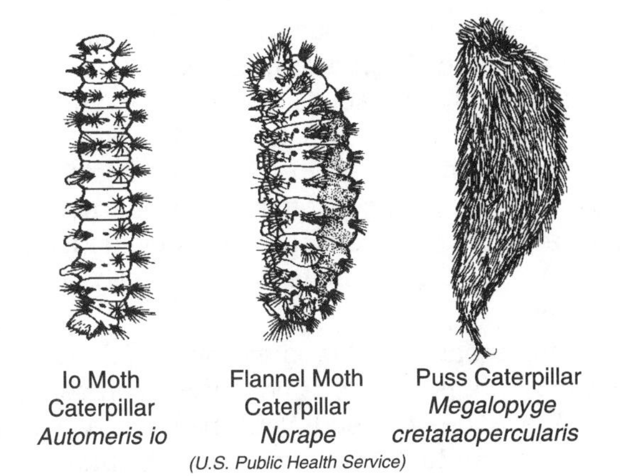

Io Moth Caterpillar *Automeris io* — Flannel Moth Caterpillar *Norape* — Puss Caterpillar *Megalopyge cretataopercularis*
(U.S. Public Health Service)

STINGING CATERPILLARS

Avoidance is the best procedure for dealing with stinging caterpillars. This requires recognition and education, especially with children. Affected areas should be promptly washed with soap and water. Cold packs may give partial relief. Severe cases, especially those involving children, should be seen promptly by a physician. Further relief may be given by use of one of several bland antiseptic ointments containing a mild local anesthetic in a petrolatum base.

Termites

Termites are social insects that live in colonies and build their nests in wood or in the ground. They attack

HOW TO DISTINGUISH ANTS FROM TERMITES

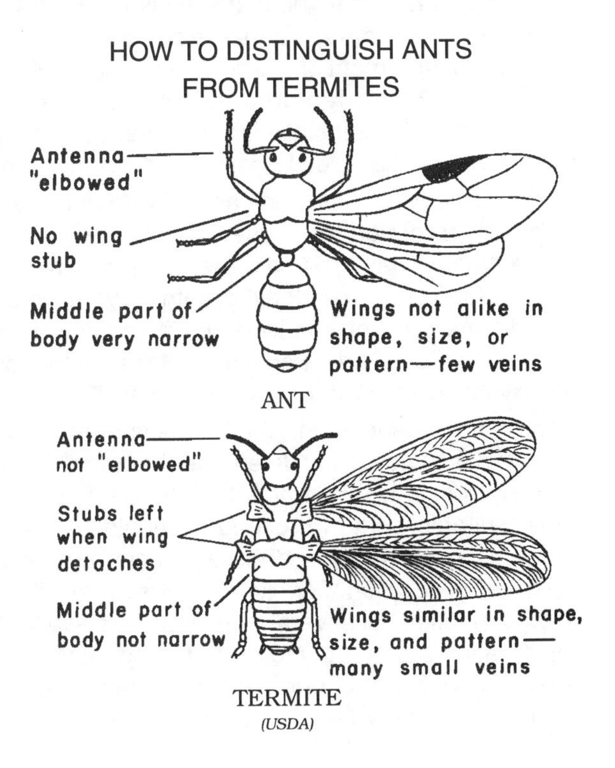

(USDA)

wood and paper because cellulose is their main source of food. Each colony is headed by a pair of functionally mature adults, called a king and queen and thousands of pale, soft-bodied workers, plus a few soldiers that protect the colony against ants and other enemies. Winged termites are most active and visible during rainy seasons when new generations of adults emerge to mate and seek sites for new colonies.

Subterranean termites form colonies deep in the ground, where needed moisture is present and attack all types of dead plant material, paper and wood in contact with soil. Most subterranean termites can also build mud tunnels, tubes or shelters over brick or concrete to reach wood several feet above the ground. They frequently bore into baseboards, moldings, door frames and porch and carport structures.

Dry-wood termites make clean tunnels or galleries in dry, sound wood and do not enter the soil for moisture. They usually cause less damage than subterranean termites but may attack floors, sills, door and window frames, rafters and rarely, furniture. From their tunnels they sometimes eliminate piles of dry fecal pellets resembling fine sawdust or tiny seeds, the main distinguishing clue.

The ideal time to protect a house from termites is while it is being built. Steps at this time should include provision for adequate drainage away from the site, removal of stumps, roots and wood scraps, insecticidal treatment of soil before footings or slabs are poured, professionally referred to as "pretreatment" and installation of screened vents in attics and crawl spaces to permit good ventilation.

The homeowner can increase his subterranean termite protection by observing a few simple precautions: Keep the under area of the home dry; don't let shrubbery block the breathing vents in the foundation; don't let sprinklers wet the stucco; avoid increasing the outside grade against the foundation; and avoid building window boxes against the house.

When it is suspected that a house is infested the first step is to positively determine that termites are actually present. When an infestation is found the most suitable control procedure should be used. Termites work slowly and there is time for a calm, considered decision. Termite control is more difficult in older houses but several techniques are available for treating such structures. The services of a licensed pest control operator are usually needed. Reputable firms will guarantee their work.

Ticks

The most common ticks encountered are the brown dog tick, *Rhipicephalus sanguineus* and the American dog tick, *Dermacentor variabilis*. The American dog tick is a carrier of Rocky Mountain spotted fever, which has not received the media attention of Lyme's disease (mentioned later), but is potentially serious.

The brown dog tick is the "domesticated" tick most commonly found in homes. It feeds on the blood of dogs, causing both irritation and loss of vigor. In homes it is objectionable by its mere presence, although neither man nor other animals are normally attacked. Consequently, the brown dog tick is not considered a disease vector.

Brown Dog Tick
(USDA)

After each feeding period the brown dog tick leaves the dog to molt, seeking shelter in places such as crevices in walls, baseboards and moldings, under the edges of rugs and in bedding used by the dog. It tends to move upward toward the ceiling for shelter.

Newly emerged adults are reddish-brown, flattened and about 1/8 inch long. As the female feeds she becomes about four times larger and bluish-brown in the swollen portion. After hatching from eggs commonly laid in indoor cracks and crevices the ticks pass through two immature feeding periods and one more adult feeding period, each separated by a resting and molting interval off the dog. Length of the life cycle varies with temperature, from a minimum of about two months in summer to a much longer period in the cooler months. Dogs do not receive this tick directly from other dogs, but from coming near objects or surfaces from which newly hatched or recently molted ticks may "climb aboard". The brown dog tick can live as long as 1-1/2 years without feeding. Dogs may thus become infested by entering homes or premises long after the departure of previous dog occupants. A tick-free family dog may be protected by confining it within an area not accessible to other dogs. Effective insecticidal control requires simultaneous treatment of the infested dog and the infested premises. For the latter, a residual insecticide is required to control the young ticks which may continue to hatch after the treatment. Infested animals should be treated according to one of the methods described in Table 22, Pest Control Suggestions for Pets, which should include a flea or tick-collar for the pet(s). Ticks are virtually impossible to control without the aid of insecticides.

Lyme disease is transmitted by the black-legged tick (formerly deer tick), *Ixodes scapularis.* The disease was first identified in Lyme, Connecticut, in 1975, with the heaviest concentrations now recognized in the Northeast, upper Midwest and along the northern California coast. Lyme disease is also found abroad in Germany and Scandinavia, as well as China and Asia to Australia. Migratory birds may partially account for the rapid spread of this disease since larval and nymphal black-legged ticks often feed on ground-feeding birds. The active months for human infection are May to August, with peak months being June and July. The most noticeable, early symptom is often but not always a gradually expanding, circular or oval-shaped red rash, resembling a "bull's eye" at the bite site. The disease not only affects humans, but a wide range of animals including livestock and pets. This tick is very small, about 1/8 inch long, dark-colored with the young (larvae) being the size of a period at the end of this sentence.

One possible explanation for the rise of Lyme disease is that populations of deer, raccoon, opossum, birds and other wildlife have rapidly expanded in the last few years and human proximity to these animals has increased as suburban development has encroached on woodlands. There are no controls for this tick in the wild.

Wasps, Hornets, Yellowjackets and Mud Daubers

Wasps include numerous species of stinging insects belonging to the same order as bees and ants. They are variable in size and color but all have constricted "wasp waists." Among the kinds found in the vicinity of homes and porches are those commonly known as paper wasps, hornets or yellow jackets and mud daubers. They may be considered beneficial in that most species are predators and capture other insects or spiders to feed their young. Adult wasps are capable of inflicting stings, temporarily painful, although some persons may develop severe allergic reactions which can be fatal. This is even more common with honeybees since more people are exposed to and stung by bees. Wasps may be attracted to garden pools or to picnic and barbecue areas where they can be a nuisance.

Wasp
(Univ. of Arizona)

Paper wasps live in colonies, which may be newly formed each year by individual overwintering fertilized females. Nests are umbrella-shaped and consist of an unprotected single circular layer of cells open at the bottom and suspended by a stalk from the eaves of houses, trees, shrubbery or other protected areas. The nests are made of "paper" of fibers from dead plants or weathered wood, moistened and formed by the mouth parts of wasps.

Hornets or yellow jackets are mostly yellow and black and live in colonies within nests built above or below ground. Nests of hornets, unlike those of paper wasps, consist of several tiers or circular layers of cells. Nests above ground are enclosed within a grayish paper covering with an entrance on the underside, as seen so often in the old comic strip, The Katzenjammer Kids.

Mud daubers construct individual cells of mud, usually in small groups side by side on walls of porches or carports. They stock these cells with paralyzed spiders which serve as food for the young.

Wasp control involves destruction of nests including the living contents of individual cells. Adult wasps may be killed by application of a fast-release, aerosol insecticide to the nest, preferably in the coolest portion of the night or at daybreak, to avoid stings. Foam-squirting aerosols are not satisfactory. Elimination of food scraps, including ripened or partially eaten fruit, from patio and barbecue areas will aid in reducing wasp numbers. A commercial, slow-acting bait is now available which foraging yellow jackets carry back to the nests where the developing young are poisoned.

Yellowjackets, hornets and sometimes wasps can become a serious annoyance around restaurants, picnic areas, swimming pools backyard or patio parties. They can be trapped with non-toxic baits, using several commercially available traps. Two very good designs are the Rescue!® trap and the Yellow Jacket Inn® picnic trap. The most effective bait for both traps is *grenadine syrup*, using roughly an ounce in a small plastic cup, the type used for catsup and other fast-food condiments. Less effective baits are raw hamburger, tuna-flavored cat food, cola, ginger ale and diluted honey. The grenadine attractant should be recharged before it is completely emptied by these wasps.

The pain from wasp stings may often be eased by applications of ice packs to the affected areas. Persons who develop allergic reactions to stings should be rushed to a physician or hospital emergency room for treatment and thereafter wear an emergency treatment information bracelet.

Waterbugs — See Cockroaches.

TABLE 20. Insect Control Suggestions for the Household With and Without Chemicals.[1]

Pest	Pesticide to Use	Non-Chemical Control
Ants	Cyfluthrin, Deltamethrin, Permethrin, Propoxur, Pyrethrins, Resmethrin. Spray or paint along ant trails, door frames, cracks, baseboards and window sills. Dust Boric Acid in spray-inaccessible areas. Set bait stations containing Abamectin, Boric acid, Fenoxycarb, Fipronil, Hydramethylnon, Methoprene, Pyriproxifen, Spinosad or Sulfuramid on ant trails and at points of entry. Read label for ant-specific products.	Pour boiling water on nest. Eliminate food sources attracting ants. Spray indoor trails with D-Limonene. Thoroughly caulk openings and cracks on outside and inside where ants enter.
Asiatic oak weevil	Cyfluthrin or Permethrin. Treat outside around foundation, basement windows and other routes of entry.	Vacuum, sweep or collect weevils by hand and destroy.
Assassin bugs	Propoxur or Deltamethrin. Spot application on bugs and on points of entry.	Vacuum or sweep with broom and destroy. Do not handle.
Bark beetles	No chemical control.	Keep reserve firewood outside. Maintain tight door and window screens.
Bedbugs	Cyfluthrin, Deltamethrin or Permethrin. Apply to springs, slats and frames and light treatment to mattress, tufts and seams. Spray or paint baseboards, cracks and crevices.	Sun mattress, springs, bedding, slats and frame. Thorough cleaning will help but not eliminate. Dust silica aerogel or diatomaceous earth into cracks and crevices of floors and baseboards.
Bees (honey)	Carbaryl, Pyrethrins, Permethrin, Propoxur or Resmethrin. Direct dusts or sprays into openings where bees have become established. Treat after dark when bees are calm and temperatures have dropped.	Spray liquid dishwashing detergent solution (2oz/gal water) into openings at night. Check with County Extension Agent for name of beekeeper who will remove colony for a fee.
Booklice or Barklice (psocids)	Propoxur, Deltamethrin or Pyrethrins. Treat damp areas where booklice have been seen.	Remove cardboard boxes, books and papers from damp storage areas. Use a fan to ventilate and dry infested areas.
Boxelder bugs	Propoxur or Deltamethrin. Spray directly on bugs. Best to spray bug clusters outdoors on tree trunks and other spots.	Vacuum or sweep with broom and destroy. Caulk cracks and other points of entry.

[1] Refer to Appendix B for trade or proprietary names of recommended insecticides.

Pest	Pesticide to Use	Non-Chemical Control
Carpenter ants	Abamectin or Boric Acid baits. Deltamethrin, Propoxur or Permethrin. Dust into wall voids through drilled 3/8-inch holes. Treat 3-5 feet on either side of where ants are entering. Colony must be reached by insecticide. Outside, treat all breaks and cracks 2 feet up and 3 feet out where ants can enter, with Propoxur or Acephate spray.	Trim bushes/trees so branches don't contact house. Correct moisture problems from any leaks or seepage. Replace rotted or water-damaged wood. Drill and dust silica aerogel as with chemical control. Treat nest also.
Carpenter bees	Carbaryl, Cyfluthrin, Propoxur or Deltamethrin. Direct spray or dust at and into wood galleries. Apply also to adjacent wooden areas and potential sites. Treat when dark and cool as for honey bees.	Dust diatomaceous earth generously into galleries as repellent. Fill holes and paint.
Carpet beetles	Bifenthrin, Cyfluthrin, Deltamethrin or Permethrin sprays. Spray carpet margins at baseboards. To avoid fabric damage, test a small area first. Where heavy infestation occurs, you may want to obtain services from a pest management operator.	Store only cleaned or well sunned clothing. Dry cleaning kills carpet beetles (and moths). Avoid accumulations of dust and lint in corners, along moldings, in hot and cold air ducts. Vacuum such areas thoroughly and frequently. Remove and destroy disposable vacuum bags.
Centipedes	Propoxur, Pyrethrins, Cyfluthrin or Permethrin. Apply directly to pests. Treat dark, moist areas in basement and garage.	Centipedes breed outside and may become pests indoors. Remove decaying grass and leaves from around house foundation. Foliage-free area around foundation makes ideal barrier. Ventilate basement & crawl-space.
Click beetles	No chemical control.	Maintain tight door and window screens. Beetles are attracted to white lights. Substitute yellow, non-attracting light bulbs.
Clothes moths	Pyrethrins, Permethrin or Resmethrin sprays, Napthalene or PDB crystals. Follow same general procedures described under carpet beetles. Store only cleaned clothing. Clothing may be protected by using a Resmethrin aerosol. A No-Pest® Strip hung in the closet also offers good protection.	Follow same procedures as given under carpet beetles. Cedar-lined closets and chests are effective in repelling moths. Remove clothing and expose to hot sunlight for 2 days.

Pest	Pesticide to Use	Non-Chemical Control
Clover mites	Dicofol, Cyfluthrin, Permethrin, Propoxur or Pyrethrins. Apply Permethrin or Pyrethrins onto mites in cracks and other areas where they hide. For outdoors,* spray house siding up to window sills and treat lawn out from house foundations 8-10 feet.	Establish a bare soil barrier about 2 ft. wide around foundation to discourage migration into dwelling.
Cluster flies or Attic flies	Permethrin, Pyrethrins or Resmethrin. Release Pyrethrins aerosol in tight enclosure, or hang No-Pest® Strip in similar infested areas. Surfaces treated with Permethrin have fairly good residual activity.	Sticky fly ribbons and the old fly swatter are about your only non-insecticidal defense. Caulk around windows and flashing on south side of buildings where flies enter.
Cockroaches	Baits are an easy method of control: Abamectin gel, Borate-baits, Hydramethylnon, Fipronil, Sulfuramid bait stations. Apply Boric Acid dust, Cyfluthrin, Deltamethrin, Permethrin, Propoxur, Resmethrin or Tetramethrin. Apply to baseboads, beneath sinks, cabinets, under and behind stove and refrigerator, openings where pipes enter walls and floors, around drains, dark-warm protected areas in basements, showers and drains and roach runways under sinks and lavatories and floor cracks.	Sanitation, removal of food sources, is the key approach. Roach traps may be prepared by setting quart or pint jars containing 2" of beer or diluted (10:1) molasses and water in which roaches drown. Sticky traps are moderately effective. Caulk entrances around pipes and wires in kitchens, laundry and other areas.
Crickets	Propoxur, Bifenthrin, Cyfluthrin, Deltamethrin or Carbaryl (bait). Treat baseboards, in closets, under stairways, around fireplaces, in basements and ground level floors, if needed and where crickets enter.	Use non-attractive yellow light bulbs outside and roach sticky-traps inside. Caulk cracks and other points of entry.
Drain flies or Moth flies	Cypermethrin, Tetramethrin or Pyrethrin aerosols. Against adult flies. Clean drains thoroughly and apply liquid or crystalline drain cleaner to kill immature stages. Apply aerosols in sink overflow openings.	Sanitation, removal of breeding site, is key approach. Clean gelatinous film from drains. Pour boiling water down drain and through overflow. Clean garbage containers regularly. Don't let wet lint accumulate under washing machine.
Earwigs	Propoxur, Resmethrin or Tetramethrin. Apply spot treatment to cracks and crevices and directly onto earwigs. Treat around and for a distance of 10 feet out from the foundation outdoors, with Propoxur or Acephate.	Remove with broom and dustpan or vacuum. Eliminate dead vegetation, leaves and excess mulch from around outside of foundation. Trap outdoors under boards or folded newspaper or dark plastic and destroy.

Pest	Pesticide to Use	Non-Chemical Control
Elm leaf beetle	Pyrethrins, Silica-gel or Boric Acid. Crack and crevice treatment.	Use vacuum, broom and dust-pan or remove by hand if they enter the home.
Firebrats	Boric Acid, Pyrethrins, Permethrin, Deltamethrin or Propoxur. Spray baseboards, window and door casings, cracks and openings in floors and walls for pipe and wire entrances.	Keep papers, cardboard boxes and books in dry storage. Reduce humidity of problem area.
Fleas	D-Limonene, Deltamethrin, Permethrin, Propoxur, Pyrethrins or Tetramethrin. Dust or spray animal's sleeping quarters and replace old bedding. Apply spray to floors, baseboards and to walls to a 1-ft height in rooms where fleas are a problem. Treat dog or cat with Carbaryl or Rotenone dust or D-Limonene aerosol outdoor resting areas. See Fleas under "Lawn and Turf", Table 17. Use flea collar on pet.	Keep pets outside at all times. Vacuum areas daily where fleas are a problem. These methods are rather ineffective. The pets are the source and must be treated.
Flies and Gnats	Propoxur, Permethrin, Pyrethrins or Deltamethrin. Use Pyrethrin aerosol when flies are present. No-Pest® Strips are an ideal choice for pesky flies. inside. Outside, treat surfaces where flies rest, walls, around doors and windows. Use cone-type or jar-type baited traps outside.	Tight screens and good sanitation are the keys. Use ribbon flypapers. Empty garbage twice weekly, rinse and invert cans. Clean up pet dung daily from yard and seal in plastic bag.
Flour, grain or Cheese mite	Pyrethrins, Resmethrin or Permethrin. Thoroughly clean, then treat infested shelves. Cover shelves with new shelf paper after spray dries.	If you are squeamish, discard infested foods and keep uncontaminated food in canisters or coffee cans with plastic lids. Otherwise place dry materials in open pan in oven for 30 minutes at 130°F, or freeze for 3-4 days and utilize in normal cooking procedures.
Ground beetles	No chemical control. Harmless incidental invaders. Outside, sprinkle Carbaryl or Deltamethrin granules around foundation, wood piles, etc.	Remove by hand, broom and dust pan or vacuum cleaner. Caulk points of entry.
Hackberry psyllids	Pyrethrins. Adults can be killed with Pyrethrins aerosol. No-Pest® Strips work equally well. Treat hackberry tree leaves with Acephate in spring.	Maintain tight door and window screens. Hack-berry trees are source. Their removal should eliminate pests.
Hornets (See also wasps)	Pyrethrins, Resmethrin, Permethrin, Deltamethrin or Propoxur. Apply spray directly to nests during coolest part of night. Repeat next night if survivors appear. Fast-release hornet aerosols are best. See Wasp and Hornet trap sources in Appendix G.	Wait until winter to remove hornet's nest. There is no safe method of removal without the help of insecticides.

Pest	Pesticide to Use	Non-Chemical Control
Horntail (pigeon tremex)	No chemical control.	Sweep up with broom and dust pan or vacuum, then destroy. They emerge from firewood, hardwood floors, studs and subflooring and do not reinfest wood in the home.
Kissing bugs (see Assassin bugs)	Propoxur or Deltamethrin. Spray outside foundation perimeter and 3 ft. of adjacent soil. Brush or spray around window and door casings.	Maintain tight screens and weather stripping. Use yellow bug lights outside. Caulk cracks in walls, foundation and around windows and doors. Remove pack rat nests where they live.
Larder beetle	Pyrethrins or Permethrin. Treat shelving with oil spray or aerosol after removing dishes, utensils and food.	Discard infested products. Wrap and store cheese and meat products at normal refrigerator temperature.
Lice (head, body and crab)	These products are available over-the-counter: Pyrethrins (Rid® Advanced Removal System, (Rid®Mousse, Pronto® Lice Shampoo), Permethrin (A-200®) Lice Control Spray, PK 7® Complete Lice Killing system, NIX® and Resmethrin Paratox® Lice Killing Kit). Most lousicides are sold on a prescription basis. In addition, head lice and nits (eggs) should be removed with the aid of a fine-toothed lice comb. Wash bedding and clothing thoroughly. Treat furniture with permethrin or pyrethrin aerosol.	For body lice only: clean underwear and outer clothing is important; laundering kills lice in 5 minutes, eggs in 10 at 125° F. water temperature. For head lice: Do not use someone else's hats, wigs, scarfs, combs or brushes.
Long-horned beetles	No chemical control.	Remove occasional beetles by hand, broom or vacuum. They usually emerge from firewood but will not infest wood in the home. Store firewood outdoors.
Millipedes	Pyrethrins, Carbaryl, Permethrin, Propoxur or Acephate. Treat areas outdoors where millipedes may be present or enter home.	Millipedes are not poisonous. Remove by hand, broom or vacuum. Around outside of foundation remove excessive mulch.

Pest	Pesticide to Use	Non-Chemical Control
Mosquitoes	(Indoors): Pyrethrins, Allethrin or Resmethrin aeresol or No-Pest® Strips. Spray according to label instructions. Strips can be hung in areas designated on label. (Outdoors): Spray yards and picnic areas, tall grass and shrubbery with Carbaryl, Pyrethrins, Acephate or Deltamethrin. Spray entry sites (screens, doors and patios). Personal repellents: Deet, Avon's "Skin So Soft®" or Permethrin (Permanone). Larvae (wigglers): *Bacillus thuringiensis* spp. *israelensis* or methoprene. Follow label directions.	Maintain tight screens and weather stripping. Use yellow non-attractive light bulbs at entrances. Remove or empty frequently any containers from premises that may hold rainwater (tires, cans). Clean out clogged roof gutters holding stagnant water. Add light-weight oil to surfaces of ponds, ditches and even animal hoof holes in mud where mosquitoes may breed. Community effort may be required. Out-of-doors, place 1/4 cup water in a white dinner plate and add 2-3 drops of Lemon Fresh Joy® dishwashing detergent. Place plate on proch or patio. Mosquitoes flock to it, then drop dead in plate or within 10 feet. Great discovery!
Pantry pests (Includes Angoumois grain moth, bean weevil, cigarette beetle, confused flour beetle, drug store beetle, foreign grain beetle, granary weevil, Indian meal moth, mealworm, Mediterranean flour moth, rice weevil, sawtoothed grain beetle and spider beetle).	Permethrin, Cyfluthrin, Propoxur, Boric Acid, Resmethrin or Pyrethrins. Empty and clean kitchen shelving, then spray shelves lightly. Direct spray into cracks and crevices. Cover shelves with new shelf paper after spray dries.	If you are squeamish, discard infested products. If not, salvage them by heating in the oven for 30 minutes at 130°F or placing them in the freeze-compartment at 0°F for 3-4 days. Store such foods in canisters with tight lids. See Pantry and Stored Products Pests narrative for carbon dioxide fumigation of larger quantites of stored grain.
Powder post beetles	Propoxur. Spray, paint or dip with dilute solution to saturate the wood where appropriate. Use Disodium Octaborate Tetrahydrate to treat infested wood as directed on the label.	Previously painted or otherwise finished wood surfaces usually remain free from infestation.

Pest	Pesticide to Use	Non-Chemical Control
Scorpions	Permethrin, Cyfluthrin, Cypermethrin, Deltamethrin, Propoxur or Malathion. Apply around the foundation and up to 1 foot above ground level on exterior walls. Use highest permissible rate. Direct application on scorpions indoors works well, of course. Scorpions are difficult to control with insecticides alone. Thus, the first control strategy is to modify the area surrounding a house.	Remove all trash, boards stones and other objects from around the home. Keep grass closely mowed near the home. Prune overhanging tree branches away from the roof. Install weather-stripping around loose fitting doors and windows. Plug weep holes in brick veneer and caulk around eaves, pipes and any other cracks into the home.
Silverfish	Propoxur, Pyrethrins, Bifenthrin, Cyfluthrin or Boric Acid. Spray or dust baseboards, door and window casings, closets, cracks and openings for pipes and wires to pass through walls and floors.	Store books and paper products in dry areas to avoid favorable habitat. Caulk openings mentioned in pesticide column.
Sowbugs and Pillbugs	Propoxur, Trichlorfon, Resmethrin or Acephate. Apply around shrubbery, plants, foundation wall, window wells and other routes of entry. Spray along doorways and basement windows.	Remove mulch, boards and excess ground cover. Caulk cracks in foundation Trap outdoors with boards, folded newspapers or dark plastic and destroy.
Spiders	Propoxur, Resmethrin, Bifenthrin, Cyfluthrin or Deltamethrin. Treat indoors with coarse droplet spray, hitting webs and probable hiding places. Outdoors, spray or dust around house foundation to reduce migration inside.	Most spiders can be kept out of the home by tight screens, weather stripping and caulking. Most are harmless and are generally beneficial. Exceptions are the black widow, brown recluse and Arizona brown spider.
Springtails (Collembola)	Resmethrin, Cyfluthrin, Pyrethrins or Permethrin. Control outdoors only by treating around foundation, especially moist areas.	Remove mulch and eliminate low moist areas around the foundation. Remove pests with vacuum. Don't overwater potted plants.
Stonefly	No chemical control.	Remove individuals by hand, broom or vacuum. They accumulate in windows in early spring, not far from the water where they breed.

Pest	Pesticide to Use	Non-Chemical Control
Termites (Subterranean)	Termite control is no longer a do-it-yourself job. Professional control is necessary. Termite control services offer several choices in baiting systems to monitor and control termites. This technology requires that the applicator know termite biology and behavior to apply the baits correctly. The hardware or garden store has several do-it-yourself kits or chemicals for termite control, but this author strongly recommends that it only be done by professionals. As for the insecticides used by professionals, those that will last the longest, up to 7 years, are bifenthrin (Talstar®), cypermethrin (Demon®, Prevail®) and fenvalerate (Tribute®) (Merchant and Gold, 2001).	Termites have not been successfully controlled with any methods other than chemical control. Metal plates installed during construction are not effective. Pre-treating the soil before slab construction is poured with one of the approved insecticides by a professional is essential. This is the ultimate in pest control prevention.
Termites (Drywood)	This calls for a professional fumigation job. There is nothing the homeowner can do other than call in a reliable pest control operator.	As with subterranean termites, there is no control without chemicals.
Ticks	Carbaryl, Permethrin, Limonene, Resmethrin, Rotenone or Propoxur. Use only Carbaryl or Rotenone dust on dogs or cats. In the home use Propoxur spray. Treat pet's sleeping areas. Replace old bedding with clean, untreated bedding. Outdoors use Carbaryl 5% G in pest areas. Spray animals with D-limonene. Personal repellents are Deet, Permethrin (Permanone®) and Avon's "Skin So Soft®".	Examine animals' heads around ears and neck daily for engorged ticks. Apply fingernail polish remover or petroleum jelly to tick an hour before removing with tweezers.
Wasps (See also Hornets)	Pyrethrins, Resmethrin, Permethrin, Deltamethrin, Propoxur or Acephate. Apply directly to nests during coolest part of night. Repeat next night if survivors appear. Remove nests when no survivors remain. Or treat with new fast-release aerosol that gives immediate knockdown. (See Wasp and Hornet Trap Sources in Appendix G.)	Wait until winter to remove nests. Otherwise, there is no safe method of removal without the help of insecticides.
Weevils	No chemical control indoors. Outdoors, Acephate, Carbaryl or Deltamethrin. Spray foundation, porches, doorways and grass 10-foot band around foundation.	Occasionally various weevils find their way into homes. These can be readily removed by hand or vacuum cleaner.

HOUSE PLANT PESTS

House plants may become infested with a number of insects and related pests, mostly feeders on plant juices. Aphids feed on stems, leaves and buds. Scale insects do not move about except when young and are recognized by their shell-like coverings, usually measuring from 1/16 to 1/8 inch in length depending on species. Whiteflies are 1/6 inch long and wedge shaped, with powdery white wings. When disturbed they resemble clouds of small snowflakes. Mealybugs are soft-bodied, waxy-white, wingless insects that may secrete honeydew which, as with aphids, may support growths of sooty mold. Spider mites are barely visible to the eye and are most commonly seen on the undersides of leaves, where they may form silk-like webbing. Fungus gnats and springtails may develop in the rich soil used for house plants. Sowbugs and pillbugs may be found in moist areas beneath potted plants. Other outdoor plant pests may also be found on house plants.

Controls for Pests of House Plants

Several methods can be used to control insects or related pests of house plants (Table 21), some utilizing chemicals, some consisting of non-chemical alternatives. What the homeowner should use varies with the pest, the number of plants involved, the size of the infestation and the personal inclination of the indoor gardener.

The simplest and easiest method, of course is to prevent the infestation from spreading from new plants to others. Examine cut flowers and new plants brought into the home to be certain they are free of all pests. New plants should be isolated for at least two weeks before placing them with other plants. During this interval you can observe the new plants and spot any infestations that develop.

Sterilized soil should be used for potting to prevent the development of infestations of soil pests such as springtails, psocids and pillbugs or sowbugs.

Home gardeners may sterilize soil in a conventional or microwave oven using the following directions. For conventional oven, place slightly moistened soil in a shallow pan and cover with aluminum foil. Seal the edges. Open a small hole and insert a meat, candy or other type thermometer in the soil. Place in a low heat—no more than 180°F. After the soil reaches this temperature, leave it for an additional 30 minutes, then remove it and let cool.

Microwave oven soil sterilization can be used for small amounts of seeding mix. Microwave approximately 2 pounds (2 pints) of dry shredded soil in an open plastic bag for 2- 1/2 minutes. This time is for a 625-watt oven; you may need to adjust the time for your oven.

Washing with soapy water and a soft brush or cloth may be all that is needed to remove aphids, mealybugs and scale insects from broad-leaved plants. The wash should be made using 2 teaspoons of a gentle, biodegradable detergent to a gallon of water. Insecticidal soaps are available to control aphids, whiteflies and mites.

When concerned with only one or two plants, you can control aphids and mealybugs by removing them with tweezers or a toothpick. Caterpillars may be removed by hand and destroyed. Cutworms, snails and slugs may be found in their hiding places during the daytime and destroyed or picked from the plants at night as they feed.

Another easy way to control a light infestation of aphids or mealybugs on one or two plants is to wet and remove the insects with a swab that has been dipped in alcohol. Use a swab made from a toothpick and a tuft of cotton or the commercial tufted swabs for that gentle touch.

Several aerosol insecticides have been registered by the EPA for use on plants within the home. They often consist of several active ingredients in combination and are usually specially formulated to kill most plant pests. Pyrethrins and resmethrin are especially effective.

When purchasing an aerosol for plant pest control, read the label on the can carefully to make certain that the spray is safe for plants. Some aerosols are designed as space sprays for flying and crawling insects in the home and contain oils or other materials that will burn or kill foliage.

Aerosol sprays for plants contain small quantities of pyrethrins and other agents. They may be used to kill pests that can be hit directly with the spray, such as aphids and whitefly adults on plants or whiteflies and fungus gnats swarming near the plants. Do not spray too close to the plants, for even the best designed aerosols will burn foliage hit with heavy spray concentrations.

THE PESTS

There are several pests of houseplants that occur and are seen in gardens, but usually do not capture the attention as they do indoors. Similarly they frequently appear as pests in the home greenhouse because of favorable environmental conditions such as temperature and humidity and because of a lack of natural control agents that occur out-of-doors. The pests described here do not make up the complete list. Those not described can be found in the garden pest section.

Cyclamen Mites

Cyclamen mites, even when adults, are too small to be seen with the naked eye. They can be viewed under a magnifying glass as oval, amber or tan-colored, glistening, semitransparent mites. The young are even smaller and milky white, while the eggs are oval and pearly white.

These mites are found mostly in protected places on young tender leaves, young stem ends, buds and flowers. They crawl from plant to plant where leaves touch; another means of spreading is the transfer of mites on hands or clothing while working and moving among them.

Damaged leaves of infested plants are twisted, curled and brittle. Buds may be deformed and fail to open. Flowers are deformed and often streaked with darker color. Blackening of injured leaves, buds and flowers is common.

Infested ivy will produce stems without leaves or with small deformed leaves. Infested African violets develop small, twisted, hairy leaves that may soon die.

To control, trim off badly injured plant parts where practicable. They can be "cooked" off infested plants by immersing infested plants, pot and all, for 15 minutes in water held at 110°F. The success of this method of treatment depends on careful control of the water temperature. In home greenhouses, control can be achieved by making 2 or 3 spray applications of an effective miticide, such as dicofol, at 10 day intervals.

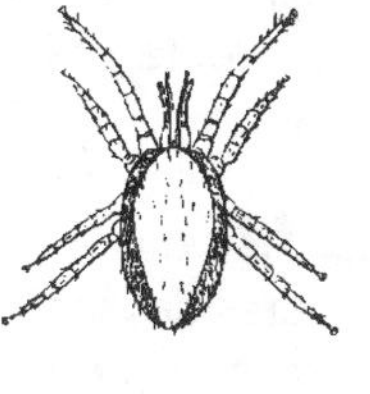
Cyclamen Mites
(USDA)

False Spider Mites

Several species of false spider mites can infest plants in the home or greenhouse. They are flat, oval, dark-red mites too small to be easily seen with the naked eye. The young and eggs are bright red. All stages of these mites are found mostly on the undersides of leaves, generally along the veins or other irregularities on the leaves.

The damage caused by feeding of these mites results in finely stippled bronze or rusty-brown areas along veins or on entire leaves. The edges of infested leaves may die or the leaves lose color and drop off. Infested plants become weakened and should be given additional attention to water and nutrition.

In the home or greenhouse mites can be controlled by making 2 or 3 applications or plant dippings with a recommended miticide, such as dicofol, at 10-day intervals.

Fungus Gnats

Fungus gnats are delicate, gray or dark gray, fly-like insects about 1/8th inch long. They are true flies, having only one pair of wings. They are attracted to light and when present in the home swarm around windows. The immature forms are white maggots that live in soil and reach a length of about 1/4 inch. These maggots are usually found in soils with surplus quantities of decaying vegetable or organic matter.

The gnats are only a nuisance, indicating the presence of maggots, likely in potted plants. The maggots cause injury to the root systems by burrowing in the soil. They may feed on the roots and crowns of plants. Severely injured plants grow poorly, appear off color and may drop leaves.

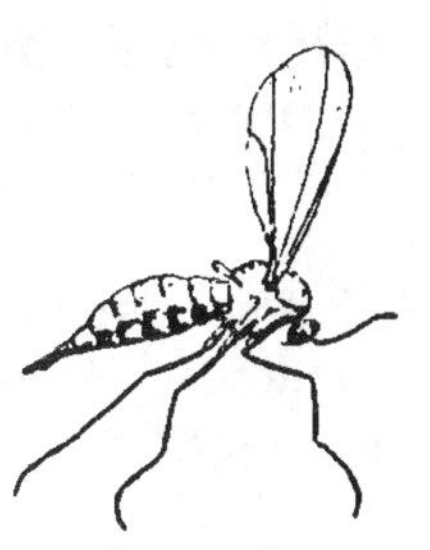
Fungus Gnats
(U.S. Public Health Service)

Maggot control can be achieved by avoiding overwatering of plants in the home and greenhouse. Soil drenches of an insecticide, such as malathion, will produce immediate control.

Mealybugs

There are several kinds or species of mealybugs that become pests on house and greenhouse plants. They are softbodied and appear as though covered with dust or flour because of their waxy coats. They grow to be about 3/16ths inch long. Some mealybugs have waxy strands or filaments extending from the rear of their bodies. They are found at rest or crawling slowly on stems, where stems and leaves join and on leaves, especially along veins on undersurfaces. Their eggs are laid in clusters enclosed in white waxy cottony or fuzzy material. Mealybugs are sometimes cared for by ants. The ground mealybug, a soil inhabitant, feeds on the roots of African violets and other plants.

Damage by mealybugs is caused by their sucking out plant juices, resulting in stunting or death of plants. Sooty mold grows on the honeydew excreted by some species of mealybugs. The ground mealybug damages the rootlets, causing the plants to grow slowly and wilt between waterings.

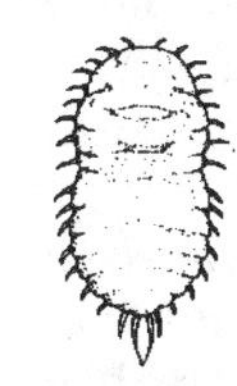
Mealybug Adult Female
(USDA)

When only one or a few plants are infested, mealybugs can be controlled by washing, by handpicking or by using an alcohol-dipped swab to remove them. Treated plants should be isolated to avoid reinfestation.

Booklice or Psocids

Psocids are softbodied, pale yellow to gray, oval insects that grow to about 1/32nd to 1/16th inch in length. Some species have wings while others are wingless. Psocids sometimes cluster in large numbers of a hundred or more. They feed on moist dead animal or organic matter, lichens and fungi.

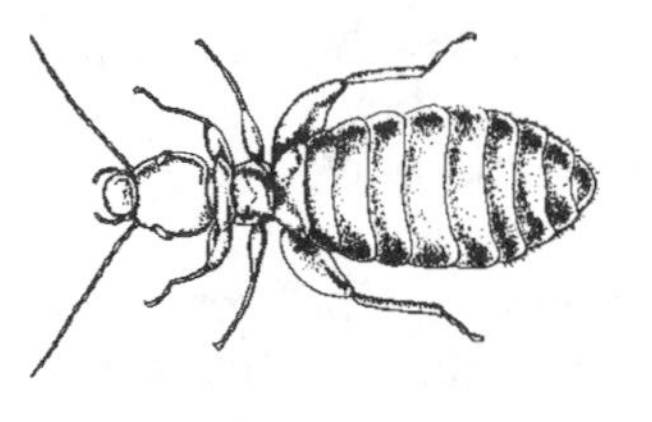

Booklouse or Psocid
(U.S. Public Health Service)

They may occur in large numbers in the soil or on pots and benches, especially in undisturbed locations in the home or greenhouse. Tiny, fast-moving psocids may sometimes be found on old books and papers stored in damp places.

Generally, psocids are only a nuisance, for though found on living plants, they are not known to feed on them. Because they are only a nuisance, control measures are unnecessary.

Scales

There are many species of scales that commonly infest several species of plants in homes or greenhouses. Scale insects have a shell-like covering or scale, that protects the body. Most species are about 1/16th to 1/8th inch in diameter, but a few species may span 3/8th inch. Some are round, others oval and some are shaped like oystershells. Color ranges from white to black, however the predominant species are browns and gray

San Jose Scale
(Union Carbide)

Some scales lay their eggs in white sacs secreted from beneath and these can be mistaken for mealybugs if not examined closely for the presence of the tiny shell-like covering. Some species infest only plant leaves, others are found on both stems and leaves, while some attack only stems.

Scales feed by sucking plant juices with their piercing-sucking mouthparts, resulting in reduced growth and stunted plants. These insects excrete honeydew as do their cousins, the aphids, which is attractive to ants. The honeydew falls on lower leaves imparting a shiny or wet appearance to the foliage and provides a source of nutrition for the growth of sooty mold.

Control can be achieved by washing with soapy water if only a few plants are involved. However, control is difficult with insecticides, requiring repeated applications. It may be best to discard heavily infested plants and purchase non-infested replacements.

Whiteflies

There are several whitefly species that can be bothersome on house plants. The adults are about 1/16th inch long, have white, wedge-shaped wings and do resemble white flies. They are not flies, however, but are related to the aphids and scales. When infested plants are moved, the adults take

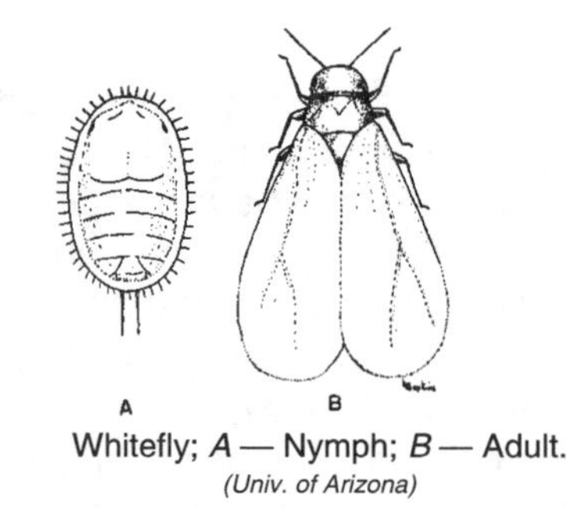

Whitefly; *A* — Nymph; *B* — Adult.
(Univ. of Arizona)

flight, resembling small snowflakes or bits of paper ash swirling in the air, seemingly with no sense of direction.

Immature whiteflies resemble scales and are mostly pale green to yellow or whitish, oval shaped and flattened. Except for the newly-hatched crawlers, the immature stages are attached to the leaves, mostly on the undersides.

Plant damage is caused by both adults and young as they suck juices from the leaves. Infested leaves become pale, then yellow and die or drop off. Surfaces of leaves may be covered with their sticky honeydew, which usually becomes blackened with sooty mold.

TABLE 21. Insect and Mite Control Suggestions for Houseplants, With and Without Chemicals.

Plant	Pest	Pesticide to Use [4]	Non-Chemical Control [3]
African violet	Mealybug	Malathion [1] or Resmethrin. Spray or dip [2] with WP formulation.	Insecticidal soap spray. Pick off by hand.
	Mites	Dicofol or Acephate. Spray or dip [2].	Insecticidal soap spray.
Aralia	Scale insects	Malathion [1] or Acephate. Dip [2]. or spray.	Insecticidal soap spray.
	Mites	Dicofol or Resmethrin. Spray or dip [2]	Insecticidal soap spray.
Begonia	Aphids	Malathion [1] or Resmethrin Spray or dip [2].	Remove by hand or with artist's brush.
	Mealybug	Malathion [1] or Resmethrin aerosol. Spray or dip [2] .	Remove by hand or with artist's brush or rotenone.
	Whitefly	Malathion [1], or Pyrethrins. Spray or dip [2].	Insecticidal soap, dip or spray.
	Mites	Dicofol or Acephate. Apply as needed.	Insecticidal soap spray.
Citrus	Scales	Malathion [1] or Acephate. Dip [2] or use artist's brush to apply to scales.	Remove by hand.
	Whitefly	Malathion [1] or Pyrethrin. Spray or dip [2].	Insecticidal soap spray.
Coleus	Mealybug	Malathion [1], Resmethrin or Rotenone. Spray or dip [2] with WP formulation.	Remove by hand or with artist's brush or rotenone.
	Whitefly	Malathion [1] or Resmethrin. Spray or dip [2].	Rotenone.
Cyclamen	Mites	Dicofol. Spray or dip [2] .	Insecticidal soap spray.
Dracaena	Mealybug	Malathion [1], Permethrin or Resmethrin aerosol. Spray or dip [2] with WP formulation.	Remove by hand or with artist's brush.
Fuchsia	Mealybug	Malathion [1], Permethrin or Resemthrin aerosol. Spray or dip [2] with WP formulation.	Remove by hand or with artist's brush.
	Whitefly	Malathion [1] or Pyrethrin Spray or dip [2].	Insecticidal soap spray, or rotenone.
Gardenia	Mealybug	Malathion [1] or Resmethrin aerosol.Spray or dip [2] with WP formulation.	Remove by hand or with artist's brush or rotenone.
	Whitefly	Malathion [1] or Pyrethrin. Spray or dip [2] .	Remove by hand or with artist's brush or rotenone.
Geranium	Whitefly	Malathion [1] or Acephate. Spray or dip [2] .	Remove by hand or with artist's brush or rotenone.
	Mites	Dicofol or Acephate. Spray or dip [2]	Insecticidal soap spray.

[1] Use premium grade formulation to avoid objectionable odor.
[2] Dip foliage and stem into solution equal to one-half the spray concentration.
[3] For a few plants with light infestations, most pests can be removed by spraying over the sink with a strong stream of water or by dipping infested parts in soapy water or by touching the insect with an artist's brush or cotton swab dipped in rubbing alcohol.
[4] Refer to Appendix B for trade or proprietary names of recommended insecticides.

Plant	Pest	Pesticide to Use [4]	Non Chemical Control [3]
Gloxinia	Aphids	Malathion [1] or Acephate. Spray or dip [2].	Remove by hand or with artist's brush.
	Mealybug	Malathion [1] or Resmethrin. Spray or dip [2] with WP formulation.	Remove by hand or with artist's brush or rotenone.
	Mites	Dicofol or Pyrethrins. Spray or dip [2].	Insecticidal soap spray.
Ivy (Boston and English)	Mites	Malathion [1], Pyrethrins or Dicofol. Spray or dip [2].	Insecticidal soap spray.
Palms	Scale	Malathion [1] or Acephate. Dip [2] or use artist's brush to apply to scales.	Remove by hand.
	Mealybug	Malathion [1], Resmethrin or Acephate. Spray or dip [2] with WP formulation.	Remove by hand or with artist's brush.
	Mites	Dicofol or Pyrethrins. Spray or dip [2].	Insecticidal soap spray.
Philodendron	Scale	Malathion [1] or Acephate. Dip [2] or spray. For small plants use artist's brush to apply to scales.	Remove by hand.
Pittosporum	Mites	Dicofol or Pyrethrins. Spray or dip [2].	Insecticidal soap spray.
Podocarpus	Mites	Dicofol or Pyrethrins. Spray or dip [2].	Insecticidal soap spray.
Roses (miniature)	Mites	Dicofol or Pyrethrins. Spray or dip [2].	Insecticidal soap spray.
Rubber Plant	Mealybug	Malathion[1], Resmethrin or Acephate. Spray or dip [2] with WP formulation.	Remove by hand or with artist's brush.
	Scale insects	Malathion [1] or Acephate. Dip [2] or spray. For small plants use artist's brush to apply to scales.	Remove by hand.
Schefflera	Scale insects	Malathion [1] or Acephate. Dip [2]. or spray. For small plants use artist's brush to apply to scales.	Remove by hand.
	Mites	Dicofol or Pyrethrins. Spray or dip [2].	Insecticidal soap spray.
Miscellaneous Plant Pests	Fungus Gnats (midges)	Propoxur, Pyrethrins, Carbaryl or Resmethrin. Apply lightly to moist soil in pots. Hang No-Pest® Strip in areas where gnats appear.	Maintain dryer flower pot soils.
	Pillbugs and Sowbugs	Propoxur, Carbaryl or Resmethrin. Apply lightly to moist soil in pots and plants.	Reduce soil moisture in pots and planters. Remove by hand and destroy.
	Springtails	Malathion [1], Pyrethrins or Resmethrin. Apply lightly to moist soil in pots and planters.	Take pots outside and expose several hours to warm, drying sun.

PETS

Can you believe that more is spent annually for pet food than for baby food in the U.S? How important is your dog or cat to you? What would you take for it? Pets are highly treasured objects of affection in the American home.

Often dogs, cats and birds as pets become so popular that they are practically considered members of the family. Because of this great recreational and companionship importance in our American way of life, it is essential that the health of these pets be protected.

Unfortunately, dogs, cats, birds and other warm-blooded pets are subject to attack by fleas, ticks, lice, flies, mites and other annoying if not harmful pests. Their attack may persist throughout the year. Infested dogs and cats scratch to relieve irritation, often rubbing off patches of their coat around the neck, shoulder and abdomen. This condition may worsen resulting in loss of appetite, weakness and increased susceptibility to disease. Birds become nervous, lose their appetite and appear old and ruffled.

With the proper selection, use and timing of insecticides, it is possible to greatly reduce if not totally eliminate these troublesome insect and mite attacks. The suggestions for control in Table 22 are up-to-date, effective, safe for the pet and his owner and registered by the Environmental Protection Agency for use on pets at the time of publication.

Nevertheless, it is possible that some of the chemicals recommended could lose their label status. **Consequently, it is the responsibility of those using the pesticides to determine the label status before use.**

Label precautions and directions are not guesswork. They are based on scientific data submitted by the manufacturer and reviewed by the Environmental Protection Agency as part of the product's registration procedures. **When controlling pests on your pets, treat only the pets listed on the pesticide label. Do not use any pesticides labeled for crops, ornamentals or livestock on pets unless the label so states; use no more of the material than the recommended dosage and only at the time recommended; and be especially cautious when treating young pets.**

Pests not described in the following section may be found elsewhere. Check the index.

Horse Bots

Horse bot flies have four stages: adult, egg, larva (bot) and pupa. The adults do not feed, since their sole purpose in life is to reproduce. They live at the most only a few weeks. There are 3 species, but their damage and life cycles are essentially the same. The eggs are attached by the adult female to hairs on the legs or near the mouth. By the licking of eggs or by normal hatching near the face, the larvae reach the mouth where they spend a period. After slight development they move on into the horse's stomach where they attach with their mouthparts to remain for about 10 to 11 months. After reaching full development they release and pass to the soil through the feces, where they pupate. In the spring or early summer the adults emerge to mate and begin the 1-generation-per-year life cycle again.

Part of a horse's stomach heavily infested with bots. Note the lesions that these bots have caused. *(USDA)*

Damage by bot flies may be indirect, in that animals under attack may inflict damage on themselves or on anyone trying to handle them. Fright and irritation caused by egg-laying adults or newly hatched bots may result in animals acting out of control. Direct damage is produced by larvae feeding on the tissues of the mouth and stomach. Infested animals often suffer from colic or other gastric disturbances. The degree of damage done by the feeding of the bots is proportional to the number present. Several hundred larvae are commonly found in one animal.

Bot control in horses should consist of prevention of infection, as well as treatment of infections with effective insecticides. Infections of bots can be prevented during the fly season by removing bot fly eggs from hairs by clipping or by applying warm rinses (120°F.) to induce hatching of the eggs and subsequent death of the young bots. Chemical control should be under the advice of a veterinarian. There are new and very effective materials available (Table 22).

Horse Bot Adult
(USDA)

Horse Flies and Deer Flies

Horse flies and deer flies belong to the same group of robust, blood-sucking flies. They vary in size from 1/4 inch to 1 inch in length. They are strong fliers, notorious pests of horses, cattle and many other warm-blooded animals. Only the females bite. The males feed on vegetable materials and cannot bite. They breed only in aquatic or semiaquatic habitats.The eggs are laid during the warm months in clusters on objects over water or marshy situations favorable for development of the larvae. Usually there is only one generation a year, overwintering as larvae in the wet soil. It completes its development in the early spring and emerges as the adult a few weeks later. The females begin seeking blood and the males feed on flowers and vegetable juices.

Horse Fly
(USDA)

Their bites cause serious annoyance and significant blood loss to domestic animals, particularly horses and cattle. Control of these biting flies can be very difficult because their developement sites may include marshy or aquatic sites, as well as relatively dry soils. Thus, their control is achieved by regular application of an approved insecticide to the affected animals.

Fleas — See description and life cycles under Household Pests, Page 153

Lice

Chewing Lice

Biting or chewing lice feed on skin scales, skin exudations and other matter on the skin and gnaw at the living epidermis. They are not blood feeders as are the sucking lice, but occasionally the irritation and scratching by the host results in bleeding.

Canaries, parakeets and other caged birds are frequently infested with a chewing louse, usually the red louse. The adults are about 1/16 inch long and reddish brown in color. Parts of the feathers, particularly the barbs and barbules, constitute a major part of the food of this pest. The irritation from the feeding of the louse causes the host to become quite restless, thus affecting its feeding habits and digestion. Young birds are particularly vulnerable. Bird lice tend to be more abundant where uncleanliness and overcrowded conditions exist. Control is best achieved with a dust, though aerosols are quite effective.

Head Louse

A Body Louse
Chicken Lice

Shaft Louse

(U.S. Public Health Service)

Dogs and cats have their very own special variety of biting or chewing louse. Dogs, particularly puppies, may suffer much irritation from the dog-biting louse, while cats, both kittens and adults, may become heavily infested with the cat louse.

Other domestic animals, cattle, horses, sheep and goats, also are occasionally plagued with biting lice. Cattle are often heavily infested on the withers, root of tail, neck and shoulders by the cattle-biting louse. Horses, mules and donkeys, but horses more particularly, may suffer from the horse-biting louse when poorly or irregularly groomed. Sheep at times may show severe infestation of the sheep-biting louse and goats are commonly infested with several species of chewing lice.

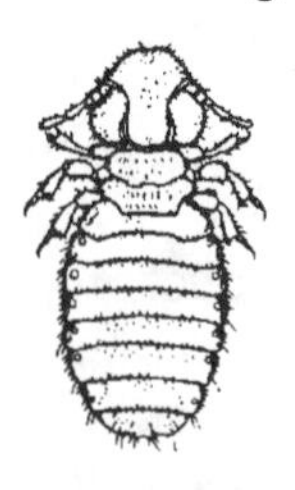
Cattle Biting Louse
(USDA)

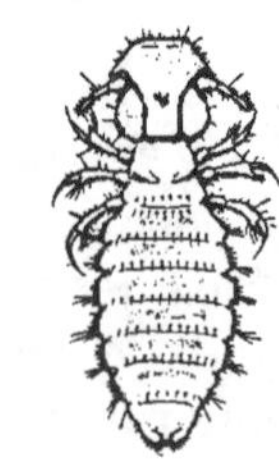
Goat Biting Louse
(USDA)

Spread from one animal to another is commonly by direct contact or by introduction of uninfested animals into quarters that have recently been vacated by infested ones. Control is dependent on sanitation practices, avoiding the introduction of infested animals into those that are uninfested and the appropriate use of recommended insecticides.

Sucking Lice

The sucking lice are totally dependent on blood meals for survival and nearly all domestic animals have their own variety of sucking louse, including humans (see index). All lice are wingless, have flattened bodies and their legs are adapted for clinging to hairs and feathers. The young resemble the adults and have the same feeding habits and nutritional requirements. The entire life cycle is spent on the host. Eggs are attached to hairs and repeat generations occur throughout the year.

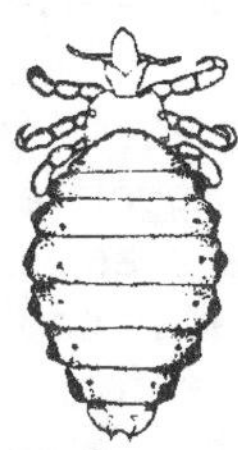

Short-nosed Cattle Louse
(U.S. Public Health Service)

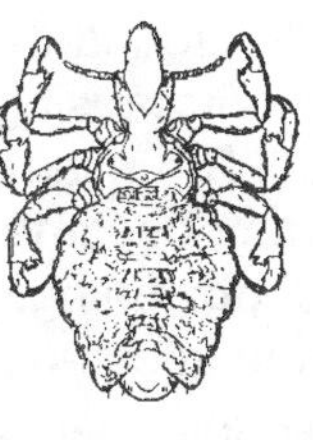

Hog Louse
(U.S. Public Health Service)

Significant weight losses and anemia may result from heavy infestations by these lice. Swine have one species, the hog louse. Cattle are infested by five species: the long-nosed cattle louse, the little blue cattle louse, the short nosed cattle louse, the cattle tail louse and the buffalo louse. Horses, mules and donkeys are infested with the horse sucking louse. Sheep may be infested with the foot louse, or the sucking body louse. Goats may be infested by the goat sucking louse.

Nutrition is very important in keeping populations of sucking lice under control. The decreased ability of poorly nourished animals to groom themselves and interference with normal seasonal hair shedding favors survival of their lice. Spread from one animal to another is only by direct contact. Control of lice requires the use of an approved insecticide with 1 or 2 repeated applications at 2 to 3-week intervals.

Mites

Bird Mites

There are several mites that are known to infest caged birds, though none are specific for any one bird. Their life cycles are similar to those described previously and most are barely visible with the naked eye. Both the adults and nymphs are blood-feeders. Some species remain on the birds throughout their lives while others feed only at night and retreat to secluded resting places within the cage during the day. Because of the two basic resting habits of the several possible species, control measures call for a good treatment both of the birds and their cages with an appropriate insecticide.

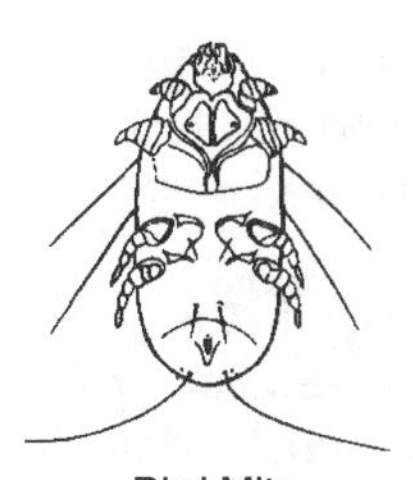

Bird Mite
(Much Enlarged)
(U.S. Public Health Service)

Mange Mites

Mange mites or sarcoptic mites are also known as itch mites. Several species attack cattle, horses, hogs, sheep, goats, dogs, rabbits and man. The mites burrow into the tender or soft areas of the skin where hair is usually sparse and often continue to spread until large portions of the body are affected. Nodules usually appear over and around the burrows. These burrows burst and ooze serum, which dries to form scabs. Intense itching causes the animals to rub and scratch, resulting in open sores which frequently are invaded by bacteria. This continual irritation causes the skin to become wrinkled and thickened as the infestation spreads. Transmission is usually by direct contact with mangy animals or with objects against which affected animals have rubbed. Overlapping generations of mites occur throughout the year at 2- to 3-week intervals.

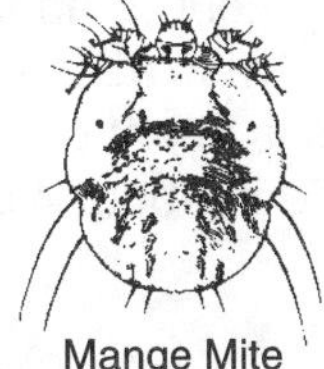

Mange Mite
(Much Enlarged)
(U.S. Public Health Service)

Follicle Mites

Follicle mites also attack domestic animals and man, causing manifestations of disease similar to those of the mange mites, by burrowing into hair follicles and oil glands. The nodules formed vary in size from a pinhead to as large as a marble and are filled with pus resulting from secondary bacterial infection. Follicle mange is only a serious pest of dogs, causing what is known as red mange. It is transmitted by direct contact and does not itch severely.

Follicle Mite
(Much Enlarged)
(U.S. Public Health Service)

Isolation of infested animals and treatment by spraying or dipping is the standard recommended procedure.

Sheep Tick or Ked

The sheep tick or ked is really a wingless fly. In no characteristic does it resemble a fly, thus its name, sheep tick. It is reddish-brown, has a saclike body, spiny and somewhat leathery. It is a widely distributed parasite of both sheep and goats. The female gives birth to a fully-grown larva that pupates in a few hours to a seed-like puparia. The pupae are most commonly found in the region of the shoulder, croup, thighs and belly of infested animals. Pupae may be found on sheep at all times of the year. The time required for development is about 3 weeks in the summer but may be twice as long in the winter. The entire life of the ked is spent on its host; when off the sheep the keds die in about 4 days.

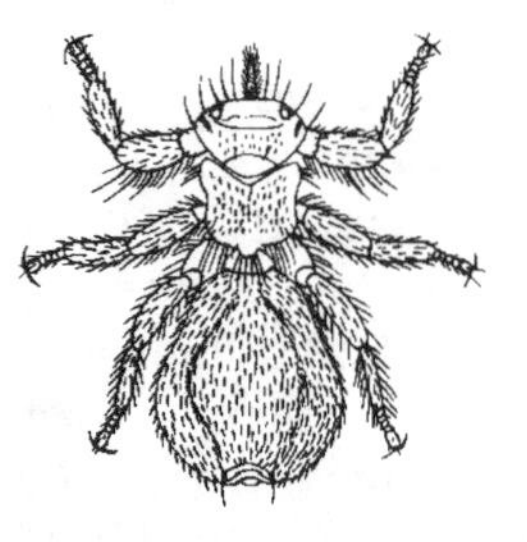

Sheep Ked
(USDA)

A few keds on the body of a sheep do not affect the animal measurably. In heavy infestations the animals rub themselves vigorously, bite the wool and scratch. Injury to lambs is especially marked by emaciation, anemia and general unthriftiness. Control is achieved only with the use of insecticides which may be applied as sprays or dusts.

Stable Flies — see Flies under Household

Wool Maggots

Wool maggots or fleece worms are the maggots of blow flies. When the wool of sheep becomes soggy from warm rains, or soiled with urine, feces, or blood from wounds or from lambing, certain blow flies are attracted and deposit their eggs in the dirty wool, usually around the rump. The maggots feed in the wet wool causing the skin to fester and wool to loosen. The inflamed flesh with the maggots tunneling in it may become infected and the sheep develop blood poisoning. Cattle and other animals may become infested if they have putrid sores.

The adult blow flies are larger than the house fly and usually are metallic bluish-green color with bronze reflections. Flies overwinter as larvae or pupae in soil beneath carcasses or in manure. The flies appear very early in spring and from that time on breeding is continuous. The fly can complete a generation from egg to egg in about 3 weeks. In warm, rainy weather they may lay eggs in the wool, where the maggots develop and drop to the ground where they complete their pupal stage.

Infestation may be prevented by protecting animals from becoming drenched in summer rains and treating sores or dirty, urine-soaked wool. When once infested the use of insecticidal control is essential.

TABLE 22. Pest Control Suggestions for Pets.[1]

Pet	Pest	Pesticide to Use	Directions
Birds	Lice, Mites (caged house birds)	Pyrethrins + synergist, aerosol. Rotenone dust (lice only). Permethrin, Allethrin or Resmethrin aerosol.	Spray parakeets, canaries and other caged birds as follows: For cage mite control, remove bird and spray entire cage thoroughly. For bird lice, leave birds in cage and spray directly on bird, holding can at least 24 inches from bird or any surface. Spray lightly, one shot 2 to 3 seconds. Treat no more than twice per week. Remove food from cage before spraying. Cages should be cleaned in hot water weekly.

[1] Refer to Appendix B for trade or proprietary names of recommended insecticides.

Pet	Pest	Pesticide to Use	Directions
Dogs	Fleas	D-limonene (shampoo, spray or dip) Propoxur 0.25% aerosol Rotenone dust or Carbaryl dust, Methoprene insect growth regulator. Lufenuron (Program®) (orally) only prescribed by a veterinarian.	Spray dog outdoors, holding aerosol 3-6 inches from dog and cover thoroughly. A single 30-60 second treatment should give protection up to one month. Repeat as needed. Treat bedding with Methoprene aerosol or spray.
		Methoprene, Cyfluthrin, Permethrin or Naled dog flea collar. Imidacloprid and Fipronil sprays for dog quarters and runs.	Gives protection up to 3 months. Some animals may be sensitive to this product. Watch for signs of irritation. Collar should be buckled loosely. Methoprene kills flea eggs, breaking the life cycle.
	Mange mite (sarcoptic)	Amitraz (Taktic®) spray.	Apply weekly for 3-4 weeks. Work into the skin with firm finger massage. Treat all areas that respond slowly. The same for minor moist fungus in ears and on feet. For ear mange mites, lay dog on side and distribute medications down into ear once daily for 1-3 days.
		Lindane (emulsifiable) (prescribed by Vet.)	Follow directions, mix sparingly. Dip or spray mature dogs once a week for 2 or more treatments as needed. Not for cats, nursing puppies or bitches with nursing pups.
	Ticks and Fleas	Carbaryl 3-5% dust Cyfluthrin, Permethrin or Propoxur collars.	Dust entire dog, including ears, legs and feet. Repeat treatment at weekly intervals. To control ticks and fleas in larger area, treat dog's sleeping quarters and surrounding area at rate of 1 lb./1000 sq. ft.
		D-limonene or Propoxur, dip.	Follow label directions for preparation of dip. Sponge on or dip animal until coat is wet. Allow animal to dry naturally.
		Pyrethrins + synergist aerosol	Hold can comfortable distance from dog and spray entire dog, avoiding the eyes. Follow label carefully.
	Lice and Fleas	Rotenone + Resmethrin shampoo.	Wet dog then shampoo thoroughly allowing lather to stand 5 minutes to kill any fleas or lice present.

Pet	Pest	Pesticide to Use	Directions
Dogs (Continued)			
	Fleas, Lice and Ticks	Pyrethrins + synergist aerosol. Methoprene insect growth regulator.	Give dog full coverage, weekly, until infestation subsides. Spray animal's bedding and living quarters regularly. Treat bedding with Methoprene aerosol or spray.
		Pyrethrins + synergist shampoo.	Mix according to label instructions and give that dog the bath of its life. Allow lather to remain 10 minutes before rinsing.
Cats	Fleas	Propoxur, Permethrin, Methoprene or Naled cat collar.	Follow label directions. Must be worn loosely. Methoprene kills flea eggs, breaking the life cycle.
		Methoprene insect growth regulator.	Treat bedding and other sleeping places with aerosol or spray.
		Pyrethrins + synergist aerosol	Treat entire pet until hair is slightly damp. Do not spray directly into face.
		Rotenone foam shampoo	Apply to palm of hand then rub into pet's coat. Dry with towel, comb then brush.
		Propoxur shampoo	Apply up to twice weekly if necessary.
	Fleas and Lice	Carbaryl 3% dust. Rotenone dust	Apply powder lightly over entire coat, while avoiding pet's eyes. Dust cat's bedding. Not to be used on kittens less than 6 weeks of age.
	Fleas, Lice and Ticks	Pyrethrins + synergist shampoo	Do not dilute with water, but sponge onto cat in original form. Rub dry with coarse towel. Not to be used on kittens less than 6 weeks of age.
Horses Ponies Donkeys and Mules	Flies (all) Gnats Mosquitoes	Permethrin ready-to-use mist spray (1%)	Spray all areas of animal with attention to face, legs and under belly. Repeat as needed.
		Oil Base Wipe-on or Spray: Tetrachlorvinphos (Rabon® 1%),Pyrethrins + synergist and repellent.	Apply by rubbing on hair coat, with attention to legs, shoulders, shanks, neck and face. Only light application required. Repeat daily as needed.
		Pyrethrins + synergist oil spray	Apply to entire animal as mist spray. Avoid coarse droplets. Wet hair but not hide.
	Mange	Permethrin	Spray animal thoroughly every 14 days.

Pet	Pest	Pesticide to Use	Directions
Horses Ponies Donkeys and Mules (Continued)			
	Horse bots	Abamectin (Eqvalan®, Zimecterin®) oral paste. Dichlorvos (Horse Wormer®) feed additive.	Check with your veterinarian.
	Lice	Pyrethrins + synergist oil spray. Permethrin spray.	Spray coat until run-off. Repeat in 3 to 5 weeks if needed.
	Horn Fly	Tetrachlorvinphos (Rabon®) or Permethrin spray.	Spray coat until run-off. Repeat in 10 days and thereafter, if needed.
Sheep Lambs and Goats	Lice and Ticks (keds)	Permethrin or Malathion sprays. Rotenone dust.	Spray animals thoroughly, wetting wool to the skin. Use about 1 gallon of finished spray per animal. Repeat at 2-week intervals, if needed. Do not treat animals less than 1 month of age.
	Wool maggots (fleece worms)	Permethrin or Pyrethrins + synergest sprays.	Spray animal thoroughly wetting wool to the skin. Repeat in 2 weeks, if needed. Don't treat lambs less than 3 months old.
Swine	Lice	Amitraz (Taktic®), Malathion, Permethrin or Pyrethrin + synergist sprays.	After spraying, keep animals out of sun and wind for a few hours. Repeat in 2-3 weeks if needed.
		Carbaryl or Malathion dust (4-5%)	Apply thoroughly to animals and pens. Repeat in 10 days and thereafter if needed.
	Mange (sarcoptic)	Amitraz (Taktic®), Permethrin or Malathion sprays.	Spray animals to point of run-off. Treat every 2-3 weeks. Check label regarding piglets under 1 month of age.
Beef Cattle	Horn Fly, Horse Fly, House Fly	Coumaphos (Co-Ral®), dust or spray or Pyrethrins or Permethrin sprays.	Use about 2 quarts per animal of water-mix and spray thoroughly.
	Horn Fly and Face Fly	Cyfluthrin, Permethrin ear tags.	Place one tag in each ear.
		Pyrethrins + synergist oil spray	Use 1-2 oz. per animal. Apply as a mist to all parts of the body, especially to the back. Apply daily as needed.

Pet	Pest	Pesticide to Use	Directions
Beef Cattle (Continued)			
	Stable Fly House Fly	Coumaphos (Co-Ral®) spray (or dust)	Apply 1-2 oz. per animal per day using hand sprayer, to all parts of the body, especially to legs. Follow label precautions.
	Lice	Amitraz (Taktic®) or Coumaphos (Co-Ral®), Malathion, Pyrethrins + Synergist or Permethrin sprays.	Use about one-half gallon of spray per animal, spraying thoroughly. Repeat in 2-3 weeks, and again if needed.
	Mange (Barn itch mite)	Coumaphos (Co-Ral®) or Amitraz (Taktic®) sprays.	Apply up to one gallon per animal, spraying thoroughly. Repeat within 10-14 days for lasting control.
Hamsters, Gerbils, White rats and White mice	Fleas	Carbaryl 3-5% dust or Rotenone dust	Apply to animals using a salt shaker with gentle rubbing. Repeat in 2-3 weeks. Frequent scratching is natural and does not necessarily indicate fleas.
	Mites	Malathion (0.5%) spray	Spray animals to point of run-off. Treat every 2-3 weeks. Avoid treating the very young or parents of the very young.
	Fleas Lice Mites	All-purpose aerosol containing: Pyrethrins .045% + Piperonyl butoxide .09% N-Octyl bicycloheptene dicarboximide .15% and inert ingredients.	This multi-action spray kills lice, fleas and mites, controls odors and helps keep pets fresh and clean. Shake can and hold upright one foot from cage and spray directly into the cage. Hamsters, gerbils and cage should be sprayed 2-3 times each week.
Snakes	Mites	Dusting sulfur	Treat the animal and its quarters with a light dusting using a salt shaker and gentle rubbing throughout the body length. Repeat as needed.

PLANT DISEASES

Some English seed they sew, as wheat and pease, but it came not to good, eather by ye badness of ye seed, or lateness of ye season, or both, or some other defects.

Gov. Wm. Bradford,
History of Plimoth Plantation
April 1621

CONTROLLING PLANT DISEASES

Most plant diseases encountered around the home are caused by certain fungi or bacteria. Other organisms that cause plant disease are viruses, rickettsias, algae, nematodes, mycoplasma-like organisms and parasitic seed plants.

There are hundreds of examples of plant diseases. These include storage rots, seedling diseases, root rots, gall diseases, vascular wilts, leaf blights, rusts, smuts, mildews and viral diseases (Figure 21). These can, in many instances, be controlled by the early and continued application of selected fungicides that either kill the pathogens or inhibit their development.

Chemical control is not the only route to follow for disease management. In some instances it is of no value at all, as in the case of virus or mycoplasma-like diseases. These are either transmitted mechanically by insect vectors feeding on diseased plants then moving to non-infected plants or by propagation methods such as grafting, rooting of diseased cuttings and the use of infected seed.

Other methods of disease control include the use of resistant varieties, planting times, cultural control and simply good plant nutrition. Gardeners should use plant varieties or cultivars which are resistant or at least tolerant to certain diseases. These are usually developed for geographical areas of the country and selected for their resistance to local disease problems.

Cultural control may involve one of several methods of altering cultural practices in favor of the plant and to the disadvantage of the disease organism. For instance, Fusarium and Verticillium wilts are carried over in the soil and in decaying plant parts from year to year. A simple crop rotation system with switching of susceptible to non-susceptible garden species will avoid the problem almost completely. The early or late planting of vegetables to avoid the normal infective periods of certain diseases is another simple way to manage diseases. Watering or sprinkling only in the mornings may avoid producing the wet afternoon environments that promote downy mildew and certain fungal fruit rots. These techniques are too numerous to give in detail.

Finally, balanced fertilization may often provide the additional plant vigor needed to outgrow a disease. This is certainly not always the case, but disease in general has a more profound effect on weak and undernourished plants than on hearty, vigorously growing specimens.

DISCOURAGING PLANT DISEASES THROUGH GOOD MANAGEMENT

Following several of the suggestions listed below will help reduce the incidence and severity of diseases in all gardens, vegetable or flower:

1. Choose a suitable location. Don't place shade-loving plants in exposed situations or sun-loving plants in the shade. Also avoid extremely wet or dry locations or use plants that are suited to these conditions. Most root diseases are favored by wet soils. Creating good drainage in garden soils can reduce the severity of these diseases.
2. Take additional time to plant a seedling with care—it pays dividends in reducing replacement and maintenance later in the growing season. Spend a little extra time working the soil into a good seed bed or planting condition.
3. Choose disease resistant varieties that are locally adapted for your area if available. Do not grow non-adapted varieties from other areas of the country unless you simply like to experiment.
4. Occasionally change the garden location and always practice a rotation within the garden plot. Many diseases, especially soil borne ones, are most serious when the same or related crops are grown in the same area year after year. Change the flower species in an area every year or two. Not only will this add variety to the landscape but it helps combat build-up of disease organisms in the soil.
5. Use disease-free seed and seedlings. Buy seed from a reputable seed company and transplants from a greenhouse operator who grows them from disease-free seed and in soil free of disease organisms. Many disease organisms can be carried to the garden on seeds, transplants or transplant soil.
6. Apply fertilizer and lime only on the basis of soil testing results or established soil fertility management practices. Exceptionally weak or vigorous plants are more susceptible to some diseases than those grown on a balanced fertility and optimum pH (soil acidity) program.
7. Weeds, particularly perennial weeds, near gardens often are initial sources of viruses in

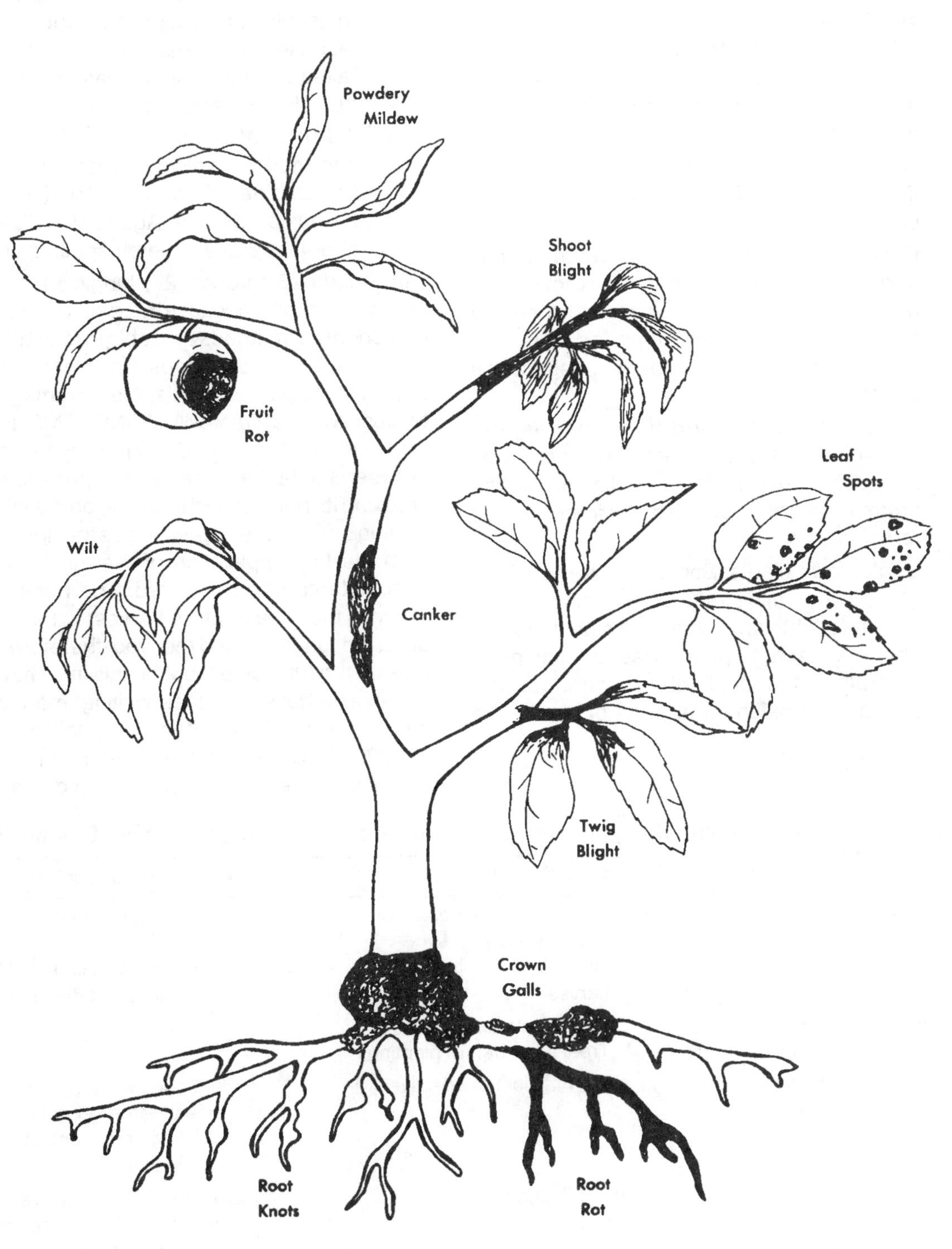

FIGURE 21. Disease Symptoms on Plants (Courtesy *Cornell University*)

the spring. Dense clusters of weeds within the garden create a microclimate that is ideal for development of fungus and bacterial diseases. Weeds also compete for water and soil nutrients.

8. Control insect pests as needed, especially leaf hoppers. Several virus diseases are transmitted by insects with piercing-sucking mouthparts.
9. Remove and destroy diseased plants when first observed. It is also unwise to compost diseased plants.
10. Provide adequate water, being sure there is good penetration below the root zone. Maintain a relatively even moisture supply throughout the growing period. Wet-dry changes result in considerable stress on plants.
11. Plow or spade under crop residue immediately after harvest. This promotes decay of organic matter and killing of disease organisms which could overwinter in crop remains.
12. Don't crowd plants. Some diseases such as the downy mildews and Sclerotinia and Botrytis blights are favored by high humidity. Thin the plants to permit free air circulation and allow the sunlight to penetrate the canopy and reach lower parts of plants and soil.
13. Remove and haul off or burn diseased branches and shoots before a disease spreads. In routine pruning, always remove diseased or sickly growth first and then prune to develop and shape the shrub or tree. Many disease organisms carry over the winter on fallen leaves; thus, diseased foliage should be collected and destroyed.
14. Additional, more localized control information on diseases of specific crops can be obtained from your local Cooperative Extension Service Agent. Usually many fact sheets are available on specific problems.

In Tables 23 through 27, fungicidal controls are emphasized, only because there simply aren't enough non-chemical methods available to match the efficacy of present-day fungicides. In the application of fungicides to garden plants, it is extremely important to observe the DAYS-WAITING-TIME to harvest. This is the waiting period required by the federal law between the last application of a pesticide to a food crop and its harvest and is shown on the label. These waiting periods are expressed as the number of days from the last application of the fungicide until the vegetable can be harvested. The number of days that you must wait before harvest varies depending upon the fungicide and the crop; therefore, you must look on the fungicide label for this information. If you observe the days-to-harvest waiting time, there should be no reason to fear that your vegetables will contain harmful fungicide residues as your family sits down to enjoy a delectable, home-grown dinner.

TABLE 23. Disease Control Suggestions for the Home Garden, With and Without Chemicals.[1]

Plant	Disease	Chemical Control	Non-Chemical Control
Asparagus	Rust	Mancozeb or Maneb. Apply at 10-day intervals, but only after harvest.	Grow resistant varieties as Viking, Mary Washington and Waltham Washington. Remove and burn plant debris in fall.
Beans (dry, snap, lima)	Seed rots	Captan, Metalaxyl, or Thiram. Treat seed before planting.	
	Bacterial blight and Anthracnose	None practical or effective.	Plant only western-grown disease-free seed. Rotate planting areas and burn bean refuse in fall.
	Mosaic	None. Manage insect vectors.	Contender, Topcrop, Tendercrop and Tendergreen are resistant varieties.
	Sclerotinia white mold	Thiophanate-methyl. Spray at green bud stage and again in 7 days.	Plant in well-drained area not planted to beans last year.

[1] Refer to Appendix D for trade or proprietary names of recommended fungicides.

Plant	Disease	Chemical Control	Non-Chemical Control
Beans (Continued)			
	Downy mildew	Maneb, Chlorothalonil. Make spray application when mildew first appears.	Water or sprinkle only in mornings.
	Powdery mildew	Dinocap. Dust or spray and again in 7-10 days.	
Beans (pole)	Rust and Halo blight	None	FM-1 is resistant.
Beets	Damping-off Seed rot	Thiram or Captan. Treat seed with or soak in Thiram solution. Use Captan or Zineb as seed treatment or apply Captan to soil before planting.	
	Curly top	None	Parma Globe is resistant.
	Cercospora leaf spot	Chlorothalonil. Apply to foliage 2-3 times at weekly intervals starting at first symptoms.	Most varieties are resistant.
	Mildew		F.M. Detroit and Dark Red are resistant. Avoid overwatering.
Cabbage Cauliflower Collards Broccoli Brussels Sprouts	Damping-off	Thiram or Captan. Dust seed before planting.	Soil temperature must be at least 55°F for planting.
	Clubroot	Terraclor® (PCNB). Add dilute solution to hole during transplanting.	Do not plant crucifers on same soil in consecutive years. Avoid low spots.
	Downy mildew	Maneb or Chlorothalonil. Apply at mid-season or on first sign of mildew and late-season treatment with Fixed Copper.	Plant hot-water-treated seed. Water or sprinkle only in mornings.
	Blackleg Black rot	Fumigation is the only chemical control. Not for home gardens.	Buy only hot-water-treated seed.
Carrots	Cercospera and Alternaria leaf spots	Maneb or Chlorothalonil. Apply when first signs of disease occur.	Use 2-3 year rotation to reduce fungus carryover.
	Carrot yellows (virus)	Control leafhoppers with recommended insecticide.	Aluminum foil mulch will repel leafhoppers.
Celery	Early or Late blight	Maneb, Thiabendazole or Chlorothalonil. Spray every 7-10 days	Soak seed at 118°F for 30 minutes.

Plant	Disease	Chemical Control	Non-Chemical Control
Corn (see sweet corn)			
Cucumbers Muskmelons Watermelons	Damping-off	Thiram or Captan. Dust seed before planting.	Control cucumber beetles which spread bacteria from plant to plant.
	Bacterial wilt	None effective.	
	Powdery mildew	Triadimefon or Thiophenate-methyl Complete coverage spray at first signs of mildew and again in 7 days.	
	Angular leaf spot	Fixed Copper. Spray at first symptoms.	Plant tolerant cucumber varieties: Pioneer, Premier, Carolina.
	Wilts (Fusarium and Verticillium)	No chemical control.	Avoid planting where problem has occurred before. Muskmelons resistant to fusarium are Iroquois, Gold Star, Harper Hybrid, Saticoy Hybrid. Summer Festival watermelon is resistant to fusarium and anthracnose.
	Curly top (virus)	No chemical control. Malathion for leafhopper control.	Transmitted by leafhoppers. Protect plants from leafhopper feeding.
	Scab	Chlorothalonil or Mancozeb. Spray at 5-7 day intervals.	Resistant cucumbers: SMR-58, Marketmore 70, Pacer, Slicemaster, and Victory.
	Mosaic	No chemical control. Manage aphids that spread virus.	Resistant cucumbers: Marketmore 70, Pacer, Slicemaster Tablegreen 65, Victory. Destroy perennial weeds.
Eggplant	Damping-off	Captan. Add to first water used on seed flats, or to hole during transplanting. Spray plants weekly, enough to run down stem.	
	Anthracnose and Phomopsis fruit rots	Maneb or Mancozeb. Spray at 10-day intervals. Begin when first fruits are 2" in diameter.	Use 4-year rotation. Plant hot-water-treated seed.
	Wilt (Verticillium)	No chemical control.	Avoid planting in wilt problem area. Rotation free of peppers, strawberries, and tomatoes. Plant resistant varieties.

Plant	Disease	Chemical Control	Non-Chemical Control
Lettuce	Downy mildew	Maneb or Metalaxyl. Apply every 7 days beginning at first signs of mildew.	Water or sprinkle only in mornings. Space head lettuce at least 10" apart for good air circulation.
	Bottom rot (Rhizoctonia) and Drop (Sclerotinia)	Vinclozolin or Iprodione. See label directions.	Rotate planting areas each year, with deep plowing. Use high beds.
Muskmelons (see Cucumbers)			
Mustard Greens	Alternaria leafspot and Downy mildew	Maneb, Metalaxyl or Mancozeb. Spray at 7-10 day intervals.	Water only in mornings. Do not sprinkle.
Onions, Garlic and Chives	Purple blotch	Maneb, Iprodione or Chlorothalonil. Spray at first signs and repeat at weekly intervals.	Plant local resistant varieties.
	Downy mildew	Maneb, Metalaxyl or Mancozeb. Apply at first signs of mildew and at 7-day intervals as needed.	Water or sprinkle only in mornings.
	Smut	Captan or Thiram. Dust seed before planting.	Smut attacks only onions grown from seed. Plant disease-free onion sets.
Parsnips	Leaf diseases	Maneb or Mancozeb. Apply when spots first appear.	Water or sprinkle only in morning. Plant on high beds.
Peas	Powdery mildew	Dinocap or Thiophanate-methyl Dust or spray at first signs of disease.	Avoid locating late plantings close to early plantings. Destroy vines as soon as harvested.
	Wilt	No chemical control.	Plant resistant varieties. Plant early and rotate planting areas each year.
	Mosaic	Control aphids early that transmit mosaic.	Thomas Laxton 60 is resistant. Eliminate weeds around garden, especially pokeweed.
Peppers	Damping-off	Captan. Add to first transplant water.	
	Bacterial spot	Fixed Copper. Several applications are effective if started before disease appears.	Purchase disease-free seed or transplants.
	Wilt (Verticillium)	No chemical control.	Avoid planting where problem has occurred before.
	Virus mosaic	Malathion. Spray to control the aphid vectors as they appear.	Remove and destroy perennial weeds and old crop residue.

Plant	Disease	Chemical Control	Non-Chemical Control
Peppers *(Continued)*			
	Anthracnose and Fruit rot	Maneb or Thiophanate-methyl. Spray at 7-day intervals through harvest. Begin when first fruits are 1" in diameter.	
Potatoes	Scab	Sulfur. Add to soil to reach pH 5.2 before planting. Treat seed pieces with Captan or Mancozeb dust before planting.	Resistant varieties: Russett Burbank, Russet Rural, Norland, Cherokee and Superior.
	Virus	Control vector aphids.	Plant certified seed potatoes.
	Early and Late blights	Maneb or Chlorothalonil. Spray every 7-10 days, beginning when plants are 6" high and continue until vines are dead. Use fixed copper when severe.	Kennebec is tolerant to late blight.
	Rhizoctonia stem canker	Captan. Dust seed pieces before planting.	Plant entire potato whole. Do not cut into pieces.
	Seed piece rot	Captan. Dust seed pieces before planting.	Plant entire potato whole. Do not cut into pieces.
	Tuber diseases		Buy certified seed potatoes. Plant uncut tubers.
Pumpkins and Squash	Damping-off and Seed rots	Thiram or Captan. Dust seed before planting.	
	Bacterial wilt	None effective.	Control cucumber beetles which spread bacteria from plant to plant.
	Powdery mildew	Triadimefon or Chlorothalonil. Apply spray or dust when mildew appears and again in 10 days.	
	Mosaic (virus)	No chemical control. Control aphid vectors and weeds, instead.	Eliminate pokeweed and wild cucumber weed hosts. Kill other perennial weeds within 150 feet. Aluminum foil mulch will repel aphids.
	Curly top (virus)	No chemical control. Control leafhopper vectors and weeds instead.	Aluminum foil mulch to repel leafhoppers.
Radishes (see Turnips)			
Rhubarb	Leaf spot	Maneb or Fixed Copper. Apply early as needed.	Remove and burn leaves in fall.
Rutabagas (see Turnips)			

Plant	Disease	Chemical Control	Non-Chemical Control
Spinach	Downy mildew	Metalaxyl or Maneb. Spray at 7-10 day intervals.	Plant resistant varieties as Hybrid 612, Early Hybrid No. 7 and Chesapeake. Water or sprinkle only in mornings.
Squash (see Pumpkins)			
Sweet corn	Smut	No chemical control.	Pick and destroy galls. Late maturing varieties are more tolerant than early varieties.
	Leaf blight	Control usually not necessary.	
Sweet potatoes	Black rot, Scurf, Foot rot	No chemical control.	Buy certified plants. Use 3- or 4- year rotation.
	Wilt, Root-knot, Soil rot	No chemical control.	Plant resistant varieties.
Tomatoes	Damping-off and Seed rot	Captan. Use in first water when transplanting. Or Metalaxyl foliage spray.	
	Early blight, Late blight and Anthracnose	Maneb, Chlorothalonil or Metalaxyl. Make first application when fruit appear and continue at 7-10 day intervals or after rains.	Rotate tomatoes with beans or sweet corn.
	Blossom end rot	No chemical control.	Maintain adequate soil moisture by watering plants if needed. Add lime to soil.
	Wilts (Fusarium and Verticillium)	No chemical control.	Plant resistant varieties and avoid planting where problem has occurred before. Campbell 1327, Jet Star, Springset, Better Boy, Heinz 1350, Supersonic are resistant varieties.
	Bacterial spot	Maneb or Fixed Copper. Apply in early flowering stage and as needed when disease appears.	Plant certified disease-free seed, preferably hot water treated.
	Septoria	Maneb, Chlorothalonil or Thiophanate-methyl. Spray at 7-day intervals.	
Turnips, Radishes and Rutabagas	Alternaria leafspot and Downy mildew	Metalaxyl, Maneb or Mancozeb. Apply at 7-day intervals. Fixed Copper. Apply to radishes at 7-day intervals.	Water only in mornings. Do not sprinkle.

IDENTIFICATION OF COMMON PROBLEMS WITH HOUSEPLANTS

SYMPTOMS	POSSIBLE CAUSES
Brown or yellow leaves	Occasional yellowing of lower leaves is natural, due to insufficient light. Temperature extremes, chills and drafts, overfeeding, low humidity, not enough fresh air, over-watering, underwatering or underfeeding can cause discolored leaves.
Yellow or white rings and spots on leaves	Splashing cold water on foliage while watering plants can result in rings and spots on the leaf surface.
Leaf drop	Usually leaf drop is caused by a sudden change in the environment; especially temperature or light intensity, cold water, dry air, transplanting shock, overwatering or overfeeding.
Bud drop	Temperatures too hot or cold during bud development, shocks due to poor humidity, drafts, or moving from the greenhouse into the home can cause bud drop.
Slow growth	During winter, slower growth is probably due to poor light or natural dormancy. Otherwise it is usually a result of overwatering or underfeeding, but can also be caused by sour, undrained soil and cramped or dry roots.
Loss of foliage, pale color, weak growth	Overwatering, lack of fertilizer, insufficient light or high heat, are causes of poor plant growth.
Yellow or brown spotted leaves	Overwatering, sun scorch from direct sunlight, drought, water left on the leaves in the hot sun, insects, or damage from pesticides can result in spotting.
Dry crumbly leaves	High heat, low humidity, overfertilizing, or pesticides can damage leaves.
Wilted leaves, new growth small	Overcrowded roots give above-ground symptoms of wilted leaves and stunted new growth.
Abnormally long stems, leaves long and pale	Insufficient light causes leggy plant growth.
Leaves curl under, new leaves are small and compact	Too much light and, on occasion chilling injury, shows as downward cupped leaves and stunted new growth.

Collapse of plant	Plants can collapse due to extreme heat or cold, gas fumes, poor drainage causing root rot, sudden change in environmental conditions or possibly bacterial or virus disease. The latter is highly unlikely, particularly if the soil is sterilized before use.
Mushy, dark stems, rotting	Too much water, particularly in cold weather, poor drainage, poor ventilation or fungal attack encouraged by damp, cold conditions can cause rot.
Swelling on leaves and corky ridges, water soaked spots that turn red or brown	Oedema occurs when the plant has absorbed too much moisture because of warm moist soil and cool moist air. To stop this, raise heat, lower humidity and place pot where soil will not heat.
Tips of leaves turn brown	Poor humidity, potash deficiency, or underwatering are causes of tip burn.
Plants bear few flowers and excess foliage	Too much nitrogen has been supplied if the plant produces excessive foliage.
Lower leaves turn yellow and remain on the plant, new leaves are small and stems are stunted	Lack of fertilizer, particularly nitrogen, causes poor plant growth.
White crust on soil	Buildup of fertilizer or salts; can be leached away with water or the plant can be repotted.
Wilting	Wilting can be caused by insufficient water, too much heat or light. If these are not to blame, there may be root damage caused by overwatering, pests, or overfertilizing.
Yellowing between the veins of young leaves, older leaves are less severely affected.	Iron or magnesium deficiency.

TABLE 24. Disease Control Suggestions for Houseplants, With and Without Chemicals. [1]

Plant	Disease	Chemical Control	Non-Chemical Control
African violet	Botrytis blight Crown rot Petiole rot Ring spot	Thiophanate-methyl or Mancozeb. Spray or dip. [3]	Water by soaking from base rather than overhead watering. Plant sanitation. [2]
Begonia	Bacterial leafspot	Maneb or Fixed Copper. Spray or dip [3] frequently.	
	Botrytis blight	Chlorothalonil or Mancozeb.	
	Powdery mildew	Triadimefon or Sulfur. Spray or dip. [3]	
Ferns	Anthracnose	Mancozeb as needed.	Plant sanitation. [2] Avoid wetting fronds.
Gardenia	Bacterial leafspot	Sulfur, if needed.	Plant sanitation. [2]
	Bud drop	No chemical control.	Avoid high night temperatures.
Geranium	Bacterial leafspot	No chemical control.	Plant sanitation. [2] Keep foliage dry.
	Botrytis blight	Maneb, Iprodione or Chlorothalonil. Spray or dip. [3]	
Gloxinia	Leaf and Stem rot	No chemical control.	Provide good soil drainage by omitting clay from potting mixture.
Ivy (Boston and English)	Leafspot disease	Sulfur or Thiophanate-methyl. Spray or dip [3].	Plant sanitation. [2]
Palms	Leafspot disease	Basic Copper. Spray as needed.	
Pittosporum	Leafspot disease	No chemical control.	Plant sanitation. [2]
Rubber plant	Anthracnose	Mancozeb.	Maintain good cultural methods.
	Leaf scorch Oedema Root rot	No chemical control.	Maintain good cultural methods. Don't overwater.

[1] Refer to Appendix D for trade or proprietary names of recommended fungicides.

[2] Remove and destroy diseased plants and plant residue.

[3] Dip foliage and stem into solution equal to one-half the spray concentration.

LAWN AND TURF

Disease free turf depends on many things. Successful disease management results from understanding there are several methods for preventing or managing infectious diseases. This total concept, referred to as integrated disease management, is extremely important and starts with the planning of the turf area.

Plant pathologists refer to the use of resistant varieties of turfgrasses as the first line of defense in disease management. This is especially true when planning and planting bluegrasses. In recent years, disease resistance has been a basic part of breeding and selection of bluegrass varieties. Most of the improved varieties have more resistance to several diseases than the "common" bluegrass varieties. They are not necessarily resistant to all diseases. For instance, Merion bluegrass is resistant to Helminthosporium melting out, but susceptible to powdery mildew. When making variety decisions, think about diseases that may be expected on the site to be seeded. For instance, shaded, damp spots would be prone to powdery mildew development, thus Merion should not be planted in those locations.

Nutrition of turfgrasses can influence the development of diseases. Some, but not all, diseases are worse when insufficient nutrition has stressed the plants and increased their susceptibility. The danger here is that homeowners will over-fertilize. This may worsen many diseases. The best advice: follow an adequate fertilization program based on needs as indicated by soil tests.

Thatch management is an important element of integrated disease prevention. Too much thatch provides places where disease-causing fungi can survive from year to year. When conditions are right, these survivors will become active and quickly create epidemic situations. Too much thatch also serves as a barrier to the proper penetration of pesticides and may prevent water from reaching the roots. Thatch may prevent grass from rooting well in the soil. Insufficient water penetration and failure of plants to root well lead to stress. Stress prevention goes hand in hand with disease management.

Even with proper thatch management, watering practices may help or hinder your disease management program. Less frequent, but thorough, waterings are preferred over frequent, light waterings. Frequent waterings promote spore production and spread of many disease causing fungi. If watering is not thorough, it can lead to shallow rooted plants more prone to stress and disease.

Information about specific diseases you are dealing with helps to better plan an integrative disease management program. The point is, that successful disease management is more than simply applying a fungicide.

Fungicides must be used correctly. Make the right diagnosis, choose the right fungicide, decide when to spray and determine the best method of application.

Correct diagnosis is most difficult with turf diseases. "Brown spots" are associated with many turf problems, including some diseases. Taking a closer look is the first step. Look further for associated symptoms such as the presence of insects or leafspots. Take your time. (Most state universities have plant disease clinics that will provide free diagnoses if turf samples are shipped to them in good condition. Check by phone before sending your sample.)

Diagnosis may turn out to be a non-infectious disease or insect problem. If so, spraying a fungicide is worthless. Remember too, that most fungicides do not work on all diseases.

The package label usually tells when and how to apply the fungicide. The first application aimed at preventing a future outbreak can be the most successful of disease management practices. Knowledge of the seasonal cycle of the disease-causing fungus and conditions that promote infection will help in making that important first application. Follow-up applications will depend on the fungicide and persistence of weather that promotes the disease.

When applying fungicides, use pressures above 25 PSI (pounds per square inch) to ensure good penetration and distribution. Hand sprayers are usually above 25 PSI when pumped up. The key to effective application is to realize that you need to provide a barrier of fungicide film over the plant parts susceptible to infections. For instance, the Helminthosporium leaf spot fungus infects leaf blades. However, the striped smut fungus infects the crown. Good penetration into the turf is required for smut prevention.

A spreader-sticker used as a spray additive helps to give better foliage coverage. Hose-on applicators that apply large volumes of water are not satisfactory when the fungicide is applied to foliage. This washes the fungicide into lower levels of the plant where it may not be the most effective for foliage diseases, such as powdery mildew. Hose-on applicators are, however, quite satisfactory for diseases that start at the crown.

TABLE 25. Disease Control Suggestions for Lawn and Turf With and Without Chemicals.[1]

Disease	Susceptible Turfgrasses	Pesticide to Use	Non-Chemical Control
Algae	All grasses	Mancozeb. When algae appear.	Reduce shade. Avoid excessive watering and improve soil drainage.
Brown Patch (Rhizoctonia)	Bentgrass Bluegrass Bermuda Fescue Ryegrass	Thiophanate-methyl, Mancozeb or Chlorothalonil. Apply at 7-day intervals in July-August.	Avoid high-nitrogen fertilizer, and increase air circulation.
Copper Spot *(Gloeocercospora sorghi)*	Bentgrass	Chlorothalonil or Triadimefon. Apply from late June to October, as needed.	
Dollar Spot (Sclerotinia)	Bentgrass Bluegrass Ryegrass Fescue Zoysia	Thiophanate-methyl, Fenarimol or Chlorothalonil. Apply from late June to October. Avoid continued use.	Mow at maximum recommended height. Increase nitrogen in soil. Maintain soil moisture. Water only in mornings. Resistant bluegrasses: Adelphi, BonnieBlue, Midnight and Victa.
Fairy rings	All grasses	Fungicides are ineffective. Requires eradication.	Provide adequate water and fertilizer. Remove infested sod and soil; replace with clean soil and reseed or sod.
Fusarium Blight	Bentgrass Bluegrass (esp. Merion) Fescue Ryegrass	Triadimefon or Fenarimol. Apply in July and August.	Water frequently but lightly during dry periods. Lime annually to soil pH above 6.2. Avoid excess nitrogen. Mow at maximum height. Remove clippings and avoid thatch.Resistant bluegrasses: Adelphi, Midnight, Mystic and Touchdown.
Leafspot and Crown Rot (Helminthosporium)	Bentgrass Bluegrass Bermuda Fescue Ryegrass	Triadimefon, Iprodione or Chlorothalonil. Apply at 7-14 day intervals, April-June more frequently in cool moist spring.	Merion Bluegrass is a resistant variety. Remove clippings, raise cutting height of mower and fertilize to maintain vigor. Resistant bluegrasses: BonnieBlue, Fylking, Midnight and Touchdown.

[1] Refer to Appendix D for trade or proprietary names of recommended fungicides.

Disease	Susceptible Turfgrasses	Pesticide to Use	Non-Chemical Control
Nematodes	All grasses	No chemical control available.	Maintain plant vigor through regular and proper fertilization.
Powdery mildew (Erysiphe)	Bermuda Bluegrass Fescue	Thiophanate-methyl or Fenarimol. Apply when mildew appears, July-September.	Reduce shading where possible. Water only in mornings. Increase air circulation by removing surrounding vegetation. Resistant bluegrasses: Bristol, Eclipse, Glade and Nugget.
Pink Patch	Red fescue Perennial ryegrasses	Triadimefon or Thiophanate-methyl.	Follow balanced fertilization program.
Pythium blight (Pythium)	Bentgrass Bluegrass Ryegrass Bermuda Fescue	Metalaxyl or Etridiazole. Hot, wet weather disease which should be treated immediately.	Maintain good growth with moderate fertilizer use and soil pH at 6.2-7.0. Avoid mowing when grass is moist.
Red Thread (Corticium)	Bentgrass Bluegrass Fescue Ryegrass	Triadimefon. Apply in May-June and again in August-September.	Maintain adequate soil fertility. Apply lime to achieve soil pH of 6.5-7.0. Water only in mornings and avoid overwatering. Resistant bluegrasses: Adelphi, Challenger, Monoply and Nassau.
Rust	Bluegrass (esp. Merion) Ryegrasses	Propiconazole, Mancozeb or Chlorothalonil. Make 2-3 applications at 12-14 day intervals in July-August.	Good soil fertility may help decrease rust problem. Resistant bluegrasses: Fylking, Park, and Sydsport.
Slime molds (Myxomycete)	All grasses	Mancozeb. Treat in August-September.	Remove affected grass by mowing and raking.
Snow molds (Fusarium and Typhula)	Bentgrass Bermuda Bluegrass Fescue Ryegrass.	Chlorothalonil or Pentachloronitrobenzene (PCNB). Apply before snow, midwinter and during spring thaw.	Avoid late fall fertilizing. Minimize thatch and accumulation by raking matted grass in spring.
Stripe smut (Ustilago)	Bentgrass Bluegrass Meadow fescue Ryegrass	Triadimefon or PCNB (Terraclor®) Apply in late fall or early spring using extra water for penetration.	Water carefully to maintain vigor. Assure adequate phosphorus fertility. Resistant bluegrasses include A-34, Bonnieblue, Sydsport and Touchdown.

Disease	Susceptible Turfgrasses	Pesticide to Use	Non-Chemical Control
Toadstools and mushrooms.	All grasses	Fenarimol or Thiram. Drench soils of trouble spots.	Maintain adequate fertilization. Remove fruiting bodies when observed and safely discard. They may be poisonous.
Yellow Patch	Bluegrasses	None	Remove excessive thatch.
Yellowing	Bermudagrass and its hybrids St. Augustine (West and Southwest U.S.)	Usually not a disease, but lack of iron.	Treat with fertilizer containing iron or iron chelates.

TABLE 26. Disease Control Suggestions for Trees, Shrubs and Woody Ornamentals With and Without Chemicals. [1]

Plant	Disease	Pesticide to Use	Non-Chemical Control
Andromeda (Pieris)	Leaf spots	Sulfur, if needed.	Remove and destroy diseased leaves.
Azalea	Leaf spots	Mancozeb or Sulfur. Apply frequently beginning in spring.	Plant sanitation.[2]
	Dieback	Bordeaux. Apply as new leaves appear.	Prune and burn infected twigs.
	Flower gall	No chemical control.	Remove galls and burn.
Barberry	Bacterial leaf spot	Bordeaux or Basic Copper Sulfate. Apply 3 times at 10-day intervals, beginning when new leaves open.	Rake and burn old leaves in fall.
	Verticillium wilt	No chemical control.	Replace with another kind of plant.
Birch	Nectria canker	Bordeaux. Apply 4 times at weekly intervals beginning in early spring.	Prune infected branches and paint stubs with wound dressing.
Boxwood	Canker	Bordeaux. Apply 4 times through growing season.	Prune infected branches, and burn old leaves.
	Leaf spots	Potassium Bicarbonate or Sulfur, if needed.	Plant sanitation [2] and fertilize to maintain vigor.

[1] Refer to Appendix D for trade or proprietary names of recommended fungicides.
[2] Remove and destroy, preferably by burning, diseased plants and plant residue. Rake and remove stems, twigs and leaves in the fall and burn.

Plant	Disease	Pesticide to Use	Non-Chemical Control
Cercis (see Red Bud)			
Clematis	Leaf spots	Thiophanate-methyl. Apply 3 times at 10-day intervals, usually May.	Plant sanitation. [2]
	Stem rot	Maneb. Apply at 2-week intervals during growing season; drench soil around crown.	
Cotoneaster	Fireblight	Bordeaux. Apply twice during flowering.	Prune dead twigs only when plant is dry.
Crab-Apple	Cedar-apple rust	Mancozeb, Triadimefon or Chlorothalonil. Apply 4-5 times at 7-10 day intervals when galls on nearby junipers produce jelly-like secretions.	Remove nearby junipers.
	Fireblight	Bordeaux. Apply twice during flowering.	Prune dead twigs only when plant is dry.
	Scab	Mancozeb or Chlorothalonil. Apply weekly beginning when buds turn pink.	Select resistant varieties.
	Powdery mildew	Sulfur or Triadimefon. Spray after periods of wet weather.	Prune to allow sun penetration into canopy.
Dogwood	Flower and Leaf Blight	Maneb or Triadimefon. Apply after periods of wet weather.	Plant sanitation. [2]
	Leaf spots	Mancozeb, Chlorothalonil or Triadimefon. Apply after periods of wet weather.	
	Crown Canker	No chemical control.	Look for borer damage and treat accordingly.
	Twig blight	No chemical control.	Plant sanitation. [2]
Dutchman's Pipe	Leaf spots	Mancozeb or Sulfur. Apply 3 times at 10-day intervals, usually May.	Plant sanitation. [2]
Euonymus	Crown gall	No chemical control.	Remove and destroy diseased plants including roots.
	Downy mildew	Chlorothalonil or Thiophanate-methyl Spray as needed.	Water only in mornings. Avoid sprinkling. Thin dense foliage.
Firethorn	Fireblight	Bordeaux. Apply twice during flowering.	Prune dead twigs. Fertilize lightly but frequently.

Plant	Disease	Pesticide to Use	Non-Chemical Control
Firethorn *(continued)*			
	Scab	Captan or Triadimefon. Apply weekly beginning when buds begin to expand.	
Hawthorn	Cedar-apple rust	(see crabapple)	
	Fireblight	Bordeaux or Fosetyl-Al. Apply twice during flowering.	Prune dead twigs only when plant is dry.
	Leaf spots	Triadimefon. Apply when leaves are young, after periods of wet weather.	Resistant varieties are Cockspur and Washington.
Ivy (Boston and English)	Leaf spot	Captan or Mancozeb. Make 3 sprays when flowers open and repeat weekly.	
	Powdery mildew	Thiophanate-methyl or Triadimefon. Spray 3 times begining in July.	Thin by pruning for air circulation.
Juniper	Twig blight	Bordeaux or Triadimefon. Apply 3 times at 14-day intervals, beginning with new growth.	Prune twigs 2 inches into live wood and burn. Select resistant varieties.
	Cedar-apple rust	Triadimefon. Spray twice at 3-week intervals in early August.	Remove galls in early spring. Plant resistant varieties.
Laurel	Leaf spot	Sulfur. Apply at 10-day intervals from beginning of new growth until leaves mature.	Remove and destroy diseased leaves.
Leucothoe	Leaf spot	Maneb, Triadimefon or Ferbam. Apply 3 times at 14-day intervals when new growth begins.	Plant sanitation. [2]
Lilac	Bacterial wilt	Bordeaux. Apply 3 times at 10-day intervals when new growth appears.	Remove and destroy wilted twigs.
	Powdery mildew	Triadimefon or Sulfur. Apply weekly when mildew appears.	
London Plane	Anthracnose	Thiophanate-methyl or Chlorothalonil. Apply 3 times when buds break and at weekly intervals	Rake and burn all leaves in fall.
	Powdery mildew	(see lilac)	

Plant	Disease	Pesticide to Use	Non-Chemical Control
Mahonia	Leaf spot	No chemical control.	Plant sanitation. [2]
	Scorch	No chemical control.	Water during dry periods and protect from wind.
Maple	Anthracnose	Maneb. Apply 3 times when buds break and at weekly intervals. Copper Oxychloride Sulfate. Apply twice-1st appearance and again 7-10 days.	Rake and burn leaves after frost.
	Scorch	No chemical control.	Water during dry periods and protect from wind.
	Tar spot	Ferbam. Apply 3-4 times beginning when buds break and at 2-week intervals. (Copper Oxychloride Sulfate. Apply twice--1st treat, repeat in 7-10 days)	
	Verticillium wilt	No chemical control.	Water during dry periods and fertilize properly.
Mountain Ash	Cytospora canker	None.	Prune diseased branches during dry weather.
	Leaf spot	Maneb or Ferbam. Apply at 10-day intervals from new growth to mature leaves.	Rake and burn dead leaves.
	Fireblight	Bordeaux. Apply twice during flowering.	Prune dead twigs and do not fertilize.
Pachysandra	Blight	Chlorothalonil or Mancozeb. Apply 3 times at weekly intervals beginning with new growth.	Clean up and thin bed.
Periwinkle (Vinca)	Blight	No chemical control.	Thin for better ventilation.
	Canker and Dieback	No chemical control.	Prune and burn infected stems.
Peach and Cherry	Leaf spot	Captan. Spray 3 times at 2-week intervals, beginning with petal fall.	Rake and burn dead leaves.
	Peach leaf curl	Dormant spray of Bordeaux in April.	
	Twig blight	Captan. Spray twice at 10-day intervals, beginning just before blossoms open.	Prune infected limbs and paint stubs with wound dressing.
	Black knot	No chemical control.	Prune galls in winter.

Plant	Disease	Pesticide to Use	Non-Chemical Control
Pieris (see Andromeda)			
Pine	Blister rust (white pine)	No chemical control.	Prune cankers on twigs and remove wild currant bushes.
	Needle blight	Bordeaux or Mancozeb. Apply twice when new growth starts and at 2-week intervals.	
	Twig blight	Bordeaux or Triadimefon. Apply 3 times, when new growth starts and at weekly intervals.	Plant sanitation. [2] Fertilize and water.
Privet	Anthracnose and Twig Blight	Maneb or Ferbam. Apply at weekly intervals until controlled.	Remove and destroy infected branches.
Pyracantha (see Firethorn)			
Quince	Crown gall	No chemical control.	Dig up and destroy infected plants and roots.
	Fireblight	Bordeaux. Apply twice during flowering	Prune dead twigs.
	Leaf spots	Mancozeb. Apply when leaves are half grown.	
	Cedar-apple rust	Ferbam or Chlorothalonil. Apply 4-5 times at 7-10 day intervals when galls on nearby junipers produce jelly-like secretions.	Remove nearby junipers.
Redbud (Cercis)	Canker	No chemical control.	Prune cankers and paint wounds with orange shellac and wound dressing.
Rhododendron	Leaf spot	Thiophanate-methyl or Ferbam. Apply at 10-day intervals from new growth to mature leaf.	Plant sanitation. [2]
	Dieback	Bordeaux. Apply 2 times at 14-day intervals. Hard to control.	Prune and burn infected plant parts.

Plant	Disease	Pesticide to Use	Non-Chemical Control
Roses	Blackspot	Thiophanate-methyl or Chlorothalonil. Apply at weekly intervals.	Blackspot-resistant rose hybrids: Hybrid teas (Duet, First Prize, Peace); Floribunda/Grandifolia (Angel Face, Goldilocks, Love); and Shrub roses (All That Jazz, Carefree Wonder).
	Botrytis blight	Chlorothalonil or Mancozeb Spray 3 times at 7-14 day intervals.	
	Crown gall	No chemical control.	Remove bushes and burn Relocate rose bed.
	Powdery mildew	Triadimefon, Sulfur or Triforine. Apply when meldew appears and at weekly intervals throughout season.	Select resistant varieties.
	Rust	Mancozeb or Sulfur. Spray 3 times at 10-day intervals, beginning May 15-June 1.	Avoid crowding. Allow air circulation between plants.
Spruce	Cytospora canker	No chemical control.	Prune diseased limb where it branches from trunk, apply wound dressing. Practice plant sanitation.
Sycamore	Anthracnose	Maneb. Apply 3 times when buds break and at weekly intervals. Copper hydroxide. Apply 2 times-at bud crack, and 7-14 days later.	Rake and burn all leaves in fall.
Vinca (see Periwinkle)			
Willow	Black canker and Twig blight	Mancozeb or Maneb. Apply 3 times at 14-day intervals. Begin when leaves are 1/4 inch long.	Prune and burn diseased twigs. Weeping willow and laurel willow are somewhat resistant.

TABLE 27. Disease Control Suggestions for Ornamental Annuals and Perennials With and Without Chemicals.[1]

Plant	Disease	Pesticide to Use	Non-Chemical Control
Ageratum	Root rot	No chemical control.	Use only sterilized potting soil.
Chrysanthemum	Leaf spot	Folpet or Captan. Spray weekly if leaf spot was severe previously.	Plant sanitation.[2]
	Powdery mildew	Triadimefon or Thiophenate-methyl. Apply after periods of wet weather.	
	Rust	Maneb. Spray at first symptoms and as needed.	Remove and bury infected leaves.
	Virus	Malathion. Treat weekly to control aphid vectors.	Avoid propagating from unknown sources. Plant sanitation.[2]
Cockscomb	Leaf spot	Maneb. Apply after periods of wet weather.	Remove and destroy badly spotted leaves.
Columbine	Crown and Root rot	No chemical control.	Sterilize soil.
	Leaf spots	Triadimefon or Captan. Apply as needed.	Plant sanitation.[2]
	Rust	Maneb. Apply when pustules first appear and as needed.	
Dahlia	Gray mold *(Botrytis)*	Maneb or Triadimefon. Apply after periods of wet weather.	Plant in areas with good air movement.
	Powdery mildew	Thiophenate-methyl or Triadimefon. Apply when mildew appears, usually August.	Burn or bury all dead plant material in fall. Avoid crowding plants.
	Stem rot and wilt	No chemical control.	Sterilize soil, or replace with fresh soil.
	Virus	Malathion. Treat weekly to control aphid vectors.	Destroy roots.
Day-Lilies	Leaf spot	Sulfur, if needed.	Plant sanitation.[2]
Delphinium (see Larkspur)			
Forget-Me-Not	Downy mildew	Triadimefon or Bordeaux. Spray as needed.	Water only in mornings. Do not sprinkle.
	Wilt	No chemical control.	Sterilize soil, or replace with fresh soil.

[1] Refer to Appendix D for trade or proprietary names of recommended fungicides.
[2] Remove and destroy, preferably by burning, diseased plants and plant residue. Rake and remove stems, twigs, and leaves in fall and burn.

Plant	Disease	Pesticide to Use	Non-Chemical Control
Hollyhock	Anthracnose	Triadimefon. Spray after periods of wet weather.	
	Leaf spots	No chemical control.	Plant sanitation.[2]
	Rust	Maneb. Apply twice weekly beginning when infection appears.	Plant sanitation.[2]
Hydrangea	Bacterial wilt	No chemical control.	Plant sanitation.[2]
	Leaf spots	Maneb or Folpet. Apply at 2-week intervals and after rains.	
	Powdery mildew	Thiophenate-methyl or Triadimefon. Apply when mildew appears and at 2- week intervals as needed.	Avoid crowding, encourage air circulation.
Impatiens	Bacterial wilt and Damping off	No chemical control.	Sterilize soil, or replace with fresh soil.
	Leaf spots	Iprodione or Captan. Apply at 2-week intervals and after rains.	
Iris	Bacterial leaf spot	No chemical control.	Plant sanitation.[2]
	Bacterial soft rot	No chemical control.	Cut rotted areas from rhizomes. Dry in direct sunlight for 1 day.
Larkspur	Bacterial blight	None.	
	Crown rot	No chemical control.	Serilize soil, or replace.
	Leaf spot	Maneb. Apply at 2-week intervals and after rains.	
	Powdery mildew	Triadimefon or Thiophanate-methyl. Apply when mildew appears, usually August.	
Lilies	Bacterial soft rot	No chemical control.	Plant sanitation.[2]
	Botrytis blight	Iprodione or Mancozeb. Apply at 10-day intervals where disease was severe previously.	Plant sanitation.[2]
	Stem canker	Terraclor® (PCNB). Dip infected bulbs.	
	Wilt	No chemical control.	Sterilize soil, or replace.
Lupine	Crown rot	No chemical control.	Sterilize soil, or replace.

Plant	Disease	Pesticide to Use	Non-Chemical Control
Lupine *(continued)*			
	Powdery mildew	Sulfur or Triadimefon. Apply when mildew appears, usually August.	Burn all dead plant material in fall.
Marigold	Blight	No chemical control.	Plant sanitation. [2]
	Leaf spots	Maneb or Captan. Apply at 2-week intervals and after rains.	
	Root rot Stem rot Wilt	No chemical control.	Sterilize soil, or replace.
Nasturtium	Bacterial leaf spots	No chemical control.	Plant sanitation. [2]
	Bacterial wilt	No chemical control.	Sterilize soil, or replace with fresh soil.
	Leaf spots	Maneb or Captan. Apply at 2-week intervals and after rains.	
Pansy	Anthracnose	Triadimefon. Apply twice at 5-day intervals when disease appears.	Remove and burn infected leaves.
	Leaf spots	Maneb or Captan. Apply at 2-week intervals and after rains.	
	Scab	Maneb. Apply after periods of wet weather.	Use sterilized soil.
	Downy mildew	Maneb or Bordeaux. Apply to densely growing beds at 2-week intervals.	Thin plants for ventilation. Water or sprinkle only in mornings.
Peony	Anthracnose	Thiophanate-methyl. Apply 3 times at 5-day intervals in early season.	Plant sanitation. [2]
	Gray mold *(Botrytis)*	Triadimefon. Apply 3 times at 2-week intervals. Begin when tips break through soil, soaking soil.	Plant sanitation. [2]
	Root rot	No chemical control.	Plant sanitation. [2]
	Leaf spots	Captan or Mancozeb. Apply several times during growing season as needed.	Plant sanitation. [2]
Petunia	Virus diseases	No chemical control.	Purchase virus-free seed and destroy infected plants.

Plant	Disease	Pesticide to Use	Non-Chemical Control
Phlox	Leaf spots	Maneb or Bordeaux. Apply after periods of wet weather.	Plant sanitation. [2]
	Powdery mildew	Triadimefon. Apply as needed.	Plant resistant varieties and use plant sanitation. [2]
Poppy	Bacterial blight	No chemical control.	Plant treated, disease-free seed. Sterilize soil.
Shasta Daisy	Leaf spots	Maneb or Captan. Apply at 2-week intervals and after rains.	Remove and burn infected parts.
Snapdragon	Anthracnose	Maneb or Thiophanate-methyl. Spray as needed, usually July.	Plant disease-free seed.
	Rust	Mancozeb. Apply at 2-week intervals until plants reach 15 inches.	Water in morning.
	Downy mildew	Maneb or Triadimefon. Apply as needed.	Water only in mornings. Do not sprinkle.
	Powdery mildew	Triadimefon. Apply as needed.	Avoid crowding, allow ventilation.
Stock	Leaf spots	Maneb or Sulfur. Apply at 2-week intervals and after rains.	
	Wilt	No chemical control.	Sterilize soil.
Sweet Alyssum	Downy mildew	Bordeaux or Maneb. Apply as needed. Buy treated seed.	Plant disease-free seed. Sterilize or replace soil.
	Wilt	Thiophanate-methyl. Soak soil before planting or after rainy periods.	Sterilize or replace soil.
Sweet Pea	Anthracnose	Triadimefon. Apply at 2-week intervals during growing season.	Plant disease-free seed.
	Root rots	Triadimefon. Soak soil.	Sterilize or replace soil.
	Leaf spots	Mancozeb. Apply at 2-week intervals and after rains.	Remove and destroy diseased leaves.
Verbena	Bacterial wilt	No chemical control.	Sterilize or replace soil.
Zinnia	Powdery mildew	Triadimefon. Apply as needed, beginning in July.	
	Root and Stem rots	No chemical control.	Sterilize or replace soil.
	Virus	No chemical control.	Plant virus-free seed, and destroy infected plants.
	Leaf spots	Maneb. Spray as needed.	

ALGAE

Controlling Algae

The algae are a group of simple, fresh-water and marine plants ranging from single-celled organisms to green pond scums and very long seaweeds. For our purposes they become problems in swimming pools, aquaria, pet drinking troughs, fish ponds, greenhouses and recirculation water systems.

Swimming pools are a good place to begin. Only chemical control will handle the day-to-day problem of algae. Like all other plants, algae need water, food, light and a certain temperature range. In swimming pools, the nutrients soon arrive: human urine, body oils, dust and trash blown in and settled on the bottom, fertilizer inadvertently sprinkled beyond the lawn into the water, an over-supply of chlorine stabilizers and finally nutritional residues of algae themselves. Summer brings the other prerequisites of light and temperature range.

Algal control is accomplished by (1) keeping the acid/alkaline range (pH) near the ideal, which is 7.4 to 7.6 and (2) maintaining a chlorine level of 0.5 to 1.5 ppm using a chlorine-based algicide that inhibits algal growth and development. This range of pH and algicide level should be monitored twice weekly to avoid algal problems.

Other algicides may be used which are effective against disease-causing bacteria. After the algae are controlled, it is also necessary to prevent the development of harmful bacteria or pollution using a disinfectant. This is achieved through the maintenance of chlorine level. Since chlorine is also a good algicide, maintaining the proper disinfectant level also serves to prevent algal growth.

That strong smell of chlorine issuing from the YMCA or other public pool and athletic foot baths is probably released from the ever-popular calcium hypochlorite or chloride of lime, not only a good algicide but an excellent disinfectant as well. It contains 70% available chlorine and is the source of most bottled laundry bleaches. In addition to this, there are three other chlorine-based inorganic salts available for pool chlorination: sodium hypochlorite (NaClO), lithium hypochlorite (LiClO) and sodium chlorite ($NaClO_2$). For a source of inorganic chlorine, this author recommends only calcium hypochlorite (70% available chlorine) in its most economical form, usually granular, purchased by the drum.

Soluble forms of copper make excellent algicides. However, they do not control disease-causing bacteria. Any of the copper-containing algicides are equally effective and longer lasting, than the chlorine materials. The copper content of swimming pool water may eventually become phytotoxic to grass, plants, shrubs and trees surrounding the pool that may be splashed or drenched occasionally. This same problem applies as well to all other algicides of greater potency.

Aquathol® (endothall) and Algae-Rhap® (copper triethanolamine) are used to control algae in ponds, but are not registered for swimming pools.

Algae Control In Aquaria and Fish Ponds Without Chemicals (Almost)

Chemicals are not the only answer for algae control in aquaria and fish ponds, as most fish enthusiasts know. Bottom-feeding scavenger fish help to reduce algae production by feeding directly on algae and on fecal material of other fish, which serves as a nutritional source for algae. Two or three species of aquatic snails serve equally well and a heavy population can keep a large aquarium immaculate. A word of caution. Snails multiply rapidly and can over-populate the tank in 2-3 months, thus requiring an occasional removal of excess young.

If the proper balance of scavengers and fish are maintained, tank-cleaning can be a seldom-if-ever affair for the proud hobbyist. However, if everything seems to suddenly go wrong and an algae "explosion" occurs, empty the aquarium or pond after transferring the fish to a temporary holding tank and give it a good scrubbing. Use a stiff bristle brush that will reach the corners and cracks and a mild detergent. After rinsing, add a 1:1 dilution of bottled laundry bleach and water to the tank and scrub thoroughly. Rinse three times, emptying the contents after each rinse and invert to drain and dry, if possible. Then refill following any intricate instructions required for certain exotic species of tropical specimens. Scrub the shells of snails with an old toothbrush and tap water to remove excess algae before returning them to the tank.

Mildew, Slime and Disease-Causing Organisms

Mildew and slimes in shower stalls, on shower curtains, in the drains and overflow openings of bathtubs, sinks and lavatories and in greenhouses and other moist situations, are various species of fungi and can be easily controlled if the right steps are taken. Fungi are microorganisms whose requirements are water, food and the proper temperature range. Unlike algae, they do not require light and consequently, appear to thrive where light, especially sunlight, is not available.

Basically we can control these mildew and slime pests by making their environment unfit for survival. It is unlikely that their water or food supply can be removed for more than a brief period, but if this is feasible, it would be the logical process. Scrubbing followed by drying out and keeping dry will prove the demise of the problem.

If this cannot be achieved, the next step is to rely on the use of disinfectants, some of which have already been discussed as algicides. These include the inorganic chlorine compounds, quaternary ammonium halides and the sodium potassium dichloroisocyanurates.

NEMATODES

Controlling Nematodes

A famous nematologist friend of mine once said, "The average home gardener won't suspect that he has nematodes unless he has rootknot on some of his plants." Nematodes are microscopic roundworms that are very common and widespread in soil, water and other habitats. Most are free-living and feed on microorganisms such as bacteria, fungi and algae. About 200 species of the more than 15,000 described species are known to be plant parasites or plant feeders. Most of these feed in or on roots and other below-ground plant parts. Some of the more common nematodes and their host plants are shown in Table 28.

The symptoms of nematode injury to plants vary. Since nematodes damage root systems the visible or above-ground symptoms are the same as those caused by a failing root system, such as lack of vigor, wilting, early leaf fall, yellowing foliage, stunting and die-back. Infected plants may have galls, decay and rough, stubby, or black and discolored roots. These symptoms are usually blamed on poor soil or on lack of water and fertilizer. Some of the more common nematodes are the root-knot nematodes, which cause the characteristic galls or knots on roots, easily seen with the naked eye.

If you suspect that nematodes are damaging your crops, send diseased plants to the nearest Extension Plant Pathologist (See Appendix H). Most effective nematicides must be applied by a certified applicator using specialized equipment.

The non-chemical methods are really the only control methods I can recommend. They include crop rotation, fallow and dry tillage and sanitation. In a rotation system, a new planting site should be chosen if an area is available and not returned to for 3-4 years. Fallow and dry tillage allows the soil to remain uncropped, while removing all vegetation including weeds during the summer and cultivating frequently. In areas of moderate to high rainfall this method is basically impractical.

Sanitation calls for pulling up and disposing of all roots of annual plantings as soon as they are harvested, or when flower blooming is finished. The roots should not be used for compost since they will merely redistribute these wormy pests. Regardless of the care taken in carrying out all other precautions, your efforts are for the most part wasted if nematodes are brought back into the garden on infested plants or tools. Thus, it is necessary to buy clean transplants.

Despite my unwillingness to recommend chemical control, this is the best method. Several types of nematicides may be injected into the soil, drenched, sprayed or sprinkled onto the soil surface, applied into furrows along the row to be planted, or applied during watering or irrigation. Some nematicides may be applied prior to planting (preplant).

If you believe it necessary to fumigate, call a professional. In all cases it is essential to follow the manufacture's label instructions carefully.

TABLE 28. Some of the More Common Nematodes and Host Plants They are Known to Attack.

Plant	Nematode [1]
Beans and peas	Root knot nematodes Pea cyst nematode Sting nematodes
Citrus	Burrowing nematode Citrus nematode Lesion nematodes
Corn	Lesion nematodes Sting nematodes Stubby root nematodes
Grapes	Dagger nematodes Lesion nematodes Root knot nematodes
Ornamentals	Depending on the plant, almost any nematode
Potatoes	Potato cyst or golden nematode Potato rot nematode Root knot nematodes
Stone fruits (Peaches, Plums, etc.)	Lesion nematodes Root knot nematodes Stubby root nematodes
Tomato	Lesion nematodes Root knot nematodes
Turf	Lesion nematodes Root knot nematodes Turf sting nematode

[1] See Table 29 for common and scientific names.

TABLE 29. The Common and Scientific Names of Some of the More Important Nematodes.

Common Name	Scientific Name
Bulb and Stem Nematodes	*Ditylenchus spp.*
Bulb or Stem Nematode	*D. dipsaci*
Potato Rot Nematode	*D. destructor*
Burrowing Nematode	*Radopholus similis*
Citrus Nematode	*Tylenchulus semipenetrans*
Cyst Nematodes	*Heterodera spp.*
Cabbage Cyst, Cabbage Root or Brassica Root Nematode	*H. cruciferae*
Pea Cyst or Pea Root Nematode	*H. goettingiana*
Potato Cyst, Potato root or Golden Nematode or Potato Eelworm	*H. rostochiensis*
Soybean Cyst Nematode	*H. glycines*
Sugar Beet Cyst or Sugar Beet Nematode or Beet Eelworm	*H. schachtii*
Dagger Nematodes	*Xiphinema spp.*
False Root Knot Nematodes	*Naccobus spp.*
Foliar or Leaf Nematodes	*Aphelenchoides spp.*
Chrysanthemum Foliar Nematode	*A. ritzemabosi*
Lance Nematodes	*Hoplolaimus spp.*
Lesion or Meadow Nematodes	*Pratylenchus spp.*
Needle Nematodes	*Longidorus spp.*
Pin Nematodes	*Paratylenchus spp.*
Ring Nematodes	*Criconemoides spp. and Criconema spp.*
Root Knot Nematodes	*Meloidogyne spp.*
Northern Root Knot Nematode	*M. hapla*
Peanut Root Knot Nematode	*M. arenaria*
Southern Root Knot or Cotton Root Knot nematode	*M. incognita*
Sheath Nematodes	*Hemicycliophora spp. and Hemicriconemoides spp.*
Spiral Nematodes	*Helicotylenchus spp. and Rotylenchus spp.*
Sting Nematodes	*Belonolaimus spp.*
Peanut Sting Nematode	*B. gracilis*
Turf Sting Nematode	*B. longicaudatus*
Stubby Root Nematodes	*Trichodorus spp.*
Stunt or Stylet Nematodes	*Tylenchorhynchus spp.*

WEEDS

Weed — a plant whose virtues
have not yet been discovered.

Ralph Waldo Emerson

THE WEEDS

When mankind ceased being the nomadic hunter and settled down periodically to become the farmer, he encountered the third plague, weeds. Until recently he controlled weeds the same old ways, year-in and year-out, for thousands of years: hand pulling, hoeing and cultivation. Also, dear gardener, those are the choices you have, with the addition of mulching and herbicides.

It is not intended that this brief introduction to weed control also include a short course in weed identification. That is much too great an assignment. Besides, there are numerous weed identification manuals available through every state Land Grant University, usually from the Cooperative Extension Service. (See Appendix H for addresses).

On the following pages are illustrated 30 of the most common and widely distributed broadleaf and grass, perennial and annual weeds in North America. In many regions of the nation most of these can be found within a short radius. These illustrations were taken from "Weeds of the North Central States," Circular 718 of the University of Illinois Agriculture Experiment Station. (Slife, Bucholtz and Kommendahl, 1960).

Dandelion *(Taraxacum officinale).*

Canada Thistle, Creeping Thistle *(Cirsium arvense).*

Cocklebur, Clotbur *(Xanthium pennsylvanicum).*

Common Ragweed *(Ambrosia artemisiifolia).*

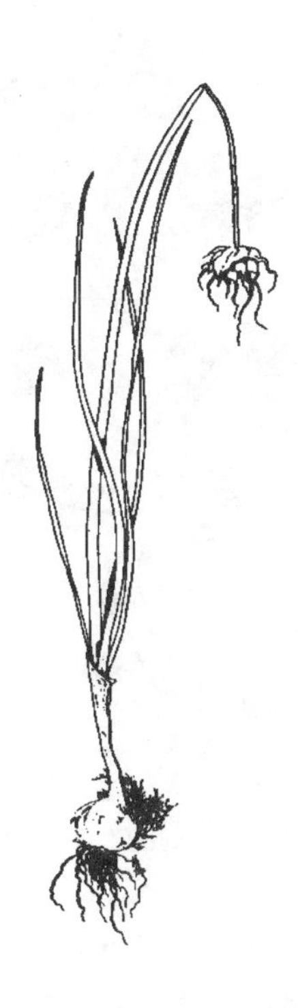

Wild Onion *(Allium canadense).*

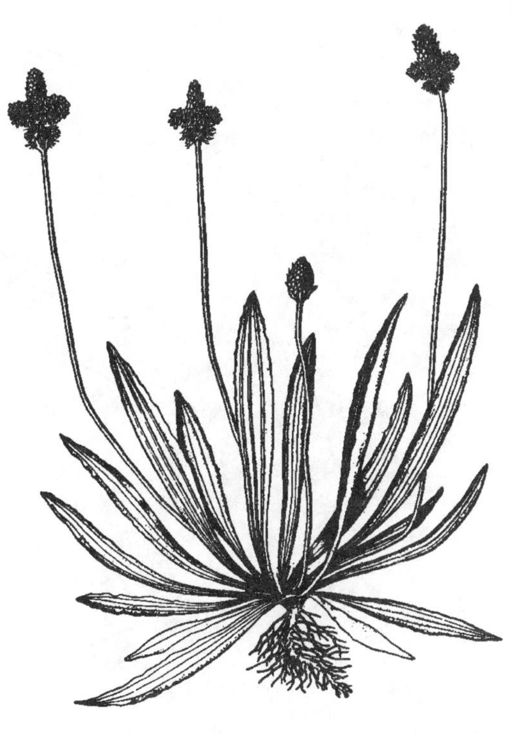

Buckhorn Plantain, Ribgrass *(Plantago lanceolata).*

Buffalobur *(Solanum rostratum).*

Prostrate Spurge, Milk purslane *(Euphorbia supina).*

Puncturevine, Caltrop *(Tribulus terrestris).*

Yellow Wood Sorrel *(Oxalis europaea).*

Shepherdspurse *(Capsella bursa-pastoris).*

Rough Pigweed, Redroot *(Amaranthus retrofleus).*

Common Chickweed *(Stellaria media).*

Black Medic, Yellow Trefoil
(Medicago lupulina)

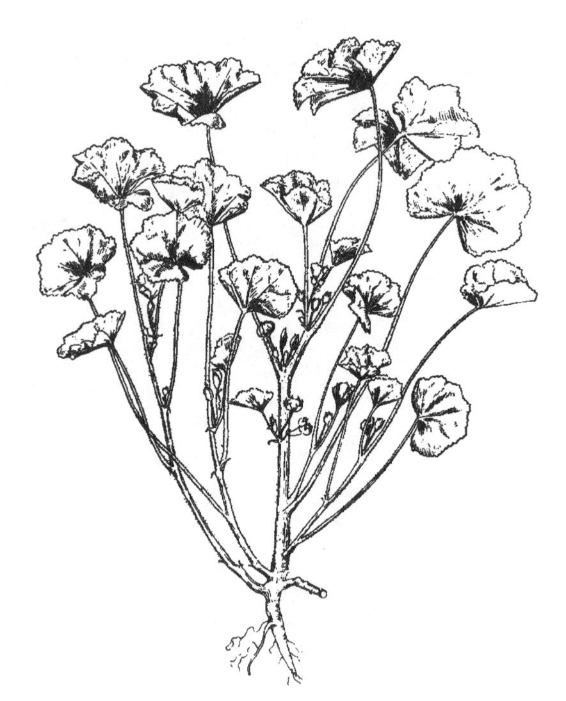

Roundleaved Mallow, Cheeses *(Malva neglecta).*

Tumbleweed, Tumble Amaranth *(Amaranthus albus).*

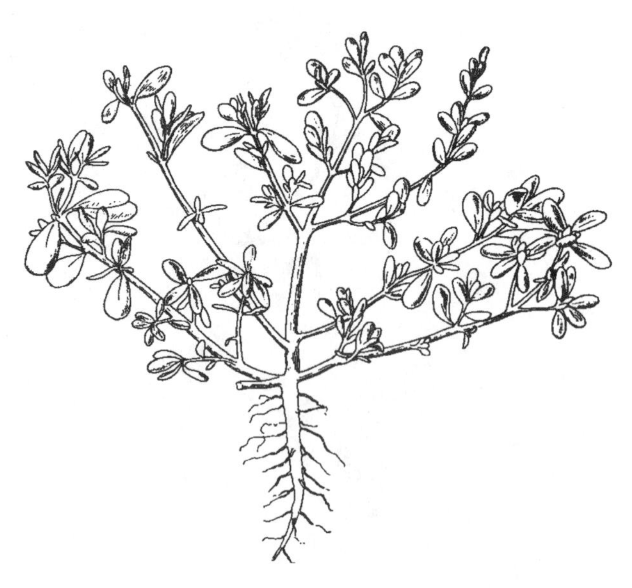

Purslane, Pusley *(Portulaca oleracea).*

Russian Thistle, Common Saltwort *(Salsola kali).*

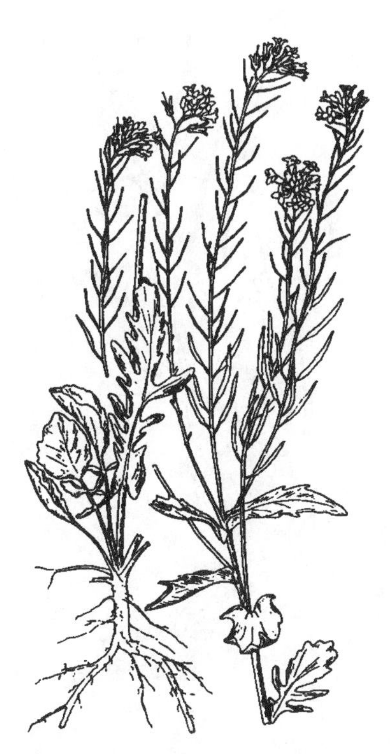

Yellow Rocket, Winter Cress *(Barbarea vulgaris).*

Lambsquarters *(Chenopodium album).*

Bermudagrass, Devilgrass *(Cynodon dactylon).*

Yellow Foxtail, Yellow Bristlegrass, Pigeongrass *(Setaria lutescens).*

Goosegrass, Yardgrass, Silver Crabgrass *(Eleusine indica).*

Nimblewill *(Muhlenbergia schreberi).*

Large Crabgrass, Large Hairy Crabgrass *(Digitaria sanguinalis).*

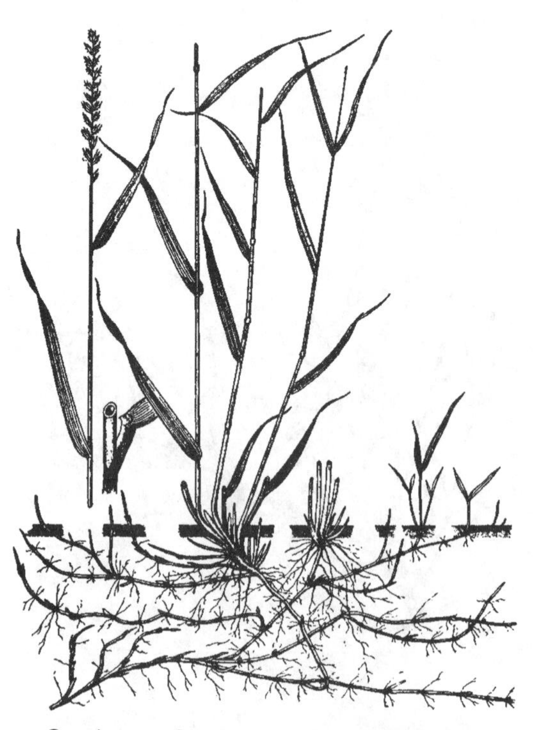

Quackgrass, Couchgrass *(Agropyron repens).*

Yellow Nutgrass *(Cyperus esculentus).*

Barnyardgrass *(Echinochloa crusgalli).*

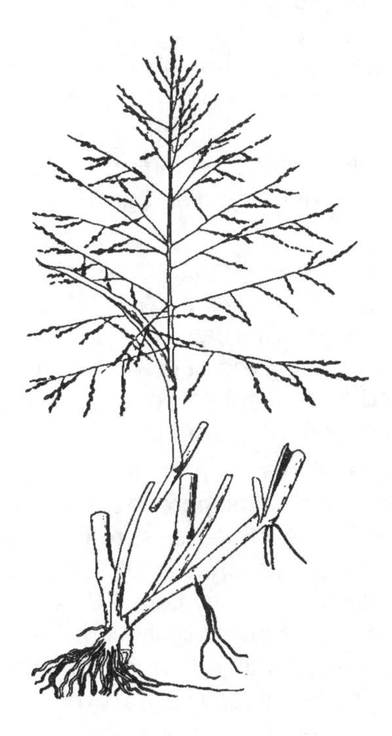

Fall Panicum, Spreading Panicgrass
(Panicum dichotomiflorum).

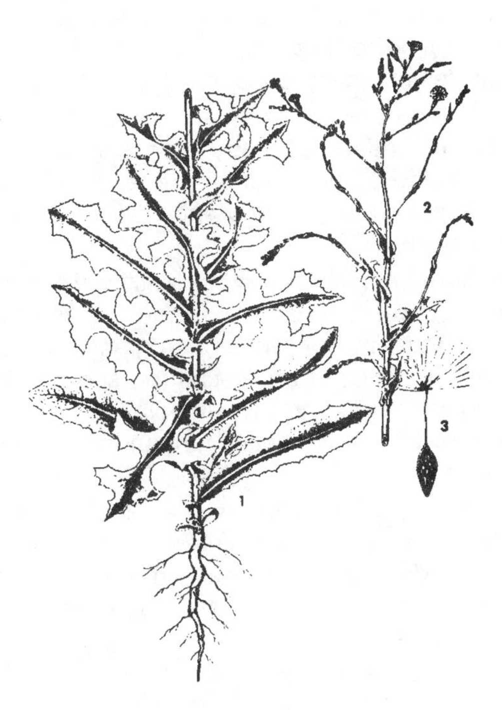

Prickly Lettuce *(Lactuca scariola).*

Controlling Weeds

Weed control with herbicides is indeed both an art and a science. The art involves the delicate placement of the materials so as not to become a threat to non-target plants. The science requires the precise selection, mixing and timing of the herbicide. The art without the science or vice versa, would result in total failure on every attempt.

Straight off, I do not recommend the use of herbicidal weed control in the home vegetable or flower garden. Most home gardeners are rank amateurs with respect to the art of herbicide employment. I am reluctant to recommend their use around woody ornamentals and trees. The benefit: risk ratio is too much out of proportion. Finally, I can suggest the use of herbicides for lawn and turf weed control. Very little art is required and the home gardener normally has a large enough target that a direct hit is generally inevitable. So, with that introduction, let's discuss briefly weed control in lawn and turf.

Several kinds of pests have a way of getting into lawn and turfgrass and weeds are no exception. Although a dense, healthy stand of grass is the most satisfactory method of controlling many weeds, it is not a fool-proof method. Broadleaf weeds and perennial or annual grass weeds are likely to show up from time to time, making it necessary to use herbicides or other means of control.

Should you have a recommended variety or a blend of recommended varieties of grasses for a lawn, maintain the proper fertility level and mow at the recommended height, you should own an almost weed-free lawn.

Certain perennials cannot be selectively controlled. These include coarse fescue, nimblewill, quackgrass, timothy and orchardgrass. To bring these grass weeds under control, spot treat and reseed or resod. Heavy infestations require a complete kill of the entire turf and reseeding or resodding.

A word of caution. The use of 2,4-D as a spray for the lawn and turf is very effective for the control of most broadleaf weeds. However the least amount of drift to other broadleaf plants, especially grapes, annual flowering plants and most vegetables, may have severe effects. So, unless you're a better artist than most home gardeners, don't spray this compound.

Weed Classification

The homeowner and gardener need to be able to recognize grass and broadleaf weeds because they differ in desirability as well as reaction to herbicides, culture and various methods of control. For weed control purposes, plants are divided into three groups - grass, broadleaf and woody.

Grass plants have one seed leaf. They generally have narrow, verticle, parallel-veined leaves and fibrous root systems. Broadleaved plants have two seed leaves and they also generally have broad, net-veined leaves and tap roots or coarse root systems. Woody plants include brush, shrubs and trees. Brush and shrubs are regarded as woody plants that have several stems and are less than 10 feet tall.

Growth Pattern vs. Weed Control

Knowing the growing habits of annuals is important in planning how and when to control weeds. Annual plants complete their life cycles from seed in less than one year. Winter annuals germinate in the fall, survive the winter, mature, develop or set seed and die in the spring or early summer when the weather grows warm and the days long. For best results with weed control, winter annuals should be controlled in the seedling stage of growth in the fall or early spring. Summer annual weeds germinate in the spring, grow, develop seed and die before or during the fall. Summer annuals should be controlled soon after germination in the seedling stage. Some weeds are specifically winter or summer annuals, while other species can germinate and grow either in the fall or spring.

Biennial weeds complete their life cycles within two years, as their name implies. The first year the biennial weed forms basal or rosette leaves and a tap root. The second year it flowers, matures, develops seed and dies. Biennial weeds should be controlled in the first year of growth for best results.

Perennial plants live more than two years and may live indefinitely. Most perennials reproduce by seed and many are able to spread and reproduce vegetatively. They are difficult to control because of their persistent root system. Seedling perennials should not be allowed to become established. An alert gardener will adapt control of established perennials to the yearly growth cycle of the dominant weed species. Perennials should be controlled during the fast growth period prior to flowering or during the regrowth period after fruiting or cutting.

Simple perennials spread by seed, crown buds and cut root segments. Most have large, fleshy tap roots.

Creeping perennials spread vegetatively as well as by seed. Grasses generally have a shallow root system compared to the deep root system of broadleaf plants.

Bulbous and tuberous perennials reproduce vegetatively from underground bulbs or tubers. Many also produce seed.

Brush, shrubs and trees may spread vegetatively as well as by seed. Woody plants can be controlled at any time of year.

When to Control with Herbicides

All weeds pass through four stages of growth: seedling, vegetative, flowering and maturity. There is a best stage with each weed for its control. If control is not obtained at the best stage of growth, the control method may require changing.

Seedling growth is the same for annuals, biennials and perennials. All start from seed. Seedling weeds are small and succulent, requiring less effort for control than at any other stage. This holds true whether the effort is derived mechanically, chemically or culturally.

For annuals, the vegetative stage is producing energy for production of stems, leaves and roots, at this stage control is still feasible but more difficult than during the seedling stage.

As an annual weed continues to develop, a chemical messenger formed by the plant tells it to change from vegetative to flowering. At this time, most of the weed's energy goes to seed production. Chemical control now, for both broadleaf and grass, is less effective at this stage.

Maturity and setting of seed in annuals complete the life cycles. Chemical control is no longer effective. The damage is done — viable seeds were produced and the cycle is ready to be repeated.

Most annual weeds can be controlled when herbicide application is made at the seedling stage. Applied later at the vegetative stage, control drops to 75% and only 40% is achieved during the flowering stage. When the herbicide is applied at plant maturity, virtually no control is obtained.

The formula for weed control described above applies also to biennials. The only distinction is that biennials go through the same stages as annuals, but require two years.

Perennial weeds are a bit more complicated. The seedling stage of growth and its control are the same as for annuals and biennials. However, the stages of growth from vegetative through maturity are different. It should be pointed out here that shoots which emerge from established roots are not seedlings.

During the vegetative stage of perennials, part of the energy used in the production of stems and leaves is derived from energy stored in the underground roots and stems. Additional energy results from photosynthesis in the leaves. Chemical control is only mediocre at best during this stage of growth but improves as the weed approaches the bud stage.

The flowering stage in perennials, as with annuals, involves a messenger which directs the plant's energy into production of flowers and seed. Food storage in the roots is initiated and continues through maturity. Chemical control is most effective just prior to flowering, referred to as the bud stage.

With perennials, only the above ground portions die each year. The underground roots and stems remain very much alive through the winter and develop new plant growth the following spring. Control with most chemicals is not feasible at this stage, despite the temptation. However, a select few herbicides will give control when applied to mature plants. The herbicide label will provide this information if it applies.

Optimum herbicidal control of perennial weeds is obtained by treating perennials during the bud and regrowth stage, thus causing the greatest drain on the underground food reserves. Treatment at early flowering usually equals that obtained at the bud stage. However, when perennials reach full flowering, control declines. Application to regrowth following this is beneficial.

Fall applications of herbicides offer one important advantage—environmental safety. In the fall, desirable plants in gardens, lawns and shrubs, have completed their growth and escape the effects of the herbicide.

Additionally, fall herbicide applications reach the underground plant parts through the natural translocation activity of the plant. In the fall, nutrients are moved from above ground parts in advance of the first killing frost and are stored over winter beneath the frost line. For this reason perennial weeds are most susceptible to herbicides in the fall.

The herbicide moves with the above ground nutrients to the underground storage organs where control is achieved.

Biennials that develop from seed the first year and overwinter in a rosette, are also controlled by fall applications. Winter annuals that germinate in the fall are also controlled by fall applications.

Fall herbicide applications provide the target weed with three stresses: (1) herbicidal effects, (2) winter effects and (3) effects of heavy nutrient demand caused by rapid spring growth. Fall application thus increases the chances of controlling the toughest of weeds, biennials and perennials.

Weed control recommendations are presented in Tables 30, 31 and 32.

Climatic Factors

Climatic factors influence weed control more than with any other form of pest control. These include temperature, humidity, precipitation and wind.

As the temperature increases, the effect of herbicide activity increases. Weed control results are the same, regardless of temperature, but the higher the temperature the faster the kill. There are exceptions. Volatile herbicides should not be applied at the higher temperatures.

Low humidity reduces penetration of the herbicide into weed tissue. When a plant is growing under humid conditions, a foliar-applied herbicide will enter the leaf more easily and rapidly than at low humidity when penetration is slow. With high humidity the weed leaf is more succulent, has a thinner outer wax layer and a thinner cuticle.

Rainfall, occurring after a foliar-applied herbicide treatment, may decrease its effectiveness. Soil-applied herbicides will be activated by rain. Heavy rains, however, may move the herbicide through and away from the target zone.

Wind can be hazardous to herbicide applications, by causing drift of the herbicide during spraying as well as movement of herbicide-laden dust particles. Wind and high temperature can also affect the weed. Hot, dry winds will cause plant stomata to close, leaf surfaces to thicken and wax layers to harden, thus making herbicide penetration more difficult.

In closing, here are a few do's and don't's to follow to achieve the best results from your efforts with herbicides: Use as low pressure as possible in your weed sprayer; don't treat clear to the edge (leave an untreated edge); use less volatile formulations of the herbicide; spray when wind speed is near zero; spray when adjacent sensitive vegetation is either mature or not present and preferable upwind; and finally, remember thy neighbor!

Mulches

The use of mulches in flower or vegetable gardens may well be your most valuable garden weed control practice. Mulches are good for many annual weeds but NOT FOR PERENNIALS. Good mulches reduce soil blowing and washing, prevent weed germination and growth, keep the soil moist and cool and generously add to the organic matter in the soil.

Leaves, grass clippings, sawdust, bark chips, straw and compost make excellent mulches and they are easy to apply. Simply spread a 3-6 inch layer of one of these organic materials on the soil surface around plants, making certain not to cover them. It is important to keep the layer deep enough to do the job. This means that it will be necessary to add more mulching material over the old layers to obtain all of the benefits of mulching.

Using lawn clippings as a mulch is a good way to dispose of a by-product that is usually hauled off with the trash. However, they may need to be mixed with other mulch materials to avoid compacting and preventing water form penetrating into the soil.

Sawdust and bark chips make better mulches if they are well rotted or if 1 to 2 cups of ammonium sulfate or sodium nitrate are added to each bushel of fresh sawdust or chips before applying. Weed-free straw is excellent but loose straw can be a fire hazard. Sometimes compost is the best mulch and it can be made from leftover plant materials from your garden.

Mulches prevent loss of moisture from the soil by evaporation. Moisture moves by capillary action to the surface and evaporates if the soil is not covered by a mulch. Sun and wind hasten this moisture loss.

You can reduce evaporation and control weeds by stirring the soil an inch or so deep, but plant roots cannot develop in this soil layer. A layer of organic material on the surface gives the same benefits and allows normal plant-root development.

The splattering action from falling raindrops is dissipated on a mulched soil. The result is less soil erosion and less soil compaction. Mulches suppress weeds, thus saving hours of back-breaking work. An occasional weed may poke through the mulch, but it is easily pulled because of the soft soil texture in which it is rooted.

Mulches prevent the soil from getting hot under intense sunlight. Many plants, including those in vegetable and flower gardens, need a cool soil surface.

Mulches, especially grass clippings and compost, add organic matter to the soil and furnish food for earthworms, which are valuable in aerating soil. The organic matter helps to keep the soil loose and easy to work. Farmers refer to this as good tilth. At the end of the growing season, the mulch can be worked into the soil to supply organic matter the following year. When mulches are used around perennials in the winter, the mulches should be removed in the spring to allow the soil to thaw and warm up.

Many organic materials, such as straw and autumn leaves, are rich in carbohydrates and low in nitrogen. It is beneficial to add nitrogen fertilizer to the material before applying it as a mulch, which causes it to break down quicker and to avoid nitrogen deficiency. One to two cups of fertilizer high in nitrogen, such as ammonium sulfate or ammonium nitrate, should be allowed for each bushel of mulch. Direct contact of fertilizer with plants will cause burning.

To provide a continuous source of the best mulch, every gardener should have a compost bin, preferably two, for making compost from organic materials. Bins can be made by attaching ordinary wire fence or boards to solid posts or with open brickwork. Each bin should be 4 to 6 feet high, 3 to 5 feet wide and any convenient length. One side of each should be removable for convenience in adding and removing the compost material. In late fall, a temporary piece of wire fence may be used to increase the height about 2 feet, which can be removed in the spring after the material settles.

Compost is not only the ideal mulch, but it is also a good fertilizer and soil conditioner when worked into the soil. Leaves, grass clippings, stems and stalks from harvested vegetables, corn husks, pea hulls and fine twigs are good material for composting. Always compost or shred leaves before using as a mulch. Raw leaves are flat and may prevent water from entering the soil. Avoid using any diseased plants, particularly the roots of plants having rootknot (nematode) in the compost or mulch.

The ideal way to make compost is to use two bins. Fill one with alternate layers of organic material 6 to 12 inches thick and of garden soil about 1 inch thick. To each layer of organic material, add chemicals at the following rate: for each tightly-packed bushel of organic materials, ammonium sulfate, 1 cup; or ammonium nitrate, one-half cup or super-phosphate, one-half cup and magnesium sulfate (Epsom salt) 1 tablespoon; or mixed fertilizer (5-10-5), 3 cups. Avoid the use of lime or ashes since their alkalinity causes loss of ammonia (NH_3) and is unneeded.

The organic material should be moistened thoroughly. Repeat this layering process until the bin is filled. Pack the material tightly around the edges but only lightly in the center so the center settles more than the edges and retains the water.

After 3 to 4 months of moderate to warm weather, commonly in June, begin turning the material by moving it from the first bin to the second. Before turning, it is best to move the material added the previous fall from the edges to the center because it has probably dried out.

In areas that have cool frosty winters, compost made from leaves in November and December can be turned the following May or June. Additional information can be obtained from your local County Cooperative Extension Agent.

TABLE 30. Weed Control in the Home Garden Without Chemicals.

Garden Plants	Weeds to Control	Herbicide to Use	Non-Chemical Control
Seeded crops (beans, beets, carrots, corn, cucumber, lettuce, potatoes, spinach, etc.)	Annual weeds (pigweed, lambsquarters purslane crabgrass foxtail)	Herbicide control is not recommended.	Give garden a good, 6" deep, cultivation before planting. Shallow cultivation, hoeing, hand pulling and black plastic mulches are effective after garden has germinated.
	Perennial and biennial weeds (quackgrass, thistles, bindweed, yellow rocket, curly dock)	Herbicide control is not recommended.	Black plastic mulch (1.5-4 mil is generally the most satisfactory.)
Transplanted crops (broccoli, cabbage, celery, melons, peppers, tomatoes)	Annual weeds (pigweed, lambsquarters, purslane, crabgrass, foxtail)	Herbicide control is not recommended.	Give garden a good 6" deep, cultivation before planting. Shallow cultivation, hoeing, hand pulling and black plastic mulches are effective after plants have begun growth.
	Perennial and biennial weed control and foliage weed treatments are the same as for seeded crops.	Herbicide control is not recommended.	Black plastic mulch.

TABLE 31. Weed Control Suggestions for Lawn and Turf.[1]

Lawn or Turf to Be Protected	Weeds	Herbicide to Use
Bentgrass (pre-emergence control)	Crabgrass Foxtail	Benefin. Apply uniformly in spring before crabgrass emergence. Dacthal. Apply in early spring before crabgrass emergence. Do not use on Cohansey or Toronto varieties.
	Annual bluegrass	Bensulide. Apply uniformly in Aug. - Sept. before annual bluegrass emergence. Dacthal. Apply in late summer and early spring. Do not use on Cohansey or Toronto varieties.
Bermudagrass turf	Spurge and most annual weeds	Dacthal. Apply to soil in spring before weeds germinate. A second treatment in mid-summer may be needed. Water after application. Do not plant winter lawns where Dacthal was used in summer or fall.
	Spurge and most broadleaf weeds.	Isoxaben or Bromoxynil plus detergent. Apply to foliage when weeds are small. Repeat if new weeds emerge. Bromoxynil may cause temporary yellowing of Bermudagrass.
	Crabgrass and most annual grass weeds.	Dacthal, Siduron, Bensulide or Benefin. Apply to soil before crabgrass germinates (Feb.-April). Water after treatment.
	Annual bluegrass	Dacthal, Bensulide or Benefin. Apply to soil before bluegrass germinates in fall (Aug.-Sept). Water after application. Do not use where winter lawns are to be planted.
	Crabgrass seedlings	Fenoxaprop-ethyl. Apply to crabgrass in June or July. Repeat if needed. May control other small seedlings.
	Established nutsedge, creeping chaffweed, woodsorrel and many perennial weeds.	Glyphosate. Apply to weeds as spot treatment. Repeat in 8 weeks or when normal growth occurs. This treatment may injure adjacent desirable plants.
Bluegrass and Fescue (pre-emergence control)	Barnyardgrass Crabgrass Foxtail	Benefin. Apply uniformly in early spring before crabgrass emergence. Bensulide. Apply uniformly in early spring before crabgrass emergence. Siduron or Oxadiazon. Apply before crabgrass emergence in early spring.
	Goosegrass	Bensulide or Oxadiazon. Same as for crabgrass. Goosegrass is more difficult to control than crabgrass, so don't expect more than 70-80% control.
Bluegrass and Fescue (post-emergence control)	Nutsedge	Bentazone. Thorough leaf coverage is essential. Delay mowing 3-5 days. Don't apply to seedling turf. Repeat application may be needed 10-14 days later.
	Crabgrass	Fenoxaprop-ethyl. Apply when crabgrass is visible, when soil moisture is adequate for rapid growth.

[1] Refer to Appendix C for trade or proprietary names of recommended herbicides.

Lawn or Turf to Be Protected	Weeds	Herbicide to Use
Seedling Turf	Broadleaf weeds	Isoxaben or Bromoxynil. Follow label instructions. Apply after grasses emerge and before broadleaf weeds are past the 3-4 leaf stage.
Turf Conversion — Converting Bermudagrass to winter lawn of ryegrass.		Cacodylic acid or Glyphosate plus detergent. Wet foliage thoroughly 2 to 6 weeks before frost. Wait 1-2 days for final seedbed preparation and planting.
Converting ryegrass to summer Bermudagrass.		Cacodylic acid plus detergent. Wet foliage thoroughly when night temperatures reach 60°F. Repeat treatment in 1 week if kill is incomplete.
Bermudagrass turf renovation.	Removal of Bermuda-grass to establish new turf.	Glyphosate. Apply to foliage when Bermudagrass has headed out. After 2 weeks spade or renovate. New seeding or sprigging may be made immediately. Avoid drift onto shrubs or trees.
		Cacodylic acid plus detergent. Apply to foliage when Bermudagrass is growing rapidly. Wait 7 days, spade or renovate and water. Wait 2-3 weeks before establishing new turf. Do not apply near trees and shrubs.
Dichondra turf	Annual grass, Wild celery, and other broadleaf weeds.	Bensulide. Apply to soil in spring and fall, followed by thorough watering.
Gravel or stone yards and desert landscape	Most annual weeds	Dacthal or Trifluralin. Apply to soil before weed seed germinate and water thoroughly.
	Annual weeds	Diquat or Cacodylic acid plus detergent. Wet foliage of small weeds thoroughly. Repeat when new weed seed germinate.
	Perennial weeds	Glyphosate. Apply to foliage when weeds are growing actively. Repeat in 8 to 12 weeks. Avoid drift onto shrubs and trees.
		Diquat or Cacodylic acid plus detergent. Wet foliage of weeds thoroughly and repeat every 1-3 weeks as long as growth continues.
Non-selective Control	Fescue Nimblewill Orchardgrass Quackgrass Timothy	Glyphosate. Wet foliage of actively growing grass in spring or summer. Repeat at 10-day intervals until complete control is obtained.
Turf, General	Black medic	2,4-D Amine. Spray or granules
	Chickweed	2,4-D Amine or Pendimethalin. Spray or granules
	Clover	Isoxaben or 2,4-D Amine. Spray or granules
	Crabgrass (pre-emergence)	Oxadiazon, Siduron, Bensulide or Benefin. Follow label directions carefully.
	Dandelion	Isoxaben or 2,4-D Amine granules.
	Plantain (narrow and broad-leaved)	Isoxaben or 2,4-D Amine spray or granules.
	Wild onion	2,4-D Amine. Spray plants twice, 5 days apart.

TABLE 32. Weed Control in Shrubs and Woody Ornamentals.[1]

Weeds to Control	Herbicide to Use
Annual weeds and perennial grass weeds from seed	Dacthal or Trifluralin. Apply to soil before weed seeds germinate and water thoroughly. Glyphosate. Apply carefully and avoid drift onto ornamentals.
Bermudagrass and nutsedge	Dichlobenil. Apply to soil and incorporate. Remove tops of weeds before or after treatment. Granular formulation is easier to use. Requires 2-3 applications per year for good control.
Nutsedge	Eptam. Apply to soil and incorporate. Remove tops before or after treatment. Granular formulation is easier to use. Requires 2-3 applications per year for good control.
All annual broadleaf and grass weeds	Diquat. Gives complete kill in 24 hrs. Apply with compressed air sprayer directly to weeds.

[1] Refer to Appendix C for trade or proprietary names of recommended herbicides.

RODENTS AND OTHER ANIMAL PESTS

Three blind mice,
see how they run!
They all ran after
the farmer's wife,
She cut off their heads with
a carving knife,
Did you ever see
such a sight in your life,
As three blind mice?
Old English
Nursery Rhyme

CONTROLLING RATS, MICE AND OTHER MAMMALS

Rodents and several other small mammals damage man's dwellings, his stored products and his cultivated crops. Among these are native rats and mice, squirrels, woodchucks, pocket gophers, hares and rabbits. Rats are notorious freeloaders and in some of the underprivileged countries where it is necessary to store grain in the open, as much as 20% may be consumed by rats before man has access to it.

The number of rodent species (order Rodentia) comprises about one-half of all mammalian species and because they are so very highly productive and widespread, they are continuously competing with man for his food.

Rats and Mice

Of these, rats and mice are the most abundant and consequently the most annoying and destructive. Rats and mice have accompanied man to probably all of the areas of the world in which he has settled, except perhaps the Arctic and Antarctic. Historically, they have been responsible for more human illnesses and deaths than any other group of mammals. Because of man's indifference and carelessness in handling food and trash-garbage, he has fostered populations of rats and mice in such close proximity to his home and place of work that they are referred to as domestic rodents.

These domestic rodents are the Norway rat, roof rat and house mouse, all members of the family Muridae, Order Rodentia (See Figure 22). The Muridae have a single pair of incisor biting teeth on each jaw and do not have canine teeth. (I really don't expect you to look to be certain!)

Norway rats *(Rattus norvegicus)* are burrowing rodents and are the most common and largest of the rats. They are generally distributed throughout the U.S. and are known by several common names: wharf rat, brown rat, house rat, sewer rat and barn rat. The adults average one pound in weight and have course, reddish brown fur and a blunt nose. The droppings or fecal pellets are large, up to 3/4 inch long. They reach sexual maturity in 3 to 5 months and have a gestation period averaging 22 days, each female averages about 20 young per year, born in 4 to 7 litters, with 8 to 12 young per litter. As in most rodents, there is considerable infant mortality.

The Norway rat lives an average of about one year. It prefers the outdoors and burrows in the ground under foundations and in trash dumps. Indoors it lives between floors and walls, in enclosed spaces of cabinets, shelving and appliances, in piles of trash and in any place concealed from view. It ranges usually no more than 100 to 150 feet. It feeds on all foods, eating from 3/4 to 1 ounce of dry food and 1/2 to 1 ounce of water per day.

The roof rat *(Rattus rattus)*, also known as the black rat and ship rat, is smaller than the Norway, is more agile and lives mostly in the warmer areas of the U.S., namely the South and to the Pacific coast and Hawaii. It is rare or absent in the colder portions of the world. The adults are smaller than the Norway rat, weighing only 8 to 12 ounces. They are black in color and have a slender, pointed nose and a tail that is longer than the body and head combined. Its droppings are up to 1/2 inch long. They reach sexual maturity in 3 to 5 months, have a gestation period averaging 22 days and average about 20 young per female per year, born in 4 to 6 litters, with 6 to 8 young per litter.

The roof rat also lives about a year. It appears to prefer above-ground dwellings, in attics, between walls and in enclosed spaces of cabinets and shelving or outdoors in trees and dense vines. It, too, ranges usually no more than 100 to 150 feet. It prefers fruits, vegetables and grains compared to the high protein preference of the Norway rat. It requires 1/2 to 1 ounce of dry food and up to 1 ounce of water per day.

The house mouse *(Mus musculus)* is the smallest of the household rodents and is found throughout the United States. It weighs 1/2 to 3/4 ounce, is dull gray and has a tail as long as its body and head combined. Its droppings are small, about 1/8 inch long. It reaches sexual maturity in 6 weeks, has a gestation period averaging 19 days and weans 30 to 35 young per female per year, born in 6 to 8 litters, with 5 to 6 young per litter. They live about 1 year and nest in any convenient, protected space. They range only up to about 30 feet. Mice will eat most anything, but prefer grain, nibble constantly and require only 1/10 ounce of food and 3/10 ounce of water per day.

Control. Rat and mouse populations are controlled by storing all food materials in rodent-proof containers, collecting and disposing of refuse and the proper storage of usable materials. Permanent removal of harborage and sources of food will eliminate existing rat and mouse populations.

Trapping is useful when poisoning fails or is too hazardous or where the odor of unrecovered dead

FIGURE 22.

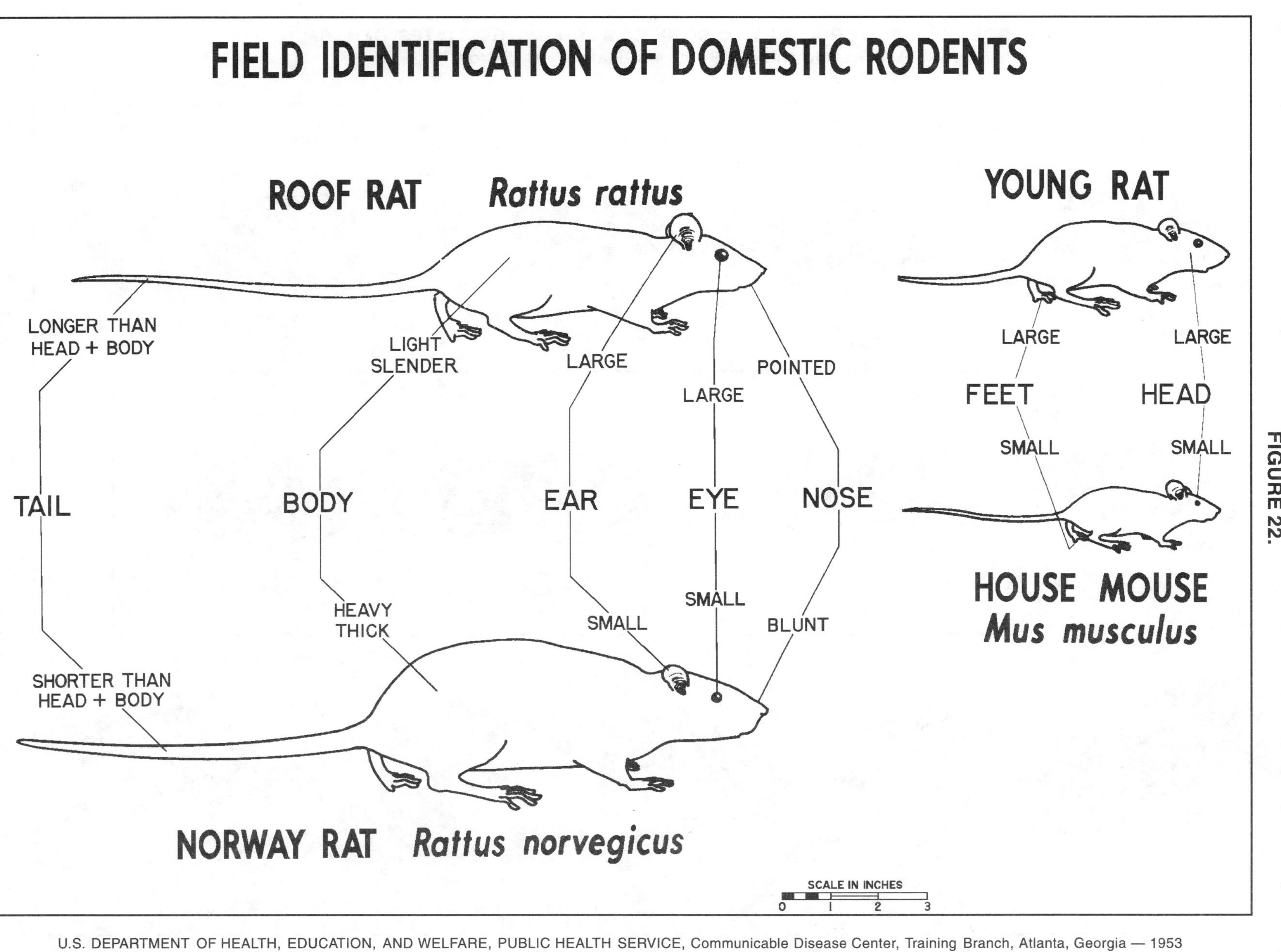

U.S. DEPARTMENT OF HEALTH, EDUCATION, AND WELFARE, PUBLIC HEALTH SERVICE, Communicable Disease Center, Training Branch, Atlanta, Georgia — 1953

FIGURE 23.

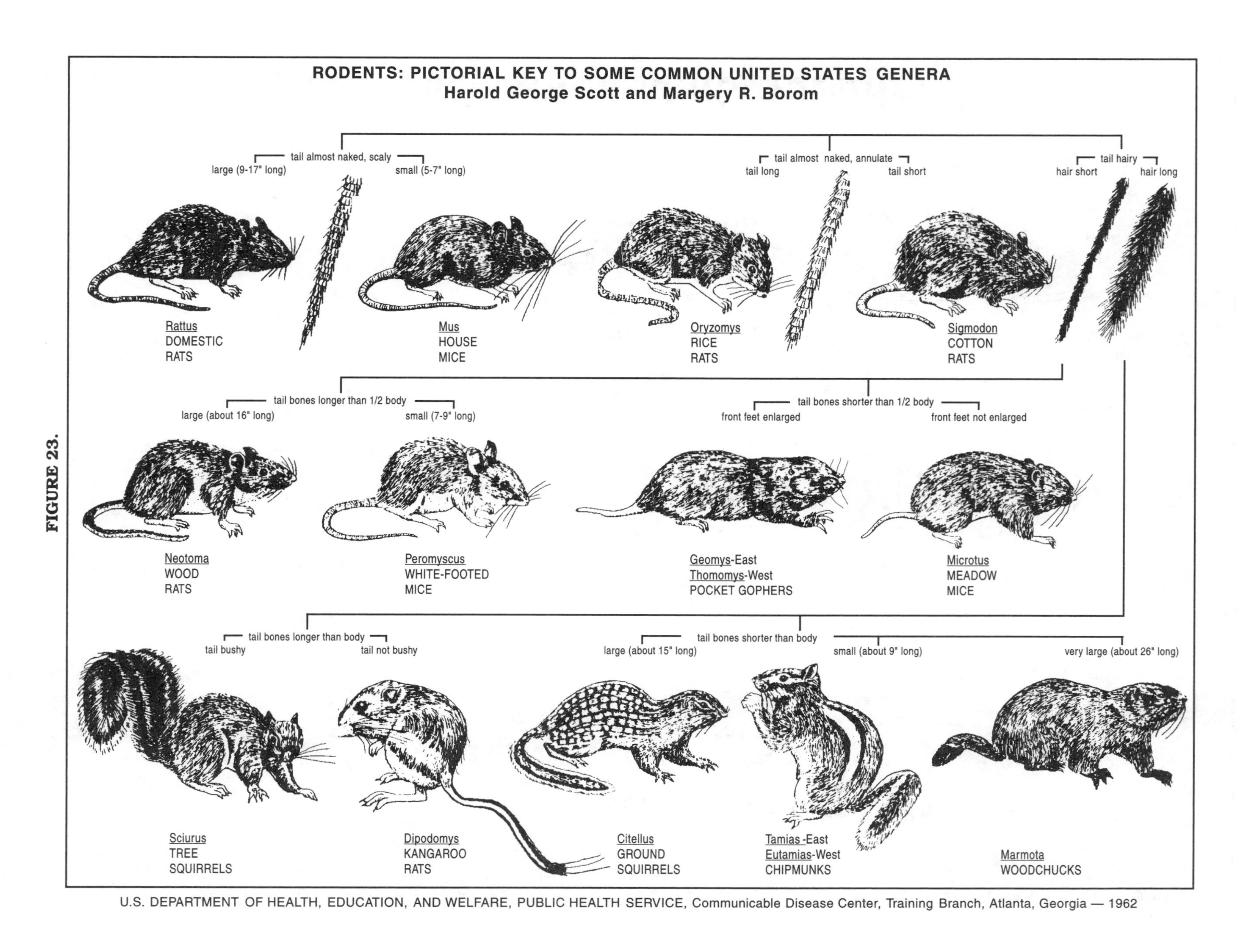
RODENTS: PICTORIAL KEY TO SOME COMMON UNITED STATES GENERA
Harold George Scott and Margery R. Borom
tail almost naked, scaly
large (9-17" long)
small (5-7" long)
tail almost naked, annulate
tail long
tail short
tail hairy
hair short
hair long
Rattus
DOMESTIC
RATS
Mus
HOUSE
MICE
Oryzomys
RICE
RATS
Sigmodon
COTTON
RATS
tail bones longer than 1/2 body
large (about 16" long)
small (7-9" long)
tail bones shorter than 1/2 body
front feet enlarged
front feet not enlarged
Neotoma
WOOD
RATS
Peromyscus
WHITE-FOOTED
MICE
Geomys-East
Thomomys-West
POCKET GOPHERS
Microtus
MEADOW
MICE
tail bones longer than body
tail bushy
tail not bushy
tail bones shorter than body
large (about 15" long)
small (about 9" long)
very large (about 26" long)
Sciurus
TREE
SQUIRRELS
Dipodomys
KANGAROO
RATS
Citellus
GROUND
SQUIRRELS
Tamias-East
Eutamias-West
CHIPMUNKS
Marmota
WOODCHUCKS
U.S. DEPARTMENT OF HEALTH, EDUCATION, AND WELFARE, PUBLIC HEALTH SERVICE, Communicable Disease Center, Training Branch, Atlanta, Georgia — 1962

rodents would be a problem. If for some reason you wish not to kill these pests, live traps are available that permit their removal and relocation.

Trapping is an art. There are many types of rat and mouse traps. One of the most effective and versatile is the snap trap, which is generally available.

Many foods make good baits: peanut butter, nut meats, doughnuts, cake, fresh crisp-fried bacon, raisins, strawberry jam and soft candies, particularly milk chocolate and gumdrops. Rats are attracted more to ground meat or fish. Surprisingly, recent studies indicate that mice aren't nearly as fond of cheese as tradition has it. Traps may also be baited by sprinkling rolled oats over and around the bait trigger. If possible, baits should be fastened by winding a short piece of thread or string around the bait and trigger. Where food is plentiful and nesting material scarce, good results can sometimes be obtained with cotton tied to the trigger.

Trap-shy individuals may be caught by hiding the entire trap under a layer of flour, dirt, sawdust, fine shavings or similar lightweight material.

The common wooden-base snap trap can be made more effective by enlarging the trigger with a piece of heavy cardboard or light screen wire. Cut the cardboard or screen in the shape of a square smaller than the limits of the guillotine wire and attach it firmly to the bait trigger. To bait the trap, smear a small dab of peanut butter in the center of the enlarged trigger or sprinkle rolled oats over the entire surface. This works well on both rat and mouse traps placed where the animals commonly run.

It is very important to place traps across the paths normally used by rats and mice. If their runways cannot be readily determined, sprinkle a light layer of talcum powder, flour or similar material in foot-square patches in likely places. Place traps in the areas where tracks appear. Because rats and mice like to run close to walls, these spots should be checked first and traps set against walls.

Use boxes or other obstacles to force the rat or mouse to pass over the trigger. Two or more traps set close together produce good results where many are present or where trap-shy individuals are a problem. Use plenty of traps rather than rely on one or two for the job. Because mice travel such short distances, traps should be placed within 10-foot intervals.

To protect other animals or small children trap boxes should be constructed so that mice and rats are forced to enter them when the boxes are placed next to walls. Where the animals travel on rafters or pipes, nail the traps in place or set them on small nipples clamped to the pipes. Leave traps in place for a few days before moving them to other locations. Check traps regularly and adjust to a fine setting.

Rats and mice are accustomed to human odors, so it is not necessary to boil traps or handle them with gloves. Neither does the scent of dead individuals warn others away.

Rat- and mouse-proofing is vital in a complete rodent control program. This consists of changing structural details to prevent entry of rodents into buildings. Openings as small as a half-inch can be entered by young rats and assuredly by mice.

There are 7 basic devices for keeping these rodents out of buildings:

1. The cuff and channel for wood doors to side and back entrances prevent rats from gnawing under or around the doors. The front doors of most establishments are less exposed to rats and are generally protected with a kick plate. Wooden door jambs can be flashed with sheet metal to protect them from rat gnawing. Because open doors provide ready entry for rodents, both the screen doors and wooden doors to food-handling establishments should be equipped with reliable self-closing devices.

2. Vents and windows can be made secure against rat entry by screening them with heavy wire mesh, preferably in a sheet-metal frame. If desired, fly screening can be incorporated into the frame also. Wooden surfaces exposed to gnawing must be covered by the frame.

3. Metal guards of suitable construction should be placed around or over wires and pipes to prevent rats from using them to gain entrance into a building.

4. Openings around pipes or conduits should either be covered with sheet metal patches or filled with concrete or brick and mortar.

5. The use of concrete for basement floors and for foundations not only prevents rat entry but also increases the value of the property.

6. Floor drains, transoms, letter drops and fan openings must receive stoppage consideration.

7. Plug any existing rodent holes with steel wool or copper mesh similar to that used in pot scrubbers.

In addition to sealing off entries, buildings should be planned or modified to avoid dead spaces such as double walls, double floors and enclosed areas under stairways. Trash and garbage piles or other materials stacked against buildings should be removed. They provide the means by which rats and mice can bypass otherwise effective stoppage measures.

Ultrasonic pest control devices simply are not effective against rats, mice, cockroaches or any other pests around the home. Manufacturer claims of such pest control or repellency with ultrasound devices cannot be substantiated by the most sophisticated research.

Repellents can be of value if used diligently against rats and mice. Of the chemical repellents mentioned later, only napthalene and paradichlorobenzene are useful in preventing rodents from entering protected spaces. The other compounds are intended to prevent eating of grains, seeds and other delectables.

Poisoning of rats and mice is by far the simplest method of control. Of the available poisons, I recommend the ready-prepared anticoagulant baits. Warfarin and Diphacinone are still available and are effective, but slow. Animals must feed daily for 5 to 10 days before they die. The effects are cumulative and the rodents finally die by internal hemorrhaging. They are slow, but very home-safe.

One-dose rat and mouse poisons are readily available to the homeowner. They are effective against Warfarin-resistant rats and kill in 2-3 days after only one feeding. These are brodifacoum and bromadiolone. You will likely find only brodifacoum at the hardware or garden supply store, because the bromadiolone baits are used primarily by professional pest control operators.

Pack Rats

Pack rats (*Neotoma spp.*) may become pests to rural and foothills homes in the Southwest U.S. Their common name is 'woodrats', but because of their amusing and sometimes annoying habit of carrying shiny and various other objects back to their nests, they have acquired the name 'pack rats', also a term used to describe the human behavior of collecting and saving objects.

These very intelligent rodents live in burrows, usually near prickly pear and cholla cactus, their main food sources. Commonly living with them and feeding on them as nymphs, are kissing bugs (*Triatoma spp.*) (See also Kissing Bugs under HOUSEHOLD—UNWANTED HOUSE GUESTS). These true, blood-sucking bugs enter homes in the summer and fall, feeding on their new human hosts, usually while they sleep.

Pack rats gnaw on wire insulation when they enter wall voids or tunnels that serve as wiring and plumbing conduits and will become a serious nuisance if not controlled. They are even known to build nests on the engines of automobiles left standing for extended periods. A single house cat can keep several colonies under control, but without a cat other methods must be used. Pack rats can be controlled by live trapping and relocation or with large snap-traps and by poisoning. Traps (live or snap) should be prebaited with apple slices for two days before setting. The apple slice should be tied on the trigger with thread, to assure their tugging at the thread will trip the trap. Walnuts or peanut butter are second choice as baits—still using the tie-down thread. Poisoning is relatively simple. Readily available bromodiolone or bromodifacoum rat and mouse baits should be sprinkled directly into their active burrows and in protected runs. Feeding on 1-2 pellets results in their death within 3 days. Imbedding 6-10 bait pellets in peanut butter and placing in protected locations has proved very effective.

Miscellaneous Animal Pests

The more common of these miscellaneous rodents are squirrels, chipmunks, rabbits, moles and bats (See Figures 23-25). They all have predictable needs and activities and can be controlled utilizing these habits.

Squirrels and Chipmunks

Squirrels and chipmunks can be repelled with RoPel® or capsaicin or the use of moth balls or moth flakes (paradichlorobenzene or napthalene) placed in the runs and holes where they enter buildings. Or they can be trapped using one of the several kinds of live-traps and released in rural areas. The best baits for squirrels are walnut meats or similar nuts, while chipmunks are attracted to rolled oats, grains of corn or peanut butter. RoPel® and capsaicin can be found in Appendix G, *Pest Control Sources*.

Rabbits

Overgrown ditches, bushy fence rows or brush piles within or near your yard or garden may be major factors contributing to the presence of rabbits. Mowing, brush cutting and general cleanup of overgrown areas and brush piles may be all that is needed for rabbit control.

The best protection from rabbit damage in a garden is a tight chicken wire fence. One-inch mesh wire 30 inches high is adequate if held firmly upright by stakes.

Rabbits can be live-trapped, using carrots or apples as bait and released in more appropriate territory. Fruit and other trees can be protected from damage by making guards of 1/4 inch hardware cloth. These guards should be formed into cylinders about 2 inches larger than the diameter of the tree trunk and long enough to protect the tree above the depth of the deepest snow expected. They should also be anchored in the soil at the base. Commercial repellents containing thiram or ziram fungicides used for plant diseases are also effective for rabbits and should be painted on the areas of tree trunks to be protected. Other repellents include blood dust and nicotine. A highly successful addition to the market is RoPel® (Burlington Scientific Corp.).

Moles

Moles are not rodents, order Rodentia. Rather they belong to another order, the Insectivora. There are several species in the U.S., but none are found in the Great Basin or Rocky Mountain regions. The eastern mole *(Scalopus aquaticus)*, hairy-tailed mole *(Parascalops breweri)* and the star-nosed mole *(Condylura cristata)* are the most common species and range from the western Great Plains to the East Coast. The eastern mole is the most common of the three and is often associated with lawn damage. On the West Coast three other species are found, but usually do not attract the attention of those found in the eastern half of the U.S.

Moles are attracted to white grubs, the larvae of several large beetles that feed on the roots of lawn grasses and also to earthworms. Moles seldom feed on plant materials; however, mice may follow mole burrows and damage bulbs and plant roots. Elimination of these grubs with soil insecticides, including the disease organism Milky Spore Disease, will usually eliminate the moles simply by removing the food attraction.

Trapping, using the old fashioned spring-activated harpoon, the choker trap and the claw-type traps, is only partially effective. The best time to trap is in the early spring when the first burrows are seen or after the first fall rains. Early spring trapping may eliminate pregnant females, thus reducing the possibility of contending with a family of moles in the summer. All burrows are not in use all the time. Find active burrows by rolling or tramping down their ridges and observe those that are raised again. These are the ones where traps should be set. If a trap is not sprung in 2 days, remove and relocate it.

Leaving the burrows raised and adding paradichlorobenzene (PDB) flakes or thiram (fungicide) to the burrow floor every 6 feet will probably prevent these beneficial nuisances from returning. Baits containing toxicants for mole control are also available. Because these baits use peanuts, grains or other non-animal food materials as carriers and because moles feed almost entirely on insects and worms, they do not readily take poisoned baits and the results are usually unsatisfactory. Insecticides that may be used by the homeowner to control the grubs include imidacloprid, trichlorfon and carbaryl. After treatment with any of these, soak the lawn surface thoroughly. Keep pets and children off the treated area until it has dried.

Mole-Med® is a commercial repellent based on castor oil and proved very successful in Michigan and Wisconsin. Available from most supply houses listed in Appendix G, *Pest Control Sources*.

Bats

House bats are flying rodents and belong to the order Chiroptera, which means "winged hand." They are an interesting and valuable component of our environment, but their presence in attics of homes, barns, garages and other structures may cause a nuisance. They often appear when their natural habitats are disturbed, such as the removal of trees with hollow limbs. On occasion they get into the eaves of attics, fouling the area with their odorous feces (guano) and urine. Their presence offers the possibility of additional pests and disease transmission, e.g., mites, bat bugs, fleas, histoplasmosis and rabies.

Although 40 species of bats are found in the U.S., there are only three species that comprise the "house bats": The big brown bat *(Eptesicus fuscus)*, the little brown bat *(Myotis lucifugus)* and the Mexican free-tailed bat *(Tadarida brasiliensis)*.

As soon as they are detected, openings in eaves and attic louvers should be sealed with 1/4 inch hardware cloth or screen, because their droppings attract new bat colonies after the original ones are broken up. Narrow cracks can be sealed with caulking compound. If a home is completely bat-proofed, make sure all bats are outside before plugging the last openings. Normally, all bats leave at about the same time. If there are several openings leave one of them unplugged for several days, then close it in the evening after all bats have left the roost.

To discourage bats from roosting in attics, scatter 3 to 5 pounds of moth flakes over the floor or hang them in mesh bags from the rafters. Floodlights in the attic or directed on outside entrances for several nights will sometimes cause bats to leave a building. The only way to permanently control bats is to keep them out—exclusion.

Skunks

Even though you may never have encountered a skunk before, you will immediately recognize both its appearance and odor. Skunks are kin to the weasels and since they are not particularly disturbed by man's presence and activities, they occasionally move into the neighborhood.

Skunks are protected by law in most states and they are frequently found to be carriers of rabies. So, avoid handeling at all costs.

Control is achieved by exclusion, that is, preventing them from returning to their sleeping or nesting quarters. Sprinkle a thin layer of flour around the hole or suspected entrance to form a tracking patch. Examine the area for skunk tracks soon after dark and when the tracks lead out of the entrance, the opening can then be safely closed off with lumber, fencing or concrete.

Skunks can be livetrapped in box traps. Use sardines or catfood for bait and insert all but the entrance door into a plastic garbage bag. After the skunk is trapped, close the bag to prevent being sprayed with its scent. Release the trapped skunk at least 5 miles from the trap site in an area where it will not cause problems.

Skunks are known for their scent, which can be ejected 6 to 10 feet. The persistence of the odor on anything touched by it is astonishing.

In Appendix F, under Deodorants, there is given the ideal deodorant recipe made of hydrogen peroxide, baking soda and household detergent. This and a second mixture are the best and most recent recommendations. Substitutes for the above to be used on inanimate objects and surfaces, but *not on pets or humans,* are chlorine laundry bleach or household vinegar, diluted 1 part in 10 parts of water, with a little household detergent added to assist in wetting and saturating the area. These also work for scrubbing floors, walls, garages, basements, outdoor furniture and the like. They can also be sprayed on contaminated soil.

The proverbial tomato juice bath is messy and doesn't work. Its supposed effectiveness is based on the small quantity of organic acid (aspartic acid) it contains. In large quantities this could be effective, but not so for tomato juice.

Deer

Deer sometimes damage gardens, orchards and ornamental shrubs and may cause extensive crop damage. Several types of fencing have been used to exclude deer from gardens and fruit tree plantings. Wire-mesh fences can be erected around small gardens and orchards without great cost. A fence must be at least 8 feet high to exclude deer effectively and may be conveniently constructed from two 4-foot widths of hog-wire fencing. Electric fences are available that have been designed specifically to exclude deer.

Materials registered as effective deer repellents include two fungicides, thiram, ziram and capsaicin derived from chili peppers. Spray trunks of trees and lower limbs generously at 2-3 week intervals. Another repellent is human hair balls, hung from tree limbs in old nylon hose or mesh. Any barber shop collection will do. This is only a "ho-hum" technique.

There are a large number of plants that are not attractive to deer, everything from certain grasses to herbaceous perennials. The University of Wisconsin has developed an excellent list of more than 200 such plants, *Plants Not Favored by Deer*, cited in the Bibliography and listed in Appendix G, Pest Control Sources and Suppliers.

Dogs and Cats

Probably nothing is more annoying than having someone's dog defecate in your front lawn, urinate on the tires of your SUV parked in the driveway or seeing a strange cat patrolling your backyard for unwary birds. Compared to some pest control problems, these are relatively simple.

Dogs and cats both respond to scolding, clapping of hands, broom-waving, hurled clods and stones and repellents. The materials registered as dog and cat repellenus are almost too numerous to describe.

Both forms of moth crystals (naphthalene and paradichlorobenzene) are very effective inside the home and out. For instance, a few crystals placed in the favorite winged-back chair will compel Fido or Kitty to seek another parking place. The same is true outside. Trees, fireplugs and other urinating points for dogs, with the exception of the tires on family vehicles, can be equally unattractive with a few crystals sprinkled around the base. Cats will avoid walking or resting on walls with moth crystals sprinkled at regular intervals. Ground cayenne pepper works even better.

If the automobile tire episodes are really important to you, try one of the commercial dog repellent aerosols containing either citral, citronella or creosote.

Other materials contained in commercial mixtures are allyl isothiocyanate, amyl acetate, anethole, bittrex, bone oil, capsaicin, citrus oil, cresylic acid, eucalyptus, geranium oil, lavender oil, lemongrass oil, menthol, methyl nonylketone, methyl salicylate, nicotine, pentanethiol, pryidine, sassafras oil and thymol. The formulated product should first be checked out with your own nose if it is to be used indoors, since some of the above are readily detectable and may be highly annoying. Better the dog or cat!

Barking dogs.

Your own dog, but more often your neighbor's dog, barks for what may seem to be mere entertainment. Barking dogs can be conditioned or trained not to bark except when excited by an intruder. My experience includes the use of (1) the "silent" dog whistle, (2) the small, purse-size compressed gas emergency alarm (Screamer™) and (3) the compressed gas air-horn, a favorite at outdoor athletic, events. The silent whistle only captures the dog's attention, while the others actually frighten the dog into silence. Used frequently at first the dog soon expects to be silenced and is thus conditioned to remain silent. Good Luck!

Garbage can/trash bag dogs.

Dogs will not overturn or disturb garbage cans or trash bags that have been sprayed lightly with a 1:1 dilution of Pine-Sol® or similar pine-scented cleaning detergent.

Geckos

Though not mammals, geckos are pests of the home in tropical and subtropical climates. They are small lizards known for their ability to scale vertical walls with their suction-cup feet. No pesticides are registered for gecko control. Doors and windows should be well-sealed and caulked, especially those near outdoor lights. Their control can be achieved with glue boards and sticky cards, used to capture mice and some rats. Catch may be improved by placing traps near lights and windows out-of-doors, where geckos prefer to congregate.

FIGURE 24.

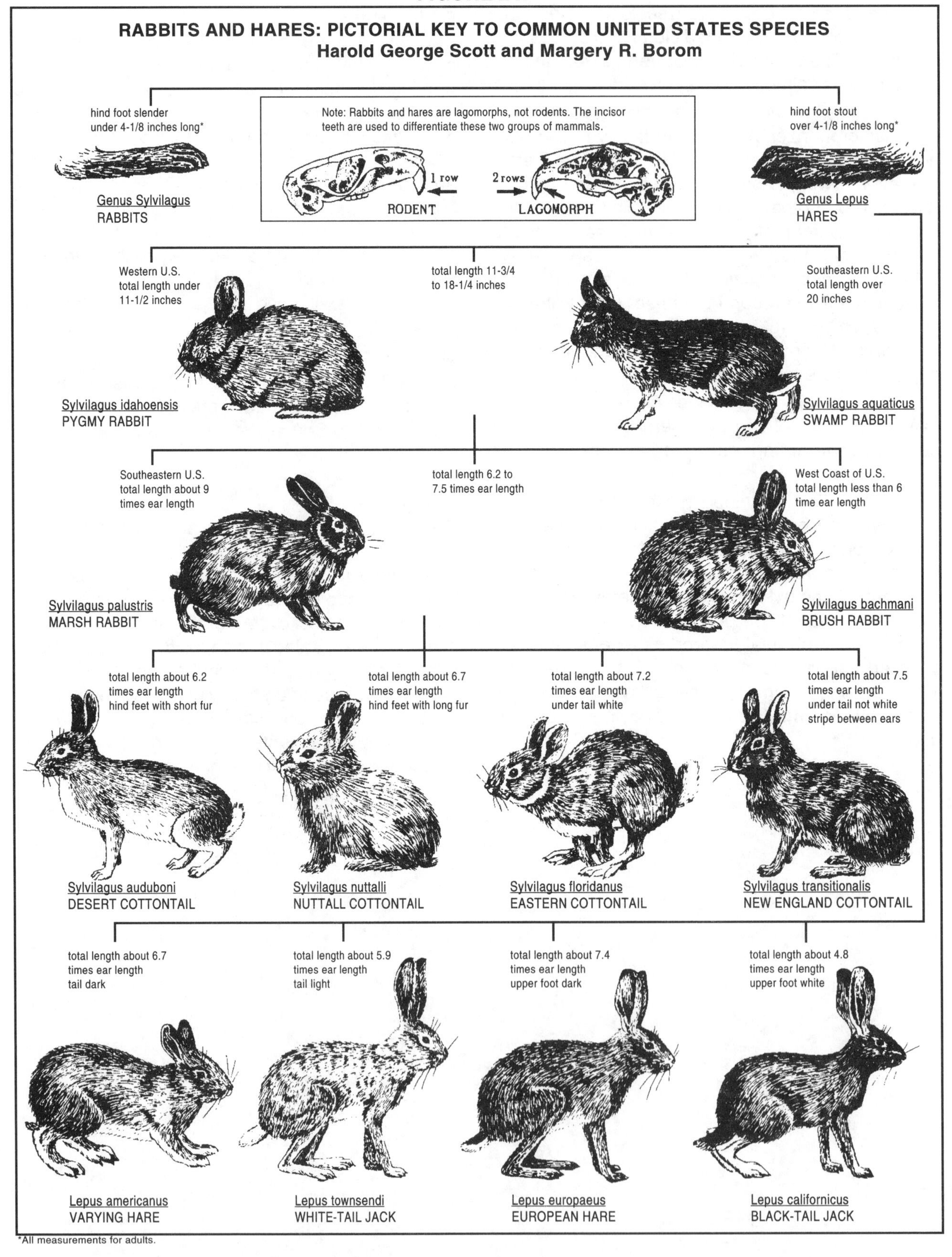

*All measurements for adults.

U.S. DEPARTMENT OF HEALTH, EDUCATION, AND WELFARE
PUBLIC HEALTH SERVICE, Communicable Disease Center, Training Branch, Atlanta, Georgia — 1962

FIGURE 25.

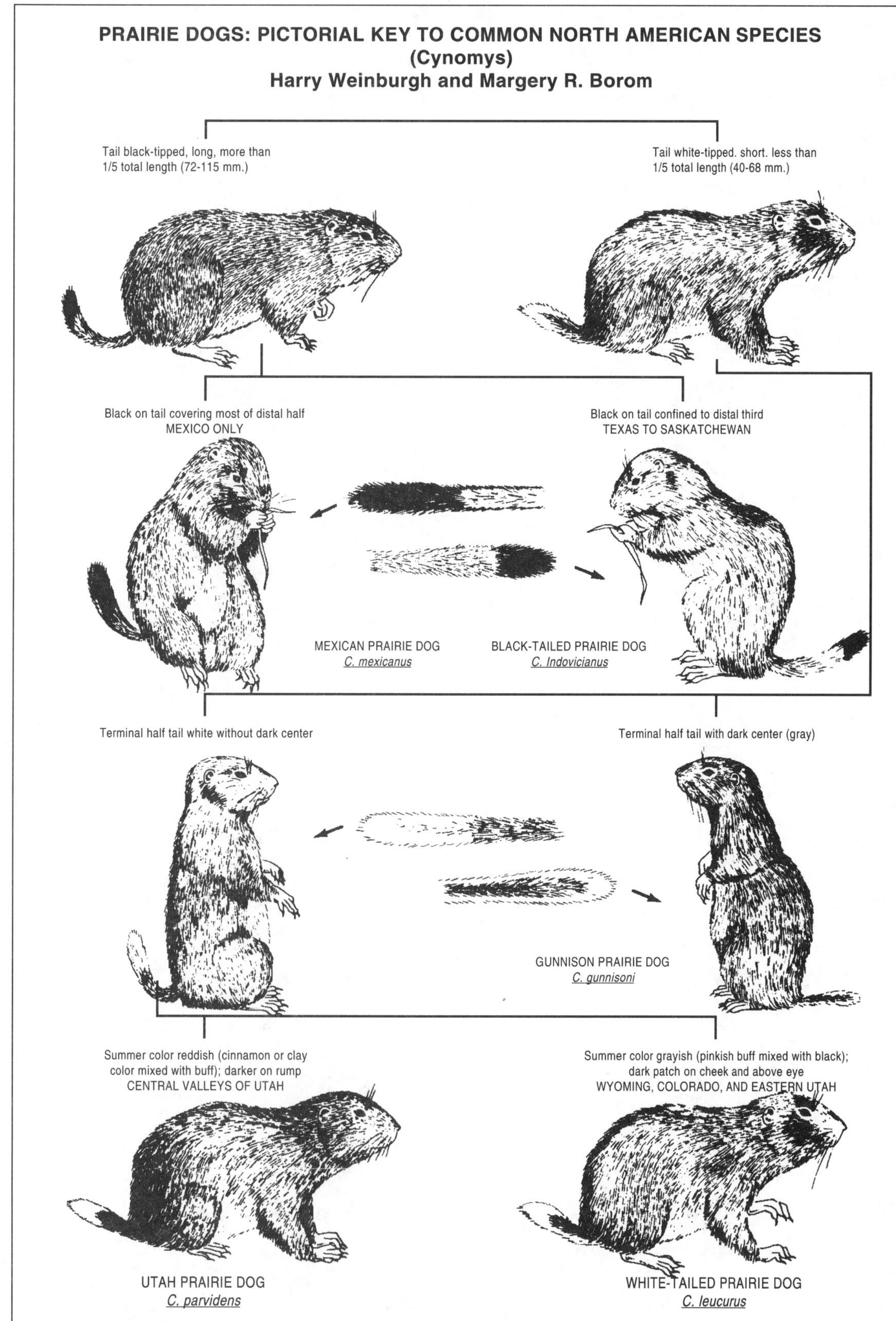

U.S. DEPARTMENT OF HEALTH, EDUCATION, AND WELFARE
PUBLIC HEALTH SERVICE, Communicable Disease Center, Training Branch, Atlanta, Georgia — 1964

BIRDS

A scarecrow in a garden of cucumbers
keepeth nothing.
The Apocrypha,
Baruch VI, 70.

BIRDS

All birds can, in one way or another at times be beneficial to mankind. They provide enjoyment and wholesome recreation for most of us regardless of where we live. Despite the fact that wild bird populations are for the most part beneficial, there are occasions when individuals of certain species can seriously compete with homeowner's interest. When these situations occur, some kinds of control measures are inevitable.

These beautiful winged creatures create pest problems singly or in small groups, but especially when in large aggregations. Most of the areas of conflict with man are: (1) destruction of agricultural foodstuffs and predation; (2) contamination of foodstuffs or defacing of buildings with their feces; (3) transmission of diseases, directly and indirectly, to man, poultry and dairy animals; (4) hazards at airports and freeways; and (5) being a general nuisance or affecting man's comfort, aesthetics or sporting values.

Crows in some areas can be particularly annoying with their shrill cries. The University of California has developed taped distress calls, that when played trick the birds into thinking one of their group is being attacked by a predator. Dispersing crows with this audiotape requires no specialized equipment and does not require federal, state or local permits. See Appendix G for obtaining copies of these tapes.

Birds also bring with them ectoparasites and the potential to transmit diseases. We are now aware that histoplasmosis, a fungal disease centering in the human respiratory tract, is associated with the bird droppings from starlings and blackbirds. Pigeons are responsible for carrying cryptococcosis, also a fungal disease and similar to histoplasmosis in its effects in humans.

Additional to the hazard represented by dust-borne fungal diseases and the nuisance of noise, birds are subject to parasites, especially mites. They multiply in the birds' nests and when the birds have departed, the hungry mites seek another available food source. If the nest is on or within a house, an apartment or office building, the mites often become pests to humans. Although many species of mites will die rather than attack an unfamiliar host, bird mites are not fussy about whose blood they consume. So, the first problem of birds nesting may be followed by a second—bird mites.

Controlling Birds

It has been my experience that an alert and aggressive cat will play a big role in repelling pest birds—and song birds, as well, unfortunately. And, there are other ways of controlling birds, mainly by repelling. How do the professionals drive nuisance birds from their roosts and perches? With frightening devices (both visual and acoustical), chemical repellents including sticky pastes on ledges and roosts, mechanical barriers, trapping and shooting, toxic baits, soporifics (stupefacients), surfactants (feather-wetting agents) and biological control (nest or roost destruction, modification of habitat and chemosterilants). However, the encouragement of natural predators or payment of bounties have not been generally successful.

You may be surprised to learn that the bird species commonly requiring control are as numerous or even more abundant today than they were years ago before controls became so widespread. The reason that the number of birds has not declined is that it is primarily the condition of food and cover that determine their density. The way man has altered his environment apparently has provided an improved habitat for many of the bird species. Our crops and the landscaping of city and suburban homeowners, provide desirable food or nesting cover not available in the past to these species.

Whether a bird is a pest depends on how many there are, where they roost and what they eat. Thus a list of pest species may or may not include your pest. Usually the avian pests would include: The English (house) sparrow, starling, common grackle, Brewer's blackbird, rusty blackbird, red-winged blackbird, feral (wild) pigeon, house finch and cow birds (See Figure 26). Other birds that may become pests on occasion are the blue jay, Steller's jay, mourning and white wing dove, swallows, martins, mocking birds, sapsuckers, flickers, gulls and pelicans. Neither list is complete, much as a list of weeds would never be complete. For every bird could, under certain circumstances, be a pest to someone.

What can we do about birds that dig up the seed which we have planted in the garden, nibble on our strawberries or attack our ripening fruit? You have about 4 choices: Repellents, stakes and flags, continuous string flagging or netting.

Repellents have been used for years in commercial planting with only moderate success and they can be used in the garden with about the same outcome.

After the garden is planted, several handfuls of naphthalene granules or flakes scattered over the seed beds will do very satisfactorily. Or the seed may be treated prior to planting with one of the chemical repellents listed later.

Scarecrows do not work. Stakes and flags do. Stakes or laths are fixed in the soil and strips of cloth or paper attached to their tops. The flag is usually tied to the end of the stake with a short string or sometimes it is tacked to the stake. For persistent birds in the garden space your stakes 15 or 20 feet apart in all directions.

The best of protective methods for garden seed and fruit eaters is the continuous string flagging. The needed materials are heavy stakes at least 4 feet long, strong cotton or sisal wrapping twine and paper, cloth or plastic streamers 2 to 2 1/2 inches wide and 20 to 24 inches long. The stakes are driven firmly into the ground and may be 50 or more feet apart. The twine is stretched from stake to stake and streamers are fastened at 5-foot intervals, making 10 streamers to each 5-foot section between stakes. It is best to install continuous string flagging ahead of attack on that treasured garden. Spiral twirlers, shiny propellers and other objects that flash in the sunlight or rustle and rattle as they spin are useful in small areas. These methods work for about a week. After that, birds grow accustomed to them and sit on them while they eat your berries.

The most effective bird-damage control method for fruit is exclusion of birds by using netting. Several types are marketed to protect fruit trees and grapes. Plastic-impregnated paper, nylon, cotton and polyethylene nettings are sold for this purpose (See Bird Barriers in Appendix G, *Pest Control Sources*). Most types are reasonably tough and can be stored and reused for many years. Life-size hawk silhouettes cut from cardboard or sheet metal suspended on string a few feet above fruit trees are also effective. Reposition daily.

For birds that roost on building window ledges and on various protruding structures of municipal buildings, churches and offices, exclusion is the most humane method of management. This exclusion process involves the use of a black plastic (polypropylene) netting, with hole sizes ranging from 1/2 inch to 7/8 inch square. The netting disappears on most surfaces and can be placed in position with clips, epoxy caulking, staples or wood lath strips. It is relatively simple to install, economical and is resistant to breakdown by sunlight. More and more, professional pest control operators are turning to this method of bird management.

For birds that assemble in your trees in large numbers during the fall and winter—these are probably grackles, blackbirds or starlings—it is best to frighten them away on their first visit. Once the area is marked with their odorous white feces, it becomes more attractive and populated.

Roost control or management now appears to be the ecologically sound method of preventing roosting by droves of birds. Recent work in Houston, Texas, by Drs. Heidi Good and Dan Johnson, indicate that removal of trees and tree trimming can prove so unattractive to blackbirds that they choose other roosts. With active roost trees, they were able to prevent roosting in particular groves for up to three years by trimming out only one-third of the canopy of all trees. Large branches are not removed unless necessary for the appearance of the tree. Trimming only the sucker growth is ineffective. Birds prefer dense bushy trees to roost in presumably because they can afford them protection from foul weather and provide many small branches on which to roost.

Recommendations that resulted from their work were to trim or remove trees before the birds arrived. This prevents the roost from ever being initiated. Until the birds find another suitable roost to use each year, it will be necessary to keep the trees trimmed to retain control. Only enough trees should be removed or pruned to create an open space within the favored site. These birds like to cluster and generally avoid roosting in isolated trees.

Carbide exploders have proven very successful, however, continued use seems to lose its effectiveness on the birds while gaining in effectiveness with the neighbors. Firecrackers can be just as effective and less costly. Firecracker ropes can be made by inserting the fuses in the strands of cotton rope which serve as a slow-burning fuse. Silhouettes of large owls placed in the trees have been reported effective in preventing alighting. Hand-clapping in combination with yelling in view of the invaders is effective. Throwing empty #3 cans containing rocks, nails or other noise-producing objects into the occupied trees will also send them flying.

Placing a cheesecloth or other veil-like covering over fruit trees during ripening satisfactorily keeps birds out of fig, cherry, plum and other fruit trees.

For birds that rest on window sills, a single strand of wire stretched across the ledge and attached to both sides, about breast high on the bird, will prevent them from landing or roosting on these ledges. Also the application of a caulk-like sticky chemical to ledges, rain gutters, roof hips, tops and gables and other resting areas is also effective.

Poisoning is effective but not selective. Invariably several song and other protected species become inadvertent victims of attempts to control pest species and consequently poisoning is not recommended. The chemical controls available to the home gardener are a few repellents, Captan, naphthalene and PDB. Methiocarb (Mesurol®) is registered as a bird repellent for cherries and blueberries and as a seed dressing. (See Avicides)

FIGURE 26.

BIRDS: PICTORIAL KEY TO SOME COMMON PEST SPECIES OF PUBLIC HEALTH IMPORTANCE
Margaret A. Parsons and Chester J. Stojanovich

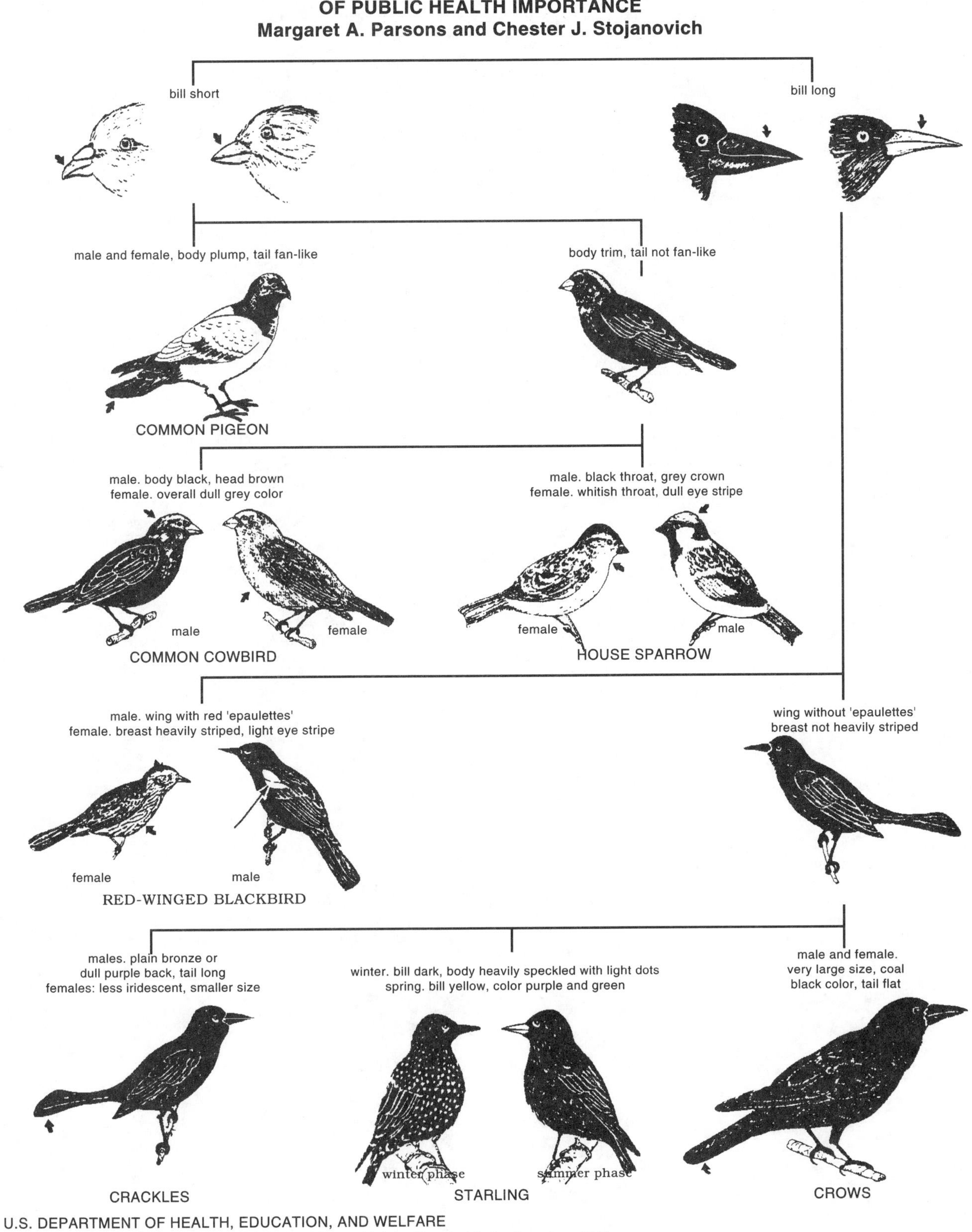

U.S. DEPARTMENT OF HEALTH, EDUCATION, AND WELFARE
PUBLIC HEALTH SERVICE, Communicable Disease Center, Training Branch, Atlanta, Georgia — 1964

CHAPTER 7

'better to be safe than sorry.
American Proverb.

SAFE HANDLING AND STORAGE OF PESTICIDES

Before any pesticide is applied in or around the home, READ THE LABEL! This is the first rule of safety in using any pesticide—read the label and follow its directions and precautions. All pesticides are safe to use, provided common-sense safety is practiced and provided they are used according to the label instructions. This especially means keeping them away from children, illiterate or mentally incompetent persons and pets.

An old cliche of industrial and safety engineers is "Safety is a state of mind". Pesticide safety is more than a state of mind. It must become a habit with those who apply pesticides around the home and certainly with those who have small children. You can control pests in your home and garden with absolute safety, if you use pesticides properly.

Pesticide Selection. When buying a pesticide, check the label. Make sure it lists the name of the pests you want to control. If in doubt, consult your County Extension Agent or other authority. Select the pesticide that is recommended by competent authority and consider the effects it may have on nearby plants and animals. Make certain that the label on the container is intact and up-to-date; it should include directions and precautions. Finally, purchase only the quantity needed for the current season. Don't buy more than you need because it's a bargain.

Pesticide Mixing and Handling. If the pesticide is to be mixed before applying, read carefully the label directions and current recommendations from the County Agent's office when available. This information can be obtained easily by telephone. Always wear rubber gloves when mixing a pesticide and stand upwind of the mixing container. Handle the pesticide in a well-ventilated area. Avoid dusts and splashing when opening containers or pouring into the sprayer. Don't use or mix a pesticide on windy days. Measure the quantity of pesticide required accurately, using the proper equipment. Over-dosage is wasteful; it won't kill more pests; it may be injurious to plants and may leave an excess residue on fruits and vegetables. Don't mix a pesticide in areas where there is a chance that spills or overflows could get into any water supply. Clean up spills immediately. Wash the pesticide off skin promptly with plenty of soap and water and change clothes immediately if they become contaminated.

Pesticide Application. Wear the appropriate protective clothing and equipment if the label calls for it. Your best protection is to wear rubber gloves, a long sleeved shirt, full-length trousers, shoes and socks. Recent studies show that 90% of exposure occurs on the hands and arms. Make certain that equipment is calibrated correctly and is in satisfactory working condition. Apply only at the recommended rate and to minimize drift, apply only on a calm day. Do not contaminate feed, food, or water supplies. This includes pet food and water bowls. Avoid damage to beneficial and pollinating insects by not spraying during periods when such insects are actively visiting flowering plants. Honey bees are usually inactive at dawn and dusk, which are good times for outdoor applications. Dusk is preferred.

Keep pesticides out of mouth, eyes and nose. Don't use the mouth to blow out clogged hoses or nozzles. Observe precisely the waiting periods specified on the label between pesticide application and harvest of fruit and vegetables. Clean all equipment used in mixing and applying pesticides according to recommendations. Don't use the same sprayer for insecticide and herbicide applications.

Wash the sprayer, protective equipment and hands thoroughly after handling pesticides. If you should ever become ill after using pesticides and believe you have the symptoms of pesticide poisoning, call your physician and take the pesticide label with you. This situation is highly unlikely, but it's always good to know in an emergency.

Pesticide Storage. The rule of thumb around the home is lock up all pesticides. Lock the room, cabinet or shed where they are stored to discourage children. Don't store pesticides where food, feed, seed or water can be contaminated. Certainly don't store them beneath the kitchen sink. Store in a dry, well-ventilated place, away from sunlight and at temperatures above freezing. If you should happen to have an operation larger than a typical homeowner's, mark all entrances to your storage area with signs bearing this caution: "PESTICIDES STORED HERE--KEEP OUT". This action would be in keeping with a commercial greenhouse or farming operation.

Pesticides should be kept only in original containers, closed tightly and labelled. Examine pesticide containers occasionally for leaks and tears. Dispose of leaking and torn containers and clean up spilled or leaked material immediately. It's a good idea to date the container when purchased. This makes the disposal of outdated materials a simple matter. Because many pesticide spray formulations are flammable, take precautions against potential fire hazards.

Disposal of Empty Containers and Unused Pesticides. Analagous to the loaded gun, empty containers are never completely empty, so don't reuse for any purpose. Instead, break glass containers, rinse metal containers three times with water, punch holes in top and bottom and leave in your trash barrels for removal to the official landfill trash dump. Empty paper bags and cardboard boxes should be torn or smashed to make unusable, placed in a larger paper bag, rolled and relegated to the trash barrel. In summary, don't leave anything tempting in the trash barrel or dump. That kid raiding your trash can may be your own!

Unwanted Pesticides. Offer to give unwanted pesticides to a responsible person in need of the materials. If this is not practicable, wrap the container in plastic and save for the next household hazardous waste collection program. Some areas may have adequate incineration facilities, approved landfills, or other approved means of disposing of unused pesticide wastes. Triple rinsed pesticide containers are considered as solid waste and will be acceptable at landfills.

Hundreds of "do" and "don't" rules for handling pesticides have been generated over the past two decades with the increased awareness of pesticide hazards. The above gleanings are from many sources and may be useful to the reader in his own home situation, around the commercial greenhouse, on the farm, in preparation for talks on safety, for inspections of schools and public buildings for safe playing and working conditions, or just for your own confident reference.

The Label

The pesticide label on the container is the single most important tool to the homeowner in using pesticides safely. The FEDERAL ENVIRONMENTAL PESTICIDE CONTROL ACT (FEPCA), which is discussed in the chapter on pesticide laws, contains three very important points concerning the pesticide label which should be further emphasized. They pertain to reading the label, understanding the label directions and following carefully these instructions.

The first provisions of FEPCA are that the use of any pesticide inconsistent with the label is prohibited and deliberate violations by growers, applicators, or dealers can result in heavy fines or imprisonment or both. The third provision is found in the general standards for certification of commercial applicators, which in essence will license them to use restricted-use pesticides, the area of label and labeling comprehension. For certification, applicators are to be tested on (a) the general format and terminology of pesticide labels and labeling, (b) the understanding of instructions, warning, terms, symbols and other information commonly appearing on pesticide labels, (c) classification of the product (unclassified or RESTRICTED USE) and (d) the necessity for use consistent with the label.

Figure 27 shows the label format for unclassified use pesticides as required by EPA to appear on all containers. This label is keyed as follows:

1. Product name
2. Company name and address
3. Net contents
4. EPA pesticide registration number
5. EPA formulator manufacturer establishment number

6A. Ingredients statement
6B. Pounds/gallons statement (if liquid)
7. Front panel precautionary statements
7A. Child hazard warning, "Keep Out of Reach of Children"
7B. Signal word—CAUTION
7D. Statement of practical treatment
7E. Referral statement
8. Side/back panel precautionary statements
8A. Hazards to humans and domestic animals
8B. Environmental hazards
8C. Physical or chemical hazards
9B. Statement of pesticide classification
9C. Misuse statement
10A. Re-entry statement
10C. Storage and Disposal block
10D. Directions for use

Emergencies Involving Pesticides

All pesticides can be used safely, provided common-sense safety is practiced and provided they are used according to the label instructions; this includes keeping them away from children and illiterate or mentally incompetent persons. Despite the most thorough precautions, accidents will occur. Below are given two important sources of information in the event of any kind of serious pesticide accident.

The first and most important source of information is for human-poisoning cases: the nearest Poison Control Center. Look it up in the telephone directory under POISON CONTROL CENTERS, or ask the telephone operator for assistance. Poison Control Centers are usually located in the larger hospitals of most cities and can provide emergency treatment information on all types of human poisoning, including pesticides. The telephone number of the nearest Poison Control Center should be kept as a ready reference by parents of small children, or employers of persons who work with pesticides and other potentially hazardous materials.

NATIONAL POISON CONTROL 800-222-1222

The second and one you will not likely use, is the CHEMTREC telephone number. From this toll-free long-distance number can be obtained emergency information on all pesticide accidents, pesticide-poisoning cases, pesticide spills and pesticide spill cleanup teams. This telephone service is available twenty-four hours a day.

CHEMTREC 800-424-9300

Specific pesticide poisoning information can also be obtained by telephone from:

The National Pesticide Information Center Hotline (800) 858-7378

Specific pesticide poisoning information can also be obtained from your nearest Certified Regional Poison Control Centers. There are presently 64 of these scattered across the U.S., with 24-hour professional services at each. Some centers are regional, serving more than one state. Following is a list of these and their direct line phone numbers, most of which have the national 800-222-1222 number. More details of this list can be accessed on the Internet at www.AAPCC.org.

CERTIFIED REGIONAL POISON CONTROL CENTERS WITH EMERGENCY TELEPHONE NUMBERS

ALABAMA, Tuscaloosa
800-222-1222
205-345-0609

ALABAMA, Birmingham
800-222-1222
205-939-9720

ALASKA, Juneau
907-465-1185

ARIZONA, Tucson
800-222-1222
520-626-6016

ARIZONA, Phoenix
800-222-1222
602-495-4884

ARKANSAS, Little Rock
800-641-3805
501-686-5540

CALIFORNIA, San Francisco
800-222-1222
[1]TTY/TDD 800-972-3323
415-502-8605

CALIFORNIA, San Diego
800-222-1222
TTY/TDD 800-972-3323
858-715-6300

CALIFORNIA, Madera
Fresno/Madera Div.
800-222-1222
TTY/TDD 800-972-3323
559-662-2300

COLORADO, Denver
Rocky Mt. Poison & Drug Ctr.
800-332-3073
TTY/TDD 303-739-1127
303-739-1100

CONNECTICUT, Farmington
800-222-1222
TTY/TDD 866-218-5372
860-679-4540

DELAWARE
Philadelphia, PA
800-222-1222
TTY/TDD 215-590-8789
215-590-2003

DISTRICT OF COLUMBIA
Washington, DC
800-222-1222 (voice & TTY)
202-362-3867

FLORIDA, Jacksonville
800-222-1222 (voice & TTY/TDD)
904-244-4465

FLORIDA, Miami
800-222-1222
305-585-5250

FLORIDA, Tampa
800-222-1222
813-844-7044

GEORGIA, Atlanta
800-222-1222
TTY/TDD 404-616-9287
404-616-9237

HAWAII
Rocky Mt. Poison & Drug Ctr.
Denver, CO
800-222-1222
808-733-9210 (Hawaii only)
303-739-1100

IDAHO
Rocky Mt. Poison & Drug Ctr.
Denver, CO.
800-860-0620 (Idaho only)
TTY/TDD 303-739-1127
303-739-1100

ILLINOIS, Chicago
800-222-1222
TTY/TDD 312-906-6185
312-906-6136

INDIANA, Indianapolis
800-222-1222
TTY/TDD 317-962-2336
317-929-2335

IOWA, Sioux City
800-222-1222
712-279-3710

KANSAS, Kansas City
800-222-1222
TTY/TDD 913-588-6639
913-588-6638

KENTUCKY, Louisville
800-222-1222
502-629-7264

LOUISIANA, Monroe
800-222-1222
318-342-3648

MAINE, Portland
800-222-1222
TTY/TDD 877-299-4447
207-842-7222

MARYLAND, Baltimore
800-222-1222
TTY/TDD 410-706-1858
410-706-7604

MASSACHUSETTS, Boston
800-222-1222
TTY/TDD 888-244-5313
617-355-6609

MICHIGAN, Detroit
800-222-1222
TTY/TDD 800-356-3232
313-745-5335

MINNESOTA, Minneapolis
Hennepin Co. Med. Ctr.
800-222-1222 (voice & TTY)
612-873-3144

MISSISSIPPI, Jackson
800-222-1222
601-984-1680

MISSOURI, St. Louis
800-222-1222
TTY/TDD 314-612-5705
314-772-8300

MONTANA
Rocky Mt. Poison & Drug Ctr.
Denver, CO
800-525-5042 (Montana only)
TTY/TDD 303-739-1127
303-739-1100

NEBRASKA, Omaha
800-222-1222
402-955-5555

NEVADA
Rocky Mt. Poison & Drug Ctr.
Denver, CO
800-446-6179 (Nevada only)
TTY/TDD 303-739-1127
303-739-1100

NEW HAMPSHIRE, Lebanon
800-222-1222
603-650-6318

NEW JERSEY, Newark
800-222-1222
TTY/TDD 973-926-8008
973-972-9280

NEW MEXICO, Albuquerque
800-222-1222 (voice & TTY)
505-272-4261

NEW YORK, Sleepy Hollow
Hudson Valley Poison Prev. Ctr.
800-222-1222
914-366-3675

NEW YORK, Mineola
Long Island Regional Ctr.
800-222-1222
TTY/TDD 516-747-3323
516-663-4574

NEW YORK, New York
New York City Poison Control Ctr.
800-222-1222
TTY/TDD 212-689-9014
212-447-8152

NORTH CAROLINA, Charlotte
800-222-1222
704-395-3795

NORTH DAKOTA, Regional Poison Ctr.
Minneapolis, MN
800-222-1222 (voice & TTY)
612-873-3144

OHIO, Columbus
800-222-1222
TTY/TDD 614-228-2272
614-722-2635

OHIO, Cleveland
800-222-1222
216-844-1573

OKLAHOMA, Oklahoma City
800-222-1222 (voice & TTY/TDD)
405-271-5062

OREGON, Portland
800-222-1222 (voice & TTY/TDD)
503-494-8600

PENNSYLVANIA, Philadelphia
800-222-1222
TTY/TDD 215-590-8789
215-590-2003

PENNSYLVANIA, Pittsburgh
800-222-1222
412-390-3300

PUERTO RICO, Santurce
800-222-1222
787-727-1000, ext. 4437

RHODE ISLAND
Regional Poison Control Ctr.
Boston, MA
800-222-1222
TTY/TDD 888-244-5313
617-355-6609

SOUTH CAROLINA, Columbia
800-222-1222
803-777-7909

SOUTH DAKOTA
Hennepin Co. Med. Ctr.
Minneapolis, MN
800-222-1222 (voice & TTY)
612-873-3144

TENNESSEE, Nashville
800-222-1222
TTY/TDD 615-936-2047
615-936-0760

TEXAS, Dallas
800-222-1222
214-589-0911

TEXAS, Temple
800-222-1222
254-724-7405

TEXAS, Amarillo
800-222-1222
806-354-1630

UTAH, Salt Lake City
800-222-1222
801-587-0600

VERMONT, Regional Poison Control Ctr.
Portland, ME
800-222-1222
TTY/TDD 207-871-2879
207-842-7222

VIRGINIA, Charlottesville
800-222-1222
434-924-0347

WASHINGTON, DC
800-222-1222 (voice & TTY)
202-362-3867

WASHINGTON, Seattle
800-222-1222
TTY/TDD 206-517-2394
206-517-2350

WEST VIRGINIA, Charleston
800-222-1222
304-347-1212

WISCONSIN, Milwaukee
800-222-1222
TTY/TDD 414-266-2542
414-266-2952

WYOMING, Regional Poison Control Ctr.
Omaha, NE
800-222-1222
402-955-555

Source: American Association of Poison Control Centers, Certified Regional Poison Centers, June 2004.
[1]TTY/TTD (Telecommunications Device for the Deaf)

FIGURE 27. Format for the Unclassified or General Use pesticide label.

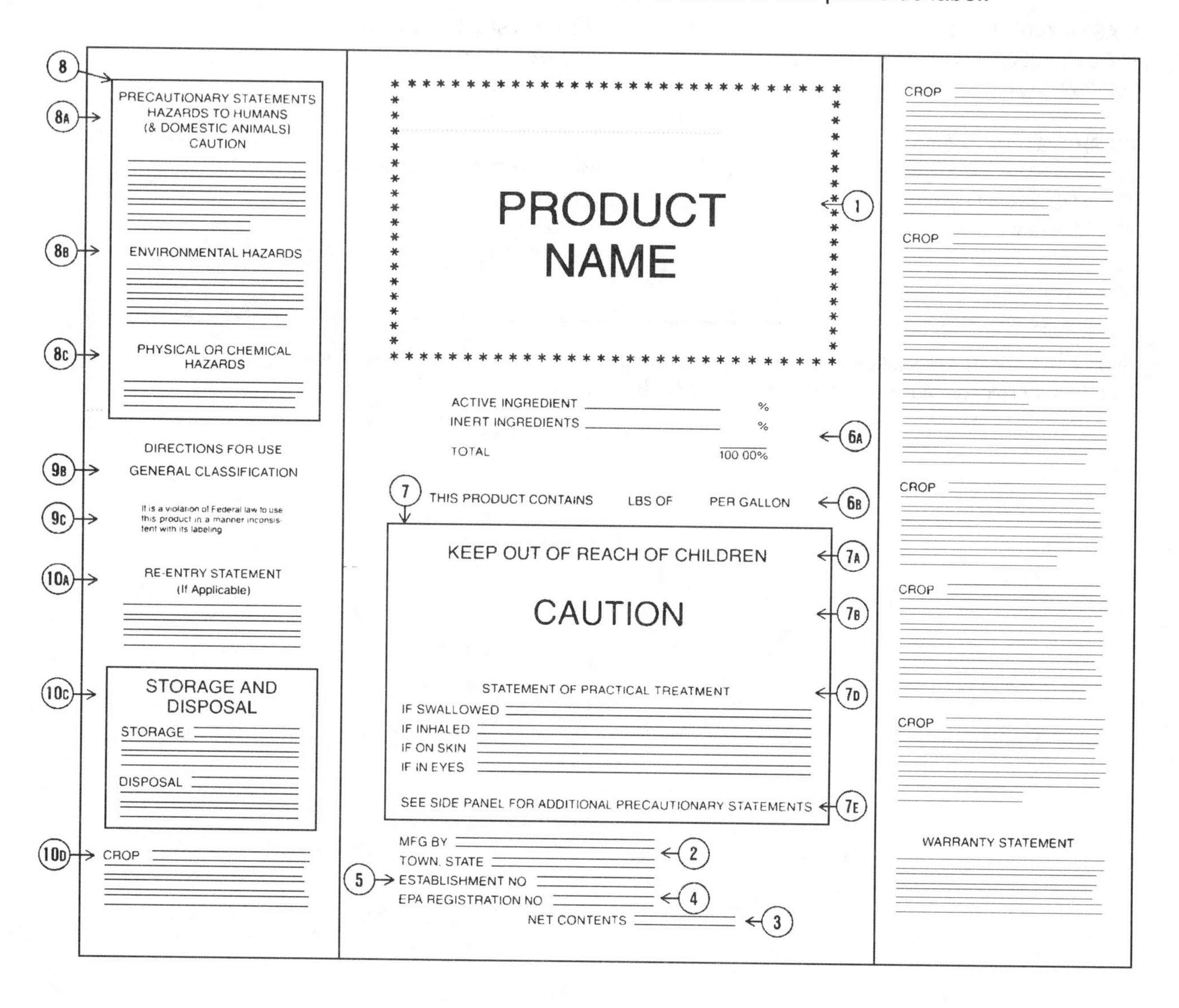

CHAPTER 8

Good laws lead to the making of better ones; bad ones bring about worse.
Jean Jacques Rousseau,
Contrat Social, Book III,
Chap. 15.

PESTICIDE LAWS

Pesticide use is controlled by both federal and state and occasionally county or city laws. Federal laws have protected the user of pesticides, his pets and domestic animals, his neighbor and the consumer of treated products for several decades. Nothing is left unprotected and after reading briefly about the laws themselves, the reader will be convinced of their omnipotence.

The first federal law, the Federal Food, Drug and Cosmetic Act of 1906, known as the Pure Food Law, required that food (fresh, canned and frozen) shipped in interstate commerce be pure and wholesome. There was not a single word in the law that pertained to pesticides.

The Federal Insecticide Act of 1910, which covered only insecticides and fungicides, was signed into law by President William Howard Taft. The act was the first to control pesticides and was designed mainly to protect the farmer from substandard or fraudulent products, for they were plentiful at the turn of the century. This was probably one of our earliest consumer protection laws.

The Pure Food Law of 1906 was amended in 1938 to include pesticides on foods, primarily the arsenicals such as lead arsenate and Paris green. It also required the adding of color to white insecticides, including sodium fluoride and lead arsenate, to prevent their use as flour or other look-alike cooking materials. This was the first federal effort toward protecting the consumer from pesticide-contaminated food, by providing tolerances for pesticide residues, namely arsenic and lead, in foods where these materials were necessary for the production of a food supply.

In 1947 the Federal Insecticide, Fungicide and Rodenticide Act (FIFRA) became law. It superseded the 1910 Federal Insecticide Act and extended the coverage to include herbicides and rodenticides and required that any of these products must be registered with U.S. Department of Agriculture before they could be marketed in interstate commerce. Basically, the law was one requiring good and useful labeling, making the product safe to use if label instructions were followed. The label was required to contain the manufacturer's name and address, name, brand and trademark of the product, its net contents, an ingredient statement, an appropriate warning statement to prevent injury to man, animals, plants and useful invertebrates and directions for use adequate to protect the user and the public.

In 1945, the Miller amendment to the Food, Drug and Cosmetic Act (1906, 1938) was passed. It provided that any raw agricultural commodity may be condemned as adulterated if it contains any pesticide chemical whose safety has not been formally cleared or that is present in excessive amounts (above tolerances). In essence, this clearly set tolerances on all pesticides in food products, for example, 10 ppm of carbaryl in lettuce or 1 ppm of ethyl parathion on string beans.

Two laws, the Federal Insecticide, Fungicide and Rodenticide Act (FIFRA) and the Miller Amendment to the Food, Drug and Cosmetic act, supplement each other and are interrelated by law in practical operation. Today they serve as the basic elements of protection for the applicator, the consumer of treated products and the environment, as modified by the following amendments.

The Food Amendment to the Food, Drug and Cosmetic Act (1906, 1938, 1954) was passed in 1948. It extended the same philosophy to all types of food additives that has been applied to pesticide residues on raw agricultural commodities by the 1954 amendment. However, this also controls pesticide residues in processed foods that had not previously fit into the 1954 designation of raw agricultural commodities. Of greater importance, however, was the inclusion of the Delaney clause, which states that any chemical found to cause cancer (a carcinogen) in laboratory animals when fed at any dosage may not appear in foods consumed by man. This has become the most controversial segment of the entire spectrum of federal laws applying to pesticides, mainly with regard to the dosage found to produce cancer in experimental animals.

The various statutes mentioned so far apply only to commodities shipped in interstate commerce. In 1959, FIFRA (1947) was amended to include nematicides, plant regulators, defoliants and desiccants as economic poisons (pesticides). (Poisons and repellents used against amphibians, reptiles, birds, fish, mammals and invertebrates have since been included as economic poisons.) Because FIFRA and the Food, Drug and Cosmetic Act are allied, these additional economic poisons were also controlled as they pertain to residues in raw agricultural commodities.

In 1964, FIFRA (1947, 1959) was again amended to require that all pesticide labels contain the Federal Registration Number. It also required caution words such as WARNING, DANGER, CAUTION and KEEP OUT OF REACH OF CHILDREN to be included on the front label of all poisonous pesticides. Manufacturers also had to remove safety claims from all labels.

Until December, 1970, the administration of FIFRA was the responsibility of the Pesticides Regulation Division of the U.S. Department of Agriculture. At that time the responsibility was transferred to the newly-designated U.S. Environmental Protection Agency (EPA). Simultaneously, the authority to establish pesticide tolerances was transferred from the Food and Drug Administration (FDA) to EPA. The enforcement of tolerances remains the responsibility of the FDA.

In 1972, FIFRA (1947, 1959, 1964) was revised by the most important pesticide legislation of this century. THE FEDERAL ENVIRONMENTAL PESTICIDE CONTROL ACT (FEPCA), sometimes referred to as 1972 FIFRA amendment. Some of the provisions of FEPCA are abstracted as follows:

1. Use of any pesticide inconsistent with the label is prohibited.
2. Deliberate violations of FEPCA by growers, applicators or dealers can result in heavy fines and/or imprisonment.
3. All pesticides will be classified into (a) General Use or (b) Restricted Use categories.
4. Anyone applying Restricted Use Pesticides must be certified by the state in which he lives.
5. Pesticide manufacturing plants must be registered and inspected by the EPA.
6. States may register pesticides on a limited basis when intended for special local needs.
7. All pesticide products must be registered by EPA, whether shipped in interstate or intrastate commerce.
8. For a product to be registered the manufacturer is required to provide scientific evidence that the product, when used as directed, will (1) effectively control the pests listed on label, (2) not injure humans, crops, livestock, wildlife or damage the total environment and (3) not result in illegal residues in food or feed.

FIFRA was further amended in 1975, 1978, 1980 and 1981. These provisions clarified the intent of the law and have a great influence on the way pesticides are registered and used.

Many of the changes in the legislation were designed to improve the registration process, which was slowed significantly by regulations resulting from the 1972 Act. The more important points are as follows:

1. Efficacy data can be waived. The EPA has the option of setting aside requirements for proving the efficacy of a pesticide before registration. This leaves the manufacturer to decide whether a pesticide is effective enough to market and final proof will depend on product performance.
2. Generic standards will be set for the active ingredients rather than for each product. This change permits EPA to make safety and health decisions for the active ingredient in a pesticide, instead of treating each product on an

individual basis. (It is easy to see how this provision will speed registration, considering that there are only about 860 active ingredients in the 21,000 formulations currently on the market.) This document will specify the information submitted in the past to register an active ingredient and identify the data now needed under re-registration.

3. Re-registration of all older products will be required. Under re-registration, all compounds are currently being reexamined to make certain the supporting data for a registered pesticide satisfies today's requirements for registration, in light of new knowledge concerning human health and environmental safety. Many pesticides were registered under the old data requirements (prior to August 1975) and registrants must submit new information to carry the product through the re-registration process.
4. Pesticides can now be given conditional registration. EPA may now grant a conditional registration for a pesticide even though certain supporting data have not been completed. That information will still be required, but it may be deferred to a later date. Conditional registration can be granted by EPA if:
   a. The uses are identical or greatly similar to those which exist on labels for already registered proucts with the same active ingredient;
   b. New uses are being added, providing a notice or Special Review has not been issued on the product or in the case of food or feed use, there is no other available or effective alternative;
   c. New pesticides have had additional data requirements imposed since the date of the original submission.
5. The use of data from one registrant can be used by other manufacturers or formulators if paid for. This new law spells out how data submitted by one registrant for an active ingredient may be used by other applicants. All data provided from 1970 on can be used for a 15-year period by other registrants, if they offer to pay "reasonable compensation" for this use. In the future, registrants will have 10 years of exclusive use of data submitted for a new pesticide active ingredient. During that time, other applicants may request and be granted permission to use the information but must obtain approval.
6. Trade secrets will be protected. A clear and detailed outline of what information is considered confidential is described. EPA may reveal data on most pesticide effects (including human, animal and plant hazard evaluation), efficacy and environmental chemistry. Four categories of data are generally to be kept confidential but may be released under certain circumstances:
   a. Manufacturing and quality control processes;
   b. Methods of testing, detecting or measuring deliberately added inert ingredients;
   c. Identity or quality of deliberately added inerts;
   d. Production, distribution, sale and inventories of pesticides.
7. The state now has primary enforcement responsibility. Under the 1978 law, the primary authority for use enforcement under federal law will be assigned to the states. Any suspected misuse of pesticides will be investigated by and acted upon by the state regulatory boards. Before the enforcement authority can be legally transferred by EPA, the state must indicate that their regulatory methods will meet or exceed the federal requirements. If a state does not take appropriate action within 30 days of alleged misuse, EPA can act. In addition, enforcement authority can be taken away from any state which consistently fails to take proper action.
8. States can register pesticides for Special Local Needs (SLN). The state authority

to register materials for Special Local Needs is increased and EPA's role is decreased. All states are given the authority automatically to register products for use within a state for special situations. Registrations can be made for new products, using already registered active ingredients. Existing product labels can be amended for new uses, including chemicals which have been subject to cancellation or suspension in the past. Only those specific uses which are cancelled or suspended may not be registered by the state for SLN's.

9. Uses inconsistent with the labeling are defined. The phrase, "to use any registered pesticide in a manner inconsistent with its labeling" is defined. Certain use practices, which previously were covered by Pesticide Enforcement Policy Statements (PEPS), are now permitted. Persons who derive income from the sale or distribution of pesticides may not make recommendations which call for uses inconsistent with labeling. Users and applicators may now:
   a. Use a pesticide for control of a target pest not named on the label.
   b. Apply the pesticide using any method not specifically prohibited on the label.
   c. Mix one or more pesticides with other pesticides or fertilizers, provided the current labeling does not actually prohibit this practice.
   d. Use a pesticide at less than labeled dosage, providing the total amount applied does not exceed that currently allowed on the labeling.

These are only the most important aspects of FEPCA that you, the interested novice, need be acquainted with.

Beyond the federal laws providing rather strict control over the use of insecticides, each state usually has two or three similar laws controlling the application of pesticides and the sales and use of pesticides. These may or may not involve the licensing of aerial and ground applicators as one group and the structural applicator or pest control applicator as another. Only the latter is of concern to the home owner, who should insist that anyone selling him pest control services be both licensed within that state and certified to apply Restricted Use pesticides.

1988 AMENDMENTS—FIFRA "LITE"

In 1988, the 100th Congress passed amendments to FIFRA, humorously referred to as FIFRA "Lite", when in fact, they were anything but trivial. These amendments strengthen the authority of EPA in several major areas, concentrating on: (1) re-registration, (2) fees, (3) expediting certain types of registration applications, (4) revised responsibilities for disposal and transportation of pesticides taken off the market by EPA, and (5) limiting the entitlement to indemnification for holders of canceled and suspended pesticides. Re-registration is the principal focus of the 1988 Amendments.

FOOD QUALITY PROTECTION ACT OF 1996

The Food Quality Protection Act (FQPA) of 1996 (P.L. 104-170) is the first important amendment to FIFRA since passage of FIFRA Lite in 1988, above.It replaces the Delaney Clause with "reasonable certainty of no harm." This no-harm standard applies to pesticide residues on both raw and processed food.

The FQPA provides for expedited registrations of biological and conventional pesticides that do one of the following: 1) Reduce the risk of pesticides to human health; 2) reduce the risk of pesticides to nontarget organisms; 3) reduce the potential for contamination of ground and surface water; or 4) broaden the adoption of Integrated Pest Management (IPM) strategies or make these strategies more available. FQPA is a very broad act and brings far-reaching changes to pesticide regulation. Among other things, it requires a screening process for determination of the estrogenic (endocrine) effects of pesticides and other substances.

There is a great deal more that some readers may want to know. You can visit Wright's PestLaw (www.pestlaw.com/pestlaw.html) or the EPA Home Page (www.epa.gov/epa.gov/epahome/index/html).

CHAPTER 9

> The single factor that determines the degree of harmfulness of a compound is the dose . . .
>
> Ted A. Loomis,
> *Essentials of Toxicology,* 1974

HAZARDS OF PESTICIDES

Because of their intended purpose, all pesticides are toxic, but their use is not necessarily a hazard. Let's get it straight at the outset. There is a marked distinction between toxicity and hazard. These two terms are not synonymous. Toxicity refers to the inherent toxicity of a compound. In other words, how toxic is it to animals under experimental conditions. Hazard is the risk or danger of poisoning when a chemical is used or applied, sometimes referred to as use hazard. The factor with which the user of a pesticide is really concerned is the use hazard and not the inherent toxicity of the material. Hazard depends not only upon toxicity but also upon the chance of exposure to toxic amounts of the material.

Webster defines the word poison as "any substance which introduced into an organism in relatively small amounts, acts chemically upon the tissues to produce serious injury or death." One can immediately spot several flaws in this definition. The "relatively small amount" statement is open to wide interpretation. For instance, many chemical agents to which man is exposed regularly could be termed poisons under this definition. An oral dose of 400 milligrams of sodium chloride per kilogram (mg/kg) of body weight, ordinary table salt, will make a person violently ill. A standard aspirin tablet contains about 5 grains of aspirin, chemically known as acetylsalicylic acid. A fatal dose of aspirin to man is in the range of 75 to 225 grains or 15 to 45 tablets. Approximately 85 deaths occur every year (about one-third are children) as a result of overdoses of aspirin. To take a third example, let us consider nicotine. A fatal oral dose of this naturally occurring alkaloid to man is about 50 milligrams (mg) or approximately the amount of nicotine contained in two unfiltered cigarettes. In smoking, however, most of the nicotine is decomposed by burning, and, thus, it is not absorbed by the smoker.

Here man is not exposed during ordinary use to amounts of salt, aspirin and nicotine which cause toxicity problems. Therefore, it is obvious that the hazard from normal exposure is very slight even though the compounds themselves would be toxic under other circumstances.

There's a better definition for the term "poison": "A chemical substance which exerts an injurious effect in the majority of cases in which it comes into contact with living organisms during normal use." The compounds mentioned above would obviously be excluded by such a definition and so would the majority of pesticides.

By necessity, pesticides are poisons, but the toxic hazards of different compounds vary greatly. As far as the possible risks associated with the use of pesticides are concerned, we can distinguish between two types: First, acute poisoning, resulting from the handling and application of toxic materials; and second, chronic risks from long-term exposure to small quantities of materials or from ingestion of them. The question of acute toxicity is obviously of paramount interest to people engaged in manufacturing and formulating pesticides and to those responsible for their application. Supposed chronic risks, however, are of much greater public interest because of their potential effect on the consumer of agricultural products.

Fatal human poisoning by pesticides is uncommon in the U. S. and is due to accident, ignorance, suicide and occasionally homicide. Fatalities represent only a small fraction of all recorded cases of poisoning, as demonstrated by these 2002 U. S. statistics (Table 33). The 18 fatalities from pesticides amounted to a mere 0.7% of the 2,497 reported deaths. In 1985, pesticide deaths amounted to 2.3% of those reported (Watson, et. al 2003). This phenomenal reduction is due to the increased safety of today's pesticides and their containers over those of two decades ago and to the increased educational program on the use and application of pesticides.

TABLE 33. Total Deaths from Poisoning (accidental, intentional, adverse reaction) Reported to Poison Control Centers in the U.S. (2002) [1].

Substance	Number of Deaths
Adhesives/glues	2
Alcohols	139
[Beverage (120), other (19)]	
Auto/aircraft/boat products	30
Bites & envenomations	6
Building & construction products	0
Chemicals	50
Cleaning substances	32
Cosmetics/personal care products	3
Deodorizers	0
Fertilizers	1
Foreign bodies/toys/misc.	9
Fumes/gases/vapors	44
Heavy metals	3
Hydrocarbons	15
Industrial Cleaners	1
Paints & stripping agents	4
Pesticides	18
[Fungicides (0), Herbicides (5), Fumigants (1) Insecticides (7), Rodenticides (5)]	
Swimming Pool/Aquarium	1
Tobacco Products	1
Other unknown non-drug substances	1
Pharmaceuticals	2130
[Analgesics (659), Antidepressants (318), Cardiovascular drugs (181), Sedative/hypnotics/ antipsychotics (364), Stimulants/street drugs (242) Others (366)]	
Plants incl. mushrooms	7
Total	2,497

[1] The Authors noted that the 291.6 million 2002 United States population, the data represented an estimated 99.8% of the human poison exposures that precipitated poison center contacts in the U.S. during 2002. Extrapolations from the number of reported poisonings to the number of actual poisonings occurring annually in the U.S. cannot be made from these data alone, as considerable variations in poison center penetrance were noted. Thus, assuming all centers reached the penetrance level of 16.0 poisonings/1000 population reported for one state, 4.6 million poisonings would have been reported to poison centers in 2002. This report included 2,380,028 human exposure cases reported by 64 participating poison centers during 2002, an increase of 4.9% compared with 2001 poisoning reports.

Source: Watson et al (2003)

Of all human poison exposure cases (these are reported incidents and not fatalities), 39.1% are children under 3 years of age. The picture for pesticides is much, much better, with only 2.1% of the reported poison exposure cases being children under 6 years of age. Somewhat of a surprise is in the category of rodenticide exposures, where 85.4% were children under the age of 6 (Watson, et. al 2003).

In 1968, 11% of all accidental poisoning deaths were children under 6 years of age, while in 1985, that figure had dropped to 6% and in 2002, it had dropped still further to 2%, a most remarkable improvement in poisoning recognition and treatment by poison control centers. Another outstanding change is that of the under 6-year-olds exposed/ poisoned. In 1968, 11% were from pesticides, while in 2003, that figure has dropped significantly to 4.1%. This is in light of the intensive poison control reporting system now available and a public that is highly sensitized to accidental poisoning in children by materials kept around the home. Finally, in 1985, 27% of the deaths attributed to accidental poisoning by pesticides were children less than 6 years of age. This figure has dropped remarkably to 2% in 2002 (Watson, et al. 2003).

Let's examine another type of data for 2002. Of the 1,227,381 children under 6 years of age reported as exposed/poisoned, pesticides accounted for only 4.1% of those exposures, while analgesics (ibuprofen, etc.) accounted for 7.4% and cosmetics and personal care products for 13.3% (Watson, et al. 2003).

Pesticides have a good safety track record and it becomes better each year, mainly through education and understandable labeling of containers.

PESTICIDE EFFECTS ON MAN

Pesticides were developed to kill unwanted organisms and are toxic materials which produce their effects by several different mechanisms. Under certain conditions they may be toxic to man and an understanding of the basic principles of toxicity and the differences between toxicity and hazard is essential. As you already know, some pesticides are much more toxic than others and severe illness may result when only a small amount of a certain chemical has been ingested, while with other compounds no serious effects would result even after ingesting large quantities. Some of the factors that influence this are related to (1) the toxicity of the chemical, (2) the dose of the chemical, especially concentration, (3) length of exposure and (4) the route of entry or absorption by the body.

Early in the development of a pesticide for further experiments and exploration, toxicity data are collected on the pure toxicant as required by the Environmental Protection Agency. These tests are conducted on test animals that are easy to work with and whose physiology, in some instances, is like that of man, for example, dog. Test animals include white mice, white rats, white rabbits, guinea pigs and beagle dogs. For instance, intravenous tests are determined usually on mice and rats, whereas dermal tests are conducted on shaved rabbits and guinea pigs. Acute oral toxicity determinations are most commonly made in rats, with the test substance being introduced directly into the stomach by tube. Chronic studies are conducted on the same two species for extended periods and the compound is usually incorporated in the animal's daily ration. Inhalation studies may involve any of the test animals, but rats, guinea pigs and rabbits are most commonly used.

These procedures are necessary to determine the overall toxic properties of the compound to various animals. From this information, toxicity to man can generally be extrapolated and eventually some micro-level portion of the pesticide may be permitted in his food as a residue, which is expressed in parts per million, that is parts of a chemical per million parts of a particular food, by weight.

Toxicologists use rather simple animal toxicity tests to rank pesticides according to their toxicity. Long before pesticides are registered with the Environmental Protection Agency and eventually released for public use the manufacturer must declare the toxicity of their pesticide to the white rat under laboratory conditions. This toxicity is defined by the LD_{50}, expressed as milligrams (mg) of toxicant per kilogram (kg) of body weight, the dose which kills 50% of the test animals to which it is administered under experimental conditions.

This toxicity value or LD_{50}, is measured in terms of oral (fed to or placed directly in the stomachs of rats), dermal (applied to the skin of rats or rabbits) and respiratory toxicity (inhaled). Using two of these tests, oral and dermal, a toxicologic ranking is shown for the organophosphate, organochlorine, carbamate and some other insecticides in Fig. 28-29. The materials on the top of the list are the most

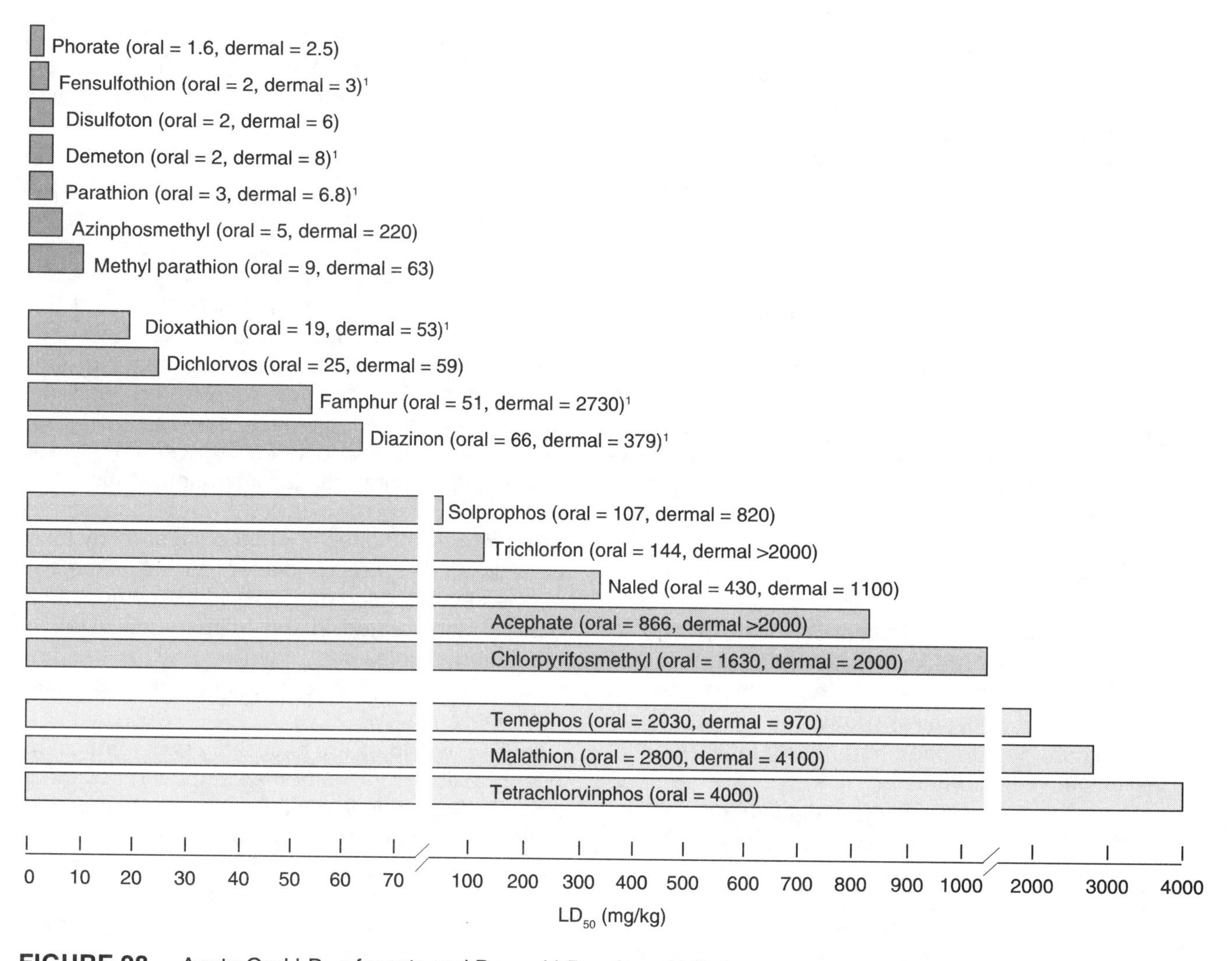

FIGURE 28. Acute Oral LD_{50}s for rats and Dermal LD_{50}s for rabbits for some organophosphate insecticides.

[1] Products are no longer registered by EPA and are used only to illustrate relative toxicities.

toxic and those at the bottom the least. The size of the dose is the most important single item in determining the safety of a given chemical and actual statistics of human poisonings correlate reasonably well with these toxicity ratings.

ESTIMATING PESTICIDE TOXICITY TO MAN

In addition to toxicity, the dose, length of exposure and route of absorption are the other important variables. The amount of pesticide required to kill a man can be correlated with the LD_{50} of the material to rats in the laboratory. In Table 34 below, for example, the acute Oral LD_{50} expressed as mg/kg dose of the technical material, is translated into the amount needed to kill a 170 lb. man. Dermal LD_{50}'s are included for a better understanding of the relationship of expressed animal toxicity to human toxicity.

In a manner of generalizing, oral ingestions are more toxic than respiratory inhalations which are more toxic than dermal absorption. Additionally, there are physical and chemical differences between pesticides which make them more likely to produce poisoning. For instance, parathion changes to the more toxic metabolite paraoxon under certain conditions of humidity and temperature. Parathion is more toxic than methyl parathion to field workers, yet there is not that great a difference in their oral toxicities. Workers' exposure is usually dermal, which explains why many more illnesses are reported in workers exposed to parathion than those

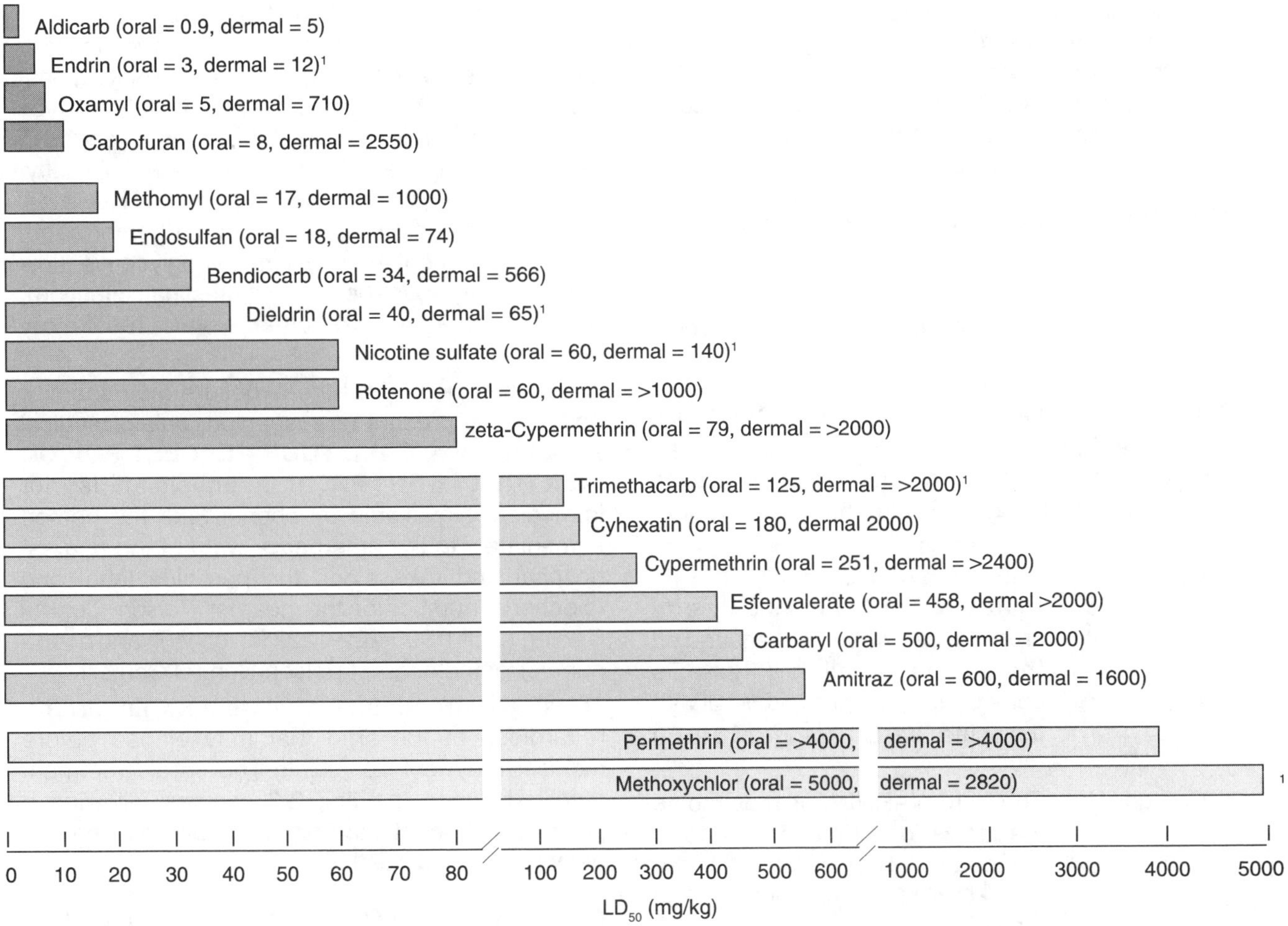

Aldicarb (oral = 0.9, dermal = 5)
Endrin (oral = 3, dermal = 12)[1]
Oxamyl (oral = 5, dermal = 710)
Carbofuran (oral = 8, dermal = 2550)

Methomyl (oral = 17, dermal = 1000)
Endosulfan (oral = 18, dermal = 74)
Bendiocarb (oral = 34, dermal = 566)
Dieldrin (oral = 40, dermal = 65)[1]
Nicotine sulfate (oral = 60, dermal = 140)[1]
Rotenone (oral = 60, dermal = >1000)
zeta-Cypermethrin (oral = 79, dermal = >2000)

Trimethacarb (oral = 125, dermal = >2000)[1]
Cyhexatin (oral = 180, dermal 2000)
Cypermethrin (oral = 251, dermal = >2400)
Esfenvalerate (oral = 458, dermal >2000)
Carbaryl (oral = 500, dermal = 2000)
Amitraz (oral = 600, dermal = 1600)

Permethrin (oral = >4000, dermal = >4000)
Methoxychlor (oral = 5000, dermal = 2820)[1]

0 10 20 30 40 50 60 70 80 100 200 300 400 500 600 1000 2000 3000 4000 5000

LD_{50} (mg/kg)

FIGURE 29. Acute Oral LD_{50}s for rats and Dermal LD_{50}s for rabbits for some organochlorine, carbamate, botanical, pyrethroid and formamidine insecticides.

[1] Products are no longer registered by EPA and are used only to illustrate relative toxicities.

TABLE 34. World Health Organization classification for estimating the acute toxicity of pesticides.

	LD_{50} for the rat (mg/kg b.w.) Oral		Dermal		Probable lethal oral dose for humans
	Solids	Liquids	Solids	Liquids	
Ia Extremely hazardous	≥5	≥20	≥10	≥40	A taste, a grain
Ib Highly hazardous	5-50	20-200	10-100	40-400	A pinch, 1 tsp
II Moderately hazardous	50-500	200-2000	100-1000	400-4000	1 tsp., 2 tbsp
III Slightly hazardous	≥501	≥2001	≥1001	≥4001	1-4 oz.
*Level at which an Acute Hazard is unlikely	≥2000	≥3000	—	—	1-1.5 pt.

*Exposures at which product is unlikely to present acute hazard in normal use. (WHO does not show toxicity of fumigants in the above system).

Source: World Health Organization

exposed to methyl parathion. So, we see that toxicity, route of absorption, dose, length of exposure and the physical and chemical properties of the pesticide contribute to its relative hazard. Hazard, then, is an expression of the potential of a pesticide to produce human poisoning.

EPA Toxicity Categories and Signal Words on Pesticide Labels.

The pesticide label contains "signal words" in bold print:

DANGER-POISON; WARNING; and CAUTION. These are significant words, since they represent an EPA category of toxicity and thus give an indication of their potential hazard (Table 35). There are four EPA groupings for pesticides:

Category I. The signal words DANGER-POISON and the skull and crossbones symbol are required on the labels of all highly toxic compounds. The pesticides fall with the acute Oral LD_{50} range of 0 to 50 mg/kg.

Category II. The word WARNING is required on the labels for all moderately toxic compounds. They all fall within the acute Oral LD_{50} range of 50 to 500 mg/kg.

Category III. The word CAUTION is required on labels for slightly toxic pesticides that fall within the acute Oral LD_{50} range of 500 to 5,000 mg/kg.

Category IV. The word CAUTION is required on labels of compounds having acute Oral LD_{50}s greater than 5,000 mg/kg. However, unqualified claims for safety are not acceptable on any label and all labels must bear the statement, "KEEP OUT OF REACH OF CHILDREN".

FIRST AID FOR PESTICIDE POISONING

The symptoms of poisoning may appear almost immediately after exposure or may be delayed for several hours depending on the chemical, dose, length of exposure and the individual. These symptoms may include, but are not restricted to headache, giddiness, nervousness, blurred vision, cramps, diarrhea, a feeling of general numbness or abnormal size of pupils. In some instances there is excessive sweating, tearing or mouth secretions. Severe cases of poisoning may be followed by nausea and vomiting, fluid in the lungs, changes in heart rate, muscle weakness, breathing difficulty, confusion, convulsions, coma or death. However, pesticide poisoning may mimic brain hemorrhage, heat stroke, heat exhaustion, hypoglycemia (low blood sugar), gastroenteritis (intestinal infection), pneumonia, asthma or other severe respiratory infection.

No matter how trivial the exposure may seem, if poisoning is present or suspected, obtain medical advice at once. **CALL YOUR NEAREST POISON CONTROL CENTER.** If a physician is not immediately available by phone, take the person immediately to the emergency ward of the nearest hospital and take along the pesticide label and telephone number of the nearest Poison Control Center. (Call 1-800-222-1222 for location and phone number of the nearest Poison Control Center)

First aid treatment is extremely important, regardless of the time that may elapse before medical treatment is available. The first aid treatment received during the first 2-3 minutes following a poisoning accident may very well spell the difference between life and death.

FIRST AID FOR CHEMICAL POISONING

If You Are Alone With the Victim

First — See that the victim is breathing; if not, give artificial respiration.

Second — Decontaminate the victim immediately by washing off any residues thoroughly. Speed is essential!

Third — Call your physician.

Note: Do *not* substitute first aid for professional treatment. First aid is only to relieve the patient before medical help is reached.

If Another Person Is With You and the Victim

Speed is essential; one person should begin first aid treatment while the other calls a physician.

The physician will give you instructions. He will very likely tell you to get the victim to the emergency room of a hospital. The equipment needed for proper

TABLE 35. EPA Labeling Toxicity Categories by Hazard Indicator.

Hazard indicators	*Toxicity categories* I (Danger—Poison)	II (Warning)	III (Caution)	IV (Caution)
Oral LD_{50}	Up to and including 50 mg/kg	From 50 to 500 mg/kg	From 500 to 5000 mg/kg	Greater than 5000 mg/kg
Inhalation LD_{50}	Up to and including 0.2 mg/liter	From 0.2 to 2 mg/liter	From 2 to 20 mg/liter	Greater than 20 mg/liter
Dermal LD_{50}	Up to and including 200 mg/kg	From 200 to 2000 mg/kg	From 2000 to 20,000 mg/kg	Greater than 20,000 mg/kg
Eye effects	Corrosive; corneal opacity not reversible within 7 days	Corneal opacity reversible within 7 days; irritation persisting for 7 days	No corneal opacity; irritation reversible within 7 days	No irritation
Skin effects	Corrosive	Severe irritation at 72 hours	Moderate irritation at 72 hours	Mild or slight irritation at 72 hours

Source. "EPA Pesticide Programs, Registration and Classification Procedures, Part II." *Federal Register 40:* 28279.

treatment is there. Only if this is impossible should the physician be called to the site of the accident.

General

1. Give mouth-to-mouth artificial respiration if breathing has stopped or is labored.
2. Stop exposure to the poison and if poison is on skin cleanse the person, including hair and fingernails. If swallowed, induce vomiting as directed below.
3. Save the pesticide container and material in it if any remains; get readable label or name of chemical(s) for the physician. If the poison is not known, save a sample of the vomitus.

Specific

POISON ON SKIN

1. Drench skin and clothing with water (shower, hose, faucet).
2. Remove clothing.
3. Cleanse skin and hair thoroughly with soap and water; rapidity in washing is most important in reducing extent of injury.
4. Dry and wrap victim in blanket.

POISON IN EYE

1. Hold eyelids open, wash eyes with gentle stream of clean running water immediately. Use copious amounts. Delay of a few seconds greatly increases extent of injury.
2. Continue washing for 15 minutes or more.
3. Do not use chemicals or drugs in wash water. They may increase the extent of injury.

INHALED POISONS (DUSTS, VAPORS, GASES)

1. If victim is in enclosed space, do not enter without air-supplied respirator.
2. Carry patient (do not let him walk) to fresh air immediately.
3. Open all doors and windows, if any.
4. Loosen all tight clothing
5. Apply artificial respiration if breathing has stopped or is irregular.

6. Call a physician.
7. Prevent chilling (wrap victim in blankets but do not overheat).
8. Keep victim as quiet as possible.
9. If victim is convulsing, watch his breathing and protect him from falling and striking his head on the floor or wall. Keep his chin up so his air passage will remain free for breathing
10. Do not give alcohol in any form.

SWALLOWED POISONS

1. CALL A PHYSICIAN OR POISON CONTROL CENTER IMMEDIATELY
2. Do not induce vomiting if :
   a. Victim is in a coma or unconscious.
   b. Victim is in convulsions.
   c. Victim has swallowed petroleum products (that is, kerosene, gasoline, lighter fluid).
   d. Victim has swallowed a corrosive poison (strong acid or alkaline products) -- symptoms: severe pain, burning sensation in mouth and throat.
3. If the victim can swallow after ingesting a corrosive poison (any material that in contact with living tissue will cause destruction of tissue by chemical action, such as lye, acids, Lysol) give the following substances by mouth. For acids or alkali: milk or water; for patients one to five years old, 2 to 4 ounces; for patients five years and older, up to 8 ounces.
4. If a noncorrosive substance has been swallowed, induce vomiting, if possible, EXCEPT if patient is unconscious, in convulsions or has swallowed a petroleum product.

CHEMICAL BURNS OF SKIN

1. Wash with large quantities of running water.
2. Remove contaminated clothing.
3. Immediately cover with loosely applied clean cloth (any kind will do).
4. Avoid use of ointments, greases, powders and other drugs in first-aid treatment of burns.
5. Treat shock by keeping victim flat, keeping him warm and reassuring him until arrival of physician.

In an EMERGENCY
Call toll-free
Poison Control Center
800-222-1222
National Pesticide
Information Center Hotline
800-858-7378
OR
Chemical Transportation
Emergency Center
(CHEMTREC)
800-494-9300
OR
Your local
Poison Control Center
OR
Call 911
Emergency Hot Line

APPENDIXES

APPENDIX **A**

Internet Pest Control Information Sources

Federal Government

- Agency for Toxic Substances and Disease Registry (ATSDR)(http://www.atsdr.cdc.gov)
- Biological Control Information (PPQ Internet)
- Organic Gardening Certification (www.usda.gov/nop)
- Organic Producers Standards (www.ofrf.org)
- U.S. Department of Agriculture Home Page (http://usda.gov/)
- U.S. Department of Agriculture, ARS Pesticide Properties Database (http://www.arsusda.gov/acsl/ppdb.html)
- U.S. Department of Health and Human Services-National Toxicology Program (http://ntp-server.niehs.nih.gov/)
- U.S. Environmental Protection Agency Home Page (http://www.epa.gov/epahome/index.html)
- U.S. Environmental Protection Agency, Office of Prevention, Pesticides and Toxic Substances (http://www.epa.gov/oppts/)
- U.S. Occupational Safety and Health Administration (http://www.osha.gov/)

State Governments and Universities

- All State Universities Cooperative Extension Service offices can be found on this U.S. map (http://www.csrees.gov/extension/index.html)
- All State Universities' websites can be accessed on this similar USDA website (http://www.csrees.usda.gov/qlinks/partners/state_partners.html)
- University of California IPM Home Page, Davis, CA (http://www.ipm.ucdavis.edu/)
- California Department of Pesticide Regulation (DPR) (http://www.dcpr.ca.gov/)
- California Environmental Protection Agency (CalEPA) (http://www.calepa.ca.gov/)
- Cornell University Pesticide Management Education Program (http://pmep.cce.cornell.edu/)
- Florida (University of) Pesticide Program (http://pdec.ifax.ufl.edu/)

- National Pesticide Telecommunications Network (NPTN), Oregon State University (http://ace.orst.edu/info/)
- Oregon State University Toxicology Network (http://ace.orst.edu/info/extoxnet)
- Purdue University Pesticide Programs (http://www.btny.purdue.edu/PPP/)
- Virginia Department of Agriculture Office of Pesticide Services (http://www.vdacs.state.va.us/index.html)
- Wisconsin Integrated Pest and Crop Management (http://ipcm.wisc.edu/)

Poison Control Centers Network

- American Association of Poison Control Centers (www.AAPCC.org)

Pesticide Information

- Insecticides (http://ipmworld.umn.edu/chapters/ware.htm)
- Herbicides (http://ipmworld.umn.edu/chapters/whitacre.htm)
- Pheromones (http://www.nysaes.cornell.edu/pheronet/)
- Biopesticides, EPA Registration (http://www.epa.gov/pesticides/biopesticides)
- Microbial Pesticides, EPA Registration (http://www.epa.gov/oppfead1/cb/ppdc/2000/regist-biopes.htm)

Pesticide Manufacturers

- Arvesta Corp. (www.arvesta.com/default.asp)
- BASF Corp. Ag Products Group (www.basf.com)
- Bayer CropScience (www.bayerus.com)
- Dow AgroSciences LLC (www.dowagro.com)
- DuPont Crop Protection (cropprotection.dupont.com)
- FMC Corp. Ag Products Group (www.ag.fmc.com)
- Gustafson LLC (www.gustafson.com)
- Monsanto Co. (www.monsanto.com)
- Sumitomo Chemical Co., Ltd. (www.sumitomo-chem.co.jp)
- Syngenta (www.syngenta-us.com)

- Valent U.S.A. Corp. (www.valent.com)

Related Organizations

- Environmental Working Group (http://www.ewg.org/)
- Friends of the Earth (http://www.foe.org/)
- Greenpeace (www.greenpeace.org/)
- National Coalition Against the Misuse of Pesticides (NCAMP) (http://www.beyondpesticides.org/)
- National Pest Control Association (www.pestworld.org)
- Sierra Club (http://www.sierraclub.org/)

Organic Gardening Websites

- Organic Gardening Certification (www.usda.gov/nop)
- Gardening: Natural/Organic (www.essortment.com/in/Gardening.Natural.Organic/)
- The Gardener's Network (www.gardenersnet.com/organic.htm)
- Suppliers of Beneficial Organisms in North America (www.cdpr.ca.gov)

Organic Gardening Book Publishers

- Sunset Publications (www.Sunset.com)
- Jerry Baker Gardening Books (www.jerrybaker.com)
- The Garden Guy, Poco Verde Landscape (www.gardenguy.com)

APPENDIX **B**

Insecticides, Insect Growth Regulators, Acaricides and Nematicides

Common, trade and chemical names, their basic manufacturer(s), general use patterns and oral and dermal LD_{50}s.

Common name, trade name and basic manufacturer(s)	*Chemical name*	*General use pattern*	*Oral LD_{50} (rat)*	*Dermal LD_{50} (rabbit)*
abamectin, avermectin, Agri-Mek®, Zephyr®, Avid® Clinch® (Syngenta)	macrocyclic lactone glycosides	For nonfood crops, mites, leafminers, animal parasiticide, cockroach & ant baits, also mosquito larvicide.	10	2000
Abate® (see temephos)				
Acaban® (see fenpyroximate)				
Acaristop® (see clofentezine)				
acephate, Orthene®, Velocity®, Pinpoint®, (Valent)	*O,S*-dimethyl acetylphosphoramidiothioate	Controls most insects on vegs, cotton, soybeans. Residual & systemic activity. Also for fire ants.	866	2000
acetamiprid Assail®, Intruder®, TriStar® (DuPont, Nippon Soda)	(E)-*N*-((6-chloro-3-pyridinyl)methyl)-*N'*-cyano-*N*-methyletha-nimidamide	Registered for wide variety of fruits, vegs, cotton & ornamentals.	314	>2000
Acramite® (see bifenazate)				
Actara® (see thiamethoxam)				
Actellifog® (see pirimiphos-methyl)				
Adept® (see diflubenzuron)				
Adios® (see cucurbitacin)				
Adjust® (see acetamiprid)				
Admire® (see imidacloprid)				
Advantage® (see carbosulfan)				
Agnique® (Henkel)	alcohol-ethoxylate	Wetting agent, dispersant used as mosquito larvicide.	Non-toxic	
Agree ® (see *Bacillus thuringiensis* spp. aizawai)				

* Chemical nomenclature used is either that of Chemical Abstracts (American Chemical Society) or International Union of Pure and Applied Chemistry (IUPAC). The author and publisher make no warranty, express or implied, regarding the accuracy of this Appendix.

Common name, trade name and basic manufacturer(s)	Chemical name	General use pattern	Oral LD_{50} (rat)	Dermal LD_{50} (rabbit)
Agri-Mek® (see abamectin)				
Agrotap® (see cartap)				
Akari®, (see fenpyroximate)				
alanycarb, Onic®, Orion®, (Otsuka)	ethyl (Z)-N-benzyl-*N*- {[methyl](1-methylthioethylideneamino-oxycarbonyl)amino]thio}-β–alaninate	Foliar spray, soil or seed treatment for wide range of pests on corn, cotton, peanuts, pome fruits, vegs, turf & ornamentals.	440	>2000
aldicarb, Temik® (Bayer CropScience)	2-methyl-2-(methylthio)propionaldehyde *O*-(methylcarbamoyl)oxime	Systemic insecticide, acaricide & nematicide, soil-applied only. For cotton, sugar beets, potatoes, pecans, oranges, ornamentals, soybeans, peanuts.	0.9	5
aldoxycarb, Award® (Bayer CropScience)	2-methyl-2-(methylsulfonyl)propanal *O*-[(methylamino)carbonyl]oxime	Nematicide, systemic insecticide, not registered.	26	1000
aldrin, Aldrex®, Aldrite® (discontinued)	1,2,3,4,10,10-hexachloro-1,4,4a,5,8,8a-hexahydro-1,4-*endo-exo*-5,8-dimethanonapthalene	No longer sold. Gives long residual control of most soil insects.	39	65
Alert® (see chlorfenapyr)				
Alfacron® (see azamethiphos)				
Alfalfa looper *(Autographa californica)*	polyhedrosis virus	Virus insecticide for caterpillars on vegetables.	Non-toxic	
allethrin, Pynamin® (Sumitomo)	*cis, trans*-(±)-2,2-dimethyl-3-(2-methylpropenyl)-cyclopropanecarboxylic acid ester with (±)-2-allyl-4-hydroxy-3-methyl-2-cyclopenten-1-one	Aerosols, for controlling flying insects in homes.	680	>2500
d-*trans*-allethrin, Bioallethrin® (Bayer CropScience)	*trans*-(+)-2,2-dimethyl-3-(2-methylpropenyl)-cyclopropanecarboxylic acid ester with (±)-2-allyl-4-hydroxy-3-methyl-2-cyclopenten-1-one	Sprays, aerosols against flying & crawling household insects.	425	4000
Alsystin® (see triflumuron)				
Altosid® (see methoprene)				
aluminum phosphide, ECO_2Fume®, Fumitoxin®, Phostoxin®, (Degesh Amer., United Phosphorus)	aluminum phosphide	Fumigant (hydrogen phosphide or phosphine, PH_3) kills insects in stored feed, grains, seeds, nuts & rodenticide.	2000 ppm gas in air	
Amazin® (see azadirachtin)				
Ambush® (see permethrin)				
Amdro® (see hydramethylnon)				
Amethopterin® (see methotrexate)				
aminocarb, Metacil® (discontinued)	4-(dimethylamino)-3-methylphenol methylcarbamate	Controls forest insects (spruce budworm, jack pine budworm, etc.).	30	275

Common name, trade name and basic manufacturer(s)	Chemical name	General use pattern	Oral LD_{50} (rat)	Dermal LD_{50} (rabbit)
amitraz, Ovasyn®, (Bayer CropScience)	*N'*-(2,4-dimethylphenyl)-*N*-[[(2,4-dimethylphenyl)-imino]methyl]-*N*-methylmethanimidamide	Controls pear psylla on pears; most mites on fruits, vegs, ornamentals; lice, ticks on livestock.	800	>200
Ammo® (see cypermethrin)				
Anagrapha falcifera NPV (Biotech)		Bioinsecticide controls celery looper on many vegetables.	Non-toxic	
Andalin® (see flucycloxuron)				
Anthio® (see formothion)				
Apache® (see cadusafos)				
Apex® (see methoprene)				
Aphistar® (see triazamate)				
ApiLife VAR® (see thymol)				
Apollo® (see clofentezine)				
Applaud® (see buprofezin)				
Asana® (see esfenvalerate)				
Aspergillus flavus AF 36		Bioinsecticide for cotton insects.	Non-toxic	
Assail® (see acetamiprid)				
Astro® (see permethrin)				
Atabron® (see chlorfluazuron)				
Atrapa® (see malathion)				
Autographa californica NPV, ACMNPV, Gusano® (BASF)		Bio-engineered baculovirus with systemic qualities.	Non-toxic	
avermectin (see abamectin)				
Avid® (see abamectin)				
Avaunt® (see indoxacarb)				
Award® (see aldoxycarb)				
Axor® (see lufenuron)				
azadirachtin, Azatin®, Neemix®, Amazin® (Amvac, Certis)	triterpenoid from neem tree extract (*Azadirachta indica*)	Botanical insect growth regulator, stops molting.	5000	>2000
azamethiphos, Snip® Alfacron® (Syngenta)	*S*-[6-chloro-oxazolo(4,*b*-5)pyridin-2-(3*H*)-on-3yl-methyl-*O,O*-dimethyl-phosphorothioate	Fly control in horse barns.	1180	>2150
Azatin® (see azadirachtin)				

Common name, trade name and basic manufacturer(s)	*Chemical name*	*General use pattern*	*Oral LD_{50} (rat)*	*Dermal LD_{50} (rabbit)*
azinphos-ethyl, Ethyl Guthion®, Crysthion® (Crystal)	*O,O*-diethyl *S*-[[4-oxo-1,2,3-benzotriazin-3(4*H*)-yl]methyl] phosphorodithioate	Controls most foliage-feeding insects on several crops., outside U.S.	13	250
azinphos-methyl, Guthion® (Bayer CropScience)	*O,O*-dimethyl *S*-[[4-oxo-1,2,3-benzotriazin-3(4*H*)-yl]methyl]phosphorodithioate	Controls most foliage-feeding insects on many crops.	9	220
Azodrin® (see monocrotophos)				
Aztec® (see tebupirimiphos)				
Bacillus cereus (MicroFlo)		Growth regulator for cotton.	>5000	>2000
Bacillus firmus, BioNem®, Biosafe® (Minrav)	bacterial isolate & nematicide	Controls root-knot nematodes in cucurbits & tomatoes		
Bacillus popilliae & Bacillus lentimorbus, Japidemic® (St Gabriel, Fairfax)	milky disease spores	Microbial insecticide for Japanese beetle larvae	Non-toxic	
Bacillus sphaericus, Vectolex® (Valent)		Microbial insecticide for mosquito larvae.	Non-toxic	
Bacillus thuringiensis Berliner		Original "Bt" registered with U.S.D.A. in 1961 for caterpillar control.	Non-toxic	
Bacillus thuringiensis spp. *aizawai*, Xen Tari®, Agree®, Design® (Certis, Valent)	serotype (H-7), spores & crystals.	For lepipdopteran larvae on corn, cotton, soybeans, fruit, vegs & tobacco.	Non-toxic	
Bacillus thuringiensis spp. *aizawai*, Mattch® (Dow Agrosciences)	encapsulated *delta* endotoxin	For most lepidopteran larvae on corn, cotton, soybeans, peanuts, vegs & many others.	Non-toxic	
Bacillus thuringiensis spp. *israelensis*, Bactimos®, Gnatrol®, VectoBac®, Teknar® (Certis, Valent)	crystalline delta endotoxin from Serotype H-14.	Larvicide for mosquitoes, aqusatic midges, black fly, & fungus gnats in greenhouses.	>5000	>2000
Bacillus thuringiensis spp. *japonensis* strain *buibui* MYX-910 (Dow Agrosciences)		Controls soil-dwelling beetle larvae in turf & ornamentals.	Non-toxic	
Bacillus thuringiensis, spp. *kurstaki*, Biobit®, DiPel®, Condor®, Crymax®, Deliver®, Javelin®, Ketch®, Thuricide® (Certis, Valent)	spores & crystalline delta endotoxin from *B. thuringiensis berliner* spp. *kurstaki* Serotype H-3a3b.	For most lepidopteran larvae on corn, cotton, soybeans, peanuts, fruits, vegs, ornamentals & tobacco.	Non-toxic	
Bacillus thuringiensis spp. *kurstaki*, MVP® (Certis)	encapsulated *delta* endotoxin.	For most lepidopteran larvae on corn, cotton, soybeans, peanuts, vegs & many others.	Non-toxic	
Bacillus thuringiensis spp. *morrisoni* (Sun Moon)	spores & crystalline *delta* endotoxin from *B. thuringiensis berliner* spp. *morrisoni* Serotype 8a8b.	For most lepidopteran larvae on corn, cotton, soybeans, peanuts, vegs & many others.	Non-toxic	

Common name, trade name and basic manufacturer(s)	Chemical name	General use pattern	Oral LD_{50} (rat)	Dermal LD_{50} (rabbit)
Bacillus thuringiensis spp. *tenebrionis,* Novodor® (Valent)		For Colorado potato beetle on potato & tomato; elm leaf beetles & other shade tree beetle larvae.	Non-toxic	
Bacillus thuringiensis spp. *tenebrionis,* MTrak® (discontinued)	encapsulated *delta* endotoxin	For most lepidopteran larvae on corn, cotton, soybeans, peanuts, vegs & many others.	Non-toxic	
Bactimos® (see *Bacillus thuringiensis*, spp. *israelensis)*				
Bancol® (see bensultap)				
Baroque® (see etoxazole)				
Barricade® (see cypermethrin)				
Baygon® (see propoxur)				
Baymix® (see coumaphos)				
Baytex® (see fenthion)				
Baythion® (see phoxim)				
Baythroid® (see cyfluthrin)				
Beauvaria bassiana, Botanigard®, Naturalis-L®, Mycotrol®, Corn Guard®		Mycoinsecticide for whiteflies, aphids, mealybugs, thrips, scales & gnats in greenhouses. Whitefly on cotton, European corn borer on corn. Fire ant & cockroach bait stations.	Non-toxic	
Beauvaria brongniarti (strain 96), Betel® (NPP)		Fungal insecticide for black vine weevil on ornamentals & chinch bugs on turf. For longicorns on citrus & mulberries.	Non-toxic	
Bee-Scent® (Scentry)		Feeding attractant added to fungicide & micronutrient sprays to blossoming crops. Not a pheromone.	Non-toxic	
Belmark® (see fenvalerate)				
bendiocarb, Ficam®,Turcam® (discontinued)	2,2-dimethyl-1,3-benzodioxol-4-yl-methylcarbamate	Contols the usual spectrum of household insects; turf, ornamentals; soil insects, ants on nonbearing fruit.	34	566
benfuracarb, Furacon® (Otsuka)	2,3-dihydro-2,2-dimethyl-7-benzofuranyl *N*-[*N*-[2-(ethylcarbonyl)ethyl]-*N*-isopropyl sulfenamoyl]-*N*-methylcarbamate	Sugarbeets, corn, potatoes.	138	>2000
bensultap, Bancol® (Takeda)	*S,S*'-2-dimethylaminotrimethylene di(benzenethiosulfonate)	Coleopteran & Lepidopteran insects in cotton, potatoes, vegs & apples.	1105	>2000
benzene hexachloride, BHC, HCH (Inquinosa)	1,2,3,4,5,6-hexachlorocyclohexane (12 to 45% gamma isomer)	Still in use internationally, but not in U.S.	125	>4000
Benzofuroline® (see resmethrin)				

Common name, trade name and basic manufacturer(s)	Chemical name	General use pattern	Oral LD_{50} (rat)	Dermal LD_{50} (rabbit)
Bestguard® (see nitenpyram)				
beta-cyfluthrin (see cyfluthrin-*beta*)				
Betel® (see *Beauvaria brongniarti*)				
BHC (see benzene hexachloride)				
Bidrin® (see dicrotophos)				
bifenazate, Floramite®, Acramite® (Crompton)	isopropyl-3-(4-methoxybiphenyl-3-yl) hydrazinoformate	Mite control on citrus, pome & stone fruits, grapes, strawberries, cotton & ornamentals.	5000	>2000
bifenthrin, Capture®, Talstar® Brigade®, Scorpion®, (FMC)	[1α,3α,(Z)]-(±)-(2-methyl[1,1'-biphenyl]-3-yl) methyl 3-(2-chloro-3,3,3-trifluoro-1-propenyl)-2,2-dimethylcyclopropanecarboxylate	Pyrethroid insecticide/miticide for wide range of crops, tree fruit & nuts, vegs, corn, ornamentals & termiticide.	375	>2000
binapacryl, Morocide® (discontinued)	2-*sec*-butyl 4,6-dinitrophenyl 3-methyl-2-butenoate	Miticide with ovicidal action, fungicide for powdery mildews. For mites on tree fruits, nuts, cotton.	136	1010
BioAct PL® (see *Paecilomyces lilacinus*)				
Biobit® (see *Bacillus thuringiensis* spp. *kurstaki)*				
Bio-Blast® (*see Metarhizium anisopliae*)				
Bioallethrin® (see *d-trans* allethrin)				
Bioline® (see *Cephalosporium lecanii*)				
BioNem® (see *Bacillus firmus*)				
bioresmethrin (see *d-trans-*resmethrin)				
Biosafe® (see *Bacillus firmus*)				
Biosafe-N® (see *Steinernema carpocapsae)*				
BioVector®355 (see *Steinernema glaseri*)				
Blex® (see pirimiphos-methyl)				
Bo-Ana® (see famphur)				
Bollex® (discontinued)	methyl α-eleostearate; methyl ester of (E,Z,E) 9,11,13-octadecatrienoic acid	Cotton boll weevil feeding deterrent; biorational control agent for	5000	—

Common name, trade name and basic manufacturer(s)	*Chemical name*	*General use pattern*	*Oral LD_{50} (rat)*	*Dermal LD_{50} (rabbit)*
		Integrated Pest Management.		
Bolstar® (see sulprofos)				
bomyl (discontinued)	dimethyl 3-hydroxy glutaconate dimethyl phosphate	Fly baits.	31	20
Bora-Care® (see sodium borate)				
boric acid, Borid®, Drax® (U.S. Borax)	inorganic salt, H_3BO_3, *ortho*-boric acid	Paste baits & dusts for cockroach, ants & fly larvae in manure.	>10,000	—
Borid® (see boric acid)				
Botanigard® (see *Beauvaria bassiana*)				
Brigade® (see bifenthrin)				
buprofezin, Applaud®, Courier®, (Nichino Amer.)	2-tert butylimino-3-isopropyl-5-phenylperhydoxo-1,3,5-thiadiazin-4-one	Persistent growth regulator for mealybugs, scales, whiteflies.	2200	>5000
Butacide® (see piperonyl butoxide)				
cadusafos, Rugby®, Apache® (FMC)	*O*-ethyl *S*, *S*-di-*sec*-butyl phosphorodithioate	Nematicide, soil insecticide, outside the U.S.	391	143
Calypso® (see thiacloprid)				
capsaicin, Hot Sauce® (Plant Biotech)	capsaicinoids	Insect irritant & synergist for other insecticides.	Irritant	Irritant
Capsyn® (Kalsec)	capsaicinoids	Synergist extracted from chili peppers.	Irritant	Irritant
Capture® (see bifenthrin)				
carbaryl, Sevin®, Sevimol® (Bayer CropScience)	1-naphthyl methylcarbamate	Has probably the greatest range of controlled pests of any insecticide; fruits, vegs, field crops, ornamentals, potatoes.	500	>2000
carbofuran, Furadan® (FMC)	2,3,-dihydro-2,2-dimethyl-7-benzofuranyl methylcarbamate	Insecticide, miticide, nematicide. Wide range of soil & foliar pests on corn, alfalfa, tobacco, peanuts, rice, sugar cane, potatoes, Christmas trees.	8	>3000
carbophenothion, Trithion® (discontinued)	*S*-[[(4-chlorophenyl)thio]methyl]-*O,O*-diethyl phosphorodithioate	Used on variety of fruit, nut, vegetable, fiber crops. Also acaricide with long residual.	6	22
carbosulfan, Advantage® Marshall® (FMC)	2,3-dihydro-2,2-dimethyl-7-benzofuranyl-[(dibutylamino)thio]methyl carbamate	Soil & foliar insects on alfalfa, citrus, corn, deciduous fruit, some nematodes. Not registered in U.S.	209	>2000
cartap, Agrotap®, Eaten®, Padan® (several)	*S,S'*-2-dimethhylaminotrimethylene bis(thiocarbanate) hydrochloride	Coleopteran & Lepidopteran insects in potatoes & vegs.	150	>2000
Carzol® (see formetanate hydrochloride)				

Common name, trade name and basic manufacturer(s)	*Chemical name*	*General use pattern*	*Oral LD_{50} (rat)*	*Dermal LD_{50} (rabbit)*
Catalyst® (see propetamphos)				
Centric® (see thiamethoxam)				
Cephalosporium lecanii, Bioline®, Mycotal® (Biotech) (formerly genus *Verticillium*)		Fungal insecticide for thrips, aphids & whiteflies on greenhouse crops.	Non-toxic	
Checkmate CM®, Checkmate DBM®, Checkmate OFM®, Checkmate OLR®, Checkmate PTB® (Suterra)	pheromane natural products	Codling moth pheremone. Diamond back moth pheremone Oriental fruit moth pheremone. Omnivorus leaf roller pheremone. Peach tree borer pheremone.	Non-toxic	
Checkmite® (see coumaphos)				
Chess® (see pymetrozine)				
Chipco Choice® (see fipronil)				
chlordane, (discontinued)	1,2,4,5,6,7,8,8-octachloro-3a,4,7,7a-tetrahydro-4,7-methanoindan	Used entirely for subterranean termite control.	283	580
chlordecone (see Kepone®)				
chlordimeform, Fundal®, Galecron® (discontinued)	*N'*-(4-chloro-o--tolyl)*N,N*-dimethylformamidine	Ovicide-insecticide for bollworm-budworm complex in cotton; ovicide-miticide for resistant mites.	170	225
chlorethoxyphos, Fortress® (Amvac)	*O,O*-diethyl *O*-1,2,2,2-tetrachloroethyl phosphorothioate	Soil insecticide for corn, vegs, potatoes, turf.	1.8	12.5
chlorfenapyr, Alert®, Phantom®, Pirate®, Pylon® (BASF)	4-bromo-2-(4-chlorophenyl)-1-ethoxymethyl-5-trifluoromethylpyrrole-3-carbonitrile	Worm complex on cotton. For corn, vegs, tree fruits, greenhouse vegs. Broad spectrum of insects controlled, Termite bait, cockroaches & ants indoors.	441	>2000
chlorfenethol, Dimite®, Qikron® (discontinued)	4,4'-dichloro-α-methylbenzhydrol	For spider mites on shrub trees, ornamentals, outside U.S.	926	—
chlorfenvinphos, Supona® (BASF)	2-chloro-1-(2,4-dichlorophenyl)vinyl diethyl phosphate	Soil insects, foliage feeders, livestock pests. Outside U.S.	12	3200
chlorfluazuron, Atabron® (ISK)	1-[3,5-dichloro-4-(3-chloro-5-trifluoromethyl-2-pyridyloxy) phenyl]-3-(2,6-difluorobenzoyl) urea	Chitin synthesis inhibitor for insecticide-resistant caterpillars in cotton & vegetables, outside U.S.	>8500	>2000
chlorobenzilate, Acaraben® (discontinued)	ethyl 4,4'-dichlorobenzilate	Mite control.	700	10,200
chloropicrin (several) (Mitsui)	trichloronitromethane	Stored grain & soil fumigant. Usually combined with other fumigants as warning agent.	lachrymatory	
chlorothiophos (discontinued)	*O*-[2,5-dichloro-4-(methylthio)phenyl]*O,O*-diethyl phosphorothioate and the 2,4,5 and 4,5,2 isomers	Controls a broad spectrum of biting, sucking & mining insects, spider mites, mosquito larvae, ticks.	7.8	50
	(*O*-chlorophenyl)glyoxylonitrile oxime *O,O*-diethyl			

Common name, trade name and basic manufacturer(s)	*Chemical name*	*General use pattern*	*Oral LD_{50} (rat)*	*Dermal LD_{50} (rabbit)*
chlorphoxim, Baythion C® (discontinued)	phosphorothioate	Controls mosquitoes, black flies.	>2500	>500
chlorpyrifos, Dursban®, Lorsban®, Lock-On® Equity®, Empire®, Lentrek® (Dow Agrosciences)	*O,O*-diethyl *O*-(3,5,6-trichloro-2-pyridyl) phosphorothioate	Ornamental pests, fire ants, turf insects; soil insects, fruit trees, cotton, horticultural crops; termite control. Not available for home use.	96	2000
chlorpyrifos-methyl, Reldan® (discontinued)	*O,O*-dimethyl *O*-(3,5,6-trichloro-2-pyridyl) phosphorothioate	Flies, mosquitoes, aquatic larvae, several foliar pests.	>3000	>3700
chromafenozide, Matric® (Nippon)	2'-tert-butyl-5-methyl-2'-(3,5-xyloyl)chromane-6-carbonhydrazide	IGR for lepidopteran pests on vegs & ornamentals	>5000	>2000
Cide-Kick® (see D-limonene)				
Cidial® (see phenthoate)				
cinnamaldehyde, Cinnamite®, Valero®, (Emerald, ProGuard)	3-phenyl-propenal	For mite control on strawberries & bush berries.	>2000	
cinnamon oil, Cinnamite® (Emerald)		For mite & aphid control on ornamentals.	Irritant	
Citation® (see cyromazine)				
Clandosan® (IGENE Biotech)	chitin (poly-D-glucosamine)	Nematicide made from crab & lobster shells.	Non-toxic	
clofentezine, Apollo®, Acaristop® (Bayer CropScience)	3,6-bis(2-chlorophenyl)-1,2,4,5-tetrazine	Acaricide, ovicide, larvicide.	3200	>1332
clothianidin, Poncho®, Clutch® (Arvesta, Bayer CropScience)	(E)-1-(2-chloro-1,3-thiazol-5-ylmethyl)-3-methyl-2-nitroguanidine	For sucking insects on fruits, vegs, pome fruits, cotton, turf, ornamentals & seed treatment.	—	—
Clutch® (see clothianidin)				
Comite® (see propargite)				
Commodore® (see *lambda*-cyhalothrin)				
Comply® (see fenoxycarb)				
Condor® (see *Bacillus thuringiensis* spp, *kurstaki)*				
Confirm® (see tebufenozide)				
Conserve® (see spinosad)				
Consult® (see hexaflumuron)				
Co-Ral® (see coumaphos)				
	O,)-diethyl-*O*-(3-chloro-4-methyl-2-oxo-2H-1-			

Common name, trade name and basic manufacturer(s)	*Chemical name*	*General use pattern*	*Oral LD_{50} (rat)*	*Dermal LD_{50} (rabbit)*
coumaphos, Baymix®, Checkmite®, Co-Ral®, Meldane® (Bayer CropScience)	benzapyran-7-yl) phosphorothioate	Systemic for control of most ectoparasites & pests on domestic animals.	16	860
Counter® (see terbufos)				
Courier® (see buprofezin)				
Cruiser® (see thiamethoxam)				
Crymax® (see *Bacillus thuringiensis* spp. *kurstaki*)				
cryolite, Kryocide® Prokil® (Cerexagri, Gowan)	sodium fluoaluminate (or sodium aluminofluoride)	Stomach insecticide usually for caterpillars on grapes, citrus, vegs.	>10,000	—
cubé (see rotenone)				
cucurbitacin, Slam®, Adios® (MicroFlo)	feeding attractant bait wih carbaryl as active ingrediant.	Buffalo gourd root powder. Feeding stimulant for corn rootworm & cucumber beetle adults.	Non-toxic	
Curacron® (see profenofos)				
Curbex® (see ethiprole)				
Curtin® (see metam-potassium)				
Cybolt® (see flucythrinate)				
Cydia pomenella granulosis virus, Virosoft CP4® (Certis, Biotepp)	bioinsecticide	For codling moth control in fruit & nut trees.	Non-toxic	
Cyflee® (see famphur)				
cyfluthrin, Baythroid®, Tempo®, Laser®, Renounce® (Bayer CropScience)	cyano(4-fluoro-3-phenoxyphenyl)methyl-3(2,2-dichloroethenyl)-2,2-dimethylcyclopropane-carboxylate	For corn, cotton, deciduous fruit, nuts, grapes, vegs & livestock quarters, stored products, turf, ornamentals.	500	>5000
cyfluthrin-*beta*, Tempo Ultra®, Responsar® (Bayer CropScience)	Reaction mixture of 2 of the diastereo-isomeric enantiomer pairs of cyfluthrin.		500	>5000
Cygon® (see dimethoate)				
gamma-cyhalothrin, Pytech® (Syngenta)	α-cyano-phenoxybenzyl 3-(2-chloro-3,3,3-trifluoroprop-1-enyl)-2,2-dimethylcyclopropanecarboxylate	Pyrethroid for cotton, soybeans, corn & vegs.	56	623
lambda-cyhalothrin, Commodore®, Demand®, Karate®, Matador®, Warrior®, Scimitar® (Syngenta)	[1α,3α(Z)]-(±)-cyano(3-phenoxyphenyl) methyl-3-(2-chloro-3,3,3-trifluoro-1-propenyl)-2,2-dimethylcyclopropane carboxylate	Controls animal ectoparasites, cattle lice & sheep keds. Vegs, cereals, nut crops, stone & pome fruits, structural pests & animal ectoparasites.	56	623
cyhexatin, Pennstyl®, (Cerexagri)	tricyclohexylhydroxystannane	Controls resistant & susceptible plant-feeding spider mites.	180	2000

Common name, trade name and basic manufacturer(s)	Chemical name	General use pattern	Oral LD_{50} (rat)	Dermal LD_{50} (rabbit)
Cymbush® (see cypermethrin)				
Cynoff® (see cypermethrin)				
cypermethrin, Ammo®, Barricade®, Cymbush®, Cynoff®, Demon®, Ripcord® (FMC, Syngenta, BASF)	(±)-a-cyano (-3-phenoxyphenyl)methyl (±)*cis, trans* 3-(2,2-dichoroethenyl)-2,2dimethylcyclopropane-carboxylate	Pyrethroid with high-level activity against broad spectrum of insect pests & crops, household aerosols.	250	>2000
zeta-cypermethrin, Fury®, Mustang® (FMC)	α,-cyano(3-phenoxyphenyl)methyl(±)-cis-trans 3-(2,2-dichlororthenyl)-2,2 dimethyl= cyclopropanecarboxylate	Broad pest spectrum on cotton, corn, soybeans, vegs, pecans, root & tuber crops, cucurbits.	79	2000
cyromazine, Citation®, Trigard® (Syngenta)	*N*-cyclopropyl-1,3,5-triazine-2,4,6-triamine	Insect growth regulator for leafminers in vegs & on livestock.	3387	>3100
Cytrolane® (see mephosfolan)				
Dacamox® (see thiofanox)				
Danitol® (see fenpropathrin)				
Danitron® (see fenpyroximate)				
dazomet, Basamid® (see Appendix E) (BASF)		Insecticide, herbicide, fungicide.		
DBCP, dibromochloropropane Nemafume®, Nemaset® (discontinued)	1,2-dibromo-3-chloropropane	Soil fumigant-nematicide.	170	1420
D-D® (see dichloropropane)				
DDD (see TDE)				
DDT (several foreign)	1,1,1-trichloro-2,2-bis(*p*-chlorophenyl)ethane	Not used in U.S. Some agricultural use in other countries, mostly in malaria eradication programs.	87	1931
DDVP (see dichlorvos)				
Deadline® (see metaldehyde)				
decamethrin (see deltamethrin)				
Decis® (see deltamethrin)				
Declare® (see methyl parathion)				
deet, Delphene®, Off® (several)	*N,N*-diethyl-*m*-toluamide	Repellent for almost all biting arthropods.	2000	—
Deliver® (see *Bacillus thuringiensis - spp. kurstaki*)				
Delphene® (see deet)				
	(*S*)-α-cyano-*m*-phenoxybenzyl (1R,3R)-3-(2,2-			

Common name, trade name and basic manufacturer(s)	Chemical name	General use pattern	Oral LD_{50} (rat)	Dermal LD_{50} (rabbit)
deltamethrin, Decis®, DeltaGard®,Suspend®, K-Othrine® (Bayer CropScience)	dibromovinyl)-2,2-dimethylcyclopropane-carboxylate	Wide range of effectivness on cotton, field, fruit, vegetable crops, pome & stone fruits & nut crops. Flying and crawling insects inside & out, lawn & turf.	128	>2000
Demand® (see *lambda-cyhalothrin*)				
demeton, Systox® (discontinued)	*O,O*-diethyl-[2-(ethylthio)ethyl]phosphorothioate and*O,O*-diethyl *S*-[2-(ethylthio)ethyl] phosphoro- thioate	Systemic insecticide applied to foliage or as soil drench for sucking insects, mites on wide variety of field, fruit, nut, vegetable crops.	2	8
demeton-methyl (see methyl demeton)				
Demon® (see cypermethrin)				
Denim® (see emamectin)				
derris (see rotenone)				
Design® (see *Bacillus thuringiensis* spp. *kurstaki x aizawai*)				
Detur® (see jojoba oil)				
dialifor, Torak® (discontinued)	*O,O*-diethyl phosphorodithioate *S*-ester with *N*-(2-chloro-1-mercaptoethyl) phthalimide	Insecticide-acaricide used on grapes, apples, citrus, pecans.	5	145
diazinon, Knox-Out®, Spectracide® (Syngenta)	*O,O*-diethyl *O*-(2-isopropyl-6-methyl-4-pyrimidinyl) phosphorothioate	Broadly used insecticide against soil insects, pests of fruit, vegs, field crops, ornamentals. No longer registered in U.S.	300	379
Dibrom® (see naled)				
dibutyl phthalate	dibutyl phthalate	Clothing impregnate for chigger repellent.	8000	—
Dicarzol® (see formetanate)				
p-dichlorobenzene, PDB (PPG)	*p*-dichlorobenzene (moth crystals)	Used as moth crystals to protect fabrics; for peachtree borers.	500	2000
dichloropropane-dichloro-propene, D-D® (discontinued)	dichloropropane-dichloropropene mixture	Soil-injected fumigant to control nema-todes, soil insects, certain diseases.	140	2100
dichloropropene, Telone II®, Telone®, Inline® (Dow Agrosciences)	1,3-dichloropropene + chloropicrin C-35 is 35% chloropicrin	Preplant fumigant applications to soil for nematode, weed, disease control in many crops & turf.	127	423 Skin burns
dichlorvos, DDVP, Vapona®, Doom® (Amvac, Syngenta)	2,2-dichlorovinyl dimethyl phosphate	Contact, stomach, fumigant insecticide for household, public health. livestock, many crop insect & mite pests, no-pest strips.	25	59
	4,4'dichloro-α-(trichloromethyl) benzhydrol			

Common name, trade name and basic manufacturer(s)	*Chemical name*	*General use pattern*	*Oral LD_{50} (rat)*	*Dermal LD_{50} (rabbit)*
dicofol, Kelthane® (Dow Agrosciences)		Wide use as acaricide on fruit, vegetable, field, ornamental crops.	575	4000
dicrotophos, Bidrin®, Ektafos® (Syngenta, Amvac)	dimethyl phosphate ester with (*E*)-3-hydroxy-*N,N*-dimethylcrotonamide	Systemic & contact insecticide, used on cotton, soybeans, ornamentals.	22	225
dieldrin, Dieldrex® Dieldrite® (discontinued)	1,2,3,4,10,10-hexachloro-6,7-epoxy 1,4,4a,5,6,7,8,8a-octahydro-1,4-endo-exo-5,8-dimethanophthalene	Formerly used for subterranean termite control. All U.S. registrations canceled in 1986.	40	65
dienochlor, Pentac® (discontinued)	bis(pentachloro-2,4-cyclopentadien-1-yl)	Miticide effective against twospotted spider mite on greenhouse ornamentals & whiteflies.	>3160	>3160
diflubenzuron, Dimilin®, Micromite®, Adept® (Crompton)	*N*-[[(4-chlorophenyl)amino]carbonyl]-2,6-difluorobenzamide	Insect growth regulator for cotton boll weevil, gypsy moth, pine tipmoth, tree fruits & nuts, animal larvicide. Termite Bait.	>4640	>10,000
Dimecron® (see phosphamidon)				
dimethoate, Roxion® (BASF)	*O,O*-dimethyl *S*-[2-(methylamino)-2-oxoethyl] phosphorodithioate	Residual fly spray for farm buildings; insect, mite control on wide variety of vegetable, field, fruit, ornamental crops. Most uses are cancelled.	250	150
Dimilin® (see diflubenzuron)				
dinitrocresol, DNOC (discontinued)	4,6-dinitro-*o*-cresol	Dormant spray for fruit trees to control mites, aphids, other pests; effective as ovicide. Not registered in U.S.	20	200
dinitrophenol	2,4-dinitrophenol	Insecticide, acaricide, fungicide as dormant fruit tree spray. Not registered in U.S.	35	700
dinocap, Mildane® (Diachem)	2-(1-methylheptyl)-4,6-dinitrophenyl crotonate	For powdery mildew & mite control on fruits, vegs, ornamentals.	980	4700
dinoseb, DNBP (discontinued)	2-*sec*-butyl-4,6-dinitrophenol	Herbicide, desiccant, dormant fruit spray for controlling eggs, larvae, adults of various mites & insects.	37	80
dinotefuran, Starkle® (Valent)	*N*-menthyl-*N'*nitro[*N''*-[(tetrahydro-3-furanyl)methyl]guanidine	Nicotinoid insecticide for broad range of pests on cotton, grapes, vegs & potatoes.	2000	>2000
dioxathion, Delnav®, Deltic® (discontinued)	2,3-*p*-dioxanedithiol *S,S*-bis(*O,O*-diethyl phosphorodithioate)	Controls insects, mites on fruit & nut trees, ornamentals; also for many livestock pests.	19	53
Dipel® (see *Bacillus thuringiensis* spp. *kurstaki)*				
Dipterex® (see trichlorfon)				
	O,O-diethyl *S*-[2-(ethylthio)ethyl]			

Common name, trade name and basic manufacturer(s)	Chemical name	General use pattern	Oral LD_{50} (rat)	Dermal LD_{50} (rabbit)
Distance® (see pyriproxyfen)	phosphorodithioate			
disulfoton, Di-Syston® (Bayer CropScience)		Systemic insecticde, acaricide used on cotton, potatoes, other crops, as seed treatment or in furrow at planting. Several home ornamental uses.	2	3.6
Di-Syston® (see disulfoton)				
DiTera® (see *Myrothecium verrucaria)*				
D-Limonene (see limonene)	botanical from tropical tree. *Lonchocarpus felipei*			
DMDP (BTG)		Nematicide for potatoes. Not registered in U.S.		
DNOC (see Appendix D)				
Doom® (see dichlorvos)				
Dragnet® (see permethrin)				
Drax® (see boric acid)				
Draza® (see methiocarb)	silica aerogel + ammonium fluosilicate			
Dri-Die® (discontinued)		Household insect control by desiccation.	Non-toxic	
Dursban® (see chlorpyrifos)				
Dylox® (see trichlorfon)				
Dynamite® (see fenpyroximate)				
Eaten® (see cartap)				
Eclipse® (see hydramethylnon))				
ECO_2Fume® (see aluminum phosphide)				
EcoTrol® (see Rosemary oil)				
Ektaphos® (see dicrotophos)				
Elsan® (see phenthoate)	analog of avermectin			
emamectin benzoate, Denim®, Proclaim® (Syngenta)		Controls caterpillars on vegetables, cucurbits & cotton.	76	>2000
Empire® (see chlorpyrifos)				
Endeavor® (see pymetrozine)				
Envidor® (see spirodiclofen)				
	6,7,8,9,10,10-hexachloro-1,5,5a,6,9,9a-			

Common name, trade name and basic manufacturer(s)	Chemical name	General use pattern	Oral LD_{50} (rat)	Dermal LD_{50} (rabbit)
endosulfan, Thiodan®, Phaser® (Bayer CropScience)	hexahydro-6,9-methano-2,4,3-benzodioaxathiepin 3-oxicide	Effective against many insect & mite pests on vegetable, fruit, forage, fiber crops, ornamentals.	18	74
endrin (discontinued)	1,2,3,4,10,10-hexachloro-6,7-epoxy-1,4,4a,5,6,7,8,-8a-octahydro-exo 1,4 -exo- 5,8-,dimethanonaphthalene	For cotton insect control & a few other crops; mouse control in orchards.	3	12
Enstar II® (see kinoprene)				
Entrust® (see spinosad)				
Enzone® (Entek)	sodium tetrathiocarbonate (32% Solution)	Nematicide, soil fungicide/insecticide.	631	>2000
EPN (discontinued)	*O*-ethyl *O*-(4-nitrophenyl) phenylphosphonothioate	Used almost exclusively for cotton insect control, outside U.S.	14	110
Equity® (see chlorpyrifos)				
esfenvalerate, Asana® XL, Hallmark® (DuPont)	(*S*)-cyano(3-phenoxyphenyl)methyl-(*S*)-4-chloro-α-(1-methylethyl) benzene acetate	Pyrethroid for vegs, cotton, soybeans, nut trees, Christmas trees.	75	>2000
Esteem® (see pyriproxifen)				
ethion, Ethanox® (discontinued)	*O,O,O',O'*-tetraethyl *S,S'*-methylene bis (phosphorodithioate)	Insecticide-acaricide for sucking & chewing pests only on citrus.	27	915
ethiprole, Curbex® (Bayer CropScience)	5-amino-1(2,6-dichloro-α,α,α-trifluoro-p-tolyl)-4-ethylsulfinylpyrazole-3-carbonitrile	Broad spectrum against chewing & sucking insects, in rice, cotton, soybeans, fruits & vegs.		
ethoprop, Mocap® (Bayer CropScience)	*O*-ethyl *S,S*-dipropyl phosphorodithioate	Soil insecticide-nematicide for root crops, corn, vegs, golf course turf.	62	26
ethyl formate	ethyl formate	Fumigant for food products.	4000	—
ethyl parathion (see parathion)				
ethylan, Perthane® (discontinued)	1,1-dichloro-2,2-bis(4-ethylphenyl)ethane	Impregnate for moth & carpet beetle control in dry cleaning, textile industry.	8170	—
ethylene dibromide, EDB (discontinued)	1,2-dibromoethane	Insecticide-nematicide fumigant for soil application; warehouse, mill, household fumigation.	146	220 ppm vapor
ethylene dichloride, EDC	1,2-dichloroethane	Fumigant for stored products.	670	1000 ppm vapor
ethylene oxide (several)	1,2-epoxyethane	Fumigant, sterilant for certain stored food products, soil.	irritant	1 ppm TLV
Etoc® (see prallethrin)				
etoxazole, Baroque®, Secure®, Tetrasan®, Zoom® (Valent)	2-(2,6-difluorophenyl)-4[4-(1,1-dimethyethyl)-2-ethoxyphenyl]-4,5-dihydrooxazole	IGR/acaricide for cotton, pome fruits, nut crops, grapes & ornamentals.	>5000	>2000
eugenol, Matran 33®	insect attractant extracted from clove oil	Attractant for cucumber & corn	Non-toxic	

Common name, trade name and basic manufacturer(s)	*Chemical name*	*General use pattern*	*Oral LD_{50} (rat)*	*Dermal LD_{50} (rabbit)*
(EcoSMART)		rootworm beetles.		
Evict® (see tefluthrin)				
Evisect® (see thiocyclam)				
Exhibit® (Biosys)	Parasitic nematodes used as soil bioinsecticide,		Non-toxic	
Extinguish® (see methoprene)				
famphur, Cyflee®, Warbex® (discontinued)	*O*-[*p*-(dimethylsulfamoyl)phenyl]*O,O*-dimethyl phosphorothioate	Systemic for louse & cattle grub control on cattle, reindeer.	35	1460
fenamiphos, Nemacur® (Bayer CropScience)	ethyl 3-methyl-4-(methylthio)phenyl (1-methylethyl)phosphoramidate	Systemic nematicide for pre- or postemergence application to soil. Not available to lay-persons.	8	72
fenazaquin, Matador® (Dow)	4-[[4-(1,1-dimethylethyl)phenyl]ethoxy]quinazoline	Stone fruits, pears, vegs, apples, citrus, grapes & ornamentals.	134	>5000
fenbutatin-oxide, Vendex®, Torque® (BASF, Griffin)	hexakis (2-methyl-2-phenylpropyl)-distannoxane	Miticide for mites on deciduous fruits, citrus, greenhouse crops.	2631	>2000
fenitrothion, Sumithion® (Sumitomo)	*O,O*-dimethyl *O*-[4-nitro-*m*-tolyl) phosphorothioate	For chewing & sucking insects on wide variety of crops; public health insect control. Outside the U.S.	250	2500
fenobucarb (see BPMC)				
fenoxycarb, Logic®, Precision®, Comply® (Syngenta)	ethyl[2-(*p*-phenoxyphenoxy)ethyl]carbamate	Insect growth regulator for ants, fleas & cockroaches. Pest control on pome fruits & nut crops.	>5,000	>2000
fenpropathrin, Danitol®, Tame® (Valent)	*(R,S)*-α-Cyano-3-phenoxybenzyl-2,2,3,3-tetramethyl-cyclo propanecarboxylate	Insecticide/acaricide with broad spectrum activity on wide range of crops, soybeans & ornamentals.	70.6	>2000
fenpyroximate, Danitron®, Acaban®, Dynamite® (Nihon Nohyaku)	(E)-1,1-dimethylethyl-4-[[[[(1,3-dimethyl-5-phenoxy-1*H*-pyrazol-4-yl) methylene] ammo] oxo] methyl] benzoate	Acaricide for fruit crops, cotton, citrus & nut crops.	480	—
fenson (discontinued)	*p*-chlorophenyl benzenesulfonate	Acaricide-ovicide for European red mite, other mites on tree fruits.	1560	2000
fensulfothion, (discontinued)	*O,O*-diethyl *O*-[*p*-methylsulfinyl)phenyl] phosphoro-thioate	Soil-applied nematicide-insecticide; limited systemic effects on foliage feeders.	2	3
fenthion, Baytex®, Lebaycid® (Bayer CropScience)	*O,O*-dimethyl*O*-[4-(methylthio)-*m*-tolyl]phosphorothioate	Mosquitoes, flies, ornamentals; livestock insect pests.	255	330
fentin hydroxide, Super Tin®, Brestanid® (Bayer CropScience)	triphenyltin hydroxide	A fungicide that has antifeeding effect on certain insects.	108	—
fenvalerate, Belmark® (see also esfenvalerate) (BASF)	cyano(3-phenoxyphenyl)methyl 4-chloro-α-(1-methylethyl)benzeneacetate	Broad spectrum activity on non-food crops, trees.	451	>5000
ferriamicide (discontinued)	mirex plus alkyl amines & ferrous chloride	Short-residual form of mirex for fire ant control.		

Common name, trade name and basic manufacturer(s)	*Chemical name*	*General use pattern*	*Oral LD_{50} (rat)*	*Dermal LD_{50} (rabbit)*
Ficam® (see bendiocarb)				
Finitron® (see sulfluramid)				
fipronil, Goblet®, Frontline®, Firestar®, Choice®, Icon®, Combat®, Regent®, Over-N-Out®, Termidor® (BASF)	(5-amino-1-(2,6-dichloro-4-(trifluoromethyl)phenyl -4-((1,R.S.)-trifluoromethyl)su-1-*H*-pyrasole-3-carbonitrile)	Soil & foliar application for corn, rice, cotton, pet, household pests, public health, lawn insects, fire ant & termite baits.	97	354
Fireban® (see tefluthrin)				
Firestar® (see fipronil)				
Firstchoice Sluggo® (Western Farm Service)	ferric phosphate	For control of slugs & snails in several crops, ornamentals & greenhouses		
Flagship® (see thiamethoxam)				
Flee® (see permethrin)				
flonicamid, F-1785® (FMC)	*N*-(cyanomethyl)-4-(trifluormethyl)-3-pyridinecarboxamide)	Controls aphids & other sucking insects on cereals, cotton, tree fruits, potatoes, vegs & ornamentals in greenhouses. Not registered in U.S.	—	—
Floramite® (see bifenazate)				
flucycloxuron, Andalin® (Crompton)	1-[α-(4-chloro-α-cyclo-propylbenzylideneamino-oxy)-ρ-tolyl]-3-(2,6-difluorobenzoyl) urea	IGR/acaricide for pome, citrus nuts, field crops, vegs.	>5,000	—
flucythrinate, Pay-Off®, Cybolt® (BASF)	(RS)-*a*-cyano-3-phenoxybenzyl (*S*)-2-(4-difluromethoxyphenyl)-3-methylbutyrate	High level of activity against a broad spectrum of insect, mite, tick pests. Outside the U.S. only.	67	>1000
Fluorguard® (see sulfluramid)				
fluvalinate, Spur® (discontinued)	*N*-[2-chloro-4-(trifluoromethyl)phenyl]-*DL*-valine (±) cyano (3-phenoxyphenyl) methyl ester	Broad spectrum pyrethroid insecticide for cotton, vegs.	260	>20,000
tau-fluvalinate, Mavrik®, Klartan® (Wellmark, Makhteshim)	(RS)-α,-cyano-3-phenoxybenzyl N-(2-chloro-α,α,α-trifluoro-p-tolyl)-D-valinate	Broad spectrum pyrethroid for most insects on cotton, cereals, fruit trees, vegs, potatoes, wine grapes, fire ants & mosquitoes.	261	>20,000
Folidol® (see methyl parathion)				
Folimat® (see omethoate)				
fonofos, Dyfonate® (discontinued)	*O*-ethyl *S*-phenyl ethyl phosphonodithioate	Soil insecticide for several species of soil pests.	8	25
Force® (see tefluthrin)				
formetanate hydrochloride, Carzol®, (Gowan)	[3-dimethylamino-(methyleneiminophenyl)]*N*-methylcarbamate hydrochloride	Insecticide-acaricide for tree fruits, citrus, alfalfa.	21	>10,200
formothion, Anthio® (discontinued)	*O,O*-dimethyl phosphorodithioate *S*-ester with *N*-formyl-2-mercapto-*N*-methylacetamide	Systemic & contact insecticide-acaricide for sucking insects, nonresistant mites, outside U.S.	365	>1000
	(RS)-S-sec-butyl O-ethyl 2-oxo-1,3-thiazolidin-3-			

Common name, trade name and basic manufacturer(s)	Chemical name	General use pattern	Oral LD_{50} (rat)	Dermal LD_{50} (rabbit)
Fortress® (see chlorethoxyfos)	yl-phosphonothioate			
fosthiazate, Nemathorin® (Syngenta)		Nematicide for peanuts, tomatoes & potatoes.	39	4970
Frontline® (see fipronil)				
FruitGuard-V® (see Indian moth granulovirus)				
Frustrate PBW® (Certis)		Pink bollworm pheremone	Non-toxic	
Fulfill® (see pymetrozine)				
Fumitoxin® (see aluminum phosphide)				
Furacon® (see benfuracarb)				
Furadan® (see carbofuran)				
Fury® (see *zeta*-cypermethrin)				
Futura® (see carbofuran)	biopesticide (insecticidal virus)			
Galaxy V4C® (Analytica)		For control of insects in brassicas	Non-toxic	
gamma-cyhalothrin, *(see* cyhalothrin-*gamma*)				
gamma BHC (see lindane)				
Gardona® (see tetrachlorvinphos)				
Gaucho® (see imidacloprid)				
Gemstar® (see *Helicoverpa zea* NPV)				
Gentrol® (see hydroprene)				
Gnatrol® (see *Bacillus thuringiensis* spp. *israelensis*)				
Goblet® (see fipronil)				
Guardian® (see *Steinernema carpocapsae*)				
Gusano® (see *Autographa californica*)				
Guthion® (see azinphos-methyl)				
Hachihachi® (see tolfenpyrad)				
Hallmark® (see esfenvalerate)	*N*-tert-butyl-*N'*-(4-chlorobenzoyl) benzohydrazine			
halofenozide, Mach 2®		Insecticide/IGR for turf insects.	>5000	>2000

Common name, trade name and basic manufacturer(s)		*General use pattern*	*Oral LD_{50} (rat)*	*Dermal LD_{50} (rabbit)*
(Dow Agrosciences)				
HCH (see benzene hexachloride)				
HCN (see hydrocyanic acid)				
Helicoverpa zea NPV, Gemstar® (Certis)		Viral insecticide for caterpillars on cotton, tomatoes & vegs.	Non-toxic	
Heliothis nuclear polyhedrosis virus, Biovirus-H®, (Biotech)	viral insecticide specific for *Heliothis*	For control of bollworm, tobacco budworm on cotton.	Non-toxic	
Helix® (see chlorfluazuron)				
hellebore	poisonous alkaloids from white hellebore (*Veratrum album*)	Home garden & orchard insecticide. No longer available.	—	—
Helper® (see thiosultap-sodium)				
heptachlor (discontinued)	1,4,5,6,7,8,8-heptachloro-3a,4,7,7a-tetrahydro-4,7-methanoindene	For residual control of subterranean termites; a few agricultural uses.	40	119
Heterorhabditis bacteriophora, Cruiser®, Heteromask® (Biologic)		Nematode parasite for turf insects (grubs).		
hexakis (see fenbutatin-oxide)				
hexaflumuron, Consult®, Trueno®, Sentricon®, Recruit® (Dow Agrosciences)	1-[3,5-dichloro-4-(1,1,2,2-tetrafluoro=ethoxy) phenyl]-3-(2,6-difluorbenzoyl) urea	IGR for broad range of insect larvae on fruit, potatoes & subterranean termites,	5000	>5000
hexythiazox, Savey®, Hexygon® (Gowan)	trans-5-(4-chlorophenyl)-N-cyclohexyl-4-methyl-2-oxothiazolodine-3-carboxamide	Miticide-larvicide for fruit, vegs & cotton.	>5000	>5000
Hirsutella thompsonii, Mycar® (discontinued)	naturally occurring disease fungus of mites	Controls certain mite species, especially citrus rust mite.	Non-toxic	
Hostathion® (see triazophos)				
Hot Sauce® (see capsaicin)				
hydramethylnon, Amdro®, Eclipse®, Siege®, Subterfuge® (BASF, Bayer CropScience)	tetrahydro-5, 5-dimethyl-2(1H)-pyrimidinone [3-[4[trifluoromethyl)phenyl]-[1-[2-(4-trifluoromethyl]phenyl]ethenyl]-2-propenylidene]-hydrazone	Stomach insecticide used for fire ant & other ant baits & as termite bait.	>5000	>2000
hydrocyanic acid, hydrogen cyanide, prussic acid, HCN		Highly toxic fumigant for buildings, premises, some stored products. Not sold in U.S.	<0.5	—
hydroprene, Mator®, GenTrol® (Wellmark)	(E,E)-ethyl 3,7,11-trimethyl-2,4-dodecadienoate	IGR for cockroaches, stored products pests & many others.	>5100	>5100
Icon® (see fipronil)				
	1-(6-chloro-3-pyridin-3-ylmethyl)-*N*-			

Common name, trade name and basic manufacturer(s)	*Chemical name*	*General use pattern*	*Oral LD_{50} (rat)*	*Dermal LD_{50} (rabbit)*
imidacloprid, Admire®, Gaucho®, Maxforce®, Pre-Empt®, Premise®, Prescribe®, Preventol®, Provado®, Merit®, Trimax® (Bayer CropScience)	nitroimidazolidin-2-ylidenamine	Systemic via soil, seed & foliage. Turf, ornamentals, fruit trees, vegs, cotton, turf. Controls most sucking insects & termiticide.	424	<5000
Imidan® (see phosmet)				
imiprothrin, Pralle® (Sumitomo)	2,5-dioxo-3-prop-2-ynyllimidazolidin-1-yl methyl (1*R*,3*R*)-2,2-dimethyl-3-(2-methylprop-1-enyl) cyclopropanecarboxylate	Pyrethroid used primarily for household insect pests, not registered in U.S.	900	>2000
Incite® (see piperonyl butoxide)				
Incline® (see dichloropropene)				
Indian Moth Granulovirus, FruitGuard-V® (AgriVir)	insecticidal baculovirus	For control of Indian meal moth in stored commodities		
indoxacarb, Steward®, Avaunt® (DuPont)	methyl-7-chloro-2,3,4a,5-tetrahydro-2-[methoxycarbonyl(-4-trifluoromethoxy=phenyl) carbamoyl]indeno[1,2-e][1,3,4]oxydiazine-4a-carboxylate	Carbamate insecticide foliar applied for cotton, vegs, potatoes, tree fruits & grapes.	268	>5000
Intruder® (see acetamiprid)				
Intrepid® (see methoxyfenozide)				
iodomethane, Midas® (Arvesta)	methyl iodide	Soil fumigant to replace methyl bromide on vegs & ornamentals.	76	
isazofos	*O*-[5-chloro-1-(1-methylethyl)-1*H*-1,2,4-triazol-3-yl] *O,O*-diethyl phosphorothioate	Nematicide & insecticide, turf insects, outside U.S.	40	>3100
isofenphos, Oftanol®, (Bayer CropScience)	1-methylethyl 2-[[ethoxy[(1-methylethyl)amino]-phosphinothioyl]oxy] benzoate	Soil insecticide for turf & ornamentals. All uses cancelled.	32	162
isopropyl formate	isopropyl formate	Fumigant for packaged dry fruits, nuts.	—	—
Japidemic® (see *Bacillus popilliae*)				
Javelin® (see *Bacillus thuringiensis* spp. *kurstaki)*				
jojoba oil, Detur® (IJO Products)	biopesticide (natural plant oil)	For control of whiteflies on several crops		
Justice® (see spinosad)				
K-Pam® (see metam-potassium)				
Kabat® (see methoprene)				
Kammo® (see D-limonene)				
	kaolinite, $Al_2O_3.2SiO_2.2H_2O$			

Common name, trade name and basic manufacturer(s)	Chemical name	General use pattern	Oral LD_{50} (rat)	Dermal LD_{50} (rabbit)
kaolin, Surround® (Englehard)		Insect & disease control, exempt from tolerance on many fruit & vegetable crops.	Non-toxic	
Karate® (see *lambda*-cyhalothrin)				
Kelthane® (see dicofol)				
Kepone®, chlordecone (discontinued)	decachlorooctahydro-1,3,4-methano-2*H*-cyclobuta-(cd)pentalen-2-one	Originally used in area-wide control programs for fire ants.	95	345
Ketch® (see *Bacillus thuringiensis*-Kurstaki)				
kinoprene, Enstar II® (Wellmark)	2-propynl (E,E)-3,7,11-trimethyl-2,4-dodecadienoate	IGR for whiteflies, aphids, scales, mealybugs, fungus gnats in greenhouses & indoor ornamental.	2130	9000
Klartan® (see *tau*-fluvalinate)				
Knack® (see pyriproxyfen)				
Knox-Out® (see diazinon)				
Kryocide® (see cryolite)				
Lagenidium giganteum, Laginex® (Agra Quest)		Fungal insecticide for caterpillar control on grasses, rice & soybeans, mosquito larvae control, in drainage ditches, ponds.	Non-toxic	
lambda-cyhalothrin (see cyhalothrin, *lambda*)				
Lannate® (see methomyl)				
Laser® (see cyfluthrin)				
Larvin® (see thiodicarb)				
Lebaycid® (see fenthion)				
Lentrek® (see chlorpyrifos)				
Lepinox® (see *Bacillus thuringiensis* spp. *kurstaki)*				
D-limonene, Cide-Kick®, Kammo® (Brewer, Helena)	1,8(9)-*p*-methadiene 1-methyl-4-isopropenyl-1-cyclohexene	A botanical insecticide effective against all external pests of pets. Excellent deodorizer.	>5000	>2000
lindane, gamma BHC (several foreign)	1,2,3,4,5,6-hexachlorocyclohexane, gamma isomer of not less than 99%	Many uses including seed treatment; moderate fumigant action. Not in the U.S.	76	500
Lock-On® (see chlorpyrifos)				
Logic® (see fenoxycarb)				
Lorsban® (see chlorpyrifos)				
	N-[[[2,5-dichloro-4-(1,1,2,3,3,3-			

Common name, trade name and basic manufacturer(s)	*Chemical name*	*General use pattern*	*Oral LD_{50} (rat)*	*Dermal LD_{50} (rabbit)*
lufenuron, Axor® (Syngenta)	hexafluoropropoxy) phenyl]amino]carbonyl]-2,6-difluorobenzamide	Controls a broad spectrum of insect pests on field crops, vegs, citrus, fruit, pet fleas & for public health insects. Not registered.	2000	>2000
Mach 2® (see halofenozide)				
magnesium phosphide, Magnaphos®, Magtoxin® (United Phosphorus)	produces phosphine (PH_3) gas.	Fumigant for grain & stored products.		
malathion, Malixol®, Atrapa® (Bayer CropScience, Griffin)	*O,O*-dimethyl phosphorodithioate of diethyl mercaptosuccinate	Controls very broad spectrum of insects in agriculture, public health, livestock, home garden & termites.	885	4,000
Mamestra configurata, Virosoft® (Biotepp)	insecticidal baculovirus	For control of Bertha amyworm in canola		
Marshall® (see carbosulfan)				
Matador® (see *lambda*-cyhalothrin, also fenazaquin)				
Mator® (see hydroprene)				
Matric® (see chromafenozide)				
Mattch® (see *Bacillus*) *thuringiensis* spp. *aizawai)*				
Mavrik® (see fluvalinate)				
Maxforce® (see imidacloprid)				
mecarbam, Afos® (discontinued)	*S*-[[(ethoxycarbonyl)methylcarbamoyl]-methyl] *O,O*-diethyl phosphorodithioate	Insecticide, acaricide-ovicide, for sucking insects, mites on fruit trees, citrus, certain vegs.	36	>1220
Meldane® (see coumaphos)				
mephosfolan, Cytrolane® (BASF)	diethyl (4-methyl-1,3-dithiolan-2-ylidene)phosphoramidate	Systemic insecticide for leaf-eating larvae, stem borers.	9	29
Meridian® (see thiamethoxam)				
Merit® (see imidacloprid)				
Mesa® (see milbemectin)				
Mesurol® (see methiocarb)				
metaldehyde, Deadline®, Snail & Slug Killer® (several)	metacetaldehyde	Molluscicide for snails & slugs	250	>5000
metam-potassium, Curtin®, K-Pam®, Sectagon K-54® (Amvac)	potassium N-methyldithiocarbamate solution	Soil fumigant	630	>1000
metam-sodium, SMDC, Vapam®, Trimaton®	sodium *N*-methyldithiocarbamate	General-purpose soil fumigant; nematicide, fungicide, herbicide.	820	800

Common name, trade name and basic manufacturer(s)	*Chemical name*	*General use pattern*	*Oral LD_{50} (rat)*	*Dermal LD_{50} (rabbit)*
(Amvac, Cerexagri)				
Metarhizium anisopliae, Bio-Blast®, Taerain®, Tick-Ex®, (Taensal, Rincon-Vitova)		Fungal insecticide for cockroaches, black vine weevil & other soilinhabiting beetles, locusts, termites, ticks.	Non-toxic	
Metarhizium anisopliae, strain ESCI (EcoSciences)		For whiteflies & aphids in greenhouses in ornamentals.		
Metarhizium anisopliae, strain F-52 (EcoSciences)		For control of coleopteran pests in ornamentals & ticks.	Non-toxic	
Metasystox® (see methyl demeton)				
Metasystox-R® (see oxydemeton-methyl)				
methamidophos, Monitor®, Tamaron® (Bayer CropScience, Arvesta)	*O,S*-dimethyl phosphoramidothioate	Controls most common insect pests on, cotton, potatoes.	30	110
methidathion, Supracide®, Ultracide® (Syngenta)	*O,O*-dimethyl phosphorodithioate *S*-ester with 4-(mercaptomethyl)-2-methoxy-Δ^2-1,3,4-thiadiazolin-5-one	Controls wide spectrum of insects in alfalfa, cotton, tree fruits, nuts.	25	1546
methiocarb, Mesurol®, Draza® (Bayer CropScience)	4-(methylthio)-3,5-xylyl methylcarbamate	Contact insecticide with bird-repellent qualities; used on ornamentals for bird-problems. Also for snail/slug control in gardens.	15	2000
methomyl, Lannate® (DuPont)	*S*-methyl *N*-[(methylcaramoyl)oxy]thioacetimidate	Controls wide range of insects in fruit crops, vegs, field crops, ornamentals.	30	>2000
methoprene, Altosid®, Apex®, Extinguish®, Kabat®, Precor® (Wellmark)	isopropyl (*E,E*)-11-methoxy-3,7,11-trimethyl-2,4-dodecadienoate	Insect growth regulator, used as mosquito larvicide.	>34,600	>3000
methotrexate, Amethopterin® (discontinued)	*N*-[4-[[(2,4-diamino-6-pteridinyl)methyl]amino] benzoyl]-L-glutamic acid	Insect chemosterilant. Historical.		
methoxychlor, Marlate® (discontinued)	1,1,1-trichloro-2,2-bis(*p*-methoxyphenyl)ethane	Many uses on fruit & shade trees, vegs, home gardens, livestock.	6000	>2000
methoxyfenozide, Intrepid®, Runner® (Dow, Bayer CropScience)	*N'*-tert-butyl-N'-(3,5-dimethylbenzoyl)-3-methoxy -2-methyl benzohydraze	Insect growth regulator for caterpillars on cotton, pome fruits, grapes, nut crops, stone fruits, vegs, & soybeans. Not registered.	>5000	>5000
methyl bromide (several)	bromomethane	Fumigant action controls all living matter; used as space, soil, stored products, drywood termite fumigant. (Extreme caution!)	200 ppm vapor	15 ppm TLV
methyl demeton, Metasystox® (discontinued)	mixture of *O,O*-dimethyl *S*(and *O*)-(2-(ethylthio)-ethyl)phosphorothioates	Controls sucking insects, mites on vegetables, fruit, hops.	64	302
methyl formate	methyl formate	Fumigant for food products.	4000	—
	O,O-dimethyl *O*-(*p*-nitrophenyl)phosphorothioate			

Common name, trade name and basic manufacturer(s)	Chemical name	General use pattern	Oral LD_{50} (rat)	Dermal LD_{50} (rabbit)
methyl parathion, Penncap M®, Declare®, Folidol®, (Bayer CropScience, Cerexagri)		Controls most cotton pests including nonresistant bollworms, tobacco budworms & most insects on many crops.	9	63
mevinphos, Phosdrin® (AmVac)	methyl (*E*)-3-hydroxycrotonate dimethyl phosphate	Contact & systemic insecticide-acaricide with broad range of use on vegetable, fruit, field, forage crops. Not registered in U.S.	3	16
mexacarbate, Zectran® (discontinued)	4-(dimethylamino)-3,5-xylyl methylcarbamate	Broad range of control for nonfood crops; ornamentals, turf, shrubs; some forest use. Snails, slugs, forestry pests & mites.	15	500
Micromite® (see diflubenzuron)				
Microthiol® (Cerexagri)	micronized sulfur	Various uses as miticide & fungicide.		
milbemectin, Milbeknock®, Mesa®, UltraFlora® (Sankyo)	A_3: (6R,25R)-5-*O*-demethyl-28-deoxy-6,28-epoxy -25-methylmilbemycin; A_4: (6R,25R)-5-*O*-demethyl-28-deoxy-6,28-3poxy-25-ethylmilbemycin B	Acaricide for citrus, cotton, stone fruits & vegs,	456	>5000
Mildane® (see dinocap)				
Midas® (see iodomethane)				
Mimic® (see tebufenozide)				
Mirex® (discontinued)	dodecachlorooctahydro-1,3,4-metheno-1*H*-cyclobuta [*cd*]pentalene	Was used to control all ant species.	235	800
Miteclean® (see pyrimidifen)				
Mocap® (see ethoprop)				
Monitor® (see methamidophos)				
monocrotophos, Azodrin® (discontinued)	dimethyl phosphate ester of (*E*)-3-hydroxy-*N*-methylcrotonamide	Controls most insect pests of cotton, tobacco, sugarcane, potatoes, peanuts, ornamentals.	8	354
Morestan® (see oxythio-quinox)				
M-Pede® (see soaps, insecticidal)				
M-Trak® (see *Bacillus thuringiensis* spp. *tenebrionis*)				
Mr. Joker® (see silafluofen)				
Mustang® (see *zeta*-cypermethrin)				

Common name, trade name and basic manufacturer(s)	*Chemical name*	*General use pattern*	*Oral LD_{50} (rat)*	*Dermal LD_{50} (rabbit)*
MVP® (see *Bacillus thuringiensis* spp. *kurstaki)*				
Myrothecium verrucaria, DiTera® (Valent)		Biorational nematicide for fruit & vegetables.		
naled, Dibrom®, Trumpet® (Amvac)	1,2,-dibromo-2,2-dichloroethyl dimethyl phosphate	Effective against chewing & sucking pests on many crops; fly control in barns, stables, poultry houses, kennels; mosquito control projects.	430	1100
naphthalene	naphthalene ("moth balls")	Fumigant for nonfood uses.	—	—
Naturalis-O® (see *Beauveria bassiana*)				
Naturalyte® (see spinosad)				
Neem oil, Triact®, Trilogy® (Certis)	triterpenoids extracted from neem kernals	Insecticide, miticide, fungicide, IGR & antifeeding agent. Exempt from residue tolerance. For fruits, vegs & ornamentals.	>5000	>2000
Neemix® (see azadirachtin)				
Nemacur® (see fenamiphos)				
Nemaslug® (see *Phasmarhabditis megidis*)				
Nemesis® (see pyriproxyfen)				
Neo-Pynamin® (see tetramethrin)				
Nexter® (see pyridaben)				
nicotine (discontinued)	3-(1-methyl-2-pyrrolidyl)pyridine	Contact, fumigant; used as greenhouse fumigant. No longer registered.	50	50
nicotine sulfate, Nico Soap® (discontinued)	3-(1-methyl-2-pyrrolidyl)pyridine sulfate	Aqueous solution with 40% nicotine equivalent; used mostly against aphids, scales on ornamentals.	60	140
nitenpyram, Bestguard® (Takeda)	*N*-[(6-chloro-3-ppyridinyl)methyl]-*N*-ethyl-*N'*-methyl-2-nitro-1,1-ethenediamine	Sucking insects on vegs,apples, fruit trees, grapes & flea control on pets. Outside U.S.	1281	>2800
Nivral® (see thiodicarb)				
Nomolt® (see teflubenzuron)				
Nosema locustae, Grasshopper Spore®, Nolo Bait® (Rincon-Vitova)	naturally occurring protozoan spores	Grasshopper control on rangeland & noncrop areas.	Non-toxic	
Novador® (see *Bacillus thuringiensis* spp. *tenebrionis*)				
	1-[3-cloro-4-(1,1,2-trifluoro-2-trifluoro-			

Common name, trade name and basic manufacturer(s)	*Chemical name*	*General use pattern*	*Oral LD_{50} (rat)*	*Dermal LD_{50} (rabbit)*
novaluron, Rimon®, Oscar®, Pedestal® (Crompton, Makhteshim)	methoxyethoxy)phenyl]-3-(2,6-difluorobenzoyl)urea	IGR for wide variety of insect pests on ornamentals & greenhouse ornamentals, cotton & pome fruits.	>5000	>2000
noviflumuron, Recruit III® (Dow Agrosciences)	(RS)-1-[3,5-dichloro-2-fluoro-4-(1,1,2,3,3,3,-hexafluoropropoxy)phenyl]-3-(2,6-difluorobenzoyl)urea	IGR & insecticide (chitin synthesis inhibitor)		
NPV (see nuclearpolyhedrosis virus)				
nuclearpolyhedrosis virus or NPV (Biotech Inter.)	biopesticide (NPV of *Helicoverpa armigera* & *Spodoptera litura*)	For control of sensitive larvae on cabbage, cotton, several vegs & roses		
Oberon® (see spiromesifen)				
Off® (see deet)				
Oftanol® (see isofenphos)				
oils (see petroleum oils)				
omethoate, Folimat® (Bayer CropScience)	*O,O*-dimethyl *S*-[2-(methylamino)-2-oxoethyl]phosphorothioate	For sucking insects, mites on fruits,vegs, hops, ornamentals, outside U.S.	25	200
Omite® (see propargite)				
Onic® (see alanycarb)				
Oracle® (see pyridaben)				
Orion® (see alanycarb)				
Orthene® (see acephate)				
Oscar® (see novaluron)				
Ovasyn® (see amitraz)				
Over-N-Out® (see fipronil)				
Ovex®, (also chlorfenson) (Nippon)	*p*-chlorophenyl *p*-chlorobenzene sulfonate	Acaricide-ovicide for cotton, fruits, nuts, ornamentals, outside U.S.	2000	—
Ovotran® (see Ovex®)				
oxamyl, Vydate® (DuPont)	methyl *N',N'*-dimethyl-*N*-[(methylcarbamoyl)oxy]-1-thiooxamimidate	Insecticide, nematicide, acaricide; controls some of each on field crops, vegs, fruits, ornamentals.	2.5	>2000
oxydemeton-methyl, Metasystox-R® (Bayer CropScience, Gowan)	*S*-[2-(ethylsulfinyl)ethyl]*O,O*-dimethyl phosphorothioate	Systemic insecticide-acaricide for vegs, fruits, field crops, flowers, shrubs, trees. Restricted Use Pesticide.	50	130
oxythioquinox, Morestan® (Bayer CropScience)	6-methyl-1,3-dithiolo(4,5-b)quinoxalin-2-one	Insecticide, acaricide, fungicide; contols mites, mite eggs, powdery mildew, pear psylla; used on fruits, vegs, citrus, ornamentals.	2500	>2000

Common name, trade name and basic manufacturer(s)	Chemical name	General use pattern	Oral LD_{50} (rat)	Dermal LD_{50} (rabbit)
Padan® (see cartap)				
Paecilomyces fumesoroseus PFR-97® (Certis)		Bionematicide for tobacco, vegs & citrus.	Non-toxic	
Paecilomyces lilacinus, BioAct PL® (Prophyta)		Mycoinsecticide for mites, whiteflies in greenhouses & on ornamentals.	Non-toxic	
paradichlorobenzene (see *p*-dichlorobenzene)	(moth crystals)			
parathion, ethyl parathion, (discontinued)	*O,O*-*diethyl O*-(4-nitrophenyl) phosphorothioate	Broad-spectrum insecticide used on wide variety of crops. Cancelled effective 12/2005.	3	6.8
parathion, methyl (see methyl parathion)				
PayOff® (see flucythrinate)				
Pedestal® (see novaloron)				
Penncap-M® (see methyl parathion)	Microencapsulated, slow-release formulation.		>270	>5400
Pennstyl® (see cyhexatin)				
permethrin, Ambush®, Astro®, Pounce®, Dragnet®, Pramex®, Prelude®, Talcord®, Flee® (Bayer CropScience, FMC, Syngenta, BASF)	3-(phenoxyphenyl)methyl (±)-cis,trans-3-(2,2-dichloroethenyl)-2,2-dimethyl cyclopropane-carboxylate (60% trans, 40% cis isomers)	Ambush® & Pounce® for cotton insects; Astro®, Pramex® for greenhouse ornamentals. Prelude® for termites.	>4000	>4000
petroleum oils, Saf-T-Side®, Stylet-Oil®, Sunspray®, Citrex®, Superior 70 Oil®, Volck Supreme Oil® (several)	long-chain hydrocarbon oils	Applied in both dormant & growing seasons as contact insecticides, acaricides & ovicides.	>5000	>2000
Phantom® (see chlorfenapyr)				
Phaser® (see endosulfan)				
Phasmarhabdtis megidis, Nemaslug® (Biobest)		Nematode culture for snail/slug control in home gardens.	Non-toxic	
d-phenothrin, Sumithrin® (Sumitomo)	(3-phenoxyphenyl)methyl 2,2-dimethyl-3-(2-methyl-1-propenyl)cyclopropanecarboxylate	Space sprays for homes, industry, aircraft.	>10,000	>10,000
phenthoate, Cidial®, Elsan® (Nissan)	ethyl α-[(dimethoxyphosphinothioyl)thio]benzene-acetate	Insecticide-acaricide; used as mosquito adulticide; controls citrus scales.	200	4000
phorate, Thimet® (BASF)	*O,O*-diethyl *S*-[(ethylthio)methyl] phosphorodithioate	Systemic insecticide used on cotton, corn, some small grains, vegs.	1.6	2.5
phosalone, Zolone® (Bayer CropScience)	*O,O*-diethyl *S*-[(6-chloro-2-oxobenzoxazolin-3-yl)methyl] phosphorodithioate	Insecticide-acaricide; used on tree fruit, nuts, citrus, potatoes, ornamentals outside U.S.	125	1500
Phosdrin® (see mevinphos)				
	2-(diethoxyphosphinylimino)-1,3-dithiolane			

Common name, trade name and basic manufacturer(s)	*Chemical name*	*General use pattern*	*Oral LD_{50} (rat)*	*Dermal LD_{50} (rabbit)*
phosfolan, Cyolane® (discontinued)		Systemic for mites, aphids, thrips, whiteflies.	8.9	23
phosmet, Imidan®, (Gowan)	*N*-(mercaptomethyl)phthalimide *S-(O,O*-dimethylphosphorodithioate)	Used for a wide variety of pests on tree fruits, nuts, grapes, blueberries including snails.	147	3160
phosphamidon, Dimecron® (discontinued)	2-chloro-3-(diethylamino)-1-methyl-3-oxo-1-propenyl dimethyl phosphate	Systemic insecticide used for sucking insects in small grains, cotton, other field crops.	15	125
phostebupirim (see tebupirimiphos)				
Phostoxin® (see aluminum phosphide)				
phoxim, Baythion® (Bayer CropScience)	phenylglyoxylonitrile oxime *O,O*-diethyl phosphorothioate	For stored products, insects in granaries, mills, ships; public health insect control, outside U.S.	1845	1126
phthalthrin (see tetramethrin)				
Pine sawfly nuclear polyhedrosis virus		Controls pine sawfly on pine & other conifers.	Non-toxic	—
Pinpoint® (see acephate)				
piperonyl butoxide, Butacide®, Incite® (Bayer CropScience)	α-[2-(2-butoxy)ethoxy]-4,5-(methylene-dioxy)-2-propyltoluene	Synergist for pyrethrins, some of the synthetic pyrethroids.	>7500	>7500
piprotal	piperonal, bis[2-(2-butoxyethoxy)ethyl]acetal	Synergist for pyrethrins, some carbamate insecticides.	4400	—
Pirate® (see chlorfenapyr)				
pirimicarb, Pirimor®, Rapid® (Syngenta)	2-dimethylamino-5,6-dimethylpyrimidin-4-yl dimethylcarbamate	Selective aphicide & for Lygus on alfalfa, lettuce, cereals, fruits & vegs.	147	>500
pirimiphos-ethyl (Jiangsu)	*O*-[2-(diethlamino)-6-methyl-4-pyrimidinyl]*O,O*-diethyl phosphorothioate	Soil insects in turf & vegs.	192	>1000
pirimiphos-methyl, Actellifog®, Blex® (Syngenta)	*O*-[2-(diethlamino)-6-methyl-4-pyrimidinyl]*O,O*-dimethyl phosphorothioate	Wide range of pests of stored products; fruit, vegetable, field crops, seed treatment.	>2000	>4592
Pirimor® (see pirimicarb)				
Platinum® (see thiamethoxam)				
Plenum® (see pymetrozine)				
plifenate, Baygon MEB® (discontinued)	2,2,2-trichloro-1-(3,4-dichlorophenyl)-ethyl acetate	Household insecticide for flies, mosquitoes, clothes moths, carpet beetles.	>10,000	>1000
Poncho® (see clothianidin)				
Pounce® (see permethrin)				
Pralle® (see imiprothrin)				
prallethrin, Etoc® (Sumitomo)	(*S*)-2-Methyl-4-oxo-3(2-propynyl)-cyclopent-2-enyl (1R)-cis, trans-chrysanthemate	Pyrethroid for outdoor pest control, kennels, food processing facilities, mosquito fogging.	640	>5000

Common name, trade name and basic manufacturer(s)	*Chemical name*	*General use pattern*	*Oral LD_{50} (rat)*	*Dermal LD_{50} (rabbit)*
Pramex® (see permethrin)				
Precision® (see fenoxycarb)				
Precor® (see methoprene)				
Pre-Empt® (see imidacloprid)				
Prefume® (see sulfuryl floride)				
Premise® (see imidacloprid)				
Prenfish® (see rotenone)				
Prentox® (see rotenone)				
Prescribe® (see imidacloprid)				
Preventol® (see imidacloprid)				
Pristine® (see acetamiprid)				
Proclaim® (see emamectin)				
profenofos, Curacron® (Syngenta)	*O*-(4-bromo-2-chlorophenyl)-*O*-ethyl-*S*-propyl phosphorothioate	Controls most caterpillar pests on cotton, soybeans.	400	472
Prokil® (see cryolite)				
promecarb, (Bayer CropScience)	3-methyl-5-(1-methylethyl)phenyl methylcarbamate	For Colorado potato beetle, caterpillars, leaf miners of fruit; corn rootworm, outside U.S.	61	>1000
propargite, Comite®, Omite® (Crompton)	2-[4-(1,1-dimethylethyl)phenoxy]cyclohexyl 2-propynyl sulfite	Acaricide used on fruit, nuts, corn, cotton, some vegs, ornamentals.	2800	4000
propetamphos, Safrotin®, Catalyst® (Wellmark)	(*E*)-1-methylethyl 3-[[(ethylamino)methoxyphosphinothioyl]oxy]-2-butenoate	Effective against cockroaches, mosquitoes, fleas.	119	2825
propoxur, Baygon® (Bayer CropScience)	*o*-isopropoxyphenyl methylcarbamate	Used against household crawling insects, flies, mosquitoes, lawn insects. Not for food crops.	50	>5000
propylene oxide (discontinued)	1,2-epoxypropane	Fumigant used to sterilize packaged products.	Irritant	100 ppm TLV
propyl isome (discontinued)	dipropyl 5,6,7,8-tetrahydro-7-methylnaptho-(2,3-*d*)-1,3-dioxole-5,6-dicarboxylate	Synergist for pyrethrins.	1500	—
Provado® (see imidacloprid)				
Pseudomonas fluorescens, Mattch® (Dow Agrosciences)		Bioinsecticide for fruits & vegs.	Non-toxic	
Pydrin® (discontinued)				
pymetrozine, Fulfill®, Chess®, Plenum®, Endeavor®	4,5-dihydro-6-methyl-4-[(3-pyridinyl-methylene)-amino]-1,2,4-triazin-3(2*H*)one	Contact/systemic for whiteflies & aphids on vegs & ornamentals as a soil	5820	—

Common name, trade name and basic manufacturer(s)	Chemical name	General use pattern	Oral LD_{50} (rat)	Dermal LD_{50} (rabbit)
(Syngenta)		drench.		
Pynamin® (see allethrin)				
Pyramite® (see pyridaben)				
Pyranica® (see tebufenpyrad)				
pyrethrins, pyrethrum	mixture of pyrethrins and cinerins	Fast knockdown natural insecticide; safe for household aerosols, pets, livestock, stored products.	1500	1800
pyridaben, Nexter®, Oracle®, Pyramite®, Sanmite (BASF)	2-tert-butyl-5-(4-tert-butylbenzylthio)-4-chloropyridazin-3(2*H*)-one	Acaricide/insecticide for whiteflies, leafhoppers, psyllids on fruit trees, vegs, field crops, ornamentals.	820	>2000
pyrimidifen, Miteclean® (Sankyo)	5-chloro-*N*-(2-[4-(2-3ethoxyethyl)-2,3-dimethylphenoxy]ethyl)-6-ethylpyrimidin-4-amine	For mites & diamondback moth on citrus, apples, pears & vegs.		
pyriproxyfen, Knack®, Distance®, Esteem®, Nemesis®, Seize® (Sumitomo)	4-phenoxyphenyl (RS)-2(2-pyridyloxy)propyl ether	Insect growth regulator for public health use, household insects, whiteflies on cotton & greenhouse ornamentals, pome & stone fruits, vegs, grapes, tropical fruits.	>5000	>2000
Pytech® (see *gamma*-cyhalothrin)				
quassia, bitterwood (discontinued)	extracts of West Indian quassia wood chips	For aphids & other horticultural pests.	—	—
quinalphos, Ekalux® (Syngenta)	*O,O*-diethyl *O*-2-quinoxalinyl phosphorothioate	Broad range of insect pests on cotton, peanuts, vegs, fruit trees, outside U.S.	65	340
Rabon® (see tetrachlorvinphos)				
Rapid® (see pirimicarb)				
Raven® (see *Bacillus thuringiensis* spp. *aizawai*)				
Raze® (see tefluthrin)				
Recruit AG® (see hexaflumuron)				
Recruit III® (see noviflumuron)				
refined petroleum distillate, Damoil®, Sunspray®, Volck® (Drexel, Sunoco, Valent)	see also 'petroleum oils', distillates, hydro-treated light paraffinic.	Mite control on bush & tree fruits		
Regent® (see fipronil)				
Reldan® (see chlorpyrifosmethyl)				
Renounce® (see cyfluthrin)				
	(5-phenylmethyl-3-furanyl)methyl 2,2-dimethyl-3-			

Common name, trade name and basic manufacturer(s)	Chemical name	General use pattern	Oral LD_{50} (rat)	Dermal LD_{50} (rabbit)
resmethrin, Benzofuroline® (Sumitomo)	(2-methyl-1-propenyl)-cyclopropanecarboxylate	Uses similar to pyrethrins but longer lasting, more effective. Household, greenhouse & landscape pests. Used in USDA meat & poultry inspection programs.	2000	2500
d-trans-resmethrin, bioresmethrin	(5-phenylmethyl-3-furanyl)methyl (1*R-trans*)-2,2-dimethyl-3-(2-methyl-1-propenyl)-cyclopropanecarboxylate	Uses similar to pyrethrins (see above) but longer lasting, more effective.	7070	>10000
Responsar® (see cyfluthrin-*beta)*				
Rimon® (see novaluron)				
Ripcord® (see cypermethrin)				
Rosemary oil, EcoTrol® (EcoSmart)	herbal biopesticide	Use on legumes & many vegs, tubers & fruits.	Eye Irritant	
rotenone, cubé, derris, Prentox®, Prenfish® (Prentiss)	1,2,12,12a-tetrahydro-2-isopropenyl-8,9-dimethoxy[1]benzopyrano[2,4-*b*] furo[2,3-*b*] [1]benzopyran-6(6a*H*)-one	Used in home gardens, on pets; fish control in ponds, lakes.	60	>1000
Roxion® (see dimethoate)				
Rugby® (see cadusafos)				
Runner® (see methoxyfenozide)				
Rutgers 612® (discontinued)	2-ethyl-1,3-hexanedi	Early insect repellent for skin, clothing.	6500	
ryania	ground stemwood of *Ryania speciosa*	Stomach & contact insecticide used in apple production; selective for worm pests, leaving beneficials.	1200	4000
ryanodine alkaloid, Ryan® (discontinuedl)	alkaloid extract from ryania.	Stomach insecticide for caterpillar control on fruit trees, citrus.	1200	
sabadilla (Dunhill)	ground seeds of *Schoenocaulon officinale*	Historic botanical insecticide useful in home gardens & citrus mites commercially.	4000	
Saf-T-Side® (see petroleum oils)				
Safrotin® (see propetamphos)				
Sanmite® (see pyridaben)				
Savey® (see hexythiazox)				
schradan, OMPA (discontinued)	octamethylpyrophosphoramide	Historic systemic insecticide-acaricide.	9	15
Scimitar® (see *lambda-*				

Common name, trade name and basic manufacturer(s)	Chemical name	General use pattern	Oral LD_{50} (rat)	Dermal LD_{50} (rabbit)
cyhalothrin)				
Scout X-TRA® (see tralomethrin)				
Scythe® (see soaps)				
Sectagon K-54® (see metam-potassium)				
Secure® (see etoxazole)				
Seize® (see pyraproxyfen)				
Sentricon® (see hexaflumuron)				
sesamex, Sesoxane® (discontinued)	2-(2-ethoxyethoxy)ethyl 3,4-(methylenedioxy) phenylacetal of acetaldehyde	Synergist for pyrethrins, allethrin.	2000	11,000
sesamin (discontinued)	2,6-bis[3,4-(methylenedioxy)phenyl]-3,7-dioxabicyclo(3.3.0)octane	Synergist for pyrethrins.	2000	11,000
sesamolin (discontinued)	6-[3,4(methylenedioxy)phenoxy]-2-[3,4-methylenedioxy)-phenyl]-3,7-dioxabicyclo-[3.3.0]octane	Synergist for pyrethrins.	2000	11,000
Sesoxane® (see sesamex)				
Sevimol® (see carbaryl)				
Sevin® (see carbaryl)				
Siege® (see hydramethylnon)				
SfNPV, *Syngrapha falcifera* nuclear polyhedrosis virus		Viral insecticide for caterpillars on vegs & cotton.	Non-toxic	
silafluofen, Mr. Joker®, Silonen® (Bayer CropScience)	(4-ethoxyphenyl)(3-4(4-fluoro-3-phenoxyphenyl) propyldimethyl silane	Outside U.S. for rice, stone/pome fruits, public health use & as termiticide.	>5000	>5000
silica aerogel, Dri Die® (discontinued)				
Silonen® (see silafluofen)				
Slam® (see cucurbitacin)				
Snail & Slug Killer® (see metaldehyde)				
Snip® (see azamethiphos)				
soaps, insecticidal, M-Pede®, Scythe® (Dow Agrosciences)	potassium salts of fatty acids.		Non-toxic	
sodium borate, Tim-Bor®, Bora-care® (U.S. Borax)	disodium octaborate tetrahydrate ($Na_2B_8O_{13}.4H_2O$)	Wood preservative against borers, termites, etc. & prevents dry, wet & white wood rotting fungi. Also for roaches, ants, silverfish, crickets & earwigs.	2550	>2000

Common name, trade name and basic manufacturer(s)	*Chemical name*	*General use pattern*	*Oral LD_{50} (rat)*	*Dermal LD_{50} (rabbit)*
Solfac® (see cyfluthrin)				
Spectracide® (see diazinon)				
spinosad, Success®, Entrust®, Spintor®, Tracer®, Conserve®, Justice®, Naturalyte® (Dow Agrosciences)	fermentation product of a soil-borne actinomycete *Sacharopolyspora spinosa.*	Controls primarily caterpillars on cotton, vegs, potatoes, tree fruits, root & tuber vegs, ornamentals & drywood termites, stored grain. Favors beneficial predators & parasites, fire ant bait.	3783	>2000
Spintor® (see spinosad)				
spirodiclofen, Envidor® (Bayer CropScience)	3-(2,4-dichlorophenyl)-2-oxo-1-oxaspirol[4.5]dec-3-en-4-yl 2,2-dimethylbutyrate	Mite control in citrus, grapes, nuts & pome/stone fruits		
spiromesifen, Oberon® (Bayer CropScience)	3-mesityl-2-oxo-1-oxaspiro[4,4]non-3-en-4-yl 3,3-dimethylbutyrate	For whiteflies & mites on cotton, vegs & ornamentals. Not reg. in U.S.		
Spodoptera exigua, Spod-X® (Certis)	beet armyworm nuclear polyhedrosis virus *SeNPV*	Virus insecticide for beet armyworm.	Non-toxic	
Spod-XLC® (see beet armyworm nuclear polyhedrosis virus)			Non-toxic	
Standak® (see aldoxycarb)				
Starkle® (see dinotefuran)				
Steinernema carpocapsae Biosafe-N®, Guardian® (Certis)		Soil-dwelling parasitic nematode.		
Steinernema glaser & *Steinernema riobravis,* BioVector®355 (Certis)		Nematodes for mole, cricket & grub control in turf.	Non-toxic	
Steward® (see indoxacarb)				
Strobane® (discontinued)	polychlorinates of camphene, pinene & related terpenes	Used almost exclusively on cotton.	220	
Stryker® (see tralomethrin)				
Stylet® (see petroleum oils)				
Subterfuge® (see hydramethylnon)				
Success® (see spinosad)				
sulfluramid, Finitron®, Raid Ant & Roach Controller II®, Fluorguard® (Makhteshim)	*N*-ethyl perfluorooctane sulfonamide	Contact & stomach poison for ants, roaches & termites.	2500	1250
sulfoxide (discontinued)	1,2-(methylenedioxy)-4-[2-(octysulfinyl)propyl]benzene	Synergist for pyrethrins, allethrin.	2000	9000
sulfuryl fluoride, Vikane®, (Dow Agrosciences)	sulfuryl fluoride	Fumigant used to control structural & household pests, especially drywood termites, walnuts & raisins. Many	—	5 ppm TLV
	O-ethyl *O*-[4-(methylthio)phenyl] *S*-propyl			

Common name, trade name and basic manufacturer(s)	*Chemical name*	*General use pattern*	*Oral LD_{50} (rat)*	*Dermal LD_{50} (rabbit)*
	phosphorodithioate	other uses as a fumigant.		
sulprofos, Bolstar® (Jiangsu)		Controls several pests on cotton, corn, tobacco, tomatoes, peanuts, alfalfa. Not marketed in the U.S.	107	820
Sumithion® (see fenitrothion)				
Sumithrin® (see *d*-phenothrin)				
Sun-Spray® (see petroleum oils)				
Super Tin® (see fentin hydroxide)				
Supona® (see chlorfenvinphos)				
Supracide® (see methidathion)				
Suspend SC® (see deltamethrin)				
Sweeper® (see methyl parathion)				
Synthrin® (see resmethrin) (discontinued name)				
Systox® (see demeton)				
Talcord® (see permethrin)				
Talstar® (see bifenthrin)				
Tamaron® (see methamidophos)				
Tame® (see fenpropathrin)				
tau-fluvalinate (see fluvalinate)				
TDE, (DDD) (discontinued)	1,1-dichloro-2,2-bis(*p*-chlorophenyl)ethane	Used in the past on many fruits, vegs.	3400	4000
tebufenozide, Mimic®, Confirm® (Dow Agrosciences)		Insect growth regulator for caterpillars on fruit crops, grapes, vegs & forestry.	>5000	>5000
tebufenpyrad, Masai®, Pyranica® (BASF)	4-chloro-*N*[[4-(1,1-dimethylethyl)phenyl]methyl]-3-ethyl-1-methyl-1*H*-pyrazole-5-carboxamide	Experimental: Acaricide for apples, citrus, vegs, cotton. Outside the U.S.	595	—
tebupirimiphos, Aztec® (Bayer CropScience)	*O*-[2-(1,1-dimethylethyl)-5-pyrimidinyl]) *O*-ethyl *O*-(1-methylethyl) phosphorothioate	Soil insects in corn.	1.3	9.4
Tedion® (see tetradifon)				
teflubenzuron, Nomolt® (BASF)	1-(3,5-dichloro-2,4-difluorophenyl)-3-(2,6-fluorobenzyoyl)-urea	Chitin synthesis inhibitor insecticide.	>5000	>2000
	[1α,3α(Z)]1-($\pm$)-(2,3,5,6-tetrafluoro-4-			

Common name, trade name and basic manufacturer(s)	*Chemical name*	*General use pattern*	*Oral LD_{50} (rat)*	*Dermal LD_{50} (rabbit)*
tefluthrin, Force®, Raze®, Fireban®, Evict® (Syngenta)	methylphenyl)-methyl 3-(2-chloro-3,3,3-trifluoro-1-propenyl)-2,2-dimethylcyclopropanecarboxylate	Pyrethroid, soil applied for corn & sugar beet insects.	22	177
Teknar® (see *Bacillus thuringiensis,* spp. *israelensis*)				
Telone II®, Telone® (see dichloropropene)				
temephos, Abate® (BASF)	*O,O'*-(thiodi-4,1-phenylene)bis(*O,O*-dimethyl phosphorothioate)	Aquatic larvicide for mosquitoes, black flies, midges.	4204	2181
Temik® (see aldicarb)				
Tempo® (see cyfluthrin)				
Tempo Ultra® (see *beta*-cyfluthrin)				
tepp (discontinued)	tetraethyl pyrophosphate	Historically used for control of noncaterpillar pests on vegs & cotton.	0.2	2
terbufos, Counter® (BASF)	*S*-[[(1,1-dimethylethyl)thio]methyl]*O,O*-diethyl phosphorodithioate	Soil insecticide, nematicide for corn rootworm, nematodes in corn, sugar beet maggots.	1.6	0.81
Tergitol® (Dow Agrosciences)	alcohol-ethoxylate	Wetting agent, dispersant used as mosquito larvicide.	Non-toxic	
Termidor® (see fipronil)				
tetrachlorvinphos, Gardona®, Rabon® (Ningbo Yihwei)	2-chloro-1-(2,4,5-trichlorophenyl)-vinyl dimethylphosphate	Wide range of insects on many crops, stored products, livestock, residual premise spray.	>2000	>2500
tetradifon, Tedion® (Crompton)	4-chlorophenyl 2,4,5-trichlorophenyl sulfone	Acaricide used on many fruits, nuts, citrus, vegs, cotton. Not registered in U.S.	>14,700	10,000
tetramethrin, phthalthrin, Neo-Pynamin® (Sumitomo)	1-cyclohexane-1,2-dicarboxyimidomethyl 2,2-dimethyl-3-(2-methylpropenyl) cyclopropanecarboxylate	Fast knockdown insecticide for aerosols, sprays; controls flying insects, garden pests, livestock insects, stored products pests.	>5000	>2,000
Tetrasan® (see etoxazole)				
tetrasul, Animert V-101® (discontinued)	4-chlorophenyl 2,4,5-trichlorophenyl sulfide	Controls spider mites, which hibernate in the egg stage on plants.	>10,800	2000
Thanite® (discontinued)	isobornyl thiocyanoacetate	Older fast-knockdown insecticide for household, pet sprays, aerosols; human louse control.	1000	6000
thiacloprid, Calypso® (Bayer CropScience)	(2Z)-3-[(6-chloro-3-pyridinyl)methyl]-1-3-thiazolidin-2-ylidenecyanamide	For most insect pests on cotton, pome fruits, vegs & potatoes outside U.S.		
	3-(2-chloro-thiazol-5-ylmethyl)5-methyl-[1,3,5]			

Common name, trade name and basic manufacturer(s)	*Chemical name*	*General use pattern*	*Oral LD_{50} (rat)*	*Dermal LD_{50} (rabbit)*
thiamethoxam, Actara®, Centric®, Cruiser®, Flagship®, Platinum®, Meridian® (Syngenta)	oxadiazinan-4-ylidene-N-nitroamine	For aphids, whiteflies, leaf hoppers, thrips, leaf miners, on cotton, citrus, fruit trees, vegs & seed treatment. Soil-applied for corn & turf. Flea control on dogs, household pests & termites.	1563	
Thimet® (see phorate)				
thiocyclam, Evisect® (Syngenta)	N,N-dimethyl-1,2,3-trithian-5-amine hydrogen oxalate	For caterpillar & beetle larvae pests on wide variety of field, veg & tree crops.	399	880
Thiodan® (see endosulfan)				
thiodicarb, Larvin®, Nivral® (Bayer CropScience)	dimethyl-*N,N'*-[thiobis[(methylimino)carbonyloxy]]-bis(ethanimidothioate)	Insecticide for control of caterpillars of various crops.	66	>2000
thiofanox, Dacamox® (Bayer CropScience)	3,3-dimethyl-1-(methylthio)-2-butanone *O*-[(methylamino)carbonyl]oxime	Systemic soil insecticide for sucking insects on cotton, soybeans, sugar beets, peanuts, potatoes, some cereals. Not registered in U.S.	9	39
thiosultap-sodium, Helper®, Pilarhope® (Sanonda)	disodium *S,S'*-(2-dimethylaminotri=methylene) di(thiosulfate)	Stomach, contact, fumigant & systemic insecticide for fruit, vegs & rice.	996	316
Thuricide® (see *Bacillus thuringiensis* spp. *kurstaki*)				
thuringiensin, DiBeta® (discontinued)	*beta*-exotoxin from *Bacillus thuringiensis*		Non-toxic	
thymol, ApiLife VAR® (Swiss Bee Res. Ctr.)		Acaricide used in Italy for control of the Varroa mite in honeybee hives.		
Tim-Bor® (see sodium borate)				
tolfenpyrad, Hachihachi® (Nihon Nohyaku)	4-chloro-3-ethyl-1-methyl-*N*-[[4-(4-methylphenoxy)phenyl]methyl]-1-*H*-pyrazole-5-carboxamide	Insecticide for use on cole & cucurbit crops, Not registered in U.S.		
Torque® (see fenbutatin oxide)				
Torus® (see fenoxycarb)				
toxaphene, Strobane-T® (discontinued)	chlorinated camphene containing 67-69% chlorine	Once used heavily on cotton; effective for grasshoppers, range caterpillars; cattle dips for mange control.	40	600
Tracer® (see spinosad)				
Tralex® (see tralomethrin)				
tralomethrin, Scout XTRA®, Stryker®, Tralex® (Bayer CropScience)	(1R, 3S) 3[(1'RS) (1',2',2',2'-tetrabromoethyl)] - 2,2-dimethylcyclopropanecarboxylic acid (S)-alpha-cyano-3-phenoxybenzyl ester	Pyrethroid for cotton, soybean & certain vegetable pests.	284	>5000
Triact® (see neem oil)				
triazamate, Aphistar® (Dow)	ethyl(3-tert-butyl-1-dimethylcarbamoyl-1H-1,2,4-triazol-5 ylthio)acetate	For cotton, vegs, apples, citrus, cereals, potatoes & ornamentals.	100	>5000
triazophos, Hostathion®	*O,O*-diethyl *O*-(1-phenyl-1H-1,2,4-triazol-3-yl) phosphorothioate	Insecticide, miticide, nematicide; very	64	1100

Common name, trade name and basic manufacturer(s)	*Chemical name*	*General use pattern*	*Oral LD_{50} (rat)*	*Dermal LD_{50} (rabbit)*
(Bayer CropScience)		broad spectrum of activity.		
trichlorfon, Dipterex®, Dylox® (Bayer CropScience)	dimethyl (2,2,2-trichloro-1-hydroxyethyl) phosphonate	For turf, ornamentals & fly control in livestock structures.	144	>2000
triflumuron, Alsystin® (Bayer CropScience)	2-chloro-N-[[[4-(trifluoromethoxy) phenyl]amino]carbonyl]benzxamide	Chitin synthesis inhibitor for biting insect larvae & especially for caterpillars.	>5000	>5000
Trigard® (see cyromazine)				
Trilogy 90 EC® (see neem oil)				
Trimax® (see imidacloprid)				
trimethacarb, Broot® (discontinued)	mixture of 3,4,5- and 2,3,5-isomers of trimethylphenyl methylcarbamate	Corn rootworm insecticide with broad potential for other crops.	125	>2000
Trimeton® (see metam sodium)				
triphenyltin chloride (see fentin chloride)				
triphenyltin hydroxide (see fentin hydroxide)				
Tristar® (see acetamiprid)				
Trueno® (see hexaflumuron)				
Trumpet® (see naled)				
Turcam® (see bendiocarb)				
Tussock moth nuclear polyhedrosis virus (discontinued)		Controls tussock moth larvae in Douglas fir.	Non-toxic	
UltraFlora® (see milbemectin)				
Valero® (see cinnamaldehyde)				
Vapam® (see metam-sodium)				
Vapona® (see dichlorvos)				
VectoBac® (see *Bacillus thuringiensis* spp. *israelinsis)*				
Vectolex® (see *Bacillus sphaericus)*				
Velocity® (see acephate)				
Vendex® (see fenbutatin-oxide)				
Verticillium lecanii (see *Cephalosporium lecanii*)				
Vidden D® (discontinued)				

Common name, trade name and basic manufacturer(s)	*Chemical name*	*General use pattern*	*Oral LD_{50} (rat)*	*Dermal LD_{50} (rabbit)*
Vikane® (see sulfuryl fluoride)				
Virosoft® (see *Mamestra configurata*)				
Vorlex® (discontinued)				
Vydate® (see oxamyl)				
Warbex® (see famphur)				
Warrior®, see *lambda*-cyhalothrin)				
XenTari® (see *Bacillus thuringiensis* spp. *aizawai*)				
X-Gnat® (Biosys)		Nematodes for control of fungus gnats in greenhouses.		
Zephyr® (see abamectin)				
zeta-cypermethrin, (see cypermethrin *zeta)*				
Zolone® (see phosalone)				
Zoom® (see etoxazole)				

* Chemical nomenclature used is either that of Chemical Abstracts (American Chemical Society) or International Union of Pure and Applied Chemistry (IUPAC). The author and publisher make no warranty, express or implied, regarding the accuracy of this Appendix.

APPENDIX **C**

Herbicides, Herbicide Safeners and Plant Growth Regulators

Common, trade and chemical names, their basic manufacturer(s), general use patterns and Oral and Dermal LD_{50}s.

Common name, trade name and basic manufacturer(s)	*Chemical name*	*General use pattern*	*Oral LD_{50} (rat)*	*Dermal LD_{50} (rabbit)*
Aatrex® (see atrazine)				
Abolish® (see thiobencarb)				
Accelerate® (see endothall)				
Accent® (see nicosulfuron)				
Acclaim® (see fenoxaprop-ethyl)				
Accord® (see quinclorac)				
acetochlor, Degree®, Harness®, Surpass®, TopNotch®, Trophy®, (Dow Agrosciences, Monsanto)	2-chloro-*N*-ethoxymethyl-6'-ethylacet-*O*-toluidide	Annual grasses, broadleaf weeds in vegs & corn.	2148	4166
Achieve® (see tralkoxydim)				
acifluorfen-sodium, Blazer®, Status® (BASF)	sodium 5-[2-chloro-4-(trifluoromethyl)-phenoxy]-2-nitrobenzoate	Selective pre-, postemergence control of broadleaf, grass weeds in soybeans, peanuts.	1300	>2000
Action® (see fluthiacet-methyl)				
acrolein, Aqualin®, Magnacide H® (Baker)	2-propenal	Aquatic weed & slime control.	46	skin burns
Afalon® (see linuron)				
Affinity® (see carfentrazone ethyl)				
Agil® (see propaquizafop)				
Aim® (see carfentrazone ethyl)				

* Chemical nomenclature used is either that of Chemical Abstracts (American Chemical Society) or International Union of Pure and Applied Chemistry (IUPAC). The author and publisher make no warranty, express or implied, regarding the accuracy of this Appendix.

Common name, trade name and basic manufacturer(s)	*Chemical name*	*General use pattern*	*Oral LD_{50} (rat)*	*Dermal LD_{50} (rabbit)*
alachlor, Lasso®, Micro-Tech®, (Monsanto)	2-chloro-2',6'-diethyl-*N*-(methoxymethyl) acetanilide	Preemergence control of annual grass, some broadleaf weeds in soybeans, corn, peanuts.	930	13,300
Alanap® (see naptalam)				
alloxydim-sodium, Fervin® (Bayer CropScience)	sodium salt of 2-(1-allyl-oxyaminobutylidene)-5,5-dimethyl-4-methoxycarbonylcyclohexane-1,3-dione	Postemergence control of grass weeds in broadleaf crops, outside U.S.	2322	>2000
Ally® (see metsulfuron-methyl)				
Alternaria cassia, Casst® (discontinued)	biopesticide	Mycoherbicide controls sicklepod & coffee senna in soybeans, peanuts & cotton.	Non-toxic	
Alternaria destruens, Smolder® (Platte)		Mycoherbicide for control of dodder.	Non-toxic	
AMADS, Wilthin® (Entek)	1-aminomethanamide dihydrogen tetraoxosulfate	Plant regulator, desiccant for seed crops.	350	>2000
Amber® (see triasulfuron)				
ametryn, Evik® (Syngenta)	2-(ethylamino)-4-(isopropylamino)-6-(methylthio)-*s*-triazine	Broadleaf, grass weeds in pineapple, sugarcane, banana; potato vine desiccant.	1100	>3100
Amex® (see butralin)				
amicarbazone, Dinamic® (Bayer CropScience)	4-amino-4,5-dihydro-*N*-(1,1,-dimethyl)-3-(1-methylethyl)-5-oxo-1H-1,2,4-triazole-1-carboxamide	For broadleaf weeds in corn, soybeans & sugarcane.	1015	>2000
amidosulfuron, Eagle®, Pursuit® (Bayer CropScience)	3-(4,6-dimethoxypyrimidin-2-yl-1(*N*-methyl-sulfonylaminosulfonyl)-urea	Selective postemergence for broad-leafs in cereals & pastures.	>5000	>5000
amitrole, Amitrol®, Weedazol® (Nufarm)	3-amino-1,2,4-triazole	Noncropland use only; for annual, perennial grass; broad-leaf weeds.	1100	>10,000
ammonium thiosulfate (Natl. Chelating)		Growth regulator for apple blossom thinning.		
Amplify®, (see cloransulam-methyl)				
ancymidol, A-Rest® (SePRO)	α-cyclopropyl-α-(*p*-methoxyphenyl)-5-pyrimidine-methanol	Plant growth regulator, reduces internode elongation for greenhouse flowers.	4500	Irritant
Ansar® (see MSMA)				
Apiro Ace® (see pyriftalid)				
Apogee® (see prohexadione-calcium)				
Appeal® (see fluthiacet-methyl)				

Common name, trade name and basic manufacturer(s)	*Chemical name*	*General use pattern*	*Oral LD_{50} (rat)*	*Dermal LD_{50} (rabbit)*
Aquacide® (see diquat)				
Aqualine® (see acrolein)				
Aquaneat® (see glyphosate)				
Aquathol® (see endothall)				
Aramo® (see tepraloxydim)				
Arelon® (see isoproturon)				
A-Rest® (see ancymidol)				
Arsenal® (see imazapyr)				
arsenic acid	*ortho* arsenic acid	Defoliant-desiccant for cotton	48	—
Assert® (see imazamethabenz-methyl)				
Assure® (see quizalofop-p-ethyl)				
asulam, Asulox® (Bayer CropScience)	methyl sulfanilylcarbamate	Postemergent grass weed control in sugarcane; bracken control in reforestation areas; crabgrass control in turf, certain ornamentals.	>4000	>1200
Asulox® (see asulam)				
atrazine (several) (Syngenta)	2-chloro-4-(ethylamino)-6-(isopropylamino)-*s*-triazine	Season-long weed control mainly in corn, sorghum.	1869	>3100
Atrimmec® (see dikegulac-sodium)				
Attribute® (see propoxycarbazone)				
Authority® (see sulfentrazone)				
AuxiGro® (see GABA)				
Avast® (see fluridone)				
Avelon® (see isoproturon)				
Avenge® (see difenzoquat)				
Axiom® (see flufenpyr-ethyl)				
azafenidin, Evolus®, Milestone® (discontinued)	2-[2,4-dichloro-5-(2-propynyloxy)phenyl]-5,6,7,8-tetrahydro-1,2,4-triazolo[4,3-a]=pyridin-3(2*H*)-one	For annual & perennial weeds in citrus, tree, nut crops, sugar cane & industrial sites.	>5000	>2000
azimsulfuron, Gulliver® (DuPont)	1-(4,6-dimethyloxypyrimidin-2-yl)-[1-methyl-1*H*-tetrazol-5-yl)pyrazol-5-ylsulfonyl]urea	Experimental: annual grasses, broadleaf weeds, sedges & perennial grasses in direct seeded rice.	>5000	>2000
Balan® (see benefin)				

Common name, trade name and basic manufacturer(s)	*Chemical name*	*General use pattern*	*Oral LD_{50} (rat)*	*Dermal LD_{50} (rabbit)*
Balance® (see isoxaflutole)				
Balfin® (see benefin)				
Banex® (see dicamba)				
Banvel® (see dicamba)				
BAP Promalin®, Perlan® (Valent)	6-benzylamino purine	PGR for apples & tree nuts. Mixture of BA + GA_4 + Ga_7	1690	>2000
Barricade® (see prodiamine)				
Basagran® (see bentazon)				
Basamid® (see dazomet)				
Baseline® (see prohexadione-calcium)				
Beacon® (see primisulfuron-methyl)				
benefin, Balan®, Balfin®, Quilan® (Dow Agrosciences)	*N*-butyl-*N*-ethyl-α-α-α-trifluoro-2,6-dinitro-*p*-toluidine	Preemergence control of annual grass, broad-leaf weeds in several crops.	>10,000	—
Benoxacor® (Syngenta)		Herbicide safener that increases tolerance of corn to metolachlor.	>5000	
bensulfuron-methyl, Londax® (DuPont)	methyl 2[[[[(4,6-dimethoxypyrimidin-2-yl)amino]carbonyl]amino]sulfonyl]methyl]benzoate	Postemergence rice herbicide. Controls broadleaf & sedge weeds, in rice.	>5000	>2000
bensulide, Betasan®, Prefar® (Gowan)	*S-(O,O-* diisopropyl phosphorodithioate) ester of *N-* (2-mercaptoethyl)benzenesulfonamide	Preemergence control of crabgrass, annual bluegrass, broadleaf weeds in dichondra, grass lawns, vegs.	270	>5000
bentazone, Basagran® (BASF)	3-isopropyl-1*H*-2,1,3-benzothiadiazin-(4)3*H*-one-2,2-dioxide	Postemergence control of broadleaf weeds in soybeans, rice, corn, peanuts.	1100	>2500
benzofenap, Yukawide® (Bayer CropScience)	2-[4-(2,4-dichloro-m-toluoyl)-1,3-dimethylpyrazol-5-yloxy]-4'-methylacetophenone	Pre- & postemergent control for grass weeds & sedge in rice.	15,000	>5000
Betanal® (see phenmedipham)				
Betanex® (see desmidipham)				
Betasan® (see bensulide)				
Beyond® (see imazamox)				
bifenox, Fox® (Bayer CropScience)	methyl 5-(2,4-dichlorophenoxy)-2-nitrobenzoate	Preemergence control of weeds in soybeans, sorghum, rice.	>6400	>20,000
Bingo® (see cinidon-ethyl)				
bispyribac-sodium, Regiment® (Valent)	sodium 2,6-bis[(4,6-dimethoxypyrimidin-2-yl) oxy]benzoate	Barnyardgrass & annual sedges plus broadleaf weeds in rice outside U.S.		

Common name, trade name and basic manufacturer(s)	*Chemical name*	*General use pattern*	*Oral LD_{50} (rat)*	*Dermal LD_{50} (rabbit)*
Blade® (see metsulfuron-methyl)				
Blazer® (see acifluorfen-sodium)				
B-Nine® (see daminozide)				
Bolero® (see thiobencarb)				
Bonzi® (see paclobutrazol)				
Broadstrike® (see flumetsulam)				
bromacil, Hyvar® (DuPont)	5-bromo-3-*sec*-butyl-6-methyluracil	Weed & brush control in noncrop area; selective weed control in citrus, pineapple & noncropland.	5200	>5000
bromoxynil, Buctril®, Connect®, Combine®, Brox®, Pardner® (Bayer CropScience)	3,5-dibromo-4-hydrobenzonitrile	Postemergence broadleaf weed control in small grains, turf noncrop areas.	190	>2000
Brox® (see bromoxynil)				
Buctril® (see bromoxynil)				
Bueno® (see MSMA)				
Busan®1236 (see metam-sodium)				
butachlor, Butanox® (Crystal)	2-chloro-2',6'-diethyl-*N*-(butoxy methyl) acetanilide	Preemergence control of annual grasses in rice.	3300	4080
butafenacil, Inspire®, Rebin® (Syngenta)	1,1-dimethyl-2-oxo-2-(2-propenyloxy)ethyl 2-chloro-5-[3,6-dihydro-3-methyl-2,6-di-oxo-4-(trifluoromethyl)-1(2*H*)-pyrimidinyl]benzoate	All weeds in grapes, nut crops, pome, stone fruits, & as a cotton defoliant.	>5000	>2000
Butanox® (see butachlor)				
butralin, Amex®, Tamex®, (Nufarm)	4-(1,1-dimethylethyl)-*N*-(1-methylpropyl)-2,6-dinitrobenzenamine	Annual grass & broadleaf weeds in cotton, soybeans. Growth regulator for small grains & tobacco sucker control.	891	>2000
butylate, Sutan® (Cedar)	*S*-ethyl diisobutylthiocarbamate	Incorporated preplant for controlling grass weeds especially nutgrass in corn.	3500	>5000
cacodylic acid, Montar®, Quick Pick® (Platte, Monterey)	hydroxydimethylarsine oxide	For killing trees in forestry, as a nonselective herbicide; cotton defoliant.	2756	—
Cadre® (see imazapic)				
calcium acid methane-asonate, Calar® (Drexel)		Postemergence weed control on turf.		
Caliber® (see simazine)				
Callisto® (see mesotrione)				

Common name, trade name and basic manufacturer(s)	*Chemical name*	*General use pattern*	*Oral LD_{50} (rat)*	*Dermal LD_{50} (rabbit)*
Caparol® (see prometryn)				
carfentrazone-ethyl, Aim®, Affinity®, Shark® (FMC)	ethyl α, 2-dichloro-5[4-(difluoromethyl)-4,5-dihydro-3-methyl-5-oxo-1*H*-1,2,4-triazol-1-yl]-4-fluorobenzenepropanoate	Postemergence broadleaf herbicide for corn, cereals, rice & cotton defoliant.	5143	>4000
Casoron® (see dichlobenil)				
CDAA, Randox® (discontinued)	*N-N*-diallyl-2-chloroacetamide	Preemergence grass, broadleaf weed control in corn, sorghum, soybeans, some vegetable crops.	700	—
CDEC, Vegadex® (discontinued)	2-chloroallyl diethyldithiocarbamate	Preemergence grass, broadleaf weeds in vegetable crops, ornamentals, shrubbery.	850	—
CeCeCe® (see chlormequat)				
Cerano® (see clomazone)				
Cerone® (see ethephon)	3-amino-2,5-dichlorobenzoic acid	Preemergence weed control in many vegetable, field crops, outside U.S.	5620	>3160
chloramben (Bayer CropScience)	methyl 2-chloro-9-hydroxyfluorene-9-carboxylate	Plant growth retardant reduces grass/lawn growth & for pineapple.	12,700	>10,000
chlorflurenol-methyl, Maintain® (Repar)	5-amino-4-chloro-2-phenyl-3-(2*H*)-pyridazinone	Pre- & postemergence weed control in sugar, red & fodder beets.	2200	>2000
chloridazon, pyrazon, Pyramin® (BASF)	ethyl-2-[[[[(4-chloro-6-methoxypirimidin 2-yl)amino]carbonyl]amino]sulfonyl]benzoate	Postemergence for annual broadleaf & yellow nutsedge in soybeans.	>4000	>2000
chlorimuron-ethyl, Classic®, (DuPont)	2-chloroethyltrimethylammonium chloride	Growth regulator for greenhouse use. Also cotton, vegs, grape, tobacco, ornamentals.	883	>4000
chlormequat chloride, Cycocel®, CeCeCe® (BASF)	isopropyl *m*-chlorocarbanilate	Preemergence herbicide for several crops; potato sprout inhibitor; growth regulator.	3,800	10,300
chlorpropham, Sprout Nip®, Unicrop CIPC® (Universal)	2-chloro-*N*[(4-methoxy-6-methyl-1,3,5 triazin-2-yl) aminocarbonyl]-benzenesulfonamide	Selective pre- & postemergence control in small grains, noncrop areas.	5,545	3,400
chlorsulfuron, Glean®, Telar® (DuPont)	2-chloro-*N*[(4-methoxy-6-methyl-1,3,5 triazin-2-yl) aminocarbonyl]-benzenesulfonamide	Selective pre-, & postemergence control in small grains, noncrop areas.	5,545	3,400
chlorthal dimethyl (see DCPA, Dacthal®)				
Chondrostoreum purpurerins, Chontrol Paste® (Myclogic)		Mycoherbicide for hardwood brush control in forests & rights-of-way.	Non-toxic	
Chopper® (see imazapyr)				
Cinch® (See S-metolachlor)				
cinidon-ethyl, Lotus®, Bingo®, Orbit® (BASF)	ethyl-(Z)-2-chloro-3-[2-chloro-5-(1,3-dioxo-4,5,6,7-tetrahydro-isoindol-2-yl) phenyl]phenyl]-acrylate	For control of broadleaf weeds in small grains, outside U.S.	>2200	>2000

Common name, trade name and basic manufacturer(s)	*Chemical name*	*General use pattern*	*Oral LD_{50} (rat)*	*Dermal LD_{50} (rabbit)*
Classic® (see chlorimuron-ethyl)				>5000
Clarity® (see dicamba)				
Clearigate®, (Applied Biochem)	copper alkanolamine complex	Algae control in irrigation ditches.	1750	—
clethodim, Envoy®, Select®, Prism® (Arvesta)	(E,E)-(±)-2[1[[3-chloro-2-propenyl=oxy]imino] propyl]5[2(ethylthio)=propyl]-3-hydroxy-2-chclohexen-1-one	For annual & perennial grasses in broadleaf crops, cotton, vegs.	2920	>5000
Clincher® (see cyhalofop-butyl)				
clodinafop-propargyl, Discover®, Horizon®, Topik® (Syngenta)	2-propynyl-(R)-2-[4-(5-chloro-3-fluoro-2-pyridyloxy)-phenoxy] propionate	Small grains, for blackgrass, wild oats, annual ryegrass, canarygrass, foxtails, etc.	1829	>2000
clofencet, (Monsanto)	2-(4-chlorophenyl)-3-ethyl-2,5-dihydro-5-oxopyridazine-4-carboxylic acid	Grain hybridizing agent.		
clomazone, Cerano®, Command®, Merit® (FMC)	2-(2-chlorophenyl)methyl-4,4-dimethyl-3-isoxazolidinone	Selective preemergence or preplant incorporated herbicide, cole crops & peas.	1369	>2000
clopyralid, Stinger®, Lontrel® Reclaim®, Transline® (Dow Agrosciences)	3,6-dichloropicolinic acid	Broadleaf weeds in small grains, pastures, rangeland, sugar beets.	>5,000	>2,000
cloquintocet-methyl (Syngenta)	1-methylhexyl [(5-chloro-8-quinolinyl)oxy]acetate	Herbicide safener for clodinafop-propargyl on wheat.	>2000	>2000
cloransulam-methyl, Amplify®, Firstrate® (Dow Agrosciences)	methyl 3-chloro-2-[[5-ethoxy-7-fluoro=[1,2,4] triazolo[1,5-c]pyrimidin-2-ylsulfonyl]amino] benzoate	Selective pre- & postemergence on soybeans for cocklebur, velvetleaf, lambsquarters & most other broad-leaf weeds.	>5000	>2000
Clove oil, Matran® (EcoSmart)	eugenol (natural plant extract)+acetic acid	Herbicide for annual grasses & broadleaf weeds. Also kills insects & repels dogs & cats		
Cobra® (see lactofen)				
Collectotrichum gloesporioides f. sp. malva, Mallet ® (Encore Tech)		Mycoherbicide for coffee weed in rice, mallow & curly indigo in soybeans.	Non-toxic	
Combine® (see bromoxynil)				
Command® (see clomazone)				
Connect® (see bromoxynil)				
Contrast® (see metribuzin)				
Corto® (see tritosulfuron)				
Cotoran® (see fluometuron)				
Cotton Pro® (see prometryn)				

Common name, trade name and basic manufacturer(s)	*Chemical name*	*General use pattern*	*Oral LD_{50} (rat)*	*Dermal LD_{50} (rabbit)*
Cotton Quik® (see ethephon)				
Credit® (see glyphosate)				
Cultar® (see paclobutrazol)				
Curbit® (see ethalfluralin)				
Cutless® (see flurprimidol)				
cyanazine, Bladex® (discontinued)	2-[(4-chloro-6-(ethylamino)-*s*-triazin-2-yl)amino]-2-methylpropionitrile	Phased out in 2001.	288	>1200
cyclanilide, Finish® (Bayer CropScience)	1-(2-4-dichloroanilinocarbonyl)= cyclopropanecarboxylic acid	Growth regulator for opening & defoliating cotton. For rice & cereals.	208	>2000
cycloate, Ro-Neet® (Cedar)	*S*-ethyl *N*-ethylthiocyclohexanecarbamate	Preplant herbicide for grass, broadleaf weeds in sugar beets, table beets, spinach.	2710	>4640
cyclohexane carboxamide, Surestem® (BASF)	1-(4-chloro-1,3-dihydro-1,3-dioxo[2*H*]-isoindol -2-yl)cyclohexanecarboxamide	Growth regulator for flowers, & ornamentals.	>5000	Eye irritant
cycloxydim, Focus®, Laser® (BASF)	2-[1-(ethoxyimino)butyl]-3-hydroxy-5-(2*H*-tetrahydrothiopyran-3yl)-2-cyclohexene-1-one	Postemergence for grass weeds in many broadleaf crops, experimental.	>5,000	>2,000
Cycocel® (see chlormequat)				
cyhalofop-butyl, Clincher® (Dow Agrosciences)	butyl (*R*)-2-[4-(4-cyano-2-fluorophenoxy) phenoxy]propionate	Controls grasses in rice.	>5000	>2000
cytokinins, Foliar TRIGGRR®, Sunburst® (Westbridge)	mixed cytokinins from seaweed or algae estracts.	Applied to both soil & foliage to increase yields of field crops, vegs & tree fruits.	Non-toxic	
2,4-D (several)	(2,4-dichlorophenoxy)acetic acid	Selective broadleaf weed control in monocots (small grains, corn, sorghum, sugarcane), noncrop areas.	375	1,500
2,4-DB (several)	4-(2,4-dichlorophenoxy)butyric acid	Selective control of certian broadleaf weeds in alfalfa, clover, soybeans, peanuts.	700	>10000
2,4-D LV6 ester		Growth regulator for potato red color.		
Dacthal® (see DCPA)				
Dagger® (see imazamethabenz-methyl)				
dalapon, Dowpon® (discontinued)	2,2-dichloropropionic acid (sodium salt)	Systemic herbicide for various grasses & rushes in crop, noncrop areas, outside U.S.	7570	10,000
daminozide, B-Nine® (Crompton)	butanedioic acid mono (2,2-dimethyl hydrazide)	Growth retardant: apples, azaleas, chrysanthemums, bedding plants, other ornamentals.	8400	>1600

Common name, trade name and basic manufacturer(s)	*Chemical name*	*General use pattern*	*Oral LD_{50} (rat)*	*Dermal LD_{50} (rabbit)*
dazomet, Basamid® (BASF)	tetrahydro-3,5-dimethyl-2*H*-1,3,5-thiadiazine-2-thione	Preplanting seedbed treatment for tobacco, turf, ornamentals; antimicrobial in glues for paper products; nematicide, slimicide.	500	>2000
DCPA, Dacthal® (Amvac)	dimethyl tetrachloroterephthalate	Preemergence herbicide for grass, broadleaf weeds in soybeans, cotton, seeded vegs.	>3000	>10,000
DEF® (see tribufos)				
Define® (see flufenacet)				
Defol® (see sodium chlorate)				
Degree® (see acetochlor)				
desmidipham, Betanex® (Bayer CropScience)	ethyl-*m*-hydroxycarbanilate carbanilate (ester)	Broadleaf weed control in sugar beets.	10,250	>2000
Detasslor® (see karetazan)				
Devrinol® (see napropamide)				
diallate, Avadex® (discontinued)	*S*-(2,3-dichloroallyl)diisopropylthiocarbamate	Soil-incorporated preemergence control of wild oats in several crops.	395	2000
diammonium glyphosate, New Touchdown® (Syngenta)	diammonium salt of glyphosate	For use on Roundup Ready® corn, cotton & soybeans.		
dicamba, Banvel®, Clarity®, Vanquish® (BASF, Syngenta)	3,6-dichloro-*o*-anisic acid	Controls annual, perennial weeds in corn, sorghum, small grains; pasture, rangeland, noncropland.	1700	>2000
dichlobenil, Casoron® (Crompton)	2,6-dichlorobenzonitrile	Selective weed control in ornamentals, orchards, vineyards; total weed control for industrial sites. Aquatic herbicide.	3160	1350
Dichlormid® (Syngenta)	*N,N*-diallyl-2,2-dichoroacetamide	Herbicide safener for Sutan® & Eptam®.	2816	>5000
dichlorprop, (several) (BASF)	2-(2,4-dichlorophenoxy)propionic acid	Systemic brush control on rights-of-way, rangeland; broadleaf weeds in cereals.	800	>4000
diclofop-methyl, Hoelon® Illoxan® (Bayer CropScience)	methyl 2-[4-(2',4'-dichlorophenoxy)-phenoxy]propanoate	For annual weed grasses (wild oats) in wheat, barley, soybeans.	563	>5000
diclosulam, Strongarm® (Dow Agrosciences)	*N*-(2,6-dichlorophenyl)-5-ethoxy-7-fluoro[1,2,4]-triazolo[1,5-c]pyrimidine sulfonamide	Preemergence for broadleaves in soybeans & peanuts.	>5000	>2000
difenzoquat, Avenge® (BASF)	1,2-dimethyl-3,5-diphenyl-1*H*-pyrazolium methyl sulfate	Postemergence for controlling wild oats in wheat, barley.	470	3540
diflufenzopyr, Distinct® (BASF)	2-[1[[[(3,5-difluorophenyl)amino]carbonyl] hydrazono] ethyl]-3-pyridinecarboxylic acid	Postemergence for annual broadleaf & perennial weeds in corn.	>5000	>5000

Common name, trade name and basic manufacturer(s)	Chemical name	General use pattern	Oral LD_{50} (rat)	Dermal LD_{50} (rabbit)
dikegulac sodium, Atrimmec® (PBI/Gordon)	sodium salt of 2,3:4,6-di-*O*-isopropylidene-α-L-xylo-2-hexalofuranosonic acid	Systemic plant growth regulator/ inhibitor for ornamentals, hedges & ground covers.	18,000	>1000
Dimension® (see dithiopyr)				
dimethenamid, Frontier® (BASF)	*(1RS,aRS)*-2chloro-N-2,4-dimethyl-3-thienyl)-*N*-(2-methoxy-1-methylethyl)-acetamide	Selective preemergence herbicide for grass weeds & nutsedge in corn & soybeans. Outside the U.S.	849	>2000
dimethenamid-P, Outlook® Frontier X-2® (BASF)	(*S*)-2-chloro-*N*-[(1-menthyl-2-methoxy) ethyl]-*N*-(2,4-dimethyl-thien-3-yl)-acetamide	Selective preemergence for major grass weeds in corn, sorghum, soybeans, vegs, potatoes & sugar beets.	695	>2000
dimethipin, Harvade®, Lintplus® (Crompton)	2,3-dihydro-5,6-dimethyl-1,4-dithiin 1,1,4,4-tetraoxide	Cotton desiccant/defoliant, enhances maturation & reduces seed moisture at harvest of rice, corn, sunflower.	1175	>120
Dinamic® (see amicarbazone)				
dinitramine, Cobex® (discontinued)	*N,N*-diethyl-2,4-dinitro-6-(trifluoromethyl)-1,3-benzenediamine	Preemergence selective herbicide, not registered in U.S.	3000	>6800
dinoseb, DNBP, Dinitro® (discontinued)	2-*sec*-butyl-4,6-dinitrophenol	General contact herbicide & desiccant.	40	75
diphenamid, Dymid®, Enide® (discontinued)	*N-N*-dimethyl-2,2-diphenylacetamide	Preemergence control of annual, broadleaf weeds in many crops.	1000	>225
dipropetryn, Sancap® (discontinued)	2-ethylthio-4-6-bis-isopropylamino-*s*-triazine	Preemergence control of pigweed, Russian thistle in cotton.	3900	>10,000
diquat, Aquacide®, Reward®, Reglone® (Syngenta)	6,7-dihydrodipyridol(1,2-α:2',1-*c*)-pyrazinediiumion	Industrial & aquatic weed control; seed crop desiccant.	231	Irritant
Distinct® (see diflufenzopyr)				
Discover® (see clodinafop-propargyl)				
dithiopyr, Dimension®, Scoop® (Dow Agrosciences)	3,5-pyridinedicarbothioic acid, 2-(difluoromethyl)-4-(2-methylpropyl)-6-(trifluoromethyl)-*S,S*-dimethyl ester	Experimental pre- & postemergence for turf, ornamentals, cotton, peanuts & soybeans.	>5000	>5000
diuron, Karmex®, Seduron® (Bayer CropScience, Crystal, Griffen)	3-(3,4-dichlorophenyl)-1,1-dimethylurea	Controls most germinating weeds in many crops; soil sterilant, general weed killer at high rates.	>5000	>5000
DNOC (Cerexagri)	4,6-dinitro-*o*-cresol	General weed killer; also insecticide, fungicide, defoliant, outside U.S.	25	200
Dormex® (see hydrogen cyanamide)				
Dropp® (see thidiazuron)				
Drive® (see quinclorac)				

Common name, trade name and basic manufacturer(s)	Chemical name	General use pattern	Oral LD_{50} (rat)	Dermal LD_{50} (rabbit)
DSMA (several) (Drexel, Crystal)	disodium methanearsonate	Postemergence grass weed control in cotton, noncrop areas.	1935	>2000
Dual Magnum® (see S-metolachlor)				
Dynam® (see oxasulfuron)				
Eagle® (see amidosulfuron)				
Ecopart® (see pyraflufen)				
EcoLyst® (Valent)	*N,N*-diethyl-2-(4-methylbenzyloxy)ethylamine hydrochloride	PGR for citrus.	531	>2000
Embark® (see mefluidide)				
endothall, Accelerate®, Hydrothol® Aquathol® Des-I-Cate® (Cerexagri)	7-oxabicyclo(2.2.1)heptane-2,3-dicarboxylic acid	Weed control in sugar beets, turf; aquatic herbicide; cotton defoliant; potato, seedcrop desiccant. Aquatic herbicide.	51	irritant
Endurance® (see prodiamine)				
Enfield® (see trifloxysulfuron)				
Enquik®, AMADS (discontinued)	sulfuric acid + urea (1:1)	Desiccant for dry beans, peas & potatoes.	—	Corrosive
Envoke® (see trifloxysulfuron)				
Envoy® (see clethodim)				
Eptam® (see EPTC)				
EPTC, Eptam®, Eradicane® (Syngenta, Cedar)	*S*-ethyldipropylthiocarbamate	Controls most grass weeds in several crops.	1367	10,000
Equinox® (see tepraloxydim)				
Eradicane® (see EPTC)				
Escort® (see metsulfuron-methyl)				
ethalfluralin, Sonalan®, Curbit® (Dow Agrosciences)	*N*-ethyl-*N*-(2-methyl-2-propenyl)-2,6-dinitro-4-(trifluoromethyl) benzenamine	For most annual grasses & many broadleaf weeds; preemergence, soil incorporated.	>10,000	>2000
ethametsulfuron-methyl, Muster® (DuPont)	methyl 2-[(4-ethoxy-6-methylamino-1,3,5-triazin-2-yl)carbamoylsulphamoy]benzoic acid	Selective postemergence for wild mustard, stinkweed, smartweed, clovers & other broadleaves.	5000	
ethephon, Ethrel®, Prep®, Cerone®, CottonQuik®, Super Boll® (Bayer CropScience, Griffen)	(2-chloroethyl) phosphonic acid	Plant growth regulator, produces ethylene; registered for many crops & turf. Controls mistletoe in dormant trees.	4229	5730
ethofumesate, Kemiron®, Nortron®, Prograss® (Bayer CropScience)	2-ethoxy-2,3-dihydro-3,3-dimethyl-5-benzofuranyl methanesulphonate	Preplant, pre- or postemergence on sugar beets for annual broadleaf, grass weeds.	>6400	>1440

Common name, trade name and basic manufacturer(s)	*Chemical name*	*General use pattern*	*Oral LD_{50} (rat)*	*Dermal LD_{50} (rabbit)*
Ethrel® (see ethephon)				
Everest® (see flucarbazone)				
Evik® (see ametryn)				
Evital® (see norflurazon)				
Evolus® (see azafenidin)				
Express® (see tribenuron-methyl)				
Facet® (see quinclorac)				
Factor® (see prodiamine)				
Falcon® (see propaquizafop)				
Far-Go® (see triallate)				
fatty acids (see Spectrum®)				
fenac, Fenatrol®, (discontinued)	(2,3,6-trichlorophenyl) acetic acid	Preemergence control of weeds in sugar cane, most weeds in non crop areas.	1760	>3160
fenoxaprop-ethyl, Acclaim®, Horizon®, Whip®, (Bayer CropScience)	(±)-ethyl 2-4-((6-chloro-2-benzoxazolyoxy)-phenoxy)propanoate	For grass weeds in broadleaf crops, rice, soybeans & turf.	2357	>2000
fenoxaprop-P-ethyl, Puma®, Ricestar®, Silverado®, Acclaim Extra® (Bayer CropScience)	(±)-ethyl 2-[4-[(6-chloro-2-benzoxazolyl) oxylphenoxy]propanoate	For annual, & perennial grass weeds in a wide variety of field crops, & vegs.	2090	>2000
fenuron	1,1-dimethyl-3-phenylurea	Weed brush killer. Outside U.S.	6400	—
fenuron TCA, Dozer® (discontinued)	1,1-dimethyl-3-phenylurea mono(trichloroacetate)	Nonselective weed, brush control in noncrop areas.	4000	—
Finale® (see haloxyfop-methyl)				
Finish® (see cyclanilide)				
FirstRate® (see chloransulam-methyl)				
Flex® (see fomesafen)				
Flexstar® (see fomesafen)				
florasulam, Primus® (Dow Agrosciences)	*N*-(2,6-difluorophenyl)-8-fluoro-5-methoxy [1,2,4]triazolo[1,5-c]pyrimidine-2-sulfonamide	Used to control broadleaf weeds in cereals & maize	>6000	>2000
fluazifop-butyl, Fusilade® (Syngenta)	butyl (RS)-2-[4-(5-trifluoromethyl-2-pyridol oxy)phenoxy]propinoate	Over-the-top control for grass weeds in broadleaf crops including cotton, soybeans & vegs, outside U.S.	2450	>2420

Common name, trade name and basic manufacturer(s)	Chemical name	General use pattern	Oral LD_{50} (rat)	Dermal LD_{50} (rabbit)
fluazifop-P-butyl, Fusilade DX®, Venture® (Syngenta)	butyl (R)-2-[4-[[5-trifluoromethyl)-2-pyridinyl]oxy]phenoxy]propanoate	Postemergence for perennial, annual grass weeds (over-the-top in broadleaved crops).	2712	>2420
flucarbazone-sodium, Everest® (Bayer CropScience)	1 *H*-1,2,4-triazole-carboxamide,4,5-dihydro-3-methoxy-4-methyl-5-oxo-N-((2-trifluoromethoxy)penyl)sulfonyl)-sodium salt	Controls annual grasses in small grains.	>5000	
fluchloralin, Basalin® (discontinued)	*N*-(2-chloroethyl)-2,6-dinitro-*N*-propyl-4-(trifluoromethyl)aniline	Preplant preemergence weed control in cotton, soybeans.	>6400	>10,000
flufenacet, Define® (Bayer CropScience)	*N*-(4-fluorophenyl) -4-(1-methylethyl)-2-[[5-(strifluoro-methyl)-1,3,4-thiadiazol-2-yl]oxy]acetaminde	Grass weeds in corn, soybeans, cottton, cereals, rice & potatoes.	1617	
flufenpyr-ethyl (Valent)	2-chloro-5-[1,6-dihydro-5-methyl-6-oxo-4-(trifluoromethyl)pyridazin-1-yl]-4-fluorophenoxyacetic acid	Controls broadleaf weeds in corn, soybeans & cotton.		
flumetralin, Prime+®, Flupro®, (Syngenta, Crompton)	2-chloro-*N*-[(2,6-dinitro-4-trifluoromethyl) phenyl]-*N*-ethyl-6-fluorobenzene-methanamine	Plant growth regulator for tobacco sucker control; herbicide for ornamentals.	3,100	—
flumetsulam, Broadstrike®, Python® (Dow Agrosciences)	*N*-(2,6-difluorophenyl)-5-methyl-[1,2,4]triazolo [1,5-a]pyrimidine-2-sulfonamide	Marketed only in premixes for corn & soybeans.	>5000	>2000
flumiclorac-pentyl, Resource® (Sumitomo)	pentyl 2-chloro-4-fluoro-5-(3,4,5,6-tetrahydrophthalimido)-phenoxyacetate	Experimental: Postemergence for velvetleaf, lambsquarters, jimsonweed & others in corn & soybeans.	3600	
flumioxazin, Valor®, Ganster® (Valent)	7-fluoro-6[3,4,5,6-tetrahydro)phthalimido] -4-(2-propynyl)-1,4-benzoxazin-3(2*H*)-one	Broadleaf weeds in soybeans, peanuts, rice, cereals, corn & other field crops.	5000	
fluometuron, Cotoran®, Meturon® (Syngenta)	1,1-dimethyl-3-(α,α,α-trifluoro-*m*-tolyl)urea	Pre-, postemergence control of weeds in cotton, sugarcane.	6416	>2000
flupyrsulfuron-methyl, Lexus® (DuPont)	2[[[[(4-6-dimethoxy-2-pyrimidinyl)amino] carbonyl]amino]-sulfonyl]-6-(trifluoro= methyl)-3-pyridinecarboxylic acid	Experimental: for annual broadleaf & grass weeds in cereals.		
fluridone, Avast®, Sonar® (SePRO, Griffin)	1-methyl-3-phenyl-5-[3-(trifluoromethyl)phenyl]-4(1*H*)-pyridinone	Aquatic herbicide.	>10,000	—
fluroxypyr-meptyl, Starane®, (Dow Agrosciences)	4-amino-3,5-dichloro-6-fluoro-2-pyridyloxacetic acid	Pyridyl herbicide selective & translocated, postemergence broadleaf weeds in pastures, cereals & corn.	>5000	>2000
flurprimidol, Cutless® (SePRO)	α-(1-methylethyl)-α-[4-(trifluoromethoxy)phenyl]-5-pyrimidinemethanol	Growth regulator for turf & golf courses also for ornamental trees.	709	—
fluthiacet-methyl, Action®, Appeal® (Syngenta)	[2-chloro-4-fluoro-5(5,6,7,8-tetrahydro-3-oxo-1*H*, 3*H*-[1,3,4]thiadiazolo[3,4-a]pyridazin-1-ylideneamino)phenylthio]acetic acid	Controls broadleaf weeds in corn, soybeans & cotton defoliant.	5000	
Focus® (see cycloxydim)				
Folex® (see tribufos)				
fomesafen, Reflex®, Flex® Flexstar® (Syngenta)	5-[2-chloro-4-(trifluoromethyl)phenoxy]-*N*-(methylsulfonyl)-2-nitrobenzamide	Selective postemergence for broadleaf weeds in soybeans.	1250	>1000

Common name, trade name and basic manufacturer(s)	Chemical name	General use pattern	Oral LD_{50} (rat)	Dermal LD_{50} (rabbit)
foramsulfuron, Option®, (Bayer CropScience)	2-[[[(4,6-dimethoxy-2-pyrimidinyl)amino] carbonyl]sulfony]-4-(formylamino)-*N,N*-dimethylbenzamide	For broadleaf & grass weeds in corn outside U.S.		
forchlorfenuron, Sitofex® (Tide)	1-(2-chloro-4-pyridyl)-3-phenylurea	PGR for apples & kiwi fruit.	4918	>2000
fosamine ammonium, Krenite® (DuPont)	ammonium ethyl carbamoylphosphonate	Fall-applied brush control agent on noncropland areas.	24,000	>4000
Fox® (see bifenox)				
Freefall® (see thidiazuron)				
Frontier® (see dimethenamid)				
Frostban® *(see Pseudomonas fluorescens/syringae)*				
Fruitone® (see NAA)				
furilazole (Monsanto)	3-dichloroacetyl-5-(2-furanyl)-2,2-dimethyl-oxazolidine	Herbicide safener for wide range of herbicides on cereals, corn & rice.	869	Irritant
Fusilade DX® (see fluazifop-P-butyl)				
GABA, AuxiGro® (Emerald)	γ-aminobutyric acid (natural substance)	PGR for beans, grapes, potatoes, surgarbeets, vegs & fruit & nut trees.		
Galigan® (see oxyfluorfen)				
Gallant® (see haloxyfop-methyl)				
Gallery® (see isoxaben)				
Ganster® (see flumioxazin)				
Garlon® (see triclopyr)				
gibberellic acid, Procone® & others (Valent, Nufarm)	(3S,3aR,4S,4aS,7S,9aR,9bR,12S)-7 12-di-hydroxy-3-methyl-6-methylene-2-2-oxoperhydro-4a,7-methano-9b,3-propenoazuleno [1,2-b]furan-4-carboxylic acid	PGR for a multitude of crops, especially vegs.	15,000	
Glean® (see chlorsulfuron)				
glufosinate-ammonium, Finale®, Liberty®, Rely® Remove® (Bayer CropScience)	(±)-2-amino-4-(hydroxymethylphosphinyl) butanoic acid	Contact herbicide for broadleaf & grass weeds in orchards, grapes, blueberries, canals, noncropland, ornamentals & Christmas trees.	2000	>2000
glyoxime, Pik-Off® (discontinued)	ethanedial dioxime	Growth regulator, enhances abscission of fruit (oranges & pineapple).	185	1580
glyphosate, Aquaneat®, Credit®, Glypro®, Rodeo®, Roundup®, Roundup Ultra®, Roundup Weather Max®, Protocol®, (Monsanto, Nufarm, Dow Agrosciences)	*N*-(phosphonomethyl)glycine, isopropylamine salt	Highly versatile translocated, non-selective herbicide–many uses on many crops.	5600	>5000

Common name, trade name and basic manufacturer(s)	Chemical name	General use pattern	Oral LD_{50} (rat)	Dermal LD_{50} (rabbit)
glyphosate-diammonium (see diammonium glyphosate)				
glyphosate-trimesium (sulfosate) Touchdown® (Syngenta)	*N*-phosphonomethylglycine trimethylsulfonium salt	Nonselective postemergence systemic for annual & perennial grass & broadleaf weeds.	750	>200
glyphosine, Polaris® (discontinued)	*N,N*-bis(phosphonomethyl)glycine	Chemical ripener for sugarcane.	3925	—
Goal® (see oxyfluorfen)				
Grammoxone® (see paraquat)				
Grandstand® (see triclopyr)				
Grasp® (see tralkoxydim)				
Gulliver® (see azimsulfuron)				
halosulfuron-methyl, Sandea®, Sempra® Manage®, Permit® (Monsanto, Gowan)	methyl 5-[[(4,6-dimethoxy-2-pyrimidinyl) amino]carbonylaminosulfony]-3-chloro-1-methyl-1*H*-pyrazole-4-carboxylate	Pre- & postemergence for annual broadleaves, nutsedge in corn, sugarcane, turf, alfalfa & vegs.	1287	>5000
haloxyfop-methyl, Gallant® (Dow Agrosciences)	methyl 2-(4-(((3-chloro-5-trifluoromethyl)-2-pyridinyl)oxy)phenoxy)propanoate	Selective postemergence for grass weeds in soybeans.	518	>5000
Harmony® (see thifensulfuron-methyl)				
Harness® (see acetochlor)				
Harvade® (see dimethipin)				
hexazinone, Velpar® (DuPont)	3-cyclohexyl-6-(dimethylamino)-1-methyl-1,3,5-triazine-2,4-(1*H*, 3*H*)-dione	Contact & residual control of weeds, also forestry on noncropland areas.	1690	>5278
Hoelon® (see diclofop-methyl)				
Homerun® (see oxaziclomefone)				
Horizon® (see fenoxaprop-ethyl, see also clodinofop-propargyl)				
Husar® (see iodosulfuron-methyl)				
hydrogen cyanamide, Dormex® (Degussa-Huls)		Growth regulator for peaches, nectarines & blueberries.		
Hydrothol® (see endothall)				
Hyvar® (see bromacil)				
Illoxan® (see diclofop-methyl)				
Image® (see imazaquin)				

Common name, trade name and basic manufacturer(s)	Chemical name	General use pattern	Oral LD_{50} (rat)	Dermal LD_{50} (rabbit)
imazamethabenz-methyl Assert®, Dagger® (BASF)	*m*-toluic acid, 6-(4-isopropyl-4-methyl-5-oxo-2-imidazolin-2-yl)-, methyl ester and *p*-toluic acid, 2-(4-isopropyl-4-methyl-5-oxo-2imidazolin-2-yl)-, methyl ester (a mix of the two methyl esters)	Selective postemergence for sunflowers, wild oats, mustards & buckwheat in small grains.	>5000	>2000
imazamox, Beyond®, Raptor® (BASF)	2-(4-dihydro-4-methyl-4-(1methylethyl)-5-oxo-1 *H*-imidazol-2-yl)-5-(methoxymethyl) -3-pyridinecarboxylic acid	Postemergence for most broadleaves in alfalfa, soybeans, IMI corn, peanuts, peas, dry beans, sunflower & rice.	>5000	>4000
imazapic, Cadre®, Plateau® (BASF)	2-(4-isopropyl-4-methyl-5-oxo-2-imidazolin-2-yl) -5-methylnicotinic acid	Pre- & postemergence broadleafs, johnsongrass, nutsedge in soybeans, sugarcane, peanuts & noncropland.	>5000	>5000
imazapyr, Arsenal®, Chopper® (BASF)	2-(4,5-dihydro-4-methyl-4-(1-methylethyl)-5-oxo-1*H*-imidazol-2-yl)-3-pyridinecarb oxylic acid with 2-propanamine (1:1)	Nonselective broad-spectrum compound with long residual activity.	>5000	>2000
imazaquin, Image®, Scepter® (BASF)	2-[4,5-dihydro-4-methyl-4-(1-methylethyl)-*S*-oxo-1*H*-imidazol-2-yl]-3-quinoline carboxylic acid	Selective pre- & postemergence for grass & broadleaf weeds in broadleaf crops.	>5000	>2000
imazethapyr, Newpath®, Pursuit® (BASF)	2-[4,5-dihydro-4-methyl-4-(1-methylethyl)-5-oxo-1*H*-imidazol-2-yl]-5- ethyl-3-pyridinecarboxylic acid	Preemergence, preplant incorporated, & post emergence for grass, broadleaf weeds in soybeans, rice & other legumes.	>5000	>2000
Inspire® (see butafenacil)				
iodosulfuron-methyl-sodium, Hussar® (Bayer CropScience)	methyl 4-iodo-2-[3-(4-methoxy-6-methyl-1,3,5,-triazin-2-yl)ureidosulfonyl]benzoate, sodium salt	Controls grass, broadleaf weeds in wheat, rye & preemerge on corn.	2678	>2000
IPC (see propham)				
isopropalin, Paarlan® (discontinued)	2,6-dinitro-*N,N*-dipropylcumidine	Preemergence control of weeds in transplant tobacco.	>5000	>2000
isoproturon, Arelon®, Strong® (Bayer CropScience)	3-(4-isopropylphenyl)1,1-dimethylurea	Most grass weeds & many annual broadleaf weeds in cereals outside U.S.	1826	<2000
isoxaben, Gallery® (Dow Agrosciences)	*N*-[3-(1-ethyl-1-methylpropyl)-5-isoxazolyl]-2,6-dimethoxy-benzamide	Preemergence for broadleaf weeds in ornamentals & turf.	>10,000	>2000
isoxadifen-ethyl (Bayer CropScience)	5-cyclopropyl-1,2-oxazol-4-yl α,α,α-trifluoro-2mesyl-p-tolyl ketone	Herbicide safener for corn.		
isoxaflutole, Balance®, (Bayer CropScience)	2-(4-chlorophenyl)-1,4-dihydro-6-methyl-4-oxonicotinic acid	Grass & broadleaves in field corn but not silage corn.	4000	
karetazan, Detasslor® (Syngenta)		For detassling hybrid corn.		
Karmex® (see diuron)				
Kemiron® (see ethofumesate)				
Kerb® (see pronamite)				

Common name, trade name and basic manufacturer(s)	Chemical name	General use pattern	Oral LD_{50} (rat)	Dermal LD_{50} (rabbit)
Krenite® (see fosamine-ammonium)				
lactic acid, Propel® (Entek)		Growth regulator for tree fruits, nuts, citrus, vegs, cotton, peanuts & many other crops.		
lactofen, Cobra®, Phoenix® (Valent)	1'-(carboethoxy) ethyl 5-[2-chloro-4-(trifluoromethyl) phenoxyl]-2-nitrobenzoate	Selective pre- & postemergence for broadleaf weeds in cereals, corn, rice, soybeans.	>5000	>2000
Laser® (see cycloxydim)				
Lasso® (see alachlor)				
lenacil, Lenazar® (Hermoo Belgium)	3-cyclohexyl-6,7-dihydro-1*H*-cyclopenta-pyrimidine-2,4(3*H*,5*H*)-dione	Weed control in sugar beets, cereal grains, outside U.S.	>11,000	>5000
Lexus® (see flupysulfuron-methyl)				
Liberty® (see glufosinate-ammonium)				
Linex® (see linuron)				
Lint Plus® (see dimethipin)				
linuron, Linex®, Lorox®, Afalon® (Bayer CropScience, Griffin)	3-(3,4-dichlophenyl)-1-methoxy-1-methylurea	Selective weed control in corn, sorghum, cotton, soybeans, wheat; annual weeds in noncrop areas.	4000	>5000
Logran® (see triasulfuron)				
Londax® (see bensulfuron-methyl)				
Lontrel® (see clopyralid)				
Lorox® (see linuron)				
Lotus® (see cinidon-ethyl)				
LPE, LPE-94® (Natra-Park)	Lysophosphatidylethanolamine (phospholipid)	PGR for apples, grapes, tomatoes, cranberrries, strawberries & cut flowers		
MAA (discontinued)	methanearsonic acid	Grass weeds in cotton, noncrop areas.	1300	—
Magnacide H® (see acrolein)				
Maintain® CF (see chlorfurenol-methyl)				
maleic hydrazide, (see MH) Royal MH-30®, Retard®, (Crompton, Drexel)		Plant growth regulator inhibits potato & onion sprouting.		
Mallet® (see *Collectotrichum* spp.)				

Common name, trade name and basic manufacturer(s)	Chemical name	General use pattern	Oral LD_{50} (rat)	Dermal LD_{50} (rabbit)
MAMA (discontinued)	monoammonium methanearsonate	Dallisgrass, nutgrass in turf.	750	—
Manage® (see halosulfuron)				
Mandate® (see thiazopyr)				
Matador® (see quizalofop-P-ethyl)				
Matran® (see clove oil)				
Matrix® (see rimsulfuron)				
Maverick® (see sulfosulfuron)				
Max® (see paraquat)				
MCPA, (several) (BASF, Crystal, Dow Agrosciences)	2'-chloro-2-(4-chloro-o-tolyloxy) acetanilide	Translocated herbicide used in small grains, rice for postemergence control of wide weed spectrum.	1160	>4000
MCPB, (several) (Nufarm)	4-(4-chloro-2-methylphenoxy)butanoic acid	Weed control in peas.	680	—
MCPP (see mecoprop)				
mecoprop, MCPP (several) (BASF, Nufarm)	2-(4-chloro-*o*-tolyl)oxypropionic acid	Turf, cereal herbicide for broadleaf weeds.	650	>4000
mefenpyr-diethyl (Bayer CropScience)	diethyl (RS)-1-(2,4-dichlorophenyl)-5-methyl-2-pyrazoline-3,5-dicarboxylate	Herbicide safener for fenoxaprop-P-ethyl	>5000	>4000
mefenpyr-methyl (Bayer CropScience)		Herbicide safener		
mefluidide, Embark® (PBI/Gordon)	*N*-[2,4-dimethyl-5-[[(trifluoromethyl)-sulfonyl]amino]phenyl]acetamide	Plant growth regulator to prevent seedhead formation in turf grasses.	>4000	>4000
Mepex®, (see mepiquat chloride)				
mepiquat chloride, Pix®, Mepichlor®, Mepex®, Topit® (BASF)	1,1-dimethyl-piperidiniumchloride	Cotton growth regulator to reduce vegetative growth; enhance grape set.	>7800	—
mepiquat pentaborate, Pix Penta® (BASF)	1,1-dimethyl-piperidinuum pentaborate	Concentrated form of cotton growth regulator.		
Merit® (see clomozone)				
Mesomaxx® (see mesosulfuron-methyl)				
merphos, (see tribufos)				
mesosulfuron-methyl, Mesomaxx® (Bayer CropScience)	2-[[[(4,6-dimethoxy-2-pyrimidinyl)amino]carbonyl] sulfonyl]-4-(formylammino)-*N,N* dimethylbenzamide	Experimentally for many broadleaf & grass weeds in cereals, outside the U.S.		
mesotrione, Callisto®, (Syngenta)	2-(4-mesyl-2-nitrobenzoyl)cyclohexane-1,3-dione	Controls broadleaf weeds and crabgrass in corn outside U.S.	>5000	>2000

Common name, trade name and basic manufacturer(s)	Chemical name	General use pattern	Oral LD_{50} (rat)	Dermal LD_{50} (rabbit)
metam-sodium, Busan® 1236, Vapam®, SMDC, Trimaton®, Turfcure® (Amvac, Cerexagri)	sodium methyldithiocarbamate	Soil fumigant for weeds, weed seeds, nematodes, fungi.	1800	1300
metamifop, DBH-129 (Dongbu Hannong)	(*R*)-2[4-(6-chloro-1,3-benzoxazol-2-yloxy]-2'-fluoro-*N*-methylpropionanilide	Herbicide. Not reg. in U.S.		
metham (see metam-sodium)				
methazole, Probe® (discontinued)	2-(3,4-dichlorophenyl)-4-methyl-1,2,4-oxadiazolidine-3,5-dione	Pre-, postemergence herbicide for cotton.	2500	>12,500
metobromuron, Patoran® (discontinued)	3-(4-bromophenyl)-1-methoxy-1-methylurea	Selective preemergence for annual grass & broadleaf weeds in broadleaf crops.	2600	irritant
S-metolachlor, Cinch®, Dual Magnum®, Pennant Magnum® (Syngenta, DuPont)	mixture of (*S*)-2-chloro-*N*-(2-ethyl-6-methylphenyl)-*N*-(2-methoxy-1-methylethyl) acetamide & (*R*)-2-chloro-*N*-(2-ethyl-6-methylphenyl)-*N*-(2-methoxy-1-methylethyl)acetamide in 80-100% to 20-0% proportion	Preemergence, preplant in corn, peanuts, soybeans, others.	2672	>2000
metribuzin, Contrast®, Sencor® (Bayer CropScience)	4-amino-6-*tert*-butyl-3-(methylthio)-1,2,4-triazin-5(4*H*)one	Controls wide spectrum of weeds in several crops.	2200	>20,000
metsulfuron-methyl, Ally®, Blade®, Escort®, Valuron® (DuPont)	2-[[[[(4-methoxy-6-methyl-1,3,5-triazin-2-yl) amino]carbonyl]amino]sulfonyl]benzoic acid	Selective pre- & postemergence for grass & broadleaf weeds in cereals, turf & pastures.	>5000	>2000
Meturon® (see fluometuron)				
MH (maleic hydrazide) (several)	1,2-dihydro-3,6-pyridazinedione	Growth retardant in trees, shrubs, grasses; sprout inhibitor for stored onions, potatoes.	>5000	>2000
Micro-Tech® (see alachlor)				
Milestone® (see azafenidin)				
Modown® (see bifenox)				
molinate, Ordram® (Syngenta)	*S*-ethyl hexahydro-1*H*-azepine-1-carbothioate	Watergrass control in rice.	369	>4640
Montar® (see cacodylic acid)				
Monument® (see trifloxysulfuron)				
monuron (discontinued)	3-(*p*-chlorophenyl)-1,1-dimethylurea	Complete weed control in noncropland areas.	3600	—
monuron TCA, Urox® (discontinued)	3-(*p*-chlorophenyl)-1,1-dimethylurea mono(trichloroacetate)	Complete weed control in noncropland areas.	2300	irritant
M-Pede® (see soaps herbicidal)				
MSMA, Ansar®, Bueno® (Drexel, KMG)	monosodium methanearsonate	Pre- & postemergence for cotton, grass weeds in turf, noncropland.	700	2500

Common name, trade name and basic manufacturer(s)	Chemical name	General use pattern	Oral LD_{50} (rat)	Dermal LD_{50} (rabbit)
Muster® (see ethametsulfuron-methyl)				
NAA, Fruitone® & others (Amvac, Platte)	1-naphthaleneacetic acid	PGR fruit thinner for apples & other fruit, prevents sprouting of root crops & fruit on ornamental trees.	>1000	Irritant
N-Serve® (see nitrapyrin)				
napropamide, Devrinol® (United Phosphorus)	2-(α-naphthoxy)-*N,N*-diethylpropionamide	Selective weed control in orchards, vineyards, several vegs & ornamentals.	>5000	>4640
naptalam, Alanap® (Crompton)	*N*-1-naphthylphthalamic acid	Broadleaf weed control in soybeans, peanuts, vine crops.	8200	>2000
neburon, Propuron® (Bayer CropScience)	1-butyl-3-(3,4-dichloropheny)-1-methylurea	Selective weed control in woody ornamental nurseries.	>11,000	—
Newpath® (see imazethapyr)				
New Touchdown® (see diammonium glyphosate)				
nicosulfuron, Accent®, (DuPont)	3-pyrimidinecarboxamide,2-[[(4,6-dimethoxy=pyrimidin-2-yl)amino-carbonyl]aminosulfonyl]-*N,N*-dimethyl	Selective postemergence Sulfonylurea herbicide.	>5000	>2000
nitralin, Planavan® (discontinued)	4-(methylsulfonyl)-2,6-dinitro-*N,N*-dipropylaniline	Weed control in cotton, soybeans, peanuts, alfalfa & vegs, outside U.S.	>2000	>2000
nitrapyrin, N-Serve® (Dow Agrosciences)	2-chloro-6-(trichloromethyl)pyridine	Inhibits nitrification in soil by *Nitrosomonas* bacteria.	1072	2830
nitrofen, Tok® (discontinued)	2,4-dichlorophenyl-*p*-nitrophenyl ether	Pre- & postemergence control of broad weed spectrum in vegs, sugar beets, rice & ornamentals.	2630	—
norflurazon, Solicam®, Evital®, Zorial Rapid®, Predict® (Syngenta)	4-chloro-5-(methylamino)-2-(α,α,α-trifluro-*m*-toyl)-3(2*H*)-pyridazinone	Preemergence control of weeds in cotton, cranberries, fruit, nut trees.	>8000	>20,000
Nortron® (see ethofumesate)				
Olympus® (see propoxycarbazone)				
Option® (see foramsulfuron)				
Orbit® (see cinidon-ethyl)				
Ordram® (see molinate)				
Organic Interceptor® (see pine extracts)				
Ornamec® (see fluazifop-P-butyl)				

Common name, trade name and basic manufacturer(s)	Chemical name	General use pattern	Oral LD_{50} (rat)	Dermal LD_{50} (rabbit)
oryzalin, Surflan® (Dow Agrosciences)	3,5-dinitro-*N⁴*,*N⁴*-dipropylsulfanilamide	Preemergence control of many weeds in soybeans, cotton, fruit, nut trees, ornamentals.	>10,000	>2000
Oust® (see sulfometuron-methyl)				
Outlook® (see dimethenamid-P)				
Outrider® (see sulfosulfuron)				
oxadiargyl, Topstar® (Bayer CropScience)	3-[2,4-dichloro-5-(2-propynyloxy)phenyl]-5-(1,1-dimethylethyl)-1,3,4-oxadiazol-2(3*H*)-one	Controls broadleaf & grassy weeds in citrus, potatoes, rice, sugarcane, sunflowers & vegs. Not reg. in U.S.	>5000	>2000
oxadiazon, Ronstar® (Bayer CropScience)	2-*tert*-butyl-4-(2,4-dichloro-5-isopropoxyphenyl)-Δ2-1,3,4-oxadiazolin-5-one	Pre-, postemergence weed control in turf & ornamentals.	>8000	>8000
oxasulfuron, Dynam®, (Syngenta)	3-(4,6-dimethyl-pyrimidin-2-yl)-1-[2-oxetan-3-yl-oxy-carbonyl)-phenylsulfonyl]urea	Broadleaf weeds in soybeans.	5000	
oxaziclomefone, Homerun®, (Bayer CropScience)	3-[1-(3,5-dichlorophenyl)-1-methylethyl]-3,4-dihydro-6-methyl-5-phenyl-2*H*-1,3-oxazin-4-one	Herbicide for rice. Not reg, in U.S.	>5000	>2000
oxyfluorfen, Galigan®, Goal® (Dow Agrosciences)	2-chloro-1-(3-ethoxy-4-nitrophenoxy)-4-(trifluoromethyl)benzene	Pre-, postemergence control of broad weed spectrum in corn, cotton, soybeans, fruit, nut trees, ornamentals.	>5000	>10,000
paclobutrazol, Bonzi®, Cultar®, Piccolo®, Trimmit® (Syngenta)	(2RS,3RD)-1-(4-chlorophenyl)-4,4-dimethyl-2-(1H-1,2,4-triazol-1-yl)pentan-3-ol	Gibberellin inhibitor growth regulator for ornamentals & trees.	2000	>1000
Paecilomyces lilacinus, BioAct PL® (Prophyta)		Bionematicide for tobacco, vegs & citrus.	Non-Toxic	
Palisade® (see trinexapec)				
Paramount® (see quinclorac)				
paraquat, Gramoxone®Max, Starfire® (Syngenta)	1,1'-dimethyl-4,4'bipyridilium ion	Contact herbicide & desiccant with broad scope of uses.	150	236
Pardner® (see bromoxynil)				
Pathfinder®II (see triclopyr)				
PCP, penta (discontinued)	pentachlorophenol	Preharvest defoliant, wood preservative, molluscicide.	50	Irritant
Peak® (see prosulfuron)				
PEBC (see pebulate)				
pebulate, PEBC, Tillam® (Cedar)	*S*-propyl butylethylthiocarbamate	Preplant control of grass, weeds in sugar beets, tobacco, tomatoes.	1120	4640
pelargonic acid, Scythe® (Mycogen)		Blossom thinner for apples & pears.		

Common name, trade name and basic manufacturer(s)	*Chemical name*	*General use pattern*	*Oral LD_{50} (rat)*	*Dermal LD_{50} (rabbit)*
pendimethalin, Prowl®, Stomp®, Pendulum® (BASF)	*N*-(1-ethylpropyl)-3,4-dimethyl-2,6 -dinitrobenzenamine	Pre-, postemergence use in corn; preplant for cotton, soybeans, tobacco.	1050	>5000
Pendulum® (see pendimethalin)				
Pennant Magnum® (see S-metolachlor)				
Perlan® (see BAP)				
perfluidone, Destun® (discontinued)	1,1,1-trifluoro-*N*-[2-methyl-4-(phenylsulfonyl)-phenyl]methanesulfonamide	Preemergence weed control in cotton.	633	>4000
Permit® (see halosulfuron methyl)				
phenmedipham, Betanal®, Spin-Aid® (Bayer CropScience)	methyl-*m*-hydroxycarbanilate-methylcarbanilate	Broadleaf weed control in sugar beets, table beets.	>8000	>4000
Phoenix® (see lactofen)				
Phytophthora palmivora DeVine® (Encore)	spores of a naturally occuring fungus	Mycoherbicide: Milkweed vine control in citrus groves.	non-toxic	
Piccolo® (see paclobutrazol)				
picloram, Tordon® (Dow Agrosciences)	4-amino-3,5,6-trichloropicolinic acid	Brush control on industrial sites, pastures, rangeland.	8200	>4000
picolinafen, Pico® (BASF)	*N*-(4-fluorophenyl)-6-[3-(trifluoromentyl]phenoxy]-2-pyridine carboxamide	Post emergence, controls broadleaf weeds in barley & wheat	>5000	>4000
Pix® (see mepiquat chloride)				
Pix Penta® (see mepiquat pentaborate)				
Plateau® (see imazapic)				
Poast® (see sethoxydim)				
Pramitol® (see prometon)				
Predict® (see norflurazon)				
Prefar® (see bensulide)				
Prime+® (see flumetralin)				
Prep® (see ethephon)				
Primatol® (see propazine)				
primisulfuron-methyl, Beacon®, Tell® (Syngenta)	3-[4,6-bis(difluoromethoxy)-pyrimidin-2-yl]-1-(2-methoxy-carbonyl phenylsulfonyl)-urea	Selective postemergence Sulfonylurea herbicide.	>5050	>2010

Common name, trade name and basic manufacturer(s)	Chemical name	General use pattern	Oral LD_{50} (rat)	Dermal LD_{50} (rabbit)
Primus® (see florasulam)				
Princep® (see simazine)				
Prism® (see clethodim)				
Procone® (see gibberellic acid)				
prodiamine, Factor®, Barricade® (Syngenta)	*N*3,*N*3-di-*n*-propyl-2,4-dinitro-6-(trifluoromethyl)-*m*-phenylenediamine	Selective preplant & preemergence for annual broadleaf & perennial grass weeds.	>15,000	—
profluralin, Tolban® (discontinued)	*N*-(cyclopropylmethyl)-α,α,α-trifluoro-2,6-dinitro-*N*-propyl-*p*-toluidine	Controls most grass, broad-leaf weeds in cotton, soybeans, sunflower & certain vegs.	10,000	>3170
Prograss® (see ethofumesate)				
prohexadione-calcium, Apogee®, Baseline® (BASF)		Growth regulator for shaping apple tree canopy, peanuts, rice, vegs & fruit.		
Promalin® (See BAP)				
prometon, Pramitol®, (Syngenta)	2,4-bis(isopropylamino)-6-methoxy-*s*-triazine	Nonselective pre-, postemergence control for most weeds in noncropland.	2980	>2000
prometryn, Caparol®, Cotton-Pro® (Syngenta)	2,4-bis(isopropylamino)-6-(methylthio)-*s*-triazine	Versatile pre-, postemergence cotton herbicide.	5233	>3100
pronamide, Kerb® (Dow Agrosciences)	3,5-dichloro(*N*-1,1-dimethyl-2-propynyl)benzamide	Pre- & postemergence weed control in legumes; also used on turf, woody ornamentals.	8350	>3160
propachlor, Prolex® (Makhteshim-Agan)	2-chloro-*N*-isopropylacetanilide	Controls most annual grass, broadleaf weeds in corn, sorghum, soybeans.	1800	>20,000
propanil, Stampede® (several) (Dow Agrosciences, Crystal)	3'-4'-dichloropropionalide	Postemergence control of grasses, certain other weeds in rice, wheat.	2000	>3900
propaquizafop, Agil®, Shogun®, Falcon® (Makhteshim-Agan)	2-isopropylidineamino-oxyethyl (R)-2-(4-(6-chloroquinoxalin-2-yloxy)phenoxy) propionate	Controls grass weeds in vegs & field crops outside U.S.	3009	>2000
propazine, Prozinex®, Primatol-P® (Griffin)	2-chloro-4,6-bis(isopropylamino)-*s*-triazine	Preemergence control of most annual weeds in milo, sorghum.	>7700	>3100
Propel® (see lactic acid)				
propham, Tuberite® (Schirm)	isopropyl carbanilate	Pre- & postemergence weed control in forage legumes, sugar beets & vegs.	5000	6800
propoxycarbazone-sodium, Attribute®, Olympus® (Bayer CropScience)	methyl 2-({[(4-methyl-5-oxo-3-pro=poxy-4,5-dihydro-1 *H*-1,2,4-triazol-1-yl)carbonyl] amino)sulfonyl)benzoate sodium salt	Annual & perennial grass weeds & some broadleaf weeds in wheat & rye.	5000	
Propuron® (see neburon)				

Common name, trade name and basic manufacturer(s)	*Chemical name*	*General use pattern*	*Oral LD_{50} (rat)*	*Dermal LD_{50} (rabbit)*
prosulfuron, Peak® (Syngenta)	1-(4-methoxy-6-methyl-1,3,5-triazin-2-yl)-3-[2-(3,3,3-trifluoropropyl) phenylsulfonyl]urea	Selective postemergence for grass & broadleaf weeds in corn & small grains.	986	>2000
Protocol® (see glyphosate)				
Prowl® (see pendimethalin)				
Prozinex® (see Propazine)				
Pseudomonas fluorescens/ syringae, Frostban® (Frost Technol.)		Biological frost damage inhibitor.	Non-toxic	
Puccinia canaliculata, Dr. Biosedge® (Tifton Innovation)		Mycoherbicide for control of yellow nutsedge.	—	—
Puccinia thalaspeos, st. Woad Warrior® (Greenville Farms)	bioherbicide; Dyers Woad rust	Registered for range grass		
Puma® (see fenoxaprop-P-ethyl)				
Pursuit® (see imazethapyr)				
pyraflufen-ethyl, Ecopart® (Nihon)	ethyl 2-chloro-5-(4-chloro-5-difluroromethoxy-1-methylprazol-3-yl)-4-flurorophenoxyacetate	Controls broadleaf weeds in cereals, potatoes & soybeans.	5000	>2000
Pyramin® (see chloridazon)				
pyrazon (see chloridazon)				
pyridate, Tough® (Syngenta)	*O*-(6-chloro-3-phenyl-4-pyridazinyl)-*S*-octyl-carbonothioate	Controls broadleaf weeds in corn, wheat & cole crops.	2000	>34,000
pyriftalid, Apiro Ace® (Syngenta)	7-[(4,6-dimethoxy-2-pyrimidinyl)thio-]-3-methyl-1(3*H*)-isobenzofuranone	Controls monocots, particularly barnyardgrass, in rice. Not reg. in U.S.	>5000	>2000
pyrithiobac-sodium, Staple® (DuPont)	sodium 2-chloro-6-[(4,6-dimethoxy= phrimidin-2-yl)thio]benzoate	Postemergence for broadleaf weeds, nutsedge & nightshade, applied over-the-top to cotton.	4000	>2000
Python® (see flumetsulam)				
Quick Pick® (see cacodylic acid)				
Quilan® (see benefin)				
quinclorac, Facet®, Accord®, Drive®, Paramount® (BASF)	3,7-Dichloro-8-quinolinecarboxylic acid	Rice herbicide for grasses & broadleaf weeds, in barley, cranberries. & pastures.	4120	>2000
quizalofop-P-ethyl, Assure II®, Co-Pilot®, Matador®, Targa® (Nissan, FMC)	ethyl (R)-2-[4-[(6-chloro-quinoxalin-2-yl)oxy]-phenoxy]propionate	For annual & perennial grass weeds in soybeans, cotton & other field crops.	1480	
quizalofop-ethyl, Assure®, Pilot® (Nissan)	2-[4-[(6-chloro-2-quinoxalinyl)oxy]-phenoxy]-propionic acid, ethyl ester	Postemergence grass weed control in soybeans & cotton, experimental.	1480	10,000
Raptor® (see imazamox)				
Rebin® (see butafenacil)				

Common name, trade name and basic manufacturer(s)	Chemical name	General use pattern	Oral LD_{50} (rat)	Dermal LD_{50} (rabbit)
Reclaim® (see clopyralid)				
Reflex® (see fomesafen)				
Regiment® (see bispyribac-sodium)				
Reglone® (see diquat)				
Remove® (see glufosinate-ammonium)				
Resource® (see flumiclorac-pentyl)				
ReTain® (Valent)	[S]-trans-2-amino-4-(2-aminoethoxy)-3-butenoic acid hydrochloride	PGR to reduce fruit drop & improve quality of apples.		
Reward® (see diquat)				
Ricestar® (see fenoxaprop-P-ethyl)				
rimsulfuron, Matrix®, TranXit®, (DuPont, Griffin)	4-((4,6-dimethoxypyrimidin)-2-yl)amino carboryl)-3-(ethylsulfonyl)-2-pyridine sulfonamide	Postemergence for field corn, potatoes, tomatoes, especially for nightshade in vegs. Turf to remove *Poa annua* & ryegrass.	>5000	>2000
Rodeo® (see glyphosate)				
Ro-Neet® (see cycloate)				
Ronstar® (see oxadiazon)				
Roundup® (see glyphosate)				
Roundup Ultra® (see glyphosate)				
Roundup Ultra Max® (see glyphosate)				
Roundup Weather Max® (see glyphosate)				
Safer® Moss/Algicide (see soaps, herbicidal)				
Sandea® (see halosulfuron-methyl)				
Saturn® (see thiobencarb)				
Scepter® (see imazaquin)				
Scoop® (see dithiopyr)				
Scythe® (see pelargonic acid & Soaps herbicidal)				
Seduron® (see diuron)				

Common name, trade name and basic manufacturer(s)	*Chemical name*	*General use pattern*	*Oral LD_{50} (rat)*	*Dermal LD_{50} (rabbit)*
Select® (see clethodim)				
Sempra® (see halosulfuron-methyl)				
Sencor® (see metribuzin)				
sethoxydim, Poast®, Vantage® (BASF)	2-[1-(ethoxyimino)butyl]-5-[2-(ethylthio)propyl]-3-hydroxy-2-cyclohexen-1-one	Systemic postemergence control of grasses in soybeans, cotton, sweet corn & many other crops.	3,200	>5,000
Shark® (see carfentrazone)				
Shogun® (see propaquizafop)				
siduron, Tupersan® (Gowan)	1-(2-methylcyclohexyl)-3-phenylurea	Weed grass control in bluegrass, rye-grass, other lawn grasses.	>7500	5500
Silgard® (see silver thiosulfate)				
Silverado® (see fenoxaprop-P-ethyl)				
silverthiosulfate, Silgard® (Gard)	silver thiosulfate	PGR protects cut flowers from ethylene effects.		
silvex (discontinued)	2-(2,4,5-trichlorophenoxy)propionic acid	For control of woody plants, broadleaf herbaceous weeds, aquatic weeds.	375	>3940
simazine, Princep®, Caliber® (Syngenta)	2-chloro-4,6-bis(ethylamino)-*s*-triazine	Most annual grasses, broadleaf weeds in corn, tree fruits, nuts, ornamentals, turf, aquatic weed & algae control.	>5000	>3100
Sinbar® (see terbacil)				
Smart Fresh®	1-methyl cyclopropene	PGR for apples, melons, tomatoes & avocados in storage.		
Smolder® (see *Alternaria destruens)*				
soaps, herbicidal, Safer®, Moss/Algicide, Scythe® (Dow)	potassium salts of fatty acids	Controls weed, moss in lawns, algae, lichens.	Low	
sodium arsenite (discontinued)	sodium aresenite	Weed control in industrial areas; tree, stump destruction.	10	—
sodium chlorate, Defol®, (Drexel)	sodium chlorate	Soil sterilant herbicide, defoliant, desiccant, harvest aid.	4950	—
Solicam® (see norflurazon)				
Sonalan® (see ethalfluralin)				
Sonar® (see fluridone)				
Spartan® (see sulfentrazone)				
Spectrum® (Coastal)	fatty acids & emulsifiers	Crop oil with spreading & wetting qualities.	Low	

Common name, trade name and basic manufacturer(s)	*Chemical name*	*General use pattern*	*Oral LD_{50} (rat)*	*Dermal LD_{50} (rabbit)*
Spike® (see tebuthiuron)				
Spin-Aid® (see phenmedipham)				
Sprout Nip® (see chlorpropham)				
Stampede® (see propanil)				
Staple® (see pyrithiobac)				
Starane® (see fluroxypyr)				
Starfire® (see paraquat)				
Status® (see acifluorfen)				
Stinger® (see clopyralid)				
Stomp® (see pendimethalin)				
Strong® (see isoproturon)				
Strongarm® (see diclosulam)				
sulfentrazone, Authority®, Spartan® (FMC, DuPont)	*N*-[2,4-dichloro-5-[4-(difluoromethyl)-4,5-dihydro-3-methyl-5-oxo-1*H*1,2,4-triazol-1-yl] phenyl]methanesulfonamide	For broadleaf, grass weeds in soybeans, sugarcane, tobacco, potatoes & turf.	2855	>2000
sulfometuron-methyl, Oust® (DuPont)	methyl 2[[[[(4,6-dimethyl-2-pyrimidinyl)amino]-carbonyl]amino]sulfonyl]benzoate	Controls bermudagrass & johnsongrass, broadleaf weeds in noncroplands.	>5000	irritant
sulfosate (see glyphosate-trimesium)				
sulfosulfuron, Maverick®, Outrider® (Monsanto)	1-(2-ethylsulfonylimidazol[1,2-a]pyridin 3-yl-sulfonyl)-3-(4,6-dimethoxyprimidin-2-yl)urea	Controls grass, broadleaf weeds in cereals, roadside & industrial sites.	5000	—
Sumagic® (see uniconazole)				
Sunburst® (see cytokinins)				
Super Boll® (see ethephon)				
Surestem® (see cyclohexane cartoxamide)				
Surflan® (see oryzalin)				
Surpass® (see acetochlor)				
Sutan® (see butylate)				
2,4,5-T (discontinued)	(2,4,5-trichlorophenoxy)acetic acid	Woody plant control on industrials sites, rangeland. Amine form used on rice, not used in U.S.	500	—
Tamex® (see butralin)				
2,3,6-TBA (discontinued)	2,3,6-trichlorobenzoic acid	Control of deep-rooted perennials.	1500	>1000

Common name, trade name and basic manufacturer(s)	Chemical name	General use pattern	Oral LD_{50} (rat)	Dermal LD_{50} (rabbit)
TCA (discontinued)	trichloroacetic acid	Soil sterilant for perennial weed grass control in noncropland.	3200	—
tebuthiuron, Spike® (Dow Agrosciences)	*N*-[5-(1,1-dimethylethyl)-1,3,4-thiadiazol-2-yl]*N,N'*-dimethylurea	Total vegetation control in noncropland areas, rangeland.	644	>200
Telar® (see chlorsulfuron)				
Tell® (see primsulfuron methyl)				
tepraloxydim, Aramo®, Equinox® (BASF)	(EZ)-(RS)-2-(1-[(2E)-3-chloroallyl=oxyimino] propyl)-3-hydroxy-5-per-hydropyran-4-ylcyclohex-2-en-1-one	Controls various grass weeds in broadleaf crops, cotton, onions, soybeans & sugar beets.	5000	—
terbacil, Sinbar® (DuPont)	3-*tert*-butyl-5-chl;oro-6-methyluracil	Selective control of annual, some perennial weeds in a number of crops.	>5000	>5000
terbucarb, terbutol, Azak® (discontinued)	2,6-di-*tert*-butyl-*p*-tolyl methyl carbamate	Selective preemergence herbicide.	>34,000	—
terbutryn, Igran® (discontinued)	2-(*tert*-butylamino)-4-(ethylamino)-6-(methylthio)-*s*-triazine	Preemergence weed control in grain sorghum; postemergence control in winter wheat, barley.	2100	>2000
thiazopyr, Visor®, Mandate® (Dow Agrosciences)	methyl 2-difluoromethyl-4-isobutyl-5-(4,5-dihydro-2-thiazolyl)-6-trifluoromethyl-3-pyridinecaboxylate	Selective for annual grass weeds, in citrus, pome fruits, nut crops & alfalfa.	>5000	>5000
thidiazuron, Dropp® Free Fall® (Griffin, Bayer CropScience)	*N*-phenyl-*N'*-(1,2,3-thiadiazol-5yl) urea	Defoliant for cotton.	>4000	>1000
thifensulfuron-methyl, Harmony®, (DuPont)	methyl 3-[[[4-methoxy-6-methyl-1,3,5-triazin -2-yl) amino-carbonyl]amino]sulfonyl]-2-thiophenecarboxylate	Selective postemergence for broadleaf weeds in corn, soybeans & safflower.	>5000	—
thiobencarb, Bolero®, Abolish®, Saturn® (Valent)	*S*-(4-chlorophenyl)methyl diethyl-carbamothioate	Pre-, postemergent herbicide for grass; broadleaf weeds in rice fields.	1300	2900
Tide® (see forchlorfenuron)				
Tillam® (see pebulate)				
Topik® (see clodinafop-propargyl)				
Topit® (see mepiquat chloride)				
Top Notch® (see acetochlor)				
Topstar® (see oxadiargyl)				
Tordon® (see picloram)				
Touchdown® (see glyphosate-trimesium)				
Tough® (see pyridate)				

Common name, trade name and basic manufacturer(s)	Chemical name	General use pattern	Oral LD_{50} (rat)	Dermal LD_{50} (rabbit)
tralkoxydim, Achieve®, Grasp® (Syngenta)	2-[1-(ethoxyimino)propyl]-3-hydroxy-5-(2,4,6-trimethylphenyl) cyclohex-2-enone	Postemergence for grass weeds in cereals.	934	—
Transline® (see clopyralid)				
TranXit® (see rimsulfuron)				
Treflan® (see trifluralin)				
triallate, Far-Go® (Monsanto)	S-(2,3,3-trichloroallyl) diisopropylthiocarbamate	Wild oat control in barley, wheat, peas, lentils.	1675	8200
triasulfuron, Amber®, Logran® (Syngenta)	2-(chloroethoxy-n-[[4-methoxy-6-methyl-1,3,5-triazin-2-yl)amino]carbonyl]benzene sulfonamide	Experimental Sulfonylurea herbicide for broadleaf weeds in small grains.	>5050	>2000
tribenuron-methyl, Express® (DuPont)	methyl 2-[[[[*N*-(4-methoxy-6-methyl-1,3,5-triazin-2-yl)methylamino]carbonyl]amino]sulfonyl] benzoate	Sulfonyl selective postemergence for broadleaf weeds in cereals.	>5000	>2000
tribufos, DEF®, Folex® (Bayer CropScience)	S,S,S-tributylphosphorothioate	Cotton desiccant & defoliant.	250	850
triclopyr, Garlon®, Grandstand®, Turflon®, Pathfinder® (Dow Agrosciences)	[(3,5,6-trichloro-2-pyridinyl)oxy]acetic acid	Systemic control of woody plants, broad-leaf weeds in rights-of-way, industrial sites, forests.	713	>2000
tridiphane, Tandem® (discontinued)	2-(3,5-dichlorophenyl)-2-(2,2,2-trichloroethyl)-oxirane	Postemergence for grasses & broadleaf weeds in corn.	1743	3536
trifloxysulfuron-sodium, Enfield®, Envoke®, Monument® (Syngenta)	1-(4,6-dimethoxypyrimidin-2-yl)-3-[3-(2,2,2,-trifluoroethoxy)-2-pyridylsulfonyl]urea	Broadleaf weed control in cotton sugarcane, citrus, stone fruits, tomatoes & turf.	>5000	>2000
trifluralin, Treflan® (Dow Agrosciences)	α,α,α-trifluoro-2,6-dinitro-*N,N*-dipropyl-*p*-toluidine	Selective preemergence incorporated herbicide registered for many crops.	>10,000	>2000
triflusulfuron-methyl, Upbeet® (DuPont)	methyl 2-[[[[[4-(dimethylamino)-6-(2,2,2-trifluoroethoxy)-1,3,5-triazin-2-yl]-amino] carbonyl]amino]sulfonyl]-e-methylbenzoate	Selective potemergence for annual & perennial broadleafs in sugarbeets.	>5000	
Triggrr® (see cytokinins)				
triketone (see sulcotrione)				
Trimaton® (see metam-sodium)				
Trimmit® (see pacelobutrazol)				
trinexapec-methyl, Palisade® Primo MAXX® (Syngenta)	ethyl 4-cyclopropyl(hydroxy)methylene-3,5-dioxocyclohexane-carboxylate	PGR for lodging in cereals, for turf to reduce mowing & as maturation promoter in sugarcane.	4460	>4000
tritosulfuron, Corto® (BASF)	1-[4-methoxy-6-(trifluoromethyl)-1,3,5-triazin-2-yl] -3-[2-(trifluoromethyl)benzenesulfonyl]urea	Experimental herbicide for use on several crops.		
Trophy® (see acetochlor)				
Tuberite® (see propham)				

Common name, trade name and basic manufacturer(s)	*Chemical name*	*General use pattern*	*Oral LD_{50} (rat)*	*Dermal LD_{50} (rabbit)*
Tupersan® (see siduron)				
Turfcure ® (see metam-sodium)				
Turflon® (see triclopyr)				
uniconazole, Sumagic® (Valent)	(E)-1-(4-chlorophenyl)-4,4-dimethyl-2-(1,2,4-triazol-1-yl)-1-penten-3-ol	Growth regulator to reduce plant height & increase flowering; Pre-plant soil treatment.	1790	>2000
Unicrop CIPC® (see chlorpropham)				
Upbeet® (see triflusulfuron-methyl)				
Valor® (see flumioxazin)				
Valuron® (see metsulfuron-methyl)				
Vanquish® (see dicamba)				
Vantage® (see sethoxydim)				
Vapam® (see metam-sodium)				
Velpar® (see hexazinone)				
Venture® (see fluazifop-butyl)				
vernolate, Vernam® (discontinued)	*S*-propyl dipropylthiocarbamate	For most common weed grasses in peanuts.	1500	>5000
Visor® (see thiazopyr)				
Weed Zol® (see amitrol)				
Whip® (see fenoxaprop-ethyl)				
Wilthin® (see AMADS)				
Xanthomonas campestris, PV *poannua*, X-Po® (Dow, EcoSoil)		Bacterial herbicide to control annual bluegrass & *Poa annua* in turf.	Non-toxic	
Yukawide® (see benzofenap)				
Zorial Rapid® (see norflurazon)				

* Chemical nomenclature used is either that of Chemical Abstracts (American Chemical Society) or International Union of Pure and Applied Chemistry (IUPAC). The author and publisher make no warranty, express or implied, regarding the accuracy of this Appendix.

APPENDIX **D**

Fungicides and Bactericides

Common, trade and chemical names, their basic manufacturer(s), general use patterns and oral and dermal LD_{50}s.

Common name, trade name and basic manufacturer(s)	*Chemical name*	*General use pattern*	*Oral LD_{50} (rat)*	*Dermal LD_{50} (rabbit)*
Abound® (see azoxystrobin)				
Acanto® (see picoxystrobin)				
acibenzolar-S-methyl, Actigard®, Blockade® (Syngenta)	*S*-methyl benzo[1,2,3]thiadiazole-7-carbothioate	Disease resistance activator for cucurbits, vegs against downy mildew, bacterial spot & blue mold.	>5000	>2000
Acrobat® (see dimethomorph)				
Actigard® (see acebenzolar-S-methyl)				
Actinovate® (see *Strepomyces lydicus*)				
Afugan® (see pyrazophos)				
Agri-Mycin® (see streptomycin)				
AgriPHAGE® (AgriPhil)	biofungicide (anti bacterial virus or bacteriophage)	For bacterial spot on tomato & peppers		
Agrobacterium radiobacter Galltroll-A®, No-Gall® (AgBioChem)		Eradicant for bacterial diseases.	Non-toxic	
Alamo® (see propiconazole)				
Algae-Rhap® (see copper-triethanolamine)				
Aliette® (see fosetyl-Al)				
Alto® (see cyproconazole)				
Amistar® (see azoxystrobin)				
Ampelomyces quisqualis AQ-10(Ecogen)		Microbial for powdery mildew on apples, cherries, cucurbits, grapes, ornamentals, vegs.	Non Toxic	
Anvil® (see hexaconazole)				
ApronXL® (see metalaxyl-M)				

* Chemical nomenclature used is either that of Chemical Abstracts (American Chemical Society) or International union of Pure and Applied Chemistry (IUPAC). The author and publisher make no warranty, express or implied, regarding the accuracy of this Appendix.

Common name, trade name and basic manufacturer(s)	*Chemical name*	*General use pattern*	*Oral LD_{50} (rat)*	*Dermal LD_{50} (rabbit)*
AQ-10® (see *Ampelomyces quisqualis*)				
Arbotect® (see thiabendazole)				
Arius® (see quinoxyfen)				
Armicarb® (see potassium bicarbonate)				
Ascend® (see TCMTB)				
Aspergillus flavus strain AF-36		Mycofungicide for aflatoxin control in cotton.		
Aspire® (see *Candida oleophila*)				
AtEze® (see *Pseudomonas chlorophis*)				
azoxystrobin, Heritage®, Abound®, Amistar®, Quadris® (Syngenta)	methyl-(E)-2-2-6-(2-cyanophenoxy)pyrimidin-4-yloxy-phenyl-3-methoxyacrylate	Bananas, grapes, peaches, peanuts, pecans, lettuce, tomatoes, berries, safflower, tropical fruits & turf.	5000	—
Bacillus licheniflormis St. 3085, 3086/1BA, Green Releaf®, Ecoguard® (Novozymes)		For dollar spot on turf, lawns & golf courses.	Non-toxic	
Bacillus pumilus GB34, QST-2808, Sonata AS® (AgraQuest, Gustafson)	biofungicide	For fireblight, scab on pome fruits, powdery mildew on apples.	Non-toxic	
Bacillus subtilis, (Sts, 6B03, MGI600, QST713) Companion®, Kodiak®, Serenade®, Subtilex®, Rhapsody® (Gustafson, AgraQuest)		Seed treatment to prevent *Rhizocotonia* & *Fusarium* diseases.	Non-toxic	
Bacillus subtilis var. *amyloiquefaciens* St. FZB-24, Taegro® (Earth Biosciences)		Root protectant for seedling & root diseases in ornamentals.	Non-toxic	
bacteriophages (see AgriPHAGE®)				
bacticin,Gallex® (AgBioChem)	mixtures of 2,4-xylenol and meta-cresol	Preventive & curative for bacterial tumors on canes, fruit & nut trees.	4640	—
Banner® (see propiconazole)				
Banol® (see propamocarb)				
Bardos® (see difenoconazole)				
basic copper sulfate (several)		Bacterial & fungal diseases on fruit, vegetable, nut, field crops.	472	Eye irritant

Common name, trade name and basic manufacturer(s)	*Chemical name*	*General use pattern*	*Oral LD_{50} (rat)*	*Dermal LD_{50} (rabbit)*
Basamid® (see dazomet)				
Bavistin® (see carbendazim)				
Baycor® (see bitertanol)				
Bayleton® (see triadimefon)				
Baytan® (see triadimenol)				
benalaxyl, Galben® (ISAGRO)	methyl *N*-(2,6-dimethylphenyl)-*N*-(phenylacetyl)-DL-alaninate	Systemic for Oomycete fungi, seed treatment.	4200	>5000
Benlate® (see benomyl)				
benomyl, Benlate® (DuPont) (discontinued)	methyl-1-(butylcarbamoyl)-2-benzamidazole carbamate	Systemic control of fruit, vegetable, nut, field crop, turf, ornamental diseases.	>10,000	>10,000
Binab T® (see *Trichoderma harzianum*)				
binapacryl, Morocide® (discontinued)	2-sec-butyl-4,6-dinitrophenyl 3-methyl-2-butenoate	Mostly powdery mildews; also a miticide.	150	750
Biocure® (see *Candida saitoana*)				
Biosave II® (see *Pseudomonas syringae*)				
biphenyl, diphenyl (discontinued)	diphenyl	Citrus wrap impregnate for rot fungi.	3280	—
bitertanol, Baycor® (Bayer CropScience)	β[(1,1'-biphenyl)-4-yloxy]-α(1,1- dimethylethyl)-1*H*-1,2,4 triazole-1-ethanol	Diseases of fruits, field crops, vegs, ornamentals, outside U.S.	>5000	>5000
Blasticidin-S® (*Streptomyces griseochromogenes*) (Kaken)	(*S*)-4[[3-amino-5[(aminoiminomethyl)methyl amino]-1-oxopentyl]amino]-1-[4-amino-2-oxo-1(2*H*)-pyrimidinyl]-1,2,3,4-tetradeoxy-β-D-erythrohex-2-enopyranuronic acid	Systemic antibiotic effective against rice blast, outside U.S.	39	3100
Blockade® (see acibenzolar-S-methyl)				
Blocker® (see PCNB)				
bluestone (see copper sulfate)				
Bordeaux mixture (several)	mixture of copper sulfate and calcium hydroxide forming basic copper sulfates	For many diseases of vegs, fruits, nuts: also acts as insecticide, repellent to some species.	Non toxic	
boscalid (see nicobifen)				
Botran® (see dicloran)				
Bravo® (see chlorothalonil)				
Break® (see propiconazole)				

Common name, trade name and basic manufacturer(s)	*Chemical name*	*General use pattern*	*Oral LD_{50} (rat)*	*Dermal LD_{50} (rabbit)*
Brestanid® (see fentin hydroxide)				
Brevibaccilus brevis	bacterial fungicide	Controls damping off & gray mold in vegs.	Non-Toxic	
bromuconazole, Condor®, Granit®, Vectra® (Bayer CropScience)	1-[(2RS,4RS; 2RS,4RS)-4-bromo-2-(2,4-dichlorophenyl)tetrahydrofurfuryl]-1*H*-1,2,4-triazole	Systemic, curative & preventive foliar, vegs, cereals, rice, fruits, turf,	365	>2000
bronopol, Bronotak® (BayerCropScience)	2-bromo-2-nitro-1,3-propanediol	Bacterial diseases of cotton, outside U.S.	180	>1600
Bronotak® (see bronopol)				
bupirimate, Nimrod® (Syngenta)	5-butyl-2-ethylamino-6-methyl-pyrimidin-4-yl-dimethylsulfamate	Systemic control of powdery mildew of fruit.	>4000	>500
Burkholderia (Pseudomonas) cepacia, Deny® (CCT)		Biofungicide, nematicide for seed treatment & transplants.		
Busan®, (see TCMTB)				
Cadminate® (discontinued)	cadmium succinate	Fungal diseases of turf.	660	>200
cadmium chloride (discontinued)	cadmium chloride	Fungal diseases of turf.	88	—
Calixin® (see tridemorph)				
Candida oleophila, Aspire® (Ecogen)		Biofungicide for post-harvest decay of citrus & pome fruits.		
Candida saitoana, Biocure® (Micro Flo)	biofungicidal fungus with lysozyme	Post-harvest fruit protectant		
captafol, Difolatan® (discontinued)	*cis-N*-(1,1,2,2-tetrachloroethyl)thio-4-cyclohexene-1,2-dicarboximide	Many diseases of fruits,vegs, nuts, seeds, outside U.S.	5000	>15,400
captan, Maestro® & many others (Drexel; Crystal, Arvesta)	*N*-[(trichloromethyl)thio]-4-cyclohexene-1,2-dicarboximide	Many diseases of fruits, vegs, nuts, seeds.	9000	—
Caramba® (see metconazole)				
Carbamate® (see ferbam)				
carbendazim, Bavistin®, Derosol® (Bayer CropScience, BASF)	2-(methoxycarbonylamino)-benzimidazole	Controls Dutch elm disease.	>10000	>2000
carboxin, Vitavax® (Crompton)	5,6-dihydro-2-methyl-*N*-phenyl-1,4-oxathiin-3-carboxamide	Systemic seed treatment for grains, peanuts, other field crops.	3820	>8000
Castellan® (see fluquinconazole)				
Cedomon® (see *Pseudomonas chloroaphis*)				

Common name, trade name and basic manufacturer(s)	*Chemical name*	*General use pattern*	*Oral LD_{50} (rat)*	*Dermal LD_{50} (rabbit)*
Ceresan®, MEMC (discontinued)	2-methoxyethylmercuric chloride	Seed treatment & bulb dip.	22	—
Charter® (see triticonazole)				
Chipco 26019® (see iprodione)				
chitosan, Elexa® (GlycoGenesys)	poly-*N*-acetyl-D-glucosamine, derived mainly from crab & shrimp exoskeletons.	For blights, mildews & gray mold. Product is also a PGR	Minimal skin irritation	
chloranil, Spergon® (discontinued)	2,3,5,6-tetrachloro-1,4-benzoquinone	Foliar diseases of fruit, vegs, ornamentals.	4000	—
chloroneb,Terraneb® (Kincade)	1,4-dichloro-2,5-dimethoxybenzene	Systemic seed treatment; in-furrow soil treatment; turf diseases.	>5000	>5000
chloropicrin	trichloronitromethane	Soil fumigant & warning agent.	250	Severe eye, Skin irritant
chlorothalonil, Daconil®, Bravo®, Echo®, Equus®, Passport®, Ole®, Tuffcide® (Syngenta, Griffin)	tetrachloroisophthalonitrile	Variety of vegetable, fruit, turf, ornamental diseases.	>10,000	>10,000
Chorus® (see cyprodinil)				
Companion® (see *Bacillus subtilis*)				
Compass® (see trifloxystrobin)				
Condor® (see bromuconazole)				
Coniothyrium minitans, Contans® (Prophyta)	biofungicidal fungus	Soil treatment for Sclerotinia, white mold, pink rot & soft water rot on vegs,	Non-toxic	—
copper ammonium carbonate, Copper-Count-N® (Mineral Research)	copper ammonium complex	Protectant for fruits, vegs tolerant to copper, turf algaecide.	low	—
copper chelates, Algimycin PLLC® (discontinued)	copper chelates of citrate and gluconate	Most common algae in various water sources.	low	—
copper hydroxide, Kocide®, Nu-Cop® (Griffin, Crystal)	copper hydroxide	Protectant for fruit, vegetable, tree, field crops.	1000	—
copper naphthenates	cupric cyclopentanecarboxylate	Rot, mildew in fabrics & wood.	low	—
copper oxides, Nordox®	cuprous oxide and cupric oxide	Various fruit, nuts, vegs & field crops.	low	—
copper oxychloride	basic copper chloride	Protectant for fruit, vegetable crops.	700	—
copper oxychloride sulfate	mixture of basic copper chloride & basic copper sulfate	Downy mildew, leaf spots of row, vine, field & tree crops.	low	—
copper quinolinolate	copper-8-quinolinolate	Treatment of fruit-handling equipment.	10,000	—

Common name, trade name and basic manufacturer(s)	*Chemical name*	*General use pattern*	*Oral LD_{50} (rat)*	*Dermal LD_{50} (rabbit)*
copper sulfate, bluestone (several)	copper sulfate pentahydrate	Mixed with lime for Bordeaux mixture; almost never used alone due to phytotoxicity. Algaecide.	300	Eye irritant
copper-triethanolamine Algae-Rhap® (Nufarm)	copper-triethanolamine complex in liquid form	Filamentous & planktonic algae in various water sources.	low	
Curalan® (see vinclozolin)				
Curamil® (see pyrazophos)				
Curzate® (see cymoxanil)				
cycloheximide, Actispray® (discontinued)	3-[2-(3,5-dimethyl-2-oxocyclohexyl)-2-hydroxyethyl] glutarimide	Mostly diseases of ornamentals, turf.	2.5	—
Cygnus® (see kresoxim methyl)				
cymoxanil, Curzate® (DuPont)	2-cyano-*N*-[(ethylamino)carbonyl]-2-(methoxyimino) acetamide	For downy mildew on grapes, late blight on tomatoes & potatoes.	1100	>3000
cyproconazole, Alto®, Sentinel® (Syngenta)	2-(4-chlorophenyl)-3-cyclopropyl-1-(1-*H*-1,2,4-triazol-1-yl) butan-2-ol	Experimental systemic/contact used on cereals, pomes, grapes, cotton, vegs. Not registered.	1,020	>2000
cyprodinil, Chorus®, Vangard® (Syngenta)	4-cyclopropyl-6-methyl-*N*-phenyl-2-pyrimidinamine	Controls *Alternaria*, *Botrytis*, *Monilinia* & more on a broad array of crops. Foliar fungicide for grapes, fruit, caneberries, vegs, cereals.	>2000	>2000
Daconil® (see chlorothalonil)				
dazomet, Basamid® (BASF)	3,5-dimethyl-(2*H*)-tetrahydro-1,3,5-thiadiazine-2-thione	Fungicide, herbicide, nematicide, insecticide, slimicide, soil fumigant. Ornamental seedbeds, lawns, potting soil, etc.	519	>2000
DCNA (see dicloran)				
Decree® (see fenhexamid)				
Denarin® (see triforine)				
Derosol® (see carbendazim)				
diallyl sulfide, Alli-up® (Platte)	diallyl sulfide	New soil fumigant for white rot in onions, garlic & leeks.		
dichlone, Quintar® (discontinued)	2,3-dichloro-1,4-naphthoquinone	Protectant for fruit, vegs, field crops, ornamentals.	1300	5000
dichlozoline, Sclex® (discontinued)	3-(3,5-dichlorophenyl)-5,5-dimethyl-2,4-oxazolidinedione	General fungicide.	3000	—
diclobutrazol, Vigil® (discontinued)	±-β-[(2,4-dichlorophenyl)methyl]α-(1,1-dimethylethyl)-1*H*-1,2,4-triazole-1-ethanol	Powdery mildews on fruit, cereals, grapes.	4000	>1000
dicloran, Botran®, DCNA (Gowan)	2,6-dichloro-4-nitroaniline	Many fruit, vegetable, field crop, greenhouse, ornamental diseases.	4000	>5000

Common name, trade name and basic manufacturer(s)	*Chemical name*	*General use pattern*	*Oral LD_{50} (rat)*	*Dermal LD_{50} (rabbit)*
diethofencarb, Diffuse® (Wangs)	isopropyl 3,4-diethoxycarbanilate	For Benzimidazole-resistant stains of fungi. Not reg in U.S.	>5000	>5000
difenoconazole, Score®, Bardos®, Dividend® (Syngenta)	1-(2-[4-(4-chlorophenoxy)-2-chlorophenyl]-4-methyl-1,3-dioxolan-2-yl-methyl)-1*H*-1.2.4-triazole	Systemic fungicide for wide variety of diseases.	2125	—
Diffuse® (see diethofencarb)				
Difolatan® (see captafol)				
dimethirimol, Milcurb® (Syngenta)	5-*n*-butyl-2-dimethylamino-4-hydroxy-6-methylpyrimidine	Systemic control of powdery mildew on cucurbits, ornamentals, outside U.S.	2350	7500
dimethomorph, Acrobat®, Forum® (BASF)	(*E*,)4-(3-(4-chlorophenyl)-3-(3,4-dimethoxyphenyl)acryloyl)morpholine	Curative & preventative foliar fungicide for grapes, potatoes, vegs, lettuce & cole crops.	3900	—
diniconazole, Reason® (Wangs)	(E)-1-(2,4-Dichlorophenyl)-4,4-dimethyl-2-(1,2,4-triazol-1-yl)-1-penten-3-ol	Broad spectrum systemic for several crops, outside U.S.	474	>5000
dinocap, Karathane®, Crotothane® (discontinued)	mixture of 2,4-dinitro-6-octylphenyl crotonate and 2,6-dinitro-4-octylphenyl crotonate	Powdery mildew on fruits, vegs, ornamentals; also a good miticide.	980	—
diphenyl (see biphenyl)				
diphenylamine, DPA No-Scald® (Cerexagri)	diphenylamine	Pre- & postharvest apple & pear scald.	300	—
dipotassium phosphate, Lexx-A-Phos® (Foliar Nutrients)		Controls several fungi in turf, ornamentals, non-bearing fruits & nuts.		
DiTera® (see *Myrothecium verrucaria*)				
Dithane® (see mancozeb)				
Dividend® (see difenconazole)				
dodemorph acetate, Meltatox® (BASF)	4-cyclododecyl-2,6-dimethylmorpholimium acetate	Powdery mildew of ornamentals & greenhouse roses.	1800	irritant
dodine, Melprex®, Syllit® (Bayer CropScience)	dodecylguanidine monoacetate	Many diseases of tree fruits.	1000	>1500
Domain® (see thiophanate methyl)				
DPA (see diphenylamine)				
Dynone® (see propamocarb)				
Eagle® (see myclobutanil)				
Echo® (see chlorothalonil)				
ECO E-RASE® (see jojoba oil)				
EcoTrol® (see Rosemary oil)				

Common name, trade name and basic manufacturer(s)	Chemical name	General use pattern	Oral LD_{50} (rat)	Dermal LD_{50} (rabbit)
EksPunge® (see potassium dihydrogen phosphate)				
Elevate® (see fenhexamid)				
Elexa® (see chitosan)				
Elite® (see tebuconazole))				
Eminent® (see tetraconazole)				
Enable® (see fenbuconazole)				
Endorse® (see polyoxin-d-zinc)				
Endura® (see nicobifen)				
Engage® (see sodium tetrathiocarbonate)				
Enzone® (see sodium tetrathiocarbonate)				
epoxiconazole, Opus® (BASF)	(2RS,3RS)-1-[3-(2-chlorophenyl)-2(4-fluorophenyl)oxiran-2-ylmethyl]-1*H*-1,2,4-triazole	Most cereal diseases, powdery mildew, sugar beets, sheath blight & blast in rice.	5000	2000
Equus® (see chlorothalonil)				
ethaboxam, Guardian® (LG Chem)	*N*-(α-cyano-2-thenyl)-4-ethyl-2-(ethylamino)-5-thiazolecarboxamide)	Controls late blight & downy mildew on fruits trees, grapes, vegs & ornamentals.	>5000	
ethirimol, Milcurb® Super (discontinued)	5-butyl-2-(ethylamino)-6-methyl-4(1*H*)-pyrimidinone	Systemic control of powdery mildew of cereals, outside U.S.	6340	>1000
etridiazole, (ethazol), Terrazole®, Koban®; Truban® (Crompton)	5-ethoxy-3-trichloromethyl-1,2,4-thiadiazole	Seeding diseases of field & vegetable crops & ornamentals.	1077	>5000
famoxadone, Famoxate® (DuPont)	5-methyl-5-(4-phenoxyphenyl)-3-(phenylamino)-2,4-oxazolidinedione	Tree fruits, grapes, vegs, potatoes, cucurbits, cereals, for late blight, downy mildew & Septoria leaf spot.		
Fandango® (see fluoxastrobin)				
fenamidone, Reason® (Bayer CropScience)	(*S*)-1-anilino-4-methyl-2-methylthio-4-phenylimidazolin-5-one	Controls early & late blights on vegs, downy mildew on grapes	2028	>2000
fenaminosulf, Lesan® (discontinued)	sodium [4-(dimethylamino)phenyl] diazenesulfonate	Seed treatment for field crops, some vegs, ornamentals, pineapple, sugar cane, outside U.S.	60	34
fenarimol, Rubigan® (Gowan)	α-(2-chlorophenyl)-α-(4-chlorophenyl)-5-pyrimidinemethanol	Powdery mildews, scab & rust of deciduous fruits, grapes & pecans.	2500	>2000
fenbuconazole, Indar®, Govern®, Enable® (Dow Agrosciences)	α-(2-(4-chlorophenyl) ethyl-α-phenyl-(1*H*-1,2,4-triazole)-1-propanenitrile	Systemic, protective & curative for wide range of diseases & crops.	>2000	>5000

Common name, trade name and basic manufacturer(s)	Chemical name	General use pattern	Oral LD_{50} (rat)	Dermal LD_{50} (rabbit)
fenhexamid, Elevate®, Decree®, Teldor® (Bayer CropScience)	*N*-(2,3-dichloro-4-hydroxyphenyl)-1-methylcyclohexanecarboxamide	For Botrytis on stone fruits, grapes, cranberries, citrus,strawberries, lettuce, Postharvest on stone & pome fruit.	>5000	>5000
fentin hydroxide, Supertin®, Brestanid® (Bayer CropScience)	triphenyltin hydroxide	Vegetable diseases, especially potato late blight.	156	1600
ferbam, Carbamate®, Trifungol® (Taminco)	tris(dimethylcarbamodithioat-*S,S'*)iron	Apple & tobacco diseases; protectant for other crops.	>4000	—
First Step® (see potassium bicarbonate)				
Flamenco® (see fluquinconazole)				
Flint® (see trifloxystrobin)				
fluazinam, Omega® (Syngenta)	3-Chloro-*N*-(3-chloro-5-trifluoromethyl-2-pyridyl)-α,α,α, trifluoro-2,6-dinitro-*p*-toluidine	Controls *Sclerotinia, Alternaria, Botrytis, Phytophthora, Plasmopara,* on tree fruits, vegs, peanuts, potatoes, apples & grapes.	5000	>2000
fludioxonil, Maxim®, Medallion®, Saphire®, Scholar® (Syngenta)	4-(2,2-difluoro-1,3-benzodioxol-4-yl)-1*H*= pyrrole-3-carbonitrile	Seed treatment for corn & soybeans for Fusarium, Helminthosporium & Rhizoctonia, for turf & ornamentals, pome fruits. Postharvest on fruit & vegs,	>5000	>2000
fluoxastrobin, HEC 5725 Fandango® (Bayer CropScience)	{2-[6-(2-chlorophenoxy)-5-fluoropyrimidin-4-yloxy] phenyl}(5,6-dihydro-1,4,2-dioxazin-3-yl) methanone *O*-methyloxime	Potential use on wheat & barley as foliar & seed treatment product. Not registered.	—	—
fluquinconazole, Vista®, Flamenco®, Castellan®, (Bayer CropScience)	3-(2,4-dichlorophenyl)-6-fluoro-2)1*H*-1,2,4-triazole-1l)-4(3*H*)-quinazolinone	Wide range of diseases on broadleaf & cereal crops, pome & stone fruits.	112	625
flusilazole, Nustar®, (discontinued)	1-[[*Bis*(4-fluorophenyl)methylsilyl]methyl]-1*H*-1,2,4-triazole	Silane systemic controls many Ascomycete, Basidiomycete & Deuteromycete fungi.	674	>2000
flutolanil, Prostar®, Moncut®, Folistar® (Bayer CropScience)	3'-isopropoxy-2-(trifluoromethyl)benzanilide	Protective systemic & curative for turf Rhizoctonia potato & peanut diseases.	10,000	>5000
Folicur® (see tebuconazole)				
Folistar® (see flutolanil)				
Fore® (see mancozeb)				
Fortress® (see quinoxyfen)				
Forum® (see dimethomorph)				
fosetyl-Al, Aliette®, Signature® (Bayer CropScience)	ethyl hydrogen phosphonate aluminum salt	Systemic for nonbearing citrus, pineapples, ornamentals, turf, bananas, grapes, macademia & bush berries.	5800	>3200

Common name, trade name and basic manufacturer(s)	*Chemical name*	*General use pattern*	*Oral LD_{50} (rat)*	*Dermal LD_{50} (rabbit)*
Fosphite® (see potassium phosphate)				
Freshgard® (see imazalil)				
Fungaflor® (see imazalil)				
Funginex® (see triforine)				
Fungo® (see thiophanate-methyl)				
furalaxyl, Fongarid® (discontinued)	methyl *N*-(2,6-dimethylphenyl)-*N*-(2-furanylcarbonyl)-DL-alaninate	Curative & preventive for soil-borne Oomycete diseases, outside U.S.	940	>3100
Galben® (see benalaxyl)				
Galltrol-A® (see *Agrobacterium radiobacter*)				
Gavel® (see zoxamide)				
Gem® (see trifloxystrobin)				
Gliocladium catenulatum, PreStop® (Kemira)		For Pythium & Rhizoctonia indoors & greenhouses.		
Gliocladium virens, SoilGard®, GlioGard® (Certis)		Mycofungicide for Pythium, Rhizoctonia seedling diseases.	Non-toxic	—
GlioGard® (see *Gliocladium virens)*				
Glyodin® (discontinued)	2-heptadecyl-2-imidazoline acetate	Wide range of tree fruit, ornamental diseases.	4600	—
Govern® (see fenbuconazole)				
Granite® (see bromuconazole)				
Green Leaf® (see *Bacillus licheniflormis*)				
harpin, Messenger® (Eden)	bacterial protein biofungicide	For control of Fusarium, Rhizoctonia & virus diseases on fruits & vegs,	Non-toxic	
Heritage® (see azoxystrobin))				
hexachlorobenzene, NoBunt® (discontinued)	1,2,3,4,5,6-hexachlorobenzene	Seed treatment for small grains.	10,000	—
hexachlorophene (discontinued)	2,2'-methylene bis (3,4,6-trichlorophenol)	Fungal & bacterial diseases of tomatoes, peppers & cucumbers; cotton seedling diseases.	2700	>10,000
hexaconazole, Anvil®, Proseed® (Syngenta)	(RS)-2-2(2,4-dichlorophenyl)-1-(1*H*-1,2,4-triazol-lyl)hexan-2-ol	Powdery mildew & blackrot on grapes.	2189	>2000
Honor® (see nicobifen)				
Horizon® (see tebuconazole)				

Common name, trade name and basic manufacturer(s)	Chemical name	General use pattern	Oral LD_{50} (rat)	Dermal LD_{50} (rabbit)
hydrogen peroxide (BioSafe)	hydrogen peroxide	Chemical wash (1%) for processed fruit & vegs.		Eye Irritant
hymexazole, Tachigaren® (Sanko)	3,5-methylisoxazol-3-ol	Soil drench & seed treatment for soil borne Pythium, damping-off, Fusarium & others.	3909	
imazalil, Freshgard®, NuZone®, Fungaflor® (Gilmore, Cerexagri)	1-[2-(2,4-dichlorophenyl)-2-(2-propenyloxy)ethyl]-1H-imidazole	Broad spectrum systemic for fruit, vegs & ornamentals.	320	4200
Indar® (see fenbuconazole)				
Input® (see prothioconazole)				
intercept® (see *Pseudomonas capacia*				
iprodione, Chipco 26-GT®, Rovral® (Bayer CropScience)	3-(3,5-dichlorophenyl)-N-(1-methyl-ethyl)-2,4-dioxo-1-imidazolidinecarboxamide	Spring & summer turf diseases, for fruits, vegs, field crops.	3500	>1000
iprovalicarb, Melody® (Bayer CropScience)	isopropyl-2-methyl-1-[(1-*p*-tolylethyl)carbamoyl]-(*S*)-propylcar=bamate	Controls early & late blight, downy mildew on potatoes, tomatoes & grapes.	>5000	>5000
jojoba oil, ECO E-RASE® (IJO Products)	plant oil of natural origin	Kills powdery mildew on vines & ornamentals. Also insecticidal on whiteflies.		
Kaligreen® (see potassium bicarbonate)				
kasugamycin (*Streptomyces kasugaenis*) (Hokko)	D-3-*O*-[2-amino-4-[(1-carboxyiminomethyl)-amino]-2,3,4,6-tetradeoxy-α-D-arabinohexopyranosyl] D-chiro-inositol	Systemic antibiotic effective against rice blast, outside U.S.	22,000	>4000
Koban® (see etridiazole)				
Kocide® (see copper hydroxide)				
Kodiak® (see *Bacillus subtilis*)				
kresoxim-methyl, Stroby®, Sovran®, Cygnus® (BASF)	methyl (E)-methoxyimino[α-(*O*-tolyloxy)-*O*-tolyl] acetate	On cereals, apples, for powdery & downey mildew, late & early blights, Vegs, also ornamentals,	>5000	>2000
Lexx-A-Phos® (see dipotassium phosphate)				
lime sulfur, (several)	calcium polysulfide	Apple scab, powdery mildews, on cherries & dormant spray for pistachios.	skin irritation	
Lynx® (see tebuconazole)				

Common name, trade name and basic manufacturer(s)	Chemical name	General use pattern	Oral LD_{50} (rat)	Dermal LD_{50} (rabbit)
macheaya extract, Qwel® (Camas Tech.)	sanguinarine chloride: [1,3]benzodioxolo[5,6-c] phenanthridinium-13-methyl chloride & chelerythrine chloride: [1,3] benzodioxolo[5,6-c]phenanthridinium-1,2-dimethoxy-12-methyl chloride	Used in greenhouses for powdery mildew; Alternaria & Septoria leaf spots on ornamentals.	960-1544	>2000
Maestro® (see captan)				
mancozeb, Dithane®, Fore®, Penncozeb®, Pentathlon®, Manzate® (Griffin, Crystal, Dow Agrosciences)	coordination product of zinc ion & manganese ethylene bisdithiocarbamate	Wide spectrum of fruit, vegetable, field crop, nut & turf diseases; also seed treatment for field crops, wheat.	>5000	>5000
maneb (several) (Griffin, Crystal)	manganese ethylenebisdithiocarbamate	Many diseases of fruits, vegs, scab on wheat, turf, but not golf courses.	7990	>5000
Manzate® (see mancozeb)				
Maxim® (see fludioxonil)				
Medallion® (see fludioxonil)				
Melprex® (see dodine)				
Melody® (see iprovalicarb)				
Meltatox® (see dodemorph acetate)				
Mertect® (see thiabendazole)				
Messenger® (see harpin)				
metalaxyl, Metax®, Rampart® (Crystal, United Phosphorus)	*N*-(2,6-dimethylphenyl)-*N*-(methoxyacetyl)alanine methyl ester	Systemic control of Pythium & Phytophthora soil-borne diseases, downy mildews, seed treatment.	669	>3100
metalaxyl-M (mefanoxam) Ridomil Gold®, Apron XL®, Meta-Mil®, Subdue Maxx®, Ultra Fluorish®, Quell® (Crompton, Sygenta)	*N*-(2,6-dimethylphenyl)-*N*-(methoxyacetyl)-D-alanine	Systemic control of Pythium & Phytophthora soil-borne diseases, downy mildews, seed treatment.	669	>3100
Meta-Mil® (see metalaxyl-M)				
metam-sodium, Vapam®, Trimaton® (AmVac, Cerexagri)	sodium *N*-methyldithiocarbamate	Soil fumigant for soil fungi, nematodes, weeds.	1800	1300
Metax® (see metalaxyl)				
metconazole, Caramba® (BASF)	(1RS,5RS;1RS,5RS)-5-(4-chlorobenzyl)-2,2-dimethyl-1-(1*H*-1,2,4-triazol-1-ylmethyl)-cyclopentanol)	Controls Fusarium, Septoria & rust on cereals, rice, fruit crops, sugar beets, apples.	661	>2000
methylisothiocyanate, Vorlex®, MIT (discontinued)	methyl isothiocyanate	Preplant soil fumigant.	175	961

Common name, trade name and basic manufacturer(s)	Chemical name	General use pattern	Oral LD_{50} (rat)	Dermal LD_{50} (rabbit)
metiram, Polyram® (BASF)	mixture of ammoniates of ethylenebis (dithiocarbamate)-zinc & ethyelnebisdithiocarbamic acid cyclic anhydrosulfides	Limited to apples, potatoes & roses.	>10,000	—
Microthiol® (see sulfur)				
Milcurb® (see dimethirimol)				
Milsana® (see *Reynoutria sachalinesis*)				
Moncut® (see flutolanil)				
Morestan® (see oxythioquinox)				
Muscador® (see *Muscador albus)*				
Muscador albus, Muscador® (AgriQuest)		Mycofumigant action on fruits & vegs.		
myclobutanil, Eagle®, Nova®, Rally®, Systhane® (Dow)	α-butyl-α-(4-chlorophenyl)-1*H*-1,2,4-triazole-1-propanenitrile	Systemic curative & protectant, for apples, grapes, sugar beets, vegs & ornamentals.	1600	>5000
MycoShield® (see terramycin)				
MycoStop®, (see *Streptomyces griseoviridis*)				
Myrothecium verrucaria, DiTera® (Valent)		Myconematicide for food & fiber crops, ornamentals & turf.		
nabam, (several)	disodium ethylenebisdithiocarbamate	Industrial applications only; not for food crops.	395	—
neem oil, Triact®, Trilogy® (Certis)	azadirachtin	Various diseases on ornamentals in nurseries & greenhouses.	Non-toxic	
nicobifen (boscalid), Endura®, Honor®, Pristine® (BASF)	2-chloro-*N*-(4'-chlorobiphenyl-2-yl)nicotinamide	For molds, rots, powdery mildews & leaf spots in vegs, pome fruits, crops, & nuts.	—	
Nimrod® (see bupirimate)				
NoBunt® (see hexachlorobenzene)				
No-Gall® (see *Agrobacterium radiobacter*)				
No-Scald® (see diphenylamine)				
Nova® (see myclobutanil)				
Nu-Cop® (see copper hydroxide)				

Common name, trade name and basic manufacturer(s)	Chemical name	General use pattern	Oral LD_{50} (rat)	Dermal LD_{50} (rabbit)
NuZone® (see imazalil)				
Ole® (see chlorothalonil)				
Omega® (see fluazinam)				
Opus® (see epoxiconazole)				
Orbit® (see propiconazole)				
oxadixyl, Recoil®, Sandofan® (discontinued)	2-methoxy-*N*-(2-oxo-1,3-oxazolidin-3yl)acet-2',6'-xylidide	Preventive & curative for potatoes, vegs, ornamentals & seed treatment.	1860	>2000
oxycarboxin, Plantvax® (Crompton)	5,6-dihydro-2-methyl-*N*-phenyl-1,4-oxathiin-3-carboxamide 4,4-dioxide	Systemic control of rusts on greenhouse flowers.	2000	>16,000
Oxytetracycline® (see terramycin)				
oxythioquinox, Morestan® (Bayer CropScience)	6-methyl-1-3-dithiolo(4,5-*b*)quinoxalin-2-one	Powdery mildews on fruits, vegs, ornamentals.	1520	>2000
Pantoea agglomerans, BlightBan® C9-1 (discontinued)		Biofungicide controls fire blight on apples & pears.		
Passport® (see chlorothalonil)				
PCNB, Terraclor®, Turfside® Blocker® (Amvac, Crompton)	pentachloronitrobenzene	Many seedling, vegetable, ornamental, turf diseases.	>12,000	—
PCP(several)	pentachlorophenol	Wood preservative.	50	105
Penncozeb® (see mancozeb)				
Pentathlon® (see mancozeb)				
peroxyacetic acid (Ecolab)		Antimicrobial for fruits, nuts, cereal grains, herbs & spices.		
phenyl phenol, Dowicide1® (discontinued)	*o*-phenyl phenol	Wax treatment or impregnated wraps for harvested fruit.	2700	Eye & skin irritant
PhosPro® (see potassium phosphite)				
phosphorus acid, Phostrol® (Agtrol)	phosphorus acid	For canker, root rot, fruit rots, on vegs, cole crops & citrus.		Eye & Skin Irritant
Phytoheath® (see sodium bicarbonate)				
picoxystrobin, Acanto® (Syngenta)	methyl (E)-2-(2-[6-trifluoromethyl)pyridin-2-yloxy-methyl]-phenyl)-3-methoxyacrylate	Controls diseases on cereals, apples & many other crops, experimentally.		
piperalin, Pipron® (SePRO)	3-(2-methylpiperidinyl)propyl 3,4-dichlorobenzoate	Powdery mildew on flowers, ornamentals.	2500	—

Common name, trade name and basic manufacturer(s)	*Chemical name*	*General use pattern*	*Oral LD_{50} (rat)*	*Dermal LD_{50} (rabbit)*
Plant Shield® (see *Trichoderma harzianum*)				
Plantvax® (see oxycarboxin)				
Polyoxin D (*Streptomyces cacaoi*) (Kaken)	5-[[2-amino-5-*O*-(aminocarbonyl)-2-deoxy-L-xylonoyl]amino]-pyrimidinyl]-β-D-allofura-1,5-dideoxy-1-[3,4-dihydro-5-(hydroxymethyl)-2,4-dioxo-1(2*H*)-nuronic acid	Systemic antibiotic effective against *Alternaria* on apples & pears. Botrytis on tomatoes & rice sheath blight.	21,000	—
polyoxin-D-zinc salt, Stopit®, Endorse® (Cleary)		Turf treatment, all types of diseases.		
Polyram® (see metiram)				
potassium bicarbonate, First Step®, Kaligreen®, Armicarb® (Cleary)		For powdery mildew on grapes, strawberries & roses; contact curative/preventive for ornamentals.		
potassium dihydrogen phosphate, EksPunge® (Lido)		For powdery mildew on fruits, vegs & ornamentals.		
potassium phosphate, Fosphite® (Griffin)	inorganic plant nutrient with fungicidal activity	For Downy mildew in vinyards & vegs.		
potassium phosphite ProPhyte®, PhosPro® (Griffin, Grow More)	potassium phosphite	Controls scab, root rot & downy mildew in cucurbits, nut crops, grapes & citrus.		Eye Irritant
Prelude® (see permethrin)				
Premis® (see triticonazole)				
Previcur® (see propamocarb hydrochloride))				
Prevex® (see propamocarb hydrochloride)				
PreStop® (see *Gliocladium catenulatum*)				
Pristine® (see nicobifen)				
Procure® (see triflumizole)				
procymidone, Sumilex® (Sumitomo)	*N*-(3',5'-dichlorophenyl)-1,2-dimethylcyclopropane-1,2-dicarboximide	Systemic eradicant & curative for fruits & vegs.	6800	>2500
Proline® (see prothioconazole)				
propamocarb hydrochloride, Banol®, Dynone®, Prevex®, Previcur®, Tattoo® (Bayer CropScience)	propyl[3-(dimethylamino)propyl]carbamate-mono-hydrochloride	Soil or foliar application to control downy mildews, on citrus, vegs, cucurbits & lettuce.	2000	>3900
ProPhyte® (see potassium phosphite)				

Common name, trade name and basic manufacturer(s)	*Chemical name*	*General use pattern*	*Oral LD_{50} (rat)*	*Dermal LD_{50} (rabbit)*
propiconazole, Alamo®, Banner®, Break®, Orbit®, Propimax®, Tilt® (Syngenta, Dow Agrosciences)	1-[[2-(2,4-dichlorophenyl)-4-propyl-1,3-dioxolan-2-yl]methyl]-1*H*-1,2,4-triazole	Broad spectrum systemic eradicant & curative. Blueberries, cereals, other crops, soybeans, vegs.	1517	>4000
Propimax® (see propiconazole)				
proquinazid, DPX-KQ926 (DuPont)	6-iodo-2-propoxy-3-propylquinazolin-4(3*H*)-one	Registration pending.		
Proseed® (see hexaconazole)				
Pro-Star® (see flutolanil)				
prothioconazole, Proline®, Input® (Bayer CropScience)	2-[2-(1-chlorocyclopropyl)-3-(2-chlorophenyl)-2-hydroxypropyl]-2,4-dihydro-1,2,4-triazole-3-thione	On cereals & oilseeds for control of Septoria- & Fusarium-spp and *Rhyncohsporium secalis*. Also an algaecide		
Pseudomonas aureofaciens strain TX-1, Spot-Less® (EcoSoil)		Bactofungicide for turfgrass disease.		
Pseudomonas cepacia Intercept® (SoilTech)		Seed treatment to prevent seedling diseases & for nematode control.	Non-toxic	
Pseudomonas chloroaphis, AtEze®, Cedomon® (EcoSoil Sys)		For Rhizoctonia & Pythium root & stem rots in greenhouse vegs.		
Pseudomonas fluorescens Dagger® (discontinued)		Cotton seedling diseases.	Non-toxic	
Pseudomonas fluorescens strain NCIB 12089, BioShield® (Anu Prods.)		Bacterial blotch on mushrooms, plus other diseases.		
Pseudomonas syringae, BioSave® (Village Farms)		Bacto fungicide for storage rots on apples, pears, cherries & potatoes.	Non-toxic	
Pseudozyma flocculosa, Sporodex® (Plant Prods)		For powdery mildew on greenhouse flowers & vegs	Non-toxic	
pyrazophos, Afugan®, Curamil® (Bayer CropScience)	ethyl 2-[(diethoxyphosphinothioyl)oxy]-5-methylpyrazolo(1,5-*a*)pyrimidine-6-carboxylate	Powdery mildew on cucurbits, fruits, cereals.	151	>2000
pyrimethanil, Scala® (Bayer CropScience)	*N*-(4,6-dimethylpyrimidin-2yl)aniline	For gray mold on pome/stone fruits, vegs, ornamentals, grapes, leaf scab on pome fruit.	4150	>5000
pyrifenox, Dorado® (discontinued)	1-(2,4-dichlorophenyl)-2-(3-pyridinyl)-ethanone *O*-methyloxime	Systemic for powdery mildews on stone fruits, grapes & vegs.	2900	—
Quadris® (see azoxystrobin)				
Quell® (see metalaxyl-M)				

Common name, trade name and basic manufacturer(s)	Chemical name	General use pattern	Oral LD_{50} (rat)	Dermal LD_{50} (rabbit)
quinoxyfen, Arius®, Fortress®, Quintec® (Dow)	5,7-dichloro-4(4-fluorophenoxy) quinoline	Systemic & preventive on cereals, tree fruits, grapes for powdery mildew & scab.	>5000	>2000
Quintec® (see quinoxyfen)				
Qwel® (see macheaya extract)				
Rally® (see myclobutanil)				
Rampart® (see metalxyl)				
Reason® (see diniconazole; also fenamidone)				
Reynoutria sachalinesis Milsana® (KHH BioSci)		Biological fungicide that raises plant's natural resistance to mildew.	Non-toxic	
Rhapsody® (see *Bacillus subtilis*)				
Ridomil Gold® (see metalaxyl-M)				
Rizolex® (see tolclofos-methyl)				
Ronilan® (see vinclozolin)				
Rosemary oil, EcoTrol®, Sporan® (EcoSmart)	biofungicide, insecticide, miticide.	Used on legumes & many vegs, tubers & fruits		
Rovral® (see iprodione)				
Rubigan® (see fenarimol)				
Saphire® (see fludioxonil)				
Scala® (see pyrimethanil)				
Scholar® (see fludioxonil)				
Score® (see difenoconazole)				
Sentinel® (see cyproconazole)				
Serenade® (see *Bacillus subtilis*)				
sodium bicarbonate, Phytoheath® (Meiji Milk)		Postharvest use on citrus for green mold during storage & transit.		
sodium borate (see sodium borate under Appendix C, Insecticides) TIM-BOR®				
sodium carbonate peroxhydrate, Terra-Cyte® (Biosafe)	mix of sodium carbonate & hydrogen peroxide	Fungicidal & algaecidal on ornamentals & other non-food plants		
sodium tetrathiocarbonate Engage®, Enzone® (Entek)	nematicide fumigant	For grapes, citrus, fruit trees & vegs.		

Common name, trade name and basic manufacturer(s)	*Chemical name*	*General use pattern*	*Oral LD_{50} (rat)*	*Dermal LD_{50} (rabbit)*
SoilGard® (see *Gliocladium virens*)				
Sovran® (see kresoxim-methyl)				
Sporan® (see Rosemary oil)				
Sporodex®, (see *Pseudozyma flocculosa*)				
Spotless® (see *Pseudomonas aureofaciens*)				
Sporodex® (see *Pseudozyma flocculosa*)				
Stopit® (see polyoxin-D-zinc)				
streptomycin, Agri-Mycin®, Streptrol® (Syngenta)	2,4-diguanidino-3,5,6-trihydroxycyclohexyl-5-deoxy-2-*O*-2-deoxy-methylamino-α-glucopyranosyl)-3-formyl pentofuranoside	Bacterial blights; cankers of pome, stone fruits, ornamentals.	9000	—
Streptomyces griseoviridis St, K61, MycoStop® (AgBio)	biofungicide	Fusarium wilt, damping off, Botrytis control in vegs, melons, cotton, cole crops.	Non-toxic	
Steptomyces lydicus, Actinovate® (Natural Indus.)	biofungicide	Downy mildew in cabbage, cucumber, grapes & late blight in tomatoes	Non-toxic	
Streptrol® (see streptomycin)				
Stroby® (see kresoxim-methyl)				
Subdue® (see metalaxyl)				
Subdue Maxx® (see metalaxyl-M)				
Subtilex® (see *Bacillus subtilis*)				
sulfur (many) Microthiol® (Cerexagri)	elemental sulfur in many formulations	Fruit rots, powdery mildews, rusts on vegetables.	Non-toxic	Skin irritant
Sumilex® (see procymidone)				
Supertin® (see fentin hydroxide)				
Syllit® (see dodine)				
Systhane® (see myclobutanil)				
Tachigaren® (see hymexazole)				
Taegro® (see *Bacillus subtilis*)				
Talent® (Luxan)	caraway oil (d-carvone)	Potato sprout inhibitor & fungicide. Not registered in U.S.		

Common name, trade name and basic manufacturer(s)	Chemical name	General use pattern	Oral LD_{50} (rat)	Dermal LD_{50} (rabbit)
Tattoo® (see propamocarb)				
TBZ (see thiabendazole)				
TCMTB, Ascend®, Busan® (Buckman; Wilbur Ellis)	2-(thiocyanomethylthio)benzothiazole)	Seed treatment for rice & seed storage diseases.	1590	Skin Irritant
tebuconazole, Folicur®, Horizon®, Elite®, Lynx® (Bayer CropScience)	(±)-2-(2,4-dichlorophenyl)-3-(1*H*-1,2,4-triazol-1yl)propyl 1,1,2,2-tetrafluorethyl ether	Systemic fungicide with curative & protective qualities on wide range of crops.	4000	>5000
Teldor® (see fenhexamid)				
Terraclor® (see PCNB)				
Terra-Cyte® (see sodium carbonate peroxyhydrate)				
Terraguard® (see triflumizole)				
terramycin, MycoShield®, Oxytetracycline® (Syngenta)	oxytetracycline hydrochloride or oxytetracycline calcium complex	Bacterial diseases of peaches, pears & ornamentals.	Low	
Terraneb® (see chloroneb)				
Terrazole® (see etridiazole)				
tetraconazole, Eminent® (ISAGRO)	(+)-2-(2,4-dichlorophenyl)-3-(1*H*-1,2,4-triazol-1-yl) propyl 1,1,2,2,tetrafluoroethyl ether	Systemic curative cereal fungicide for rust, powdery mildew & Septoria on peanuts, sugarbeets.	1031	>2000
thiabendazole, TBZ, Arbotec®, Mertect® (Syngenta)	2-(4'-thiazoyl)benzimidazole	Systemic for blue & green molds; soybean pod, stem blight; others.	3100	>2000
Thin-X® (discontinued)	fatty acids	Blossom thinning for fruit trees.		
thiophanate-methyl, Domain®, Fungo®, Topsin-M® (Crystal; Cleary)	dimethyl [(1,2-phenylene)bis-(imino carbonothioyl)bis-(carbamate)	Systemic, controls wide spectrum of diseases of fruits, vegs, field crops, grapes, potatoes, turf, greenhouses.	7500	>10,000
thiram (several) (Cerexagri)	tetramethylthiuramdisulfide	Seed treatment; diseases of fruits, vegs, turf.	1000	>5000
Tilt® (see propiconazole)				
TIM-BOR® (see sodium borate under Appendix C, Insecticides)				
tolclofos-methyl, Rizolex® (Sumitomo)	*O*-(2,6-dichloro-4-methylphenyl) *O,O*-dimethyl phosphorothioate	Non-systemic for seedling diseases, (*Rhizoctonia*, *Sclerotinia* & *Typhula*) on many crops & turf.	5000	>5000
Topshields® (see *Trichoderma harzianum*)				
Topsin E® (see thiophanate)				
Topsin M® (see thiophanate-methyl)				

Common name, trade name and basic manufacturer(s)	*Chemical name*	*General use pattern*	*Oral LD_{50} (rat)*	*Dermal LD_{50} (rabbit)*
Triact® (see neem oil)				
triadimefon, Bayleton® (Bayer CropScience)	1-(4-chlorophenoxy)-3,3-dimethyl-1-(1*H*-1,2,4-triazol-1-yl)-2-butanone	Systemic control of powdery mildews of fruits, vegs, cereals; rusts of cereals, coffee.	363	>1000
triadimenol, Baytan® (Bayer CropScience)	β-(4-chlorophenoxy)-α-(1,1-dimethylethyl)-1*H*-1,2,4-triazole-1-ethanol	Systemic control of cereal diseases; smuts, powdery mildew, rust.	700	>5000
Trichoderma harzianum strain KRLAG2, T-22®, Topshield®, Trichodex®, Plant Shield® (BioWorks)		Mycofungicide for *Pythium* seedling diseases. Downy & powdery mildew on greenhouse flowers & vegs.	Non-toxic	
Trichoderma harzianum strain ATCC20476, + *T polysporum* A TCC20475, Binab T® (Binab)		Mycofungicide for wood rot organisms, Fusarium & Botrytis.	Non-toxic	
Trichodex® (see *Trichoderma harzianum*)				
tridemorph, Calixin® (BASF)	2,6-dimethyl-4-tridecyl-morpholine	Variety of field crop, vegetable, fruit ornamental diseases.	825	1350
triflumizole, Terraguard®, Procure® (Crompton)	(E)-4-chloro-α,α,α-trifluoro-*N*-(-1-imidazol-1-yl-2-propoxyethylidene-*o*-toluidine	Systemic for powdery mildew, scab & rust on deciduous fruits, cherries, strawberries, vegs & cereals.	715	>5000
trifloxystrobin, Compass®, Gem®, Flint®, Twist® (Bayer CropScience)	benzeneacetic acid, (E,E)-α-(methoxyimino)-2-[[[[1-[3-trifluoromethyl)phenyl]ethylidene]amino] oxyl]methyl],methyl ester	Controls powdery mildew & leaf spots on cereals, grapes, pome fruits, almonds, tree fruits, vegs, potatoes, sugar beets, turf & ornamentals.	>16,000	>10,000
triforine, Funginex®, Denarin® (BASF)	*N,N'*-[piperazinediyl-bis(2,2,2-trichloroethylidene)]-bis-(formamide)	Systemic control of several fungal diseases of fruits, vegs, cereals, ornamentals.		
Trifungol® (see ferbam)				
Trilogy® (see neem)				
Trimaton® (see metam sodium)				
triphenyltin acetate (see fentin acetate)				
triphenyltin chloride (see fentin chloride)				
triphenyltin hydroxide (see fentin hydroxide)				
triticonazole, Charter®, Premis® (Bayer CropScience)	(±)-(E)-5-(4-chlorobenzylidene)-2,2-dimethyl-1-(1*H*-1,2,4-triazol-1ylmethyl)cyclopentanol	Seed treatment for cereals & corn, controls turf diseases on golf courses & turf farms.	>2000	>2000
Truban® (see ethazol)				
Tuffcide® (see chlorothalonil)				

Common name, trade name and basic manufacturer(s)	Chemical name	General use pattern	Oral LD_{50} (rat)	Dermal LD_{50} (rabbit)
TurfSide® (see PCNB)				
Twist® (see trifloxystrobin)				
Ultra Fluorish® (see metalaxyl-M)				
VacciPlant® (Agrimar)	natural product from brown algae	Used for fire blight in apples & pears		
validamycin, Validacin® (Takeda)	1L-(1,3,4/2,6)-2,3-dihydroxy-6-hydroxymethyl-4-[(1S,4R,5S,6S)-4,5,6-trihydroxy-3-hydroxymethyl=cyclohex-2-enylamino]cyclohelyl-β-D-glucopyranoside	Antibiotic for rice, seed potatoes & Rhizoctonia damping-off.	>20,000	—
Vangard® (see cyprodinil)				
Vapam® (see metam-sodium)				
Vectra® (see bromuconazole)				
vinclozolin, Ronilan®, Curalan® (BASF)	3-(3,5-dichlorophenyl)-5-ethenyl-5-methyl-2,4-oxazolidinedione	*Botrytis*, *Sclerotina* & *Monilia* spp. in fruits, vegs, ornamentals.	5000	>2000
Vista® (see fluquinconazole)				
Vitavax® (see carboxin)				
Xanthomonas compestris spp. *vesicatoria* (AgriPhi)		Bactofungicide for bacterial diseases of tomatoes, peppers & other vegs.		
zineb, Cuprothex®, Tritoftorol® (Bayer CropScience, Cerexagri)	zinc ethylenebis (dithiocarbamate)	Many diseases of fruits, vegs.	>5200	>10,000
ziram (several) (Cerexagri)	zinc dimethyldithiocarbamate	Many diseases of vegs, some fruits.	1400	>6000
zoxamide, Gavel®, Zoxium® (Dow Agrosciences)	3,5-dichloro-4-(3-chloro-1-ethyl-1-methyl-2-oxopropyol)-*p*-toluamide	For downy mildew, early & late blights, Phytophthora on vegs, grapes, ornamentals.	>5000	>2000

* Chemical nomenclature used is either that of Chemical Abstracts (American Chemical Society) or International Union of Pure and Applied Chemistry (IUPAC). The author and publisher make no warranty, express or implied, regarding the accuracy of this Appendix.

APPENDIX **E**

Rodenticides

Common, trade and chemical names, their basic manufacturer(s) and their Oral LD_{50}s to rats.

Common name, trade name and basic manufacturer(s)	*Chemical name**	*Oral LD_{50} (rat)*
alpha-chloralose (see chloralose)		
Alfamat® (see chloralose)		
Alfa-Z® (see chloralose)		
antu (discontinued)	α-naphthylthiourea	6
barium carbonate	barium carbonate	
Boot Hill® (see) brodifacoum)		630
brodifacoum,** Talon®, Ratak Plus®, Weather Blok® (Syngenta)	3-[3-[4'-bromo(1,1'-biphenyl)-4-yl]-1,2,3,4-tetrahydro-1-naphthalenyl]-4-hydroxy-*2H*-1-benzopyran-2-one	0.27
bromadiolone, Maki®, Boothill®, Hawk®, Contrac® (LiphaTech, Motomco)	3-[3-[4'-bromo(1,1'-biphenyl)-4-yl]-3-hydroxy-1-phenylpropyl]-4-hydroxy-*2H*-1-benzopyran-2-one	1.13
bromethalin, Vengeance®, Fastrac®, Gladiator®, Rampage® (Bayer CropScience, Motomco)	*N*-methyl-2,4-dinitro-*N*-(2,4,6-tribromophenyl)-6-(trifluoromethyl)benzenamine	2.0
Caid® (see chlorophacinone)		
chloralose, alpha-chloralose, Alfamat® (RIMI)	glucochloral $C_6H_{11}O_6Cl_3$	200
chlorophacinone, Rozol®, Caid® (LiphaTech)	2-[(*p*-chlorophenyl)phenylacetyl]-1*H*-indene-1,3(2*H*)-dione	3.1
cholecalciferol, AZ Rat®, Muritan® (Bayer, KI Ambiente)	9,10-secocholesta-5,7,10(19)trien-3 betanol; activated 7-dehydrocholesterol	Bait low Tox.
Compound 1080	sodium fluoroacetate or sodium monofluoroacetate (Used only in government predator control projects).	0.22
Compound 1081, Rhodex® (RIMI)	fluoroacetamide	5.75
Contrac® (see bromadiolone)		
coumachlor (Syngenta)	3-(α-acetonyl-4-chlorobenzyl)-4-hydroxycoumarin (not sold in U.S.)	900
coumafuryl, Fumarin® (discontinued)	3-(α-acetonylfurfuryl)-4-hydroxycoumarin	25

* Chemical nomenclature used is either that of Chemical Abstracts (American Chemical Society) or International Union of Pure and Applied Chemistry (IUPAC). The author and publisher make no warranty, express or implied, regarding the accuracy of this Appendix.

Common Name, trade name and basic manufacturer(s)	*Chemical name*		*Oral LD_{50} (rat)*
coumatetralyl, Racumin® (Bayer)	3-(α-tetralyl)-4-hydroxycoumarin (not sold in U.S.)		16.5
Cov-R-Tox® (see warfarin)			
crimidine, Castrix® (discontinued)	2-chloro-4-dimethylamino-6-methylpyrimidine		1.25
D-Cease® (see difethialone)			
Dethdiet® (see red squill)			
difenacoum, Ratak®, Sorexa® (Sorex)	3-(3-1,1'-biphenyl-4yl-1,2,3,4-tetrahydro-1-naphthylenyl)-4-hydroxy-2*H*-1-benzopyran-2-one	(Not sold in U.S.)	1.8
difethialone, Frap®, Generation®, D-Cease® (LiphaTech)	3-[1R,S,3RS:1RS,3RS)-3-(4-bromodyshenyl-4-yl)-1,2,3,4-tetrahydro-1-naphthyl]4-hydroxy-1-benzothi-*m*-2-one	(Used only by professionals)	0.4
Diphacin® (see diphacinone)			
diphacinone, Diphacin®, Havoc Chunks®, Tomcat®, Promar®, Ramik® (HACCO, Motomco)	2-diphenylacetyl-1,3-indandione		1.86-2.88
endrin (discontinued)	1,2,3,4,10,10-hexachloro-7-epoxy-1,4,4a,5,6,7,8,8a-octahydro exo-1,4-exo-5,8-dimethanonaphthalene		10
Fastrac® (see bromethalin)			
flocoumafen, Storm®, Stratagem® (BSAF)	4-hydroxy-3-[1,2,3,4-tetrahydro-3-[4-(4-trifluoromethylbenzyloxy)phenyl]-1-naphthyl] coumarin (not sold in U.S.)		0.25
fluoroacetamide (see Compound 1081)			
Frap® (see difethialone)			
Generation® (see difethialone)			
Gladiator® (see bromethalin)			
Havoc Chunks® (see diphacinone)			
Hawk® (see bromadiolone)			
isovaleryl (see Valone®)			
Maki® (see bromadiolone)			
Muritan® (see cholecalciferol)			
norbormide, Raticate® (discontinued)	5-(α-hydroxy-a-2-pyridyl-benzyl)-7-(α-2-pyridyl-benzylidene)-5-norbornene-2,3-dicarboximide		5.3
nux vomica (see strychnine)			
phosacetim, Gophacide® (discontinued)	*O,O*-bis(4-chlorophenyl)acetimidoylphosphoramidothioate		3.7
phosphorus (see yellow phosphorus)			
pindone, Pival® (discontinued)	2-pivaloylindane-1,3-dione		280
Prolin® (discontinued)	warfarin plus sulfaquinoxaline		

Common Name, trade name and basic manufacturer(s)	*Chemical name*	*Oral LD_{50} (rat)*
Promar® (see diphacinone)		1000
Prozap® (see zinc phosphide)		
pyrinuron, Vacor® (discontinued)	1-(3-pyridylmethyl)-3-(4-nitrophenyl)urea	4.75
Quintox® (see cholecaliciferol)		
Racumin® (see coumatetralyl)		
Ramik® (see diphacinone)		
Rampage® (see cholecalciferol)		
Ratak® (see difenacoum)		
Ratak Plus® (see brodifacoum)		
Ratol® (see zinc phosphide)		
Red Quill® (see scilliroside)		
red squill, Dethdiet®, (Syngenta)	(From plant: *Urginea maritima)* scilliroside = (3b,6b-6-acetyloxy-3-(β-D-glucopyranosyloxy)- 8,14-dihydroxybuta-1,20,22-trienolide)	0.7
Rhodex® (see fluoroacetamide)		
Rozol® (see chlorophacinone)		
scilliroside (see red squill) (Kwisda) Red Quill®	For Norway rats.	
sodium cyanide, Cymag® (discontinued)	sodium cyanide	0.5
sodium fluoroacetate (see compound 1080)		
Sovex® (see difenacoum)		
Storm® (see flocoumafen)		
Stratagem® (see flocoumafen)		
strychnine alkaloid, nux vomica (several baits)	alkaloid from tree, *Strychnos nux-vomica* (only for gopher and mole control in U.S.)	1 to 30
Talon® (see brodifacoum)		
Tomcat® (see diphacinone)		
thallium sulfate, Ratox® (discontinued)	thallium sulfate (for use by government agencies only)	16
Tox-Hid® (see warfarin)		
Valone® (discontinued)	2-isovaleryl-1,3-indandione	50
Vengeance® (see bromethalin)		

Common Name, trade name and basic manufacturer(s)	*Chemical name*	*Oral LD_{50} (rat)*
warfarin, Cov-R-Tox®, Kaput®, Rodex®, Tox-Hid® (Hacco)	3-(a-acetonylbenzyl)-4-hydroxycoumarin	Varies greatly from 1 to 186 (in daily doses)
warfarin, sodium salt	water soluble sodium salt of warfarin	Same as warfarin
Weather Blok® (see brodifacoum)		
yellow phosphorus (discontinued)	elemental phosphorus	<6
zinc phosphide, Ratol®, Prozap®, ZP Rodent Bait® (several) (HACCO, Motomco)	zinc phosphide	45

* Chemical nomenclature used is either that of Chemical Abstracts (American Chemical Society) or International Union of Pure and Applied Chemistry (IUPAC). The author and publisher make no warranty, express or implied, regarding the accuracy of this Appendix.

** Brodifacoum has been reformulated as the new Talon® with Bitrex (denatonium benzoate) which acts as a human taste deterrent. Bitrex is the most bitter substance known to man, but apparently is tasteless to rodents. Its purpose is to prevent the accidental ingestion by humans, but especially children.

PESTICIDES THAT CAN BE MADE AT HOME

APPENDIX F

INSECTICIDES

Detergents. Add a tablespoon of household detergent to a plastic bottle that has a pistol pump spray handle, and fill with water. When ants or any other undesirable crawling pests appear, a squirt of the detergent water kills them. If they were sprayed in the house, they can be wiped up with a sponge.

Diatomaceous Earth. There are several home remedies for snails and slugs that do not include metaldehyde baits. Diatomaceous earth is very effective, easily applied, and nontoxic. Snails and slugs have a very soft underside, and they move by gliding across their own thin film of slime that they produce as they crawl. As they move over diatomaceous earth they puncture their tender undersides and die. Surround the garden or area to be protected with a 12-24 inch band of diatomaceous earth. Snails stop moving almost immediately. Ground oyster shells also work, but not quite as well. To prevent either of these slimy pests from traveling up trees and shrubs, attach thin *copper strips* around the trunks.

Boric Acid. An excellent cockroach bait is made by mixing equal parts of boric acid powder, powdered sugar, and cornmeal. The bait should be placed in small containers such as bottle caps or individual condement serving cups and placed in protected places, e.g. beneath refrigerator, cabinets, stove, etc.

Pharaoh ants are attracted to and killed by a mix of mint apple jelly and boric acid (2 tablespoons of boric acid powder blended into 10 ounces of the jelly). Another Pharaoh ant control, involving boric acid, is made by blending together 3 tablespoons of smooth peanut butter, 1 teaspoon of brown sugar, 1 teaspoon of boric acid powder and 1/4 teaspoon of salt. Serve in bottle caps.

All ants entering homes can be attracted to and killed with a mixture of equal parts of boric acid and confectioner's sugar, or a syrup concoction made by dissolving 1 cup of sugar and 4 teaspoons boric acid powder in 3 cups of water and saturating cotton wads placed in small dishes or bottle caps.

Plaster of Paris. Plaster of Paris mixed with equal parts of powdered sugar makes another successful cockroach poison bait. The plaster of Paris, once moistened in the gut of the roach, begins to harden. Guess what!

Nicotine extract. Soak one or two shredded, cheap cigars, or a standard plug of chewing tobacco in 1 gallon of water overnight at room temperature. Remove tobacco parts and add 1 teaspoon of household detergent. Used as a foliar spray against aphids and other small insects it is usually as effective as sprays made commercially. CAUTION: This nicotine extract is very toxic to all living animals, including man. It may also transmit tobacco virus to tomatoes and other sensitive plants.

Lime-sulfur concentrate. Use 1 pound of unslaked lime (quicklime), 2 pounds of sulfur and 1 gallon of water for each batch. Make a paste of the sulfur with some of the water. In a well-ventilated room or out-of-doors heat the water and lime until the lime starts to dissolve, then add the sulfur paste. Bring to a boil and simmer for about 20-30 minutes or until the material becomes dark amber. Add water as necessary to maintain original level. When the free sulfur has all disappeared the reaction is complete, and the mixture should be strained and stored in jars.

This is the *concentrate*. To use as a dormant spray, dilute 1 part of concentrate to 9 parts of water. As a summer foliage spray, dilute 1 part to 49 parts of water. Lime sulfur sprays will stain the skin and clothing. Do not apply to vegetables to be canned. Jars that contain trace amounts of elemental sulfur are known to explode.

Dormant oil spray concentrate. Heat 1 quart of good lubricating oil (10 wt) free of additives, white oil such as mineral oil is preferred, with 1 pint of water and 4 ounces of potash fish-oil soap. Yellow kitchen bar soap can be used as a substitute for fish-oil soap. Heat until the boiling point is reached. Pour back and forth until blended. For deciduous trees, mix 5 teaspoons of concentrate with 1 quart of water. For evergreens, use 2 teaspoons per quart. For leftover concentrate, reheat and mix again before using.

Kerosene emulsion concentrate. Dissolve 2 cubic inches of kitchen grade bar soap in 1 pint of water. Add 1 quart of kerosene and beat with egg beater or electric mixer until a thick cream results. For dormant deciduous, trees dilute 1 part of concentrate with 7 parts of water. For ordinary mid-summer growth, dilute with 10 parts water, and for a weak spray, dilute with 15 parts water.

Bordeaux mixture. Add 3 ounces of copper sulfate (bluestone) to 3 gallons of water and dissolve thoroughly. Add 5 ounces of hydrated lime and mix completely. Use without further dilution. This fungicide-insecticide can be stored but is corrosive to metal containers. While primarily a fungicide, it is also very repellent to many insects such as flea beetles, leafhoppers and potato psyllid, when sprayed over the leaves of plants.

Wormwood. Use dry or fresh wormwood leaves. Cover with water and bring to a boil. Dilute with 4 parts of water and spray on plants immediately. Wormwood sprays are purported to kill slugs, aphids and crickets.

Quassia. Soak 2 ounces of quassia wood chips in 1 gallon water for 2-3 days. Simmer for 2-3 hours over low heat. Remove chips and mix with 2 ounces of soft soap. It's not the best but will control aphids.

Soaps. Soap dilutions have been used for control of soft-bodied insects, such as aphids, since 1787, when this control method first appeared in writing. Undoubtedly it had been used long before that. Most often these soaps were derived from either plants (cottonseed, olive, palm, or coconut) or from animal fat, such as lard, whale oil, or fish oil. Vegetable or plant-derived soaps are more effective than those derived from petroleum. Commercial soaps today vary greatly in composition and purity, therefore vary widely in effectiveness.

If in doubt, you should try true soap suds from a known brand of inexpensive laundry soap against aphids on only a few plants. If this proves successful, the practice could be expanded. Old-fashioned homemade soap may be prepared using inexpensive waste lard or tallow, lye, water and borax (optional). Six pounds of fat and a can of lye will make 7 pounds of soap, or 12 to 15 pounds of soft soap.

To make, strain 3 quarts of heated-to-melting fat. Dissolve a can of commercial lye in one quart of cold water. When this is dissolved and the water is still warm, stir into it the warm fat and add 1 cup of ammonia and 2 tablespoons of borax which have been dissolved in a half-cup of warm water. If fragrance is desired add a teaspoon of citronella oil. Stir thoroughly to mix and allow to cool. The soap will harden and can then be cut up into bars.

If soft soap is preferred, add 3 quarts of hot water where the oil of citronella is called for. This of course will prevent hardening and can be spooned or poured after cooling.

(See also SOAPS in Appendix G).

HERBICIDES

Borax. For grass and weeds growing between stones or bricks on walks or terraces, sprinkle 20 Mule Team Borax® powder and sweep into cracks. One application every other year should be adequate.

Corn gluten meal. Working as both a natural herbicide and fertilizer (10% nitrogen), a corn by-product identified as corn gluten inhibits the germination of crabgrass and other annual weeds. It was discovered at Iowa State University in 1992, and has become available commercially, marketed for crops as well as home gardens. It is not a complete substitute for synthetic products, but it is a viable alternative for those who don't want to use traditional preemergence herbicides.

Petroleum oils. A non-selective, all-purpose weed killer can be made using kerosene, paint thinner, or diesel fuel. This is an effective contact herbicide for all vegetation and should be used with caution. Though only temporary in effect, it is fast-acting and probably the safest of the materials to use around the home. Especially good for driveways, alleys and paths. Caution: May leave oily residue on concrete and stones.

Copper sulfate (bluestone). Copper in concentrated form is lethal to all plant life. To make an all-purpose weed killer that will also sterilize the soil on which it falls for 2-3 years, dissolve 8 ounces of copper sulfate in 1 gallon of water. This is corrosive to metal and sprayers should be rinsed thoroughly after use.

FUNGICIDES

Bordeaux mixture. See instructions for making under "Insecticides". It is particularly effective against downy mildew and several diseases of deciduous fruits when applied as a dormant spray.

Sulfur. Dusting sulfur, flotation or colloidal sulfur and wettable sulfur are all available for fungicidal use. Sulfur is quite effective against powdery mildews and most effective when temperature is above 70°F. Caution: Do not use when temperatures exceed 90°F, or on susceptible plants. Do not apply to vegetables to be canned. Jars are known to explode that contain trace amounts of elemental sulfur.

Lime-sulfur. Primarily a dormant fungicide for deciduous fruit trees. (See "Insecticides" for recipe).

ALGAECIDES

Copper sulfate. (Bluestone) Copper is lethal to all algae. Where other plants are not involved and there is no need to control bacteria, copper sulfate can be effectively used to inhibit algae growth at the rate of 1 to 2 ounces per 100 gallons of water.

INSECT REPELLENTS FOR PLANTS

Garlic and onion. In a blender, place onion, garlic, horseradish, peppers, mint or any other aromatic herbs. Add enough water to blend into a frappe. Dilute with an equal volume of water and let stand overnight at room temperature. Filter through an old stocking, add 1 teaspoonful of detergent and spray on plants to be protected. This is only a repellent to a few species and will not kill any insect.

Horseradish. See "Garlic and Onion".

Peppers. See "Garlic and Onion".

Mint. See "Garlic and Onion".

None of these are very good and are not recommended.

INSECT REPELLENTS FOR HUMANS

Skin Lotion. Avon Home Products' *Skin-So-Soft®* is one of the better mosquito and biting fly repellents, though it was not manufactured for that purpose. It also helps relieve itching caused by insect bites and dry skin.

Vitamin B Complex. Daily "mega-doses" of vitamin B complex are purported to repel biting insects.

Anise oil. A fragrant oil with slight repellency to gnats, biting flies and mosquitoes.

Oil of citronella. One of the better natural repellents against mosquitoes and biting flies, with a pleasant fragrance.

Asefetida. A terrible smelling herb that has absolutely no insect repellent qualities. (It does keep other humans away!)

Oil of Pennyroyal. A pleasant, short-life repellent for mosquitoes and biting flies. Pennyroyal herb rubbed fresh on skin is reported also to be effective for a short period. The fragrance resembles mint.

Camphor. Dissolve 2 squares of camphor in 1 pint of olive oil.

INSECT REPELLENTS FOR ANIMALS

Brewer's Yeast-Garlic Powder. Ticks can be controlled on dogs without the need for dipping. Add two tablespoons of brewer's yeast-garlic powder mix daily to the dog's food for an 80+ pound animal. Scale the dose down for smaller animals. This mix is available at most health food stores.

Tansy. Dried tansy leaves rubbed into a pet's fur are reputed to drive off fleas.

Oil of Lavender. This is purported to drive fleas off dogs when rubbed into their fur. Dogs probably won't be any happier for your efforts, however!

INSECT REPELLENTS FOR THE HOME

Tansy. Dried tansy leaves are supposed to keep ants off kitchen shelves and protect stored woolens and furs from moths by its repellent action.

Camphor. Camphor crystals have long been used for moth protection of woolens and furs in closed containers as chests and drawers.

Cedar. Still an old favorite is the pungent fragrance of cedar-lined chests and closets which are only moderately effective in repelling clothes moths from furs and woolens stored therein.

Oil of lavender. Supposed to drive fleas and sand fleas from houses in the summer by sprinkling on carpets and floors. Of doubtful effectiveness.

Diatomaceous earth. When sprinkled heavily on dung heaps (chicken, cow, horse and dog) diatomaceous earth apparently repels flies so that egg laying is prevented, thus the usual reproduction is prevented.

Oil of Pennyroyal. The oil or the fresh herb rubbed on kitchen shelves will act as a short-lived ant repellent.

ANIMAL REPELLENTS

Moles and gophers. Blend 2 ounces of castor oil with 1 ounce of liquid detergent. Add an equal volume of water and blend again. Fill a 2-quart sprinkling can with warm water and add 2 tablespoons of the mix, stir, and sprinkle over infested areas. Puncture mole tunnels and treat sparingly.

Gophers and ground squirrels. These burrowing animals can be forced to leave for more pleasant quarters. Pour 1 cup of standard laundry bleach (the active ingredient is 7% calcium hypochlorite) into the burrow. Within an hour, toss in 6-10 moth balls (naphthalene), then cover the opening with soil. There is no hazard to pets because of the strong repellency of this method.

Squirrels. Mix ground red (cayenne) pepper with Vaseline or any other petroleum jelly and smear on the surfaces you don't want the squirrels to climb, e.g. bird feeder poles, metal bands around selected trees.

Mice and Rats. Ground cayenne pepper sprinkled at likely entry points and along their runs, identified by their fecal pellets, has proved very effective.

Another remedy for stopping these rodents from gnawing on any object is to coat or paint on a film of Tabasco pepper sauce. The principle ingredient in Tabasco sauce is the pepper, *Capsicum frutescens* var. *tabasco*, which is the ultimate of peppers. The active ingredient is capsaicin, described by the Merck Index as "a powerful irritant that causes intense pain in humans and experimental animals".

To prevent rodents from entering and exiting through their old familiar gnawed openings in wood or drywall, pack the opening with steel wool or copper mesh similar to that used for scrubbing pots and pans.

Dogs and rabbits in shrubs and flower beds. Red or cayenne pepper powder, or napthalene flakes sprinkled on the dogs' urinating areas and where rabbits have gnawed or eaten foliage, work equally well. Tobacco dust, lightly sprinkled over the area to be protected every 5-7 days acts as a good repellent. Also, tansy, rue and rosemary interplanted in the flower bed will offer some repellency and may make the area generally undesirable for these visitors.

Dogs and cats on furniture. Dogs and cats can be prevented from sleeping in your best arm chairs or other locations by placing 2-3 layers of newspaper over the entire sitting surface. An alternate is to place one or two naphthalene moth balls or a few moth crystals (paradichlorobenzene) in their normal resting place. Another that works, but I do not recommend, is to give the fabric, but not wood surfaces, a very light spray of nicotine.

Dogs, raccoons, opossums, skunks in garbage bags or trash cans. These 4-legged pests will not disturb trash bags or trash cans that have been sprayed with a 1:1 dilution of Pine-Sol® or similar pine-scented cleaning detergent.

A second guaranteed repellent is a mild ammonia solution and a teaspoon of cayenne powder sprayed liberally on the tied trash bags or lids of trash cans.

Cats. To prevent cats from defecating or urinating in flower or vegetable gardens, work about a 1/4 inch layer of chicken fertilizer into the soil. The odor is not noticeable to humans.

Since cats are territorial, borax soap sprinkled around the outer perimenter of flower beds and shrubs has been effective. Pay special attention to the areas where cats enter and exit.

Fill a mason or mayonnaise jar with water. Fold aluminum foil into a strip 1 1/2 inches wide by the length of the jar and lower it into the jar. Then add 2-3 drops of laundry bleach to prevent the growth of algae. Place the jars around the problem area. Reflections appear to frighten cats away, but this, like the others, is not a permanent solution.

Deer. Deer are very persistent creatures when hungry and consequently are difficult to keep away. They may be repelled from fruit trees and areas by dusting the subject targets with blood meal or animal blood procured from a slaughter house.

Blood meal diluted with water and sprayed on dormant trees is reported to be successful in controlling deer damage. Effective control requires frequent applications, sometimes as often as weekly. Hanging kerosene- or creosote-soaked rags, or perforated cans of moth balls or crystals around the garden or orchard perimeter sometimes works, at least for a short period. Rags should be retreated at monthly intervals. Hanging an old nylon stocking containing 2-3 handfuls of human hair from trees and shrubs has been reported as an effective deer-damage control. Fresh hair is required several times per season.

Bars of Dial® soap and small hotel soap bars, with wrappers removed and suspended chest high with string, from the limbs of fruit trees, is used extensively in the Northern and Eastern parts of the U.S. and is reported to be an effective deterrent.There are a large number of plants not preferred by deer. A source of this information is found under *Deer* in Chapter 6, *Rodents and Other Animal Pests.*

Rabbits. An effective rabbit repellent can be made by dissolving 7 pounds of tree rosin in 1 gallon of denatured alcohol. Allow this mixture to stand in a warm place for 24 hours. It should be stirred occasionally to dissolve the rosin. The mixture can be painted or sprayed on dry tree trunks in the fall. For spraying, thin the mixture by adding another gallon of denatured alcohol. Trees should be treated 2 feet higher than snow is expected to drift. This formula is not effective against *snowshoe hares.*

BIRD REPELLENTS

There are bird repellents, but none that can be made by the home gardener. (See Controlling Birds in the **BIRDS** chapter.)

SKUNK DEODORANTS

To remove the skunk scent from pets that have been "greeted" by a skunk, give them a good bathing in the following mixture for pets: 1 quart of 3% hydrogen peroxide (from the drugstore), 1/4 cup of baking soda (sodium bicarbonate), and 1 teaspoon of liquid soap or detergent. The bath should be followed with a tap water rinse. Guaranteed! (Tomato juice is a waste of time. It simply doesn't work). This same mixture can be used as a spray-on or wipe-on mixture for surfaces contacted by skunks. With a fabric, such as carpeting, treat a small area, to see initial effects, e.g. bleaching, staining. If no objectionable effect is visible after 8 hours, then treat the entire contaminated area.

A second skunk deodorizing elixir is composed of 1 cup of sodium perborate, sold in the super market under the proprietary name of Snowy Bleach®, 3 tablespoons of liquid detergent, in 1 gallon of water. Wash the animal gently for about 10 minutes while pouring the mixture repeatedly over all body parts, but avoiding the eyes. Rinse the animal thoroughly, followed with a bath of detergent only, preferably a shampoo grade. Rinse the animal twice and dry quickly with bath towels. Another guaranteed deodorant!

BENEFICIAL PREDATORS AND PARASITES

Abbreviations used:

PM	=	predatory mites
PN	=	parasitic nematodes
SP	=	stored product parasites & predators
AP	=	aphid parasites & predators
WP	=	whitefly parasites & predators
PG	=	greenhouse parasites & predators
SM	=	scale & mealybug parasites & predators
EP	=	egg parasites
LP	=	moth & butterfly larval parasites
FP	=	filth fly parasites
OP	=	other parasites
GP	=	general predators

United States

ARBICO, Inc.
P.O. Box 4247 CRB
Tucson, AZ 85738-1247
520/825-9785; FAX 520/825-2038
ARBICO@aol.com
PM, PN, SP, AP, WP, PG, SM, EP, LP, FP, GP, WC

Beneficial Insectary
9664 Tanqueray Ct.
Redding, CA 96003
800/477-3715, 530/226-6300
FAX 530/226-6310
bi@insectary.com
PM, AP, WP, PG, EP, FP, GP

Beneficial Insects Co.
244 Forrest Street
Fort Mill, SC 29715
803/547-2301
PM, PN, SP, AP, WP, PG, SM, EP, FP, GP

Bio Ag Services
4218 West Muscat
Fresno, CA 93706
559/268-2835: FAX 559/268-7945
AP, WP, PG, EP, GP

Biofac Crop Care, Inc.
P.O. Box 87
Mathis, TX 78368
800/233-4914, 361/547-3259
FAX 361/547-9660
PM, SP, AP, WP, PG, EP, LP, FP, GP

BioTactics, Inc.
20780 Warren Rd.
Perris, CA 92570
909/943-2819; FAX 909/943-8080
www.benemite.com
PM, WP, PG, EP, GP

Bonide Products
6301 Sutliffe Rd.
Oriskany, NY 13424
315/736-8231; FAX 315/736-7582
www.bonideproducts.com
PM, WP, PG, EP, GP

Buena Biosystems
P.O. Box 4008
Ventura, CA 93007
805/525-2525; FAX 805/525-6058
bugdude@msn.com
PN, AP, WP, SM, FP, GP

Burpee Seed Company
300 Park Avenue
Warminster, PA 18974
800/888-1447; FAX 800/487-5530
http://garden.burpee.com
AP, WP, PG, EP, FP, GP

CropKing Inc.
5050 Greenwich Rd.
Seville, OH 44273
330/769-2002; FAX 330/769-2616
cropking@cropking.com
PM, PN, SP, AP, WP, PG, SM, EP, FP, GP

Gardens Alive!
5100 Schenley Place
Lawrenceburg, IN 47025
812/537-8652; FAX 812/537-8660
PM, PN, AP, WP, PG, EP, FP, GP

Gardener's Supply Co.
128 Intervale Rd.
Burlington, VT 05401
800/955-3370; FAX 800/551-6712
info@gardeners.com
PN, AP, WP, PG, EP, FP, GP

Great Lakes IPM
10220 Church Road NE
Vestaburg, MI 48891
800/235-0285, 989/268-5693
FAX 989/268-5311
PM, PN, SP, AP, WP, PG, SM, EP, LP, FP, GP

Hydro-Gardens, Inc.
P.O. Box 25845
Colorado Springs, CO 80936
719/495-2266, 800/634-6362
FAX 719/495-2266
PM, PN, AP, WP, PG, SM, EP, GP

IPM Laboratories, Inc.
P.O. Box 300
Locke, NY 13092-0300
315/497-2063; FAX 315/497-3129
PM, PN, AP, WP, PG, SM, EP, FP, PG

Nature's Control
P.O. Box 35
Medford, OR 97501
541/541-6033, 800/698-6250
FAX 541/899-9121
www.naturescontrol.com
PM, PN, AP, WP, PG, SM, EP, FP, GP

Rincon-Vitova Insectaries Inc.
P.O. Box 1555
Ventura, CA 93002-1555
805/643-5407, 800/248-2847
FAX 805/643-6267
PM, PN, AP, WP, PG, SM, EP, FP, GP

San Jacinto Environmental Supplies
2221-A West 34th Street
Houston, TX 77018
713/957-0909; FAX 713/957-0707
PM, PN, AP,WP, PG, SM, EP, GP

Territorial Seed Company
P.O. Box 157
Cottage Grove, OR 97424
541/942-9547; FAX 541/942-9881
TSC@ordata.com
PN, AP, WP, PG, GP

Tip Top Bio-Control
P.O. Box 7614
Westlake Village, CA 91359-7614
805/375-1382; FAX 805/375-3180
PN, AP, WP, SM, FP, GP

Worm's Way, Inc.
7850 North Highway 37
Bloomington, IN 47404-9477
800/274-9676 (IN); 800/283-9676 (FL)
800/284-9676 (MA); FAX 800/316-1264
jwray@wormsway.com
PM, PN, AP, WP, PG, SM, EP, GP

Canada

Coast Agri Ltd.
RR #2, 464 Riverside Road South
Abbotsford, BC
CANADA V2S 7N8
604/864-9044; FAX 604/864-8418
wporter@coast-agri.com
PM, AP, WP, PG, SM, EP, FP, GP

Natural Insect Control
RR #2, Stevensville, Ontario
CANADA L0S IS0
905/382-2904; FAX 905/382-4418
PM, PN, SP, AP, WP, PG, SM, EP, LP, FP, GP

Westgro Sales Inc.
7333 Progress Way
Delta, BC
CANADA V4G 1E7
604/940-0290; FAX 604/940-0258
westgro@westgro.com
PM, PN, AP, WP, PG, SM, EP, FP, GP

Mexico

CREROB-Matamoros
Sendero Nacional Km. 1
Apartado Postal No. 550
Matamoros, Tamaulipas
MEXICO
891/2-12-02
AP, WP, EP, GP

CREROB-Cuidad Obregón
Bvd. Rodolfo Elias Calles No. 711 Pte.
Colonia Sochiloa
Cuidad Obregón, Sonora C.P. 85150
MEXICO
641/6-54-07, 641/7-58-51
AP, WP, PG, EP, GP

MOSQUITO FISH--*Gambusia affinis*

Chico Game Fish Farm
971 East Avenue
Chico, CA 95926
530/343-1849

J & J Aquafarms
P.O. Box 922
Sanger, CA 93657
209/857-0477; FAX 209/262-2324
jjaquafarms@juno.com

WEED CONTROLS

Common Weeds

Bio Collect
5481 Crittenden St.
Oakland, CA 94601
510/436-8052; FAX 510/532-0288
Biological weed controls

Biological Control of Weeds
1418 Maple Dr.
Bozeman, MT 59715
406/586-5111; FAX 406/586-1679
Biological weed controls
www.bio-control.com

Aquatic Weeds—Chinese grass carp *(Ctenopharyngodon idella)*

J. M. Malone & Son Enterprises
P.O. Box 158
Lonoke, AR 72086
501/676-2800; FAX 501/676-2910

Keo Fish Farm, Inc.
P.O. Box 166
6444 Highway 165 N
Keo, AR 72083
501/842-2872; FAX 501/842-2156

SNAIL CONTROLS—Predatory decollate snail *(Rumina decollata)*

ARBICO, Inc.
P.O. Box 4247 CRB
Tucson, AZ 85738-1247
520/825-9785; FAX 520/825-2038
ARBICO@aol.com

EARTHWORMS

Burpee Seed Company
300 Park Avenue
Warminster, PA 18974
800/888-1447; FAX 800/487-5530
http://garden.burpee.com

Tip Top Bio-Control
P.O. Box 7614
Westlake Village, CA 91359-7614
805/375-1382; FAX 805/375-3180

Worm's Way Inc.
7850 North Highway 37
Bloomington, IN 47404-9477
800/274-9676 (IN); 800/283-9676 (FL)
800/284-9676 (MA); FAX 812-331-0854
jwray@wormsway.com

FROGS

Nu-Tex Frog Farm
P.O. Box 4029
Corpus Christi, TX 78400

BOTANICAL INSECTICIDES
(Several garden supply houses market their own brand-name botanicals)

Abamectin (avermectins)

Peaceful Valley Supply (Avid®)
P.O. Box 2209
110 Springhill Dr.
Grass Valley, CA 95945
530/272-4769; FAX 530/272-4794
www.groworganic.com

Planet Natural (Ascend®)
1612 Gold Ave
Bozeman, MT 59715
800/289-6656; FAX 406/587-0223
www.planetnatural.com

Limonene (d-limonene)

Helena Chemical Co. (Kammo®)
225 Schilling Blvd, Ste. 300
Collierville, TN 38017
901/761-0050; FAX 901/756-9947
www:helenachemical.com

Orange Guard (Orange Oil®)
7 Trampa Canyon Rd.
Carmel Valley, CA 93924
888/659-3217; FAX 831/659-5128
www.orangeguard.com

Neem (azadirachtin)

Peaceful Valley Supply
(Safer's Bioneem®, Neemix®)
P.O. Box 2209
110 Springhill Dr.
Grass Valley, CA 95945
530/272-4769; FAX 530/272-4794
www.groworganic.com

Harmony Farm Supply (Azatin®)
3244 Gravenstein Hwy., No. B
Sebastopol, CA 95472
707/823-9125; FAX 707/823-1734
www.harmonyfarm.com

Nicotine sulfate (discontinued)
(See Appendix F, Pesticides That Can Be Made At Home)

Pyrethrins

Monterey Lawn & Garden Products Inc.
(Bug Buster O®, Pyganic®)
3654 S. Willow Ave.
P.O. Box 35000
Fresno, CA 93745
559/499-2100; FAX 559/499-1015
www.montereychemical.com

Nichols Garden Nursery
(pyrethrum powder)
1190 Old Salem Rd. NE
Albany, OR 97321-4580
800/422-3985; FAX 541/967-8406

Rotenone (cubé)

Bonide Products
6302 Sutliffe Rd.
Oriskany, NY 13424
315/736-8231; FAX 315/736-7582
www.bonideproducts. corn

Peaceful Valley Supply
P.O. Box 2209
110 Springhill Dr.
Grass Valley, CA 95945
530/272-4769; FAX 530/272-4794
www.groworganic.com

Ryania (formulators)

AgriSystems International
125 W. Seventh St.
Wind Gap, PA 18091
610/863-6700

Dunhill Chemical
642 S. Duggan
Azusa, CA 91702
626/815-1663; FAX 626/815-1687

Sabadilla (formulators)

Harmony Farm Supply
3244 Gravenstein Hwy, No. B
Sebastopol, CA 95472
707/823-9125; FAX 707/823-1734
www. harmonyfarm.com

Peaceful Valley Supply (Veratran®)
110 Springhill Dr.
Grass Valley, CA 95945
530/272-4769; FAX 530/272-4794
www.groworganic.com

Cinnamaldehyde

Emerald BioAgriculture
(Cinnamite®, Valero®)
3125 Sovereign Dr., Ste. B
Lansing, MI 48909-0519
517/882-7536; FAX 517/882-7521

Eugenol

Bioganic Crop Protection (Matran 33®)
EcoSMART Technologies, Inc.
318 Seabord Lane, Ste. 208
Franklin, TN 37067
888/326-7233
www.BIOGANIC.com

Jojoba Oil

IJO Products, LLC (Detur®, E-Rase®)
4672 W. Jennifer, Ste. 103
Fresno, CA 93722
559/221-6048; FAX 559/221-6049
www.ijoproducts.com

Rosemary Oil

Bioganic Crop Protection (Hexacide®, EcoTrol®)
(See Eugenol, above)

Bacillus lentimorbus / Bacillus popillae
(Milky disease spores for beetle grubs)

St. Gabriel Labs (Milky Spore®)
14044 Litchfield Rd.
Orange, VA 22960
800/801-0061, 540/672-0866
FAX 540/672-0052
www.milkyspore.com

North Country Organics (Milky Spore®)
P.O. Box 372
Bradford, VT 05033
802/222-4277; FAX 802/222-9661
www.noorganics.com

Bacillus thuringiensis spp. aizawai
(Caterpillar larvicide)

Certis (Agree®)
9145 Guilford Rd., Ste. 175
Columbia, MD 21046
800/250-5020, 301/604-7340
FAX 301/604-7015
www.certisusa.com

Valent USA (Xentari®, Florbac®)
P.O. Box 8025
Walnut Creek, CA 94596-8025
800/624-6094, 925/256-2700
FAX 925/256-2776
www.valent.com

Bacillus thuringiensis spp. israelensis
(For mosquito & blackfly larvae)

Valent BioSciences (Vectobac®)
870 Technology Way
Liberty, IL 60048
800/323-9597; FAX 847/968-4780
www.valentbiosciences.com

Peaceful Valley Supply (Bactimos®)
P.O. Box 2209
110 Springhill Dr.
Grass Valley, CA 95945
530/272-4769; FAX 530/272-4794
www.groworganic.com

Wellmark International (Teknar®)
1100 E. Woodfield Rd., Ste. 500
Schaumberg, IL 60173
800/248-7763; FAX 800/426-7473

Bacillus thuringiensis spp. kurstaki
(Caterpillar larvicide)

Certis (Condor®, Crymax®)
(See above under *Bacillus thuringiensis aizawai)*

Valent USA (Biobit®, Foray®)
(See above under *Bacillus thuringiensis aizawai)*

Peaceful Valley Supply (Caterpillar Attack®, Javelin®)
(See above under *Bacillus thuringiensis israelensis)*

Bacillus sphaericus (for *Culex* mosquito larvae)

Valent BioSciences (VectoLex®, Spherimos®)
(See above under *Bacillus thuringiensis spp. aizawai)*

Metarhizium anisopliae (Fungus for meadow spittlebug & coffee leafminer)

Bio Pre
Geerweg 65
2461 TT Langeraar
The Netherlands
31-172-539333; FAX 31-172-537859
www.biopre.nl

Nosema locustae
(Protozoan spore pathogen for grasshoppers, locusts, Mormon crickets)

M & R Durango (Nolo Bait®)
P.O. Box 886
Bayfield, CO 81122
800/526-4075, 970/259-3521
FAX 970/259-3857

Gardens Alive! (Semispore Bait®)
5100 Schenley Place
Lawrenceburg, IN 47025
812/537-8650; FAX 812/537-8660

Steinernema carpocapsae
Steinernema feltiae
Seinername scapterisci
(Nematode parasites for caterpillars, beetle larvae, some flies and other soil-dwelling insects) (See any of the vendors under BENEFICIAL PREDATORS AND PARASITES with 'PN' in their listing)

Verticillium lecanii
(Fungus for aphid, whitefly and thrips control in greenhouses)

Koppert USA (Mycotal®, Vertalec®)
28465 Beverley Rd.
Romulus, MI 48174
800/928-8827; FAX 734/641-3793
www.koppertonline.com

INSECT PHEROMONE TRAPS
(Several garden supply houses market their own brand-name traps)

Trécé, Inc. (Pherocon® Monitoring Systems)
P.O. Box 129
Adair, OK 74330
918/785-3061; FAX 918/785-3063
custserv@earthlink.net

Scentry Biologicals (NoMate®, Scentry®)
610 Central Ave.
Billings, MT 59102
800/735-5323; FAX 406-245-2790
www.scentry.com

Certis (Decoy®, Magnet®)
(See above under *Bacillus thuringiensis spp. aizawai)*

Pacific Biocontrol Corp. (BP Rope®, Isomate® traps)
14615 NE 13th Court, Ste. A
Vancouver, WA 98685-1451
800/999-8805; FAX 360/571-2248
www.pacificbiocontrol.com

Hercon Environmental Products (Hercon® Lure-N-Kill, Disrupt®)
Aberdeen Road
Emigsville, PA 17318
800/717-764-1192; FAX 717/767-1016

BIOFUNGICIDES

Agrobacterium radiobacter

AgBioChem, Inc. (Galltrol-A®, Gallex®)
3 Fleetwood Court
Orinda, CA 94563
925/254-0789; FAX 530/527-6288

***Streptomyces griseoviridis* K61**
(For Fusarium, damping off, Pythium, etc.)

Hydro-Gardens (Mycostop®)
P.O. Box 25845
Colorado Springs, CO 80936
800/634-6362; FAX 719/495-2366
www.hydro-gardens.com

OTHER CONTROL MATERIALS

Bordeaux mixture (basic producers)

Nufarm USA (Comac®)
1333 Burr Ridge Rd., Ste 125
Burr Ridge, IL 60527
708/754-3330, 913/402-1094
FAX 866/241-0612
www.ag.us.nufarm.com

Cerexagri, Inc.
1713 S. California Ave.
Monrovia, CA 91016-0120
626/358-1838; FAX 626/359-7248

Boric acid

U.S. Borax, Inc. (Borid®, Drax®)
26877 Tourney Rd.
Valencia, CA 91355-1847
661/287-5400; FAX 661/287-5455
www.borax.com

Peaceful Valley Supply (Roach Powder®) (See above under *Bacillus thuringiensis spp. israelensis)*

Seabright Labs. (Roach Free®)
P.O. Box 8647
Emeryville, CA 94662
800/284-7363, 510/655-3126
FAX 510-654-7982

Cryolite

Cerexagri, Inc. (Kryocide®)
(See above under Bordeaux Mixture)

Gowan Company (Prokil®)
P.O. Box 5569
Yuma, AZ 85366-5569
800/883-1844, 928/783-8844
FAX 928/343-9255

Diatomaceous Earth

Celite Corp. (Celite®, Diafil®, Kenite®)
137 W. Central Ave.
Lompoc, CA 93434-0519
800/527-7315; FAX 805/735-5699

Eagle-Picher Minerals, Inc. (Celatom®, Crop Guard®, All Gone®)
9785 Gateway Dr.
Reno, NV 89511
800/366-7607, 775/824-7651
FAX 775/824-7694
www.minerals.epcorp.com

Lime-sulfur

Best Sulfur Products (BSM Lime Sulfur®)
5427 E. Central Ave.
Fresno, CA 93725
559/485-0114; FAX 559/264-1715
office@bestsulfurproducts.com

Oils, summer, dormant, superior
Many formulators-check your local garden/orchard center

Silica Aerogel (basic producer)

Cabot Corp. (Cab-O-Sil®)
700 E. US Hway 36
Tuscola, IL 61953-9643
800/222-6745, 217/253-3370
www.cabot-corp.com

Target Specialty Products
(Drione® w/pyrethrin)
1155 Mabury Rd.
San Jose, CA 95133-1029
800/767-0719; FAX 408/287-2004

Silica, Fumed (basic producer)

Cabot Corp. (Cab-O-Sil®)
(See above under Silica Aerogel)

Degussa Corp. (Aerosil®)
379 Interpace Parkway
P.O. Box 677
Parsippany, NJ 07054-0677
800/237-6745, 973/541-8515
FAX 973/541-8501

Soaps, Insecticidal

Dow AgroSciences (M-Pede®)
9330 Zionsville Rd.
Indianapolis, IN 46268-1054
800/256-3726, 800/745-7476
317/337-4385; FAX 800/905-7326
www.dowagro.com

St. Gabriel Labs (Concern®)
(See above under *Bacillus lentimorbus)*

Peaceful Valley Supply (K-Neem®)
(See above under Boric Acid)

Soaps, Herbicidal

Dow AgroSciences (Scythe®)
(See above under Soaps, Insecticidal)

Monterey Lawn & Garden Products Inc. (QuiK Weed®)
(See above under Pyrethrum)

Gardens Alive! (WeedAside®)
(See above under *Nosema locustae)*

PEST CONTROL DEVICES

Adhesive for Crawling Insects

Tanglefoot Co. (Tangle-Trap®, Sticky Trapping Systems®)
314 Straight Ave., SW
Grand Rapids, MI 49504
616/459-3139; FAX 616/459-4140
www.tanglefoot.com

ARBICO (Yellow Sticky Traps & Tapes®)
(See above under Snail Controls)

Seabright Labs. (Sticky Traps®, Stickem Special®)
(See above under Boric Acid)

Animal Barriers & Repellents

Dogs & Cats

Animal Repellents, Inc.
(Halt Dog Repellent®)
P.O. Box 510
Orchard Hill, GA 30266
800/241-5064; FAX 770/227-9190
www.halt.com

Dr. T's Nature Products (RoPel®)
P.O. Box 682
Pelham, GA 31779
800/299-6288; FAX 299/294-3027

Farnam (Get Off My Garden®)
301 W. Osborne
Phoenix, AZ 95013
800/234-2269; FAX 941/435-3737

Deer

Plants Not Favored by Deer
University of Wisconsin Bulletin (A3727)
45 N. Charter St.
Madison, WI 53715
887/947-7827
www.uwex.edu/ces/pubs/

Gardens Alive! (Hinder Deer & Rabbit Repellent®)
(See above under Soaps, Herbicidal)

Deer Busters (Deer Away®)
9735A Bethel Rd.
Frederick, MD 21702
888/422-3337; FAX 301/694-9254

Harmony Farm Supply (National Deer Repellent®, Not Tonite Deer®)
(See above under Sabadilla)

Gophers & Moles

Gardens Alive! (Mole-Med®)
(See above under Deer)

Gardens Alive (Gopher-Med®)

Rabbits

Gardens Alive! (Hinder Deer & Rabbit Repellent®)
See above under Soaps, Herbicidal)

Deer Busters (Deer Away®)
(See above under Deer)

Envirodyne (Green Screen®)
P.O. Box 357
Manistee, MI 49660
800/968-9453; FAX 231/723-2514

Raccoons

Peaceful Valley Supply (Get Away®)
(See above under Boric Acid)

Dr. T's Nature Products (Repel®)
(See above under Dogs & Cats)

ANIMAL TRAPS

Kness Mfg. Co., Inc.
(KAGE-ALL® live traps)
Highway 5 South
P.O. Box 70
Albia, IA 52531
800/247-5062

BioQuip Products, Inc. (Havahart®, Trap Away®, Victor®, Tomahawk®, Intruder®)
2321 Gladwick St.
Rancho Dominguez, CA 90220
310/667-8800; FAX 310/667-8808

BIRD BARRIERS AND REPELLENTS

Crow Taped Distress Calls
Catalog No. 21561
Division of Agriculture & Natural Resources
University of California
Oakland, CA 94612-3560
800/994-8849

J.T. Eaton & Co.
(4 The Birds® transparent repellent)
1393 E. Highland Rd.
Twinsburg, OH 44087
800/321-3421; FAX 330/425-8353

Rid-A-Bird, Inc.
(Scare-Eye® balloons, Rid-A-Bird® perches for English sparrows, starlings, pigeons)
Box 436
1217 W. 3rd St.
Wilton, IA 52778-9985
800/432-4737

Gardener's Supply Co.
(Reemay® lightweight polyester fabric & Bird Netting to protect fruit trees)
(See above, Animal Barriers & Repellents)

Bird-X Inc.
(Bird-X® transparent repellent,BirdGard® Electronic Repeller, Bird-Proof® gel repellent, Irri-Tape® holographic irridescent foil, Pigeon Hawk® replica, Terror-Eyes® balloons Bird-Net® barrier, Nixalite® steel needle roosting strips, Bird-Lite® flashing lite repeller, Ava-Alarm® outdoor sonic repeller.)
300 N. Elizabeth
Chicago, IL 60607
800/662-5021; FAX 312/226-2480
www.bird-x.com

Reed-Joseph Intl. Co., (Scare-Away Reflecting Tape®)
P.O. Box 894
800 Main St.
Greenville, MS 38701
800/647-5554; FAX 662/335-8850
www.reedjoseph.com

Hot Foot America (Hot Foot Gel®)
P.O. Box 1339
Sausalito, CA 94966
800/533-8421, 415/789-5135
FAX 415/789-0564

Tanglefoot Co. (Tanglefoot Bird Repellent®)
(See above under Adhesive for Crawling Insects)

Seabright Labs. (Stickem Special®)
(See above under Adhesive for Crawling Insects)

INSECT TRAPS

BioQuip Products, Inc. (Ultra violet light traps)
(See above under Animal Traps)

Gilbert Industries
(Electrocuting light traps)
5611 Krueger Dr.
Jonesboro, AR 72401-9102
800/643-0400, 870/932-6070
FAX 870/932-5609

American Biophysics
(Mosquito Magnet®)
2240 S. County Trail
East Greenwich, RI 02818
401/884-3500; FAX 401/884-6688
www.mosquitomagnet.com

Mosquito Wizard (Megacatch®)
4800 District Blvd.
Vernon, CA 90058
866/339-4927

WASP AND HORNET TRAPS

Seabright Labs. (Yellowjacket Inn®)
(See above under Adhesive for Crawling Insects)

Peaceful Valley Supply (SureFire®)
(See above under Boric Acid)

Woodstream (Safer Pheromone Trap®)
(See below under Rodent Barriers)

RODENT BARRIERS, REPELLENTS & TRAPS

Allen Special Products, Inc.
(STUF-FIT® copper mesh hole repellent for rats, mice, bats)
P.O. Box 605
Montgomeryville, PA 18936
800/848-6805; FAX 215/997-6654

Bird-X Inc. (Transonic®IXL sonic pest repeller) (See above, under Birds)

Kness Mfg. Co. Inc. (SNAP-E®, BIG SNAP-E®, mouse & rat traps; STICK-ALL® Depot & Glueboard adhesive traps; KETCH-ALL® multiple catch no-bait mousetrap)
(See above under Animal Traps)

Enforcer Products, Inc.
(Rat & Mouse Glue Traps®)
P.O. Box 1060
Cartersville, GA 30120
800/241-5656, 770/386-0801
FAX 770/386-1659

Woodstream (Victor Holdfast®)
69 N. Locust St.
Lititz, PA 17543-0327
800/800-1819, 717/626-2125
FAX 717/626-1912
www.woodstreampro.com

Burlington Scientific Corp. (RoPel®)
71 Carolyn Blvd.
Farmingdale, NY 11735-1527
631/694-4700; FAX 631/694-9177
www.ropel.com

DIRECTORY OF STATE COOPERATIVE EXTENSION OFFICES

APPENDIX H

One of your best sources of sound pest control information is your local County Extension Agent located in nearly every county of the nation. They can be found by looking in the white pages of your telephone directory for Cooperative Extension Service, listed under the local County Government. Because of Extension's universal availability, there will be an office nearby. With your computer, they can be found by logging onto this U.S. Department of Agriculture website: ***http://www.csrees.gov/extension/index.html.*** Click on your state map, then the appropriate county. County information lists the Extension addresses, phone and FAX numbers and their website. A second website: ***http:www.csrees.usda.gov/qlinks/partners/state_partners.html*** will provide similar information.

College of Agriculture
Auburn University
Auburn, Alabama 36849

College of Agriculture and Land Resources Management
University of Alaska
Fairbanks, Alaska 99775-0500

College of Agriculture and Life Sciences
University of Arizona
Tucson, Arizona 85721

Dale Bumpers College of Agricultural, Food and Life Sciences
University of Arkansas
Fayetteville, Arkansas 72701

Division of Agriculture and Natural Resources
University of California
Oakland, California 94612-3560

College of Agriculture and Environmental Sciences
University of California
Davis, California 95616

College of Natural and Agricultural Sciences
University of California
Riverside, California 92521

College of Agricultural Sciences
Colorado State University
Fort Collins, Colorado 80523

College of Agriculture and Natural Resources
University of Connecticut
Storrs, Connecticut 06269

College of Life Sciences
University of the District of Columbia
Washington, DC 20008

College of Agriculture and Natural Resources
University of Delaware
Newark, Delaware 19717-1303

College of Agricultural and Life Sciences
University of Florida
Gainesville, Florida 32611

College of Agricultural and Environmental Sciences
University of Georgia
Athens, Georgia 30602-7503

College of Natural and Applied Sciences
University of Guam
Mangilao, Guam 96923

College of Tropical Agriculture and Human Resources
University of Hawaii
Honolulu, Hawaii 96822

College of Agricultural and Life Sciences
University of Idaho
Moscow, Idaho 83843

College of Agricultural, Consumer and Environmental Sciences
University of Illinois
Urbana, Illinois 61801

School of Agriculture, Consumer and Family Sciences
Purdue University
West Lafayette, Indiana 47907

College of Agriculture
Iowa State University of Science and Technology
Ames, Iowa 50011

College of Agriculture
Kansas State University
Manhattan, Kansas 66506

College of Agriculture
University of Kentucky
Lexington, Kentucky 40546

College of Agriculture
Louisiana State University
Baton Rouge, Louisiana 70893

College of Natural Sciences, Forestry and Agriculture
University of Maine
Orono, Maine 04469-0163

College of Agriculture and Natural Resources
University of Maryland
College Park, Maryland 20742

College of Natural Resources and the Environment
University of Massachusetts
Amherst, Massachusetts 01003

College of Agriculture and Natural Resources
Michigan State University
East Lansing, Michigan 48824

College of Agricultural, Food and Environmental Sciences
University of Minnesota
St. Paul, Minnesota 55108

College of Agriculture and Life Sciences
Mississippi State University
Mississippi State, Mississippi 39762

College of Agriculture, Food and Natural Resources
University of Missouri
Columbia, Missouri 65211

College of Agriculture
Montana State University
Bozeman, Montana 59717

Institute of Agriculture and Natural Resources
University of Nebraska
Lincoln, Nebraska 68583

College of Agriculture, Biotechnology and Natural Resources
University of Nevada-Reno
Reno, Nevada 89557

College of Life Sciences and Agriculture
University of New Hampshire
Durham, New Hampshire 03824

Cook College
Rutgers State University
New Brunswick, New Jersey 08903

College of Agriculture and Home Economics
New Mexico State University
Las Cruces, New Mexico 88003

College of Agriculture and Life Sciences
Cornell University
Ithaca, New York 14853

College of Agriculture and Life Sciences
North Carolina State University
Raleigh, North Carolina 27695

College of Agriculture, Food Systems and Natural Resources
North Dakota State University
Fargo, North Dakota 58105

College of Food, Agricultural and
Environmental Sciences
The Ohio State University
Columbus, Ohio 43210

College of Agricultural Sciences and
Natural Resources
Oklahoma State University
Stillwater, Oklahoma 74078

College of Agricultural Sciences
Oregon State University
Corvallis, Oregon 97331

College of Agricultural Sciences
Pennsylvania State University
University Park, Pennsylvania 16802

College of Agricultural Sciences
University of Puerto Rico
Mayaguez, Puerto Rico 00681-5000

College of the Environment and Life Sciences
University of Rhode Island
Kingston, Rhode Island 02881

College of Agriculture, Forestry and Life
Sciences
Clemson University
Clemson, South Carolina 29634

College of Agriculture and Biological Sciences
South Dakota State University
Brookings, South Dakota 57007

College of Agricultural Sciences and
Natural Resources
University of Tennessee
Knoxville, Tennessee 37901

College of Agriculture and Life Sciences
Texas A&M University
College Station, Texas 77843

College of Agriculture
Utah State University
Logan, Utah 84322

College of Agriculture and Life Sciences
University of Vermont
Burlington, Vermont 05405

University of the Virgin Islands
R.R. #2, P.O. Box 10,000
Kingshill, St. Croix
Virgin Islands 00850

College of Agriculture and Life Sciences
Virginia Polytechnic Institute and State
University
Blacksburg, Virginia 24061

College of Agriculture, Human and
Natural Resources
Washington State University
Pullman, Washington 99164

College of Agriculture and Forestry
West Virginia University
Morgantown, West Virginia 26506

College of Agriculture and Life Sciences
University of Wisconsin
Madison, Wisconsin 53706

College of Agriculture
University of Wyoming
Laramie, Wyoming 82071

CONVERSION OF U.S. MEASUREMENTS TO THE METRIC SYSTEM [1]

APPENDIX I

The metric system originated in France in 1790 and spread throughout Europe, Latin America, and the East during the 19th century. With the exception of the United States (where it is used consistently in medicine and the scientific realm) and a few former British associated areas, the metric system is the official language of measurements.

Simply described, the metric system is a decimal system of weights and measures in which the gram (.0022046 pound), the meter (39.37 inches) and the liter (61.025 cubic inches) are the basic units of weight, length, and volume respectively. Most names of the various other units are formed by the addition of the following prefixes to these three terms, namely:

milli—(one thousandth), as 1 millimeter = 1/1000 meter
centi—(one hundredth), as 1 centimeter = 1/100 meter
deci—(one tenth), as 1 decimeter = 1/10 meter
deca or deka—(ten), as 1 decameter = 10 meters
hecto—(one hundred), as 1 hectometer = 100 meters
kilo—(one thousand), as 1 kilometer = 1,000 meters

Distance:

1 inch = 2.54 cm.; 12 inches = 1 foot; 3 feet = 1 yard; 1 yard = .91 (meter) M.
1,760 yards (5,280 feet) = 1 mile; 1 mile = 1,619 M.
Therefore, 1 (kilometer) Km = .62 miles or 1 mile = 1.62 Km.
Accordingly, for example, 50 Km. = 31 miles (50 miles x .62 miles per Km. = 31).

Area:

1 acre (43,560 square feet) = 4,047 square meters = 0.405 (hectare) Ha.
1 square mile (a section of land) = 640 acres or 259 Ha.
1 Ha. = 2.47 acres. To convert acres into Ha. simply multiply acres by 0.405.
Accordingly, for example, 200 acres = 81 Ha. (200 acres x 0.405 Ha. per acre = 81).

Volume:

1 quart (58 cubic inches) = .9463 liter. One liter = 1.0567 quarts.
4 quarts = a liquid gallon = 231 cubic inches or 3.78 (liter) L.
1 cubic yard (36 x 36 x 36 inches) = .76 (cubic meter) M^3.
1 bushel (dry measure) = 2,150 cubic inches = 35.2 liters.

Weight:

1 ounce = 28.35 grams; 16 ounces = 1 pound (#); 100# = 45.35 Kg; 2,000# = 1 short ton; 1 short ton = .907 metric ton.
1 gram (g) = 1,000 milligrams (mg)
1 mg = 1,000 micrograms (ug)
1 ug = 1,000 nanograms (ng)
1 ng = 1,000 picograms (pg)
1 pg = 1,000 femtograms (fg)
1 Kilogram (Kg) = 2.205 pounds; 1,000 Kg. = 2,204.6 pounds or metric ton; 1 Cwt (England) = 112#; 20 Cwt = 2240# = long ton.

[1] Courtesy of George W. Ware, Sr.

Temperature:

Temperature in the U.S. is measured in Fahrenheit (F), while under the metric system in Centigrade (C). Freezing is 32°F and boiling 212°F — a difference of 180 degrees F, compared with freezing at 0° Centigrade and boiling at 100° C — a difference of 100 degrees (See comparison below). Accordingly, 1 degree F = 5/9 degree C, and 9 degrees F = 5 degrees C.

Approximate F and C comparisons are: 50°F = 10°C; 60°F = 16°C; 70°F = 21°C; 80°F = 27°C; 90°F = 32°C; 100° F = 38°C etc.

System	Freezing Point	Boiling Point	Difference Between Freezing & Boiling
(F) Fahrenheit	32°F	212°F	180 degrees F
(C) Centigrade	0°C	100°C	100 degrees C

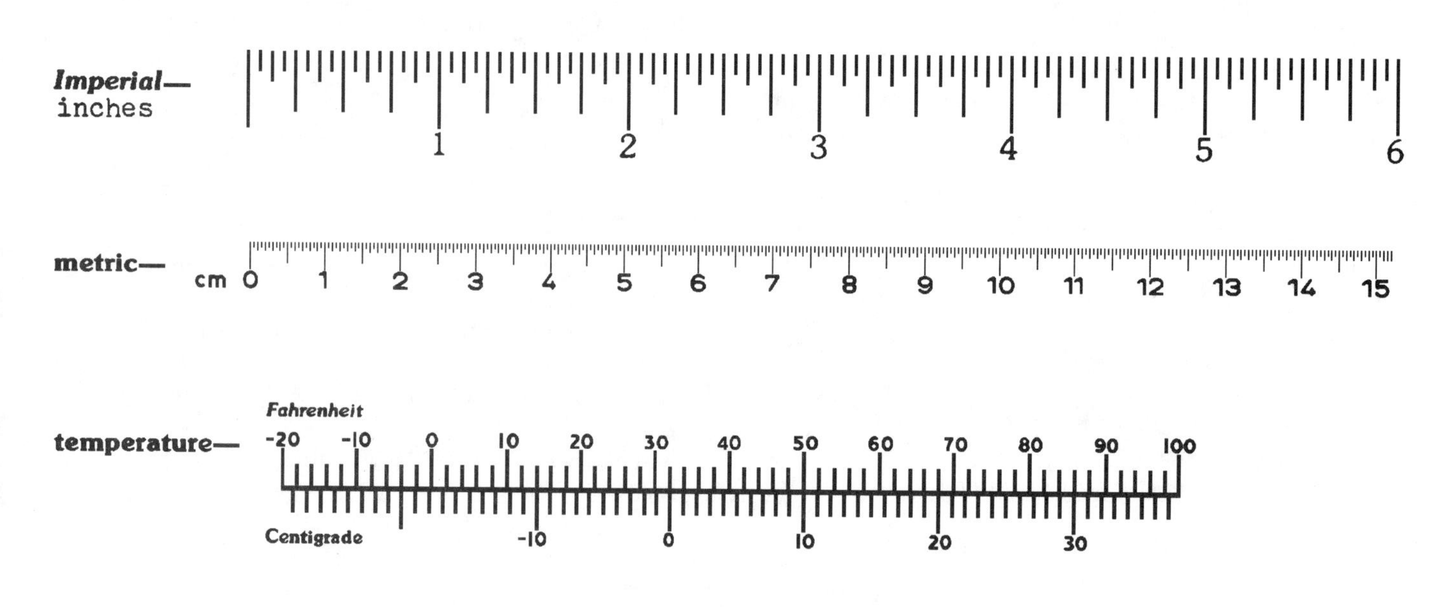

Visual Scales for Converting Simple Distances and Temperatures to The Metric System

GLOSSARY

Abscission — Process by which a leaf or other part is separated from the plant.

Absorption — Process by which pesticides are taken into tissues, namely plants, by roots or foliage (stomata, cuticle, etc).

Acaricide (miticide) — An agent that destroys mites and ticks.

Acetylcholine (ACh) — Chemical transmitter of nerve and nerve-muscle impulses in animals.

Activator — Material added to a pesticide to increase toxicity.

Active ingredient (a.i.) — Chemicals in a product that are responsible for the pesticidal effect.

Acute toxicity — The toxicity of a material determined at the end of 24 hours to cause injury or death from a single dose or exposure.

Additive, pesticide — See Adjuvant.

Adjuvant — An ingredient that improves the properties of a pesticide formulation. Includes wetting agents, spreaders, emulsifiers, dispersing agents, foam suppressants, penetrants and correctives.

Adsorption — Chemical and/or physical attraction of a substance to a surface. Refers to gases, dissolved substances, or liquids on the surface of solids or liquids.

Adulterated pesticide — A pesticide that does not conform to the professed standard or quality as documented on its label or labeling.

a.i. — Active ingredient in a pesticide product.

Algaecide — Chemical used to control algae and aquatic weeds.

Annual — Plant that completes its life cycle in one year, i.e., germinates from seed, produces seed, and dies in the same season.

Antagonism — (Opposite of synergism) decreased activity arising from the effect of one chemical on another.

Antibiotic — Chemical substance produced by a microorganism that is toxic to other microorganisms.

Anticoagulant — A chemical which prevents normal bloodclotting. The active ingredient in some rodenticides.

Antidote — A practical treatment, including first aid, used in the treatment of pesticide poisoning or some other poison in the body.

Antimicrobials — Or more recently, biocides. Defined by EPA as pesticides used to reduce growth of microorganisms or to protect water and industrial systems from these organisms.

Antitranspirant — A chemical applied directly to a plant which reduces the rate of transpiration or water loss by the plant.

Apiculture — Pertaining to the care and culture of bees.

Atropine (atropine sulfate) — An antidote used to treat organophosphate and carbamate poisoning.

Attractant, insect — A substance that lures insects to trap or poison-bait stations. Usually classed as food, oviposition and sex attractants.

Auxin — Substance found in plants, which stimulates cell growth in plant tissues.

Avicide — Lethal agent used to repel or destroy birds.

Bacillus thuringiensis (Bt)— A naturally-occurring soil bacterium that occurs world-wide and produces a toxin specific to certain insects (e.g., moths, beetles, mosquitoes and blackflies).

Bactericide — Any bacteria-killing chemical.

Bacteriostat — Material used to prevent growth or multiplication of bacteria.

Bentonite — A colloidal native clay (hydrated aluminum silicate) that has the property of forming viscous suspensions (gels) with water; used as a carrier in dusts to increase a pesticide's adhesion.

Biennial — Plant that completes its growth in 2 years. First year it produces leaves and stores food; the second year it produces fruit and seeds.

Biocide — Antimicrobials, defined by EPA as pesticides used to reduce growth of microorganisms or to protect water and industrial systems from these organisms.

Biological control agent — Any biological agent that adversely affects pest species.

Biomagnification — The increase in concentration of a pollutant in animals as related to their position in a food chain, usually referring to the persistent, organochlorine insecticides and their metabolites.

Biopesticides — Include biochemical pest control agents, microbial pest control agents and transgenic plants with pesticidal activity.

Biorational pesticides — Biological pesticides such as bacteria, viruses, fungi and protozoa; includes pest control agents and chemical analogues of naturally occurring biochemicals (pheromones, insect growth regulators, etc.).

Biota — Animals and plants of a given habitat.

Biotechnology – The science and art of genetically modifying an organism's DNA, so that transformed individuals can express new traits that enhance survival (e.g., insect or disease resistance, herbicide, resistance) or modify quality (e.g., oil, amino acids).

Biotic insecticide — Microorganisms, known as insect pathogens that are applied in the same manner as conventional insecticides to control pest species.

Biotype — Subgroup within a species differing in some respect from the species such as a subgroup that is capable of reproducing on a resistant variety.

Blast — Plant disease similar to blight.

Blight — Common name for a number of different diseases on plants, especially when collapse is sudden - e.g. leaf blight, blossom blight, shoot blight.

Botanical pesticide — A pesticide produced from naturally-occurring chemicals found in some plants. Examples are nicotine, pyrethrum, strychnine and rotenone.

Broadcast application — Application over an entire area rather than only on rows, beds, or middles.

Broad-spectrum insecticides — Nonselective, having about the same toxicity to most insects.

Bt-corn — Genetically engineered hybrids that carry the *Bacillus thuringiensis* toxin to control European corn borer and corn earworm.

Calibrate, calibration — To determine the amount of pesticide that will be applied to the target.

Canker — A lesion on a stem.

Carbamate insecticide — One of a class of insecticides derived from carbamic acid.

Carcinogen — A substance that causes cancer in animal tissue.

Carrier — A inert material that serves as a diluent or vehicle for the active ingredient or toxicant.

CAS Number — The specific identification number assigned to a chemical by the Chemical Abstracts Service of the American Chemical Society, e.g. the CAS No. For 2,4-D is 94-75-7.

Causal organism — The organism (pathogen) that produces a given disease.

Certified applicator — Commercial or private applicator qualified to apply Restricted Use Pesticides as defined by the EPA.

CFR —See Code of Federal Regulations.

Chelating agent — Certain organic chemicals (i.e. ethylenediaminetetraacetic acid) that combine with metal to form soluble chelates and prevent conversion to insoluble compounds.

Chemical name — Scientific name of the active ingredient(s) found in the formulated pesticide. The name is derived from the chemical structure of the active ingredient.

Chemophobia — A pronounced fear or dislike of chemicals, particularly those that have detectable odors. *See also* Multiple Chemical Sensitivity.

Chemotherapy — Treatment of a diseased organism, usually plants, with chemicals to destroy or inactivate a pathogen without seriously affecting the host.

CHEMTREC — A toll-free, long-distance, telephone service that provides 24-hour emergency pesticide information (800-424-9300).

Chlorosis — Loss of green color in foliage.

Cholinesterase (ChE) — An enzyme of the body necessary for proper nerve function that is inhibited or damaged by organophosphate or carbamate insecticides taken into the body by any route.

Chronic toxicity — The toxicity of a material determined usually after several weeks of continuous exposure.

Code of Federal Regulations (CFR) — Regulations pertaining to pesticides and other chemicals, published as they are passed in the *Federal Register* and by the EPA in chapters or in sections.

Common pesticide name — A common chemical name given to a pesticide by a recognized committee on pesticide nomenclature. Many pesticides are known by a number of trade or brand names but have only one recognized common name. Example: The common name for Sevin® insecticide is carbaryl.

Compatible (Compatibility) — Two materials that can be mixed together with neither affecting the action of the other.

Concentration — Content of a pesticide in a liquid or dust, for example lbs/gallon or percent by weight.

Contact herbicide — Phytotoxic by contact with plant tissue rather than as a result of translocation.

Cry proteins — Several proteins that comprise the crystal found in spores of *Bacillus thuringiensis.* Activated by enzymes in the insect's midgut, these proteins attack the cells lining the gut, cause gut paralysis and subsequent death.

Cultivar — An accepted term for a variety of a man-made selection of a particular plant.

Cumulative pesticides — Those chemicals. which tend to accumulate or build up in the tissues of animals or in the environment (soil, water).

Curative pesticide — A pesticide which can inhibit or eradicate a disease-causing organism after it has become established in the plant or animal.

Cuticle — Outer covering of insects or leaves. Chemically they are quite different.

Days-to-Harvest — The least number of days between the last pesticide application and the harvest date, as set by law. Same as "harvest intervals".

Deciduous — Plants that lose their leaves during the winter.

Decontamination — The removal or breakdown of any pesticide chemical from any surface or piece of equipment.

Deflocculating agent — Material added to a spray preparation to prevent aggregation or sedimentation of the solid particles.

Defoliant — A chemical that initiates abscission.

Deoxyribonucleic acid (DNA) — Double-stranded molecule, consisting of paired nucleotide units grouped into genes and associated regulatory sequences. These genes serve as blueprints of protein construction from amino-acid building blocks.

Dermal toxicity — Toxicity of a material as tested on the skin, usually on the shaved belly of a rabbit; the property of a pesticide to poison an animal or human when absorbed through the skin.

Desiccant — A chemical that induces rapid desiccation of a leaf or plant part.

Detoxify — To make an active ingredient in a pesticide or other poisonous chemical harmless and incapable of being toxic to plants and animals.

Diapause — A period of arrested developement or suspended animation, usually used in reference to insects.

Diatomaceous earth — A whitish powder prepared from deposits formed by the silicified skeletons of diatoms. Used as diluent in dust formulations.

Diluent — Component of a dust or spray that dilutes the active ingredient.

Dioxin — Common name for tetrachlorodioxin, more precisely, 2,3,7,8-tetrachlorodibenzo-*p*-dioxin, also referred to as TCDD. A highly toxic compound.

Disinfectant — A chemical or other agent that kills or inactivates disease producing micro-organisms in animals, seeds, or other plant parts. Also commonly referred to chemicals used to clean or surface sterilize inanimate objects (see also Biocide).

DNA — *See* Deoxyribonucleic acid.

Dormant spray — Chemical applied in winter or very early spring before treated plants have started active growth.

Dose, dosage — Same as rate. The amount of toxicant given or applied per unit of plant, animal, or surface.

Drift, spray — Movement of airborne spray droplets from the spray nozzle beyond the intended contact area.

EC_{50} — The median effective concentration (ppm or ppb) of the toxicant in the environment (usually water) which produces a designated effect in 50% of the test organisms exposed.

Ecdysone — Hormone secreted by insects essential to the process of molting from one stage to the next.

Ecology — Derived from the Greek *oikos*, "house or place to live". A branch of biology concerned with organisms and their relation to the environment.

Economic threshold — The density of a pest at which control measures should be initiated to prevent an increasing pest population from reaching the economic injury level.

Ecosystem — The interacting system of all the living organisms of an area and their nonliving environment.

Ectoparasite — A parasite feeding on a host from the exterior or outside.

ED_{50} — The median effective dose, expressed as mg/kg of body weight, which produces a designated effect in 50% of the test organisms exposed.

Emulsible (Emulsifiable) Concentrate — Concentrated pesticide formulation containing organic solvent and emulsifier to facilitate emulsification with water.

Emulsifier — Surface active substances used to stabilize suspensions of one liquid in another, for example, oil in water.

Emulsion — Suspension of miniscule droplets of one liquid in another.

Endocrine disruptors — Hormone-mimicking substances, that may have an effect in humans similar to that of estrogen or other such endocrine effect. Their presence in food, water or other environmental media and their potential risk to humans and wildlife are of scientific concern.

Endoparasite — A parasite that enters host tissue and feeds from within.

Enhanced seed — Seed that has been improved through traditional breeding or genetic engineering to resist pest and diseases, tolerate certain herbicides or improve yield.

Environment — All the organic and inorganic features that surround and affect a particular organism or group of organisms.

Environmental Protection Agency (EPA) — The Federal agency responsible for pesticide rules and regulations and all pesticide registrations.

EPA — The U.S. Environmental Protection Agency.

EPA Establishment Number — A number assigned to each pesticide production plant by EPA. The number indicates the plant at which the pesticide product was produced and must appear on all labels of that product.

EPA Registration Number — A number assigned to a pesticide product by EPA when the product is registered by the manufacturer or his designated agent. The number must appear on all labels for a particular product.

Eradicant — Applies to fungicides that are used to eliminate a pathogen from its host or environment.

Exterminate — Often used to imply the complete extinction of a species over a large continuous area such as an island or a continent.

FDA — See Food and Drug Administration.

FEPCA — The Federal Environmental Pesticide Control Act of 1972.

FIFRA — The Federal Insecticide, Fungicide and Rodenticide Act of 1974.

Filler — Diluent in powder form.

Fixed coppers — Insoluble copper fungicides where the copper is in a combined form. Usually finely divided, relatively insoluble powders.

Flowable — A type of pesticide formulation in which a very finely ground solid particle is mixed in a liquid carrier. (See Formulations, Page 19).

Foaming agent — A chemical which causes a pesticide preparation to produce a thick foam. This aids in reducing drift.

Fog treatment — The application of a pesticide as a fine mist for the control of pests.

Foliar treatment — Application of the pesticide to the foliage of plants.

Food and Drug Administration (FDA) — Federal agency responsible for purity and wholesomeness of food; safety of drugs, cosmetics and food additives; and enforcement of pesticide tolerances (residues) in food, as set by the EPA.

Food chain — Sequence of species within a community, each member of which serves as food for the species next higher in the chain.

Food Quality Protection Act (FQPA) — Passed by Congress in August 1996, it established new standards for evaluating the health and safety of pesticides "for reasonable certainty of no harm." FQPA is concerned primarily with pesticide residues and combinations of residues on raw food crops.

Formulation — Way in which basic pesticide is prepared for practical use. Includes preparation as wettable powder, granular, emulsifiable concentrate, etc.

FQPA — *See* Food Quality Protection Act.

Full coverage spray — Applied thoroughly over the crop to a point of runoff or drip.

Fumigant — A volatile material that forms vapors which destroy insects, pathogens and other pests.

Fungistatic — Action of a chemical that inhibits the germination of fungus spores while in contact.

Gallonage — Number of gallons of finished spray mix applied per 1000 square feet, acre, tree, hectare, square mile, or other unit.

General Use Pesticide — A pesticide which can be purchased and used by the general public without undue hazard to the applicator and environment as long as the instructions on the label are followed carefully. (See Restricted Use Pesticide).

Gene "stacking" — Combining such traits in seed as resistance to herbicides, diseases, insects and/or poor growing conditions, without sacrificing yields.

Germicide — A substance that kills germs (microorganisms). (Antiquated term).

Granular — A dry formulation of pesticide and other components in discrete particles generally less than 10 millimeters in diameter.

Growth regulator — Organic substance effective in minute amounts for controlling or modifying (plant or insect) growth processes.

Harvest intervals — Period between last application of a pesticide to a crop and the harvest as permitted by law.

Herbicide modifier, safener — Chemical used with herbicides to change herbicidal properties by a physiological mechanism. Includes safeners, synergists and extenders, but not surfactants that may modify activity by a chemical or physical mechanism.

Hormone — A product of living cells that circulates in the animal or plant fluids and that produces a specific effect on cell activity remote from its point of origin.

Host — Any plant or animal attacked by a parasite.

Host plant resistance — Plants that are resistant to attack by insects, diseases, nematodes or birds.

Hydrolysis — Chemical process of (in this case) pesticide breakdown or decomposition involving a splitting of the molecule and addition of a water molecule.

Hyperplasia — Abnormal increase in the number of cells of a tissue.

Hypertrophy — Abnormal increase in the size of cells of a tissue.

Incompatible — Two or more materials which cannot be mixed or used together.

Inert ingredients — The inactive materials in a pesticide formulation, which would not prevent damage or destroy pests if used alone.

Ingest — To eat or swallow.

Ingredient statement — That portion of the label on a pesticide container which gives the name and amount of each active ingredient and the total amount of inert ingredients in the formulation.

Inhalation toxicity — To be poisonous to man or animals when breathed into lungs.

Insect growth regulator (**IGR**) — Chemical substance which disrupts the action of insect hormones controlling molting, maturity from pupal stage to adult and others.

Insect pest management — The practical manipulation of insect (or mite) pest populations using any or all control methods in a sound ecological manner.

Integrated control — The integration of the chemical and biological methods of pest control.

Integrated pest management (IPM) — A management system that uses all suitable techniques and methods in as compatible a manner as possible to maintain pest populations at levels below those causing economic injury.

Intramuscular — Injected into the muscle.

Intraperitoneal — Injected into the viscera, but not into the organs.

Intravenous — Injected into the vein.

Invert emulsion — One in which the water is dispersed in oil rather than oil in water. Usually a thick, salad-dressing-like mixture results.

IPM — *See* Integrated Pest Management.

Juvenoid —Insect growth regulator that affects embryonic, larval and nymphal development.

Kg or kilogram — A unit of weight in the metric system equal to 2.2 pounds.

Label — All printed material attached to or part of the container.

Labeling — Supplemental pesticide information which complements the information on the label, but is not necessarily attached to or part of the container.

LC_{50} — The median lethal concentration, the concentration which kills 50% of the test organisms, expressed as milligrams (mg), or cubic centimeters (cc), if liquid, per animal. It is also the concentration expressed as parts per million (ppm) or parts per billion (ppb) in the environment (usually water) which kills 50% of the test organisms exposed.

LD_{50} — A lethal dose for 50% of the test organisms. The dose of toxicant producing 50% mortality in a population. A value used in presenting mammalian toxity, usually oral toxicity, expressed as milligrams of toxicant per kilogram of body weight (mg/kg).

Leaching — The movement of a pesticide chemical or other substance downward through soil as a result of water movement.

Low volume spray — Concentrate spray, applied to uniformly cover the crop, but not as a full coverage to the point of runoff.

Material Safety Data Sheets (MSDS) — OSHA Form 20, designed to contain all pertinent information on a particular chemical that has been determined to pose toxic, fire or reactivity hazards. Includes appropriate safety measures for use/handling. Intended as information for employees and employers on hazardous materials.

mg/kg (**milligrams per kilogram**) — Used to designate the amount of toxicant required per kilogram of body weight of test organism to produce a designated effect, usually the amount necessary to kill 50% of the test animals.

Microbial insecticide — A microorganism applied in the same way as conventional insecticides to control an existing pest population.

Mildew — Fungus growth on a surface.

Minimum tillage — Recent agricultural practices that utilize minimum cultivation for seedbed preparation and may reduce labor and fuel costs; may also reduce damage to soil structure.

Miscible liquids — Two or more liquids capable of being mixed in any proportions, and that will remain mixed under normal conditions.

M.L.D. — Median lethal dose, or the LD_{50}.

Mode of Action — The sum of anatomical, physiological and biochemical interactions and responses that result in toxic action of a chemical in an organism (e.g., the mode of action of *Bt* is the disruption of cells lining the midgut of insect pests).

Molluscicide — A chemical used to kill or control snails and slugs.

Mosaic — Leaf pattern of yellow and green or light green and dark green produced by certain virus infections.

MSDS — *See* Material Safety Data Sheets.

Multiple Chemical Sensitivity— Loosely defined as a seeming hypersensitivity by certain individuals to a broad array of chemicals, usually those easily detected by smell. Doctors are in the midst of a large debate on whether this condition is a genuine medical problem or the result of some overly stressed imaginations. There is the possibility that it is a new class of disease, with, as yet, no clearly defined symptoms or diagnosis.

Mutagen — Substance causing genes in an organism to mutate or change.

Mycoplasma — A microorganism intermediate in size between viruses and bacteria possessing many virus-like properties and not visible with a light microscope.

Necrosis — Death of tissue, plant or animal.

Negligible residue — A tolerance which is set on a food or feed crop permitting an ultra-small amount of pesticide at harvest as a result of indirect contact with the chemical.

Nematicide — Chemical used to kill nematodes.

Nonselective herbicide— One that is generally toxic to plants without regard to species- toxicity may be a function of dosage or method of application.

Nuclear polyhedrosis virus (NPV) — A disease virus of insects, cultured commercially and sold as a biological insecticide.

Oncogenic — The property to produce tumors (not necessarily cancerous) in living tissues. (See carcinogenic.)

Oral toxicity — Toxicity of a compound when given by mouth. Usually expressed as number of milligrams of chemical per kilogram of body weight of animal (white rat) when given orally in a single dose that kills 50% of the animals. The smaller the number, the greater the toxicity.

Organochlorine insecticide — One of the many chlorinated insecticides e.g. DDT, dieldrin, chlordane, BHC, lindane, etc., no longer registered by EPA.

Organophosphate — Class of insecticides (also one or two herbicides and fungicides) derived from phosphoric acid esters, e.g. malathion, diazinon, etc.

Organotins —A classification of miticides and fungicides containing tin as the nucleus of the molecule.

Ovicide — A chemical that destroys an organism's eggs.

Pathogen — Any disease-producing organism or virus.

Penetrant — An additive or adjuvant which aids the pesticide in moving through the outer surface of plant tissues.

Perennial — Plant that continues to live from year to year. Plants may be herbaceous or woody.

Persistence — The quality of an insecticide to persist as an effective residue due to its low volatility and chemical stability, e.g. certain organochlorine insecticides.

Persistent herbicide — Herbicide that, when applied at the recommended rate, will harm susceptible crops planted in normal rotation after harvesting the treated crop or that interferes with regrowth of native vegetation in noncrop sites for an extended period of time.

Pesticide — An "economic poison" defined in most state and federal laws as any substance used for controlling, preventing, destroying, repelling, or mitigating any pest. Includes fungicides, herbicides, insecticides, nematicides, rodenticides, desiccants, defoliants, plant growth regulators, etc.

Pesticide resistance management (PRM) — Broadly defined as alternating pesticides in the control of an organism to provide exposure to other modes of action, thereby extending the useful lives of selected pesticides.

Pheromones — Highly potent insect sex attractants produced by the insects. For some species laboratory-synthesized pheromones have been developed for trapping purposes.

Physical selectivity — Refers to the use of broad-spectrum insecticides in such ways as to obtain selective action. This may be accomplished by timing, dosage, formulation, etc.

Physiological selectivity — Refers to insecticides that are inherently more toxic to some insects than to others.

Phytotoxic — Injurious to plants.

Piscicide — Chemical used to kill fish.

Plant pesticides — Transgenic plants containing *Bacillus thuringiensis* endotoxin genes, e.g., *Bt*-cotton, *Bt*-potato, *Bt*-corn.

Plant regulator (Growth regulator) — A chemical that increases, decreases, or changes the normal growth or reproduction of a plant.

Poison Control Center — Information source for human poisoning cases, including pesticides, usually located at major hospitals.

Postemergence — After emergence of the specified weed or crop.

ppb — Parts per billion (parts in 10^9 parts) is the number of parts of toxicant per billion parts of the substance in question.

ppm — Part per million (parts in 10^6 parts) is the number of parts of toxicant per million parts of the substance in question. They may include residues in soil, water or whole animals.

ppt — Parts per trillion (parts in 10^{12} parts) is the number of parts of toxicant per trillion parts of the substance in question.

Predacide — Chemical used to poison predators.

Preemergence — Prior to emergence of the specified weed or planted crop.

Preplant application — Made before the crop is planted.

Preplant soil incorporated — Herbicide applied and tilled into the soil before seeding or transplanting.

Protectant — Fungicide applied to plant surface before pathogen attack to prevent penetration and subsequent infection.

Protective clothing — Clothing to be worn in pesticide-treated fields under certain conditions as required by federal law, e.g. reentry intervals.

Protopam chloride (2-PAM) — An antidote for certain organophosphate pesticide poisoning, but not for carbamate poisoning.

Rate — Refers to the amount of active ingredient applied to a unit area regardless of percentage of chemical in the carrier (dilution).

Raw agricultural commodity — Any food in its raw and natural state, including fruits, vegetables, nuts, eggs, raw milk and meats.

Reentry intervals — Waiting interval required by federal law between application of certain hazardous pesticides to crops and the entrance of workers into those crops without protective clothing.

Registered pesticides — Pesticide products which have been approved by the Environmental Protection Agency for the uses listed on the label.

Repellent (insects) — Substance used to repel ticks, chiggers, gnats, flies, mosquitoes and fleas.

Residue — Trace of a pesticide and its metabolites remaining on and in a crop, soil, or water.

Resistance (insecticide) — Natural or genetic ability of an organism to tolerate the poisonous effects of a toxicant.

Restricted Use Pesticide — One of several pesticides designated by the EPA that can be applied only by certified applicators, because of their inherent toxicity or potential hazard to the environment.

RNA — Ribonucleic acid.

Rodenticide — Pesticide applied as a bait, dust, or fumigant to destroy or repel rodents and other animals, such as moles and rabbits.

Ropewick applicator — A rope saturated with a foliage-applied translocated herbicide solution that is wiped across the surface of weed foliage. The rope utilizes forces of capillary attraction and conducts the herbicide from a reservoir.

Rust — A disease with symptoms that usually include reddish-brown or black pustules; a group of fungi in the Basidiomycetes.

Safener — Chemical that reduces the phytotoxicity of another chemical.

Safer pesticides — Those designated as "safer" (or "reduced risk") by EPA that have favorable characteristics affecting health or environmental risks, resistance management and integrated pest management.

Scientific name — The one name of a plant or animal used throughout the world by scientists and based on Latin and Greek.

Secondary pest — A pest which usually does little if any damage but can become a serious pest under certain conditions, e.g. when insecticide applications destroy its predators and parasites.

Selective insecticide — One which kills selected insects, but spares many or most of the other organisms, including beneficial species, either through differential toxic action or the manner in which insecticide is used.

Selective pesticide — One which, while killing the pest individuals, spares much or most of the other fauna or flora, including beneficial species, either through differential toxic action or through the manner in which the pesticide is used (formulation, rate, timing, placement, etc.)

Semiochemicals — Chemicals emitted by plants or animals that modify the behavior of other similar or different organisms. The three types are: pheromones (chemically messaging within a species), allomones (emitted by one species but affect another to benefit the emitter), kairomones (emitted by one species to the benefit of another receptor species).

Senescence — Process or state of growing old.

Sex lure — Synthetic chemical which acts as the natural lure (pheromone) for one sex of an insect species.

Signal word — A required word which appears on every pesticide label to denote the relative toxicity of the product. The signal words are either "Danger—poison" for highly toxic compounds, "Warning" for moderately toxic, or "Caution" for slightly toxic.

Slimicide — Chemical used to prevent slimy growth, as in wood pulping processes for manufacture of paper and paperboard.

Slurry — Thin, watery mixture, such as liquid mud, cement, etc. Fungicides and some insecticides are applied to seeds as slurries to produce thick coating and reduce dustiness.

Smut — A fungus with sooty spore masses; a group of fungi in the Basidiomycetes.

Soil application — Application of pesticide made primarily to soil surface rather than to vegetation.

Soil persistence — Length of time that a pesticide application on or in soil remains effective.

Spore — A single to many-celled reproductive body in the fungi that can develop a new fungus colony.

Spot treatment — Application to localized or restricted areas as differentiated from overall, broadcast, or complete coverage.

Spreader — Ingredient added to spray mixture to

improve contact between pesticide and plant surface.

Sterilize — To treat with a chemical or other agent to kill every living thing in a certain area.

Sticker — Ingredient added to spray or dust to improve its adherence to plants.

Structural pests — Pests which attack and destroy buildings and other structures, clothing, stored food and manufactured and processed goods. Examples: Termites, cockroaches, clothes moths, rats, dry-rot fungi.

Stupefacient or soporific — Drug used as a pesticide to cause birds to enter a state of stupor so they can be captured and removed or to frighten other birds away from the area.

Subcutaneous toxicity — The toxicity determined following its injection just below the skin.

Surfactant — Ingredient that aids or enhances the surface-modifying properties of a pesticide formulation (wetting agent, emulsifier, spreader).

Suspension — Finely divided solid particles or droplets dispersed in a liquid.

Sustainable agriculture — Farming systems that emphasize environmentally benign practices while also sustaining yield and net farm income, e.g., crop rotation, pest resistant varieties, recycling animal manures and use of biologically based pest control methods.

Swath — The width of the area covered by a sprayer or duster making one sweep.

Synergism — Increased activity resulting from the effect of one chemical on another.

Synthesize — Production of a compound by joining various elements or simpler compounds.

Systemic — Compound that is absorbed and translocated throughout the plant or animal.

Tank mix — Mixture of two or more pesticides in the spray tank at time of application. Such mixture must be cleared by EPA.

Target — The plants, animals, structures, areas, or pests to be treated with a pesticide application.

Temporary tolerance — A tolerance established on an agricultural commodity by EPA to permit a pesticide manufacturer or his agent time, usually one year, to collect additional residue data to support a petition of a permanent tolerance; in essence, an experimental tolerance (See Tolerance).

Teratogenic — Substance that causes physical birth defects in the offspring following exposure of the pregnant female.

Tolerance — Amount of pesticide residue permitted by federal regulation to remain on or in a crop. Expressed as parts per million (ppm).

Tolerant — Capable of withstanding effects.

Topical application — Treatment of a localized surface site such as a single leaf blade, on an insect, etc., as opposed to oral application.

Toxic — Poisonous to living organisms.

Toxicant — A poisonous substance such as the active ingredient in pesticide formulations that can injure or kill plants, animals, or microorganisms.

Toxin — A naturally occurring poison produced by plants, animals, or microorganisms. Examples: The poison produced by the black widow spider, the venom produced by snakes and the botulism toxin.

Trade name (Trademark name, proprietary name, brand name) — Name given a product by its manufacturer or formulator, distinguishing it as being produced or sold exclusively by that company.

Transgenic plants —Plants whose genetic composition has been altered to include selected genes from other plants or species by methods other than those of traditional plant breeding, introducing resistance to herbicides, insects and bacterial, fungal and viral pathogens.

Translocation — Transfer of food or other materials such as herbicides from one plant part to another.

Trivial name — Name in general or common-place usage; for example, nicotine.

Ultra low volume (ULV) — Sprays that are applied at 0.5 gallon or less per acre or sprays applied as the undiluted formulation.

Unclassified Pesticide — Formerly General Use Pesticide. A pesticide that can be purchased and used by the general public without undue hazard to the applicator and environment as long as the instructions on the label are followed carefully (See also Restricted Use Pesticide).

Vector — An organism, as an insect, that transmits pathogens to plants or animals.

Vermin — Pests, usually rats, mice, or insects.

Viricide — A substance that inactivates a virus completely and permanently.

Virustatic — Prevents the multiplication of a virus.

Volatilize — To vaporize.

Weed — Plant growing where it is not desired.

Wettable powder — Pesticide formulation of toxicant mixed with inert dust and a wetting agent which mixes readily with water and forms a short-term suspension (requires tank agitation).

Wetting agent — Compound that causes spray solutions to contact plant surfaces more thoroughly.

Winter annual — Term usually applies to weeds. Plant that starts germination in the fall, lives over winter, and completes its growth, including seed production, the following season.

BIBLIOGRAPHY

AAPCC (2004) American Association of Poison Control Centers: Certified Regional Poison Control Centers. (www.aapcc.org)

American Phytopathological Society (1998) Compendium of Plant Disease Series: Turfgrass, Greenhouse Crops, Vegetable Gardens, Apple & Pear, Citrus, Ornamental Foliage, Roses, Grapes, Deciduous Trees, Strawberries. APS. St. Paul, MN.

Andrews E (1996) Disposing of Hazardous Wastes from the Home. University of Wisconsin Cooperative Extension Bull. G3453. Madison. 4 pp.

Anonymous (2003) Chemical & Engineering News (Mar. 17) p. 7.

Anonymous (2004) 2004 Directory of Least-Toxic Pest control Products. The IPM Practitioner. Bio-Integral Resource Center (BIRC), Berkeley, CA. 52 pp.

Atkinson TH, Koehler PG, Patterson RS (l990) The Nicaraguan Cockroach-Our Newest Immigrant. Pest Management. pp 12-14 (Oct.)

Becker R, Welty C (1994) Integrated Pest Management for the Home Vegetable Garden. Ohio State University Cooperative Extension. HYG-2205-94. Columbus. 4 pp.

Carruth LA (l975) Vegetable Garden Pests. University of Arizona Cooperative Extension Bull. A-81. Tucson. 18 pp.

Carruth LA, Olton GS (1976) Household Pests. University of Arizona Cooperative Extension Bull. A-72. Tucson. 24 pp.

Caslick JW, Decker DJ (1981) Control of Wildlife Damage in Homes and Gardens. Cornell University Cooperative Extension. Ithaca, NY. 18 pp.

Centers for Disease Control and Prevention (1997) Plague Update. General news release.

Corbett JR, Wright K, Baillie AC (1984) The Biochemical Mode of Action of Pesticides. Academic Press, New York. 382 pp.

Craven, SR (2000) Mole Control. University of Wisconsin, Cooperative Extension Bull. G32000. Madison. 4 pp.

Crop Protection Handbook (2004) Meister Media Worldwide. Willoughby, OH. Vol. 90, 510 pp.

Davidson RN, Lyon WF (1987) Insect Pests of Farm, Garden and Orchard. 8th Ed. Meister Publishing, Willoughby, OH. 618 pp.

Drees BM, Summerlin B (2003) House-infesting Ants and their Management. Texas A & M University Cooperative Extension Bull. L-2061. College Station. 13 pp.

Ettlinger SR (1990) The Complete Illustrated Guide to Everything Sold in Garden Centers (except the plants). Macmillan, New York. 368 pp.

Donaldson D, Kiely T, Grube A (2002) Pesticides Industry Sales and Usage, 1998 and 1999 Market Estimates. US EPA, Washington, DC. EPA-733-R-02-001, 34 pp.

Flint ML (1998) Pests of the Garden and Small Farm. 2nd Ed. University of California Press, Berkeley. 276 pp.

Furrer JD, Mitich LW, Williams Jr. JL, Gaussoin RE, Shearman RC (1992) Lawn Weeds and Their Control. North Central Regional Extension Publ. 26. University of Nebraska Cooperative Extension, Lincoln. 24 pp.

Gleason M, Lewis D (1997) Tree Fruits—Insect and Disease Management for Backyard Fruit Growers in the Midwest. Iowa State University Cooperative Extension Bull. IDEA 3/May. Ames. 40 pp.

Gold RE, Glenn GJ (2001) Drywood Termites. Texas A & M University Extension Bull. L-1782. College Station. 6 pp.

Gold RE, Glenn GJ (1999) Carpenter Ants. Texas A & M University Extension Bull. L-1783. College Station. 5 pp.

Gold RE, Howell Jr HN, Glenn GJ (I999) Subterranean Termites. Texas A&M University Cooperative Extension Bull. B-6080. College Station. 8 pp.

Hahn J, Pellitteri P, Lewis D (1997) Common Spiders in and Around Homes. University of Minnesota Cooperative Extension Bull. FO-1033-B. Minneapolis. 4 pp.

Hahn J, Pellitteri P, Lewis D (1996) Wasp and Bee Control. University of Minnesota Cooperative Extension Bull. FO-3732-S. Minneapolis. 4 pp.

Hamman PJ, Gold RE (1994) Cockroaches-Recognition and control. Texas A&M University Cooperative Extension Bull. B-1458. College Station. 7 pp.

Hayden RA, Ellis MA, Foster RE (1993) Midwest Tree Fruit Handbook. Purdue University Cooperative Extension Bull. 506A. Lafayette, IN. 46 pp.

Heimann MF, Stevenson WR, Worf GL (1996) Powdery Mildew of Ornamentals. University of Wisconsin Cooperative Extension Bull. A2404. Madison. 3 pp.

Herbicide Handbook (2002) Weed Science Society of America, 8th Ed. (Vencill WK, ed), Lawrence, KS. 493 pp.

Hoelscher CE, Patrick CD, Robinson JV (2000) Managing External Parasites of Texas Livestock and Poultry. Texas A & M University Extension Bull B-1306. College Station. 23 pp.

Horst RK (1998) Westcott's Plant Disease Handbook, 6th Ed. Van Nostrand Reinhold, New York. 953 pp.

Howell H (2000) Bed Bugs. Texas A & M University Cooperative Extension Bull. L-1742. College Station. 3 pp.

Hunter CD (1997) Suppliers of Beneficial Organisms in North America. California Dept. of Pesticide Regulation, Environ. Monitoring and Pest Management Br., Sacramento. 32 pp.

Illinois Agrinews (2003) Jan. 24 issue.

Jackman JA (1999) Scorpions. Texas A & M University Extension Bull. L-1678. College Station. 5 pp.

Jackson WB (1992) How Many Rats Are There? Pest management. pp 12-15 (Aug.)

Johnson WT, Lyon HH (1991) Insects That Feed on Trees and Shrubs. 2nd Ed. Cornell University Press, Ithaca, NY. 560 pp.

Jull LG (2001) Plants Not Favored by Deer. University of Wisconsin, Cooperative Extension Bull. A3727. Madison. 4 pp.

Klass C, Snover KL (2003) Pest Management Around the Home, Part II. 2003-2004 Pesticide Guidelines. Cornell University Cooperative Extension Misc. Bull. S74. Ithaca, NY. 94 pp.

Klass C, Snover KL (2000) Pest Management Around the Home, Part 1. Cultural Methods. Cornell University Cooperative Extension Misc. Bull. S74. Ithaca, NY. 103 pp.

Klassen CD, Amdur MO, Doull J (1986) Casarett and Doull's Toxicology-The Basic Science of Poisons, 3rd Ed. Macmillan, New York. 974 pp.

Klotz J, Bennett G (1992) Carpenter Ants. The Urban Pest Management Challenge. Pest Control. pp 44-50 (May).

Koehler PG, Patterson RS, Webb JC (1986) Efficacy of Ultrasound for German Cockroach and Oriental Rat Flea Control. Journal of Economic Entomology. 79(4):1027-1031.

Koval CF, Pellitteri P (1992) Woody Ornamental Pest Control Guide. Urban Phytonarian Series. University of Wisconsin Cooperative Extension A3134. Madison. I I pp.

Kunerth J (1992) A New Invader. *Aedes albopictus.* Pest Management 28 (April).

Lyon WF (1987) Safe Pesticides for Pets. Ohio State University Cooperative Extension Bull. 586. Columbus. 18 pp.

Lyon WF (1989) Spiders in and Around the House. Home, Yard and Garden Facts, HYG-2060-89. Ohio State University Cooperative Extension. Columbus. 4 pp.

Lyon WF (1981) Wood's Cockroach. Home, Yard and Garden Facts, HYG-2119-91. Ohio State University Cooperative Extension. Columbus. 2 pp.

Lyon WF, Steel JA (I 994) Household and Structural Pest Management. Ohio State University Cooperative Extension Bull. 512. Columbus. 20 pp.

Mahr DL, McManus PS, Roper TR (2002) Apple Pest Management for Home Gardeners. University of Wisconsin Cooperative Extension Bull. A2179. Madison. 4 pp.

Mahr DL, McManus PS, Roper TR (2002) Apricot, Cherry, Peach and Plum Pest Management for Home Gardeners. University of Wisconsin Cooperative Extension Bull. A2130. Madison. 2 pp.

Mahr DL, Heimann MF, Roper TR (1994) Strawberry Pest Management for Home Gardeners. University of Wisconsin Cooperative Extension Bull. A2127. Madison. 2 pp.

Mahr DL, Roper TR, Jeffers SN (1993) Raspberry Pest Management for Home Gardeners. University of Wisconsin Cooperative Extension Bull. A2128. Madison. 4 pp.

Mahr DL, Jeffers SN, Roper TR (1994) Grape Pest Management for Home Gardeners. University of Wisconsin Cooperative Extension Bull. A2129. Madison. 2 pp.

McManus PS, Mahr DL, Roper TR, Flashinski RA (2004) 2004 Commercial Tree Fruit Spray Guide. University of Wisconsin Cooperative Extension Bull. A3314. Madison. 58 pp.

McManus PS, Heimann MF (1997) Apple, Pear and Related Trees Disorder: Fire Blight. University of Wisconsin Cooperative Extension Bull. Al 616. Madison. 4 pp.

Merchant M (2001) Geckos in homes. Texas A & M University, Cooperative Extension Fast Sheet Ent-1019. College Station. 3 pp.

Merchant M, Teel P (2001) Biting Mites in Homes. Texas A&M University Cooperative Extension FastSheet Ent-1025. College Station. 3 pp.

Merchant M (2000) Honey Bee Swams and Their Control. Texas A & M University Cooperative Extension FastSheet Ent-1003. College Station. 3 pp.

Merchant M, Gold RE (2001) Choosing a Termite Treatment Chemical. Texas A & M University Cooperative Extension FastSheet Ent-1005. College Station. 4 pp.

Merchant M, Robinson J (1999) Controlling Fleas. Texas A&M University Cooperative Extension Bull. L-1738. College Station. 8 pp.

Miller R L, Roach K (l 992) Insect Pests of Christmas Trees. Ohio State University Cooperative Extension Bull. 619. Columbus. 22 pp.

Moore GC, Olson JK (2004) Human Lice. Texas A & M University Cooperative Extension Bull. L-1315. College Station. 6 pp.

National Pest Control Association, Inc. (1990) Wildlife-Transmitted Diseases. Tech. Release ESPC 043249. Dunn Loring, VA. 6 pp.

National Pest Control Association, Inc. (1992) Caulking Materials for Insect Pest Management. Tech. Release ESPC 039817. Dunn Loring, VA. 5 pp.

Olsen LG, Whalon ME, Hamilton JG (1993) Insect Traps for Home Fruit Insect Control. North Central Regional Publ. 359. Michigan State University Cooperative Extension. East Lansing. 7 pp.

Parrella MP, Heinz KM, Nunnery L (1992) Biological Control Through Augmentative Releases of Natural Enemies: A Strategy Whose Time Has Come. American Entomologist. 172-179 (Fall)

Pellitteri PJ (l999) Cluster flies in the home. University of Wisconsin Cooperative Extension Bull. A2090. Madison. 2 pp.

Pesticide Manual (2000) A World Compendium, 12th Ed. Tomlin CDS (ed). British Crop Protection Council. Farnham, Surrey, UK. 1250 pp.

Pound WE, Rimelspach JW, Shetlar DJ, Street JR (1990) Control of Turfgrass Pests. Ohio State University Cooperative Extension Bull. L-187. Columbus. 17 pp.

Pratt HD (l966) Pictorial Keys to Arthropods, Reptiles, Birds and Mammals of Public Health Significance. U.S. Dept. HEW. Communicable Disease Center. Training Branch. Atlanta, GA. 80 pp.

Rasche RE (1992) Mapping Malaria. Pest Management 30 (April).

Robinson JV (2000) Horticultural Oils and Pest Control. Texas A & M Cooperative Extension Bull. L-5350. College Station. 2 pp.

Robinson JV, Meola R (1997) Suggestions for Indoor and Outdoor Flea Control. Texas A & M Cooperative Extension Bull. UC-034. College Station. 4 pp.

Scott HG (1972) Household and Stored-Food Insects of Public Health Importance and Their Control. DHEW Publ. (HSM) 72-8122. U.S. Dept. HEW, Center for Disease Control. Atlanta, GA. 28 pp.

Sinclair WA, Lyon HH, Johnson WT (1991) Diseases of Trees & Shrubs. Cornell University Press. Ithaca, NY. 574 pp.

Slife WA, Lyon HH, Kommendahl T (l 960) Weeds of the North Central Regional States. North Central Regional Publ. 36. Circ. 718, University of Illinois, Agric. Expt. Station. Urbana. 165 pp.

Smith EH, Whitman RC (2000) Field Guide to Structural Pests. National Pest Control Association. Dunn Loring, VA. (l 6 chapters, pp. unnumbered).

Smith RL (1982) Venomous Animals of Arizona. University of Arizona Cooperative Extension Bull. 8245. Tucson. 134 pp.

Stewart JW (1997) Common Insect and Mite Pests on Humans. Texas A&M Cooperative Extension Bull. UC-013. College Station. 7 pp.

Swiader JM, Ware GW (2002) Producing Vegetable Crops, 5th Ed. Interstate Publishers, (now Prentice Hall, Upper Saddle River, NJ). 658 pp.

Tashiro H (1991) Turfgrass Insects of the United States and Canada. Cornell University Press. Ithaca, NY. 472 pp.

The IPM Practitioner (2004) 2004 Directory of Least-Toxic Pest Control Products. BIo-Integral Resource Center (BIRC), Berkeley, CA. 52 pp.

Thomson WT (2002) Agricultural Chemicals, Book II-Herbicides. Thomson Publications, Fresno, CA. 309 pp.

Thomson WT (2001) Agricultural Chemicals, Book I-Insecticides. Thomson Publications, Fresno, CA. 248 pp.

Thomson WT (2000) Agricultural Chemicals, Book IV-Fungicides. Thomson Publications, Fresno, CA. 225 pp.

Thomson WT (2000) Agricultural Chemicals, Book III-Misc. Agricultural Chemicals. Thomson Publications, Fresno, CA. 189 pp.

Timm RM (1987) Prevention and Control of Wildlife Damage. University of Nebraska Cooperative Extension. Lincoln. 418 pp. + supplements.

Ware GW, Whitacre DM (2004) The Pesticide Book, Meister Media Worldwide, Willoughby, OH. 496 pp.

Watson WA, Litovitz TL, Rodgers GC Jr., Klein-Schwartz W, Youniss J, Rose SR, Borys D, May ME (2003) 2002 Annual Report of the American Association of Poison Control Centers Toxic Exposure Surveillance System. American Journal of Emergency Medicine 21(5):353-411.

Whitson T (ed) (1991) Weeds of the West. Western Society of Weed Science. University of Wyoming Cooperative Extension Bull. Laramie. 54 pp.

Worf GL, Heimann MF (l994) Common Foliage Diseases of Shade Trees in Wisconsin. University of Wisconsin Cooperative Extension Bull. A2509. Madison. 6 pp.

Wyman JA, Pellitteri PJ (2001) Managing Insects in the Home Vegetable Garden. University of Wisconsin Cooperative Extension Bull. A2088. Madison. 20 pp.

Aldrich SR (ed) (l981) Organic and conventional Farming Compared (Report No. 84, Council for Agricultural Science and Technology (CAST). Iowa State University, Ames. 18 pp.

Anonymous (2004) 2004 Directory of Least-Toxic Pest Control Products. The IPM Practitioner. Bio-Integral Resource Center (BIRC), Berkeley, CA. 52 pp.

Baker, Jerry (1999) Old-Time Garden Wisdom, Jerry Baker, Wixom, MI. 367 pp.

Better Homes & Gardens (1987) Step-By-Step Successful Gardening. Meredith Books, Des Moines, IA. 256 pp.

Better Homes & Gardens (2003) Bird Gardens. Meredith Books,Des Moines, IA. 143 pp.

Carr A (1979) Rodale's Color Handbook of Garden Insects. Rodale Press, Emmaus, PA. 241 pp.

Christie GD (1992) Non-Toxic Bait Trapping for Yellow Jackets. Pest Control. pp 30-32 (May).

Cuthbertson, Yvonne (2001) Beginners Guide to Herb Gardening. Guild of Master Craftsman Publications, East Sussix, U.K. 168 pp.

Denckla, Tanya LK. (2003) The Gardener's A-Z Guide to Growing. Storey Publications, North Adams, MA. 485 pp

Encyclopedia of Natural Insect and Disease Control (1984) Roger B. Yepsen, ed., Rodale Press, Emmaus, PA. 534 pp.

Ettlinger SR (1990) The Complete Illustrated Guide to Everything Sold in Garden Centers (except the plants). Macmillan, New York. 368 pp.

Fedor, John (2001) Organic Gardening for the 21st Century. Reader's Digest, Pleasantville, NY. 288 pp.

Flint ML (1998) Pests of the Garden and Small Farm. 2nd Ed. University of California Press, Berkeley. 276 pp.

Gussow, Joan Dye (2001) This Organic Life-confessions of a suburban homeowner. Chelsea Green Publications Co., White River Junction, VT. 273 pp.

Horst RK (1998) Westcott's Plant Disease Handbook, 6th Ed. Van Nostrand Reinhold, New York. 953 pp.

Hunter CD (1997) Suppliers of Beneficial Organisms in North America. California Dept. of Pesticide Regulation, Environ. Monitoring & Pest Management Br., Sacramento. 32 pp.

Irish, Mary (2002) Gardening in the Desert Southwest, Month-By-Month, Cool Springs Press, Nashville, TN. 319 pp.

Klass C, Snover KL (2000) Pest Management Around the Home, Part 1. Cultural Methods. Cornell University Cooperative Extension Misc. Bull. S74. Ithaca, NY. 103 pp.

Lewis, Eleanore (2001) Better Homes and Gardens Perennial Gardens, Meredith Books, Des Moines, IA. 143 pp.

Lovejoy, Ann (2001) Organic Gardening Design School, Rodale Press, Emmaus, PA. 280 pp.

Lyon WF, Miller RL (1989) Home Remedies-Pest Detection, Attraction, Repellents and Control Practices. Ohio State University Cooperative Extension Home, Yard & Garden Facts #2145. Columbus. 2 pp.

Lyon WF, Steel JA (1994) Household and Structural Pest Management. Ohio State University Cooperative Extension Bull 512. Columbus. 20 pp.

Lyon WF, Steel JA (1995) Mosquito Pest Management. Ohio State University Cooperative Extension Bull. 641. Columbus. 10 pp.

Olkowski W, Daar S, Olkowski H (1991) Common Sense Pest Control. Taunton Press, Newton, CT. 715 pp.

Olsen LG, Whalon ME, Hamilton JG (1992) Insect Traps for Home Fruit Insect Control. North Central Regional Extension Publ. 359. Michigan State University. East Lansing. 32 pp.

Organic Gardener's Handbook of Natural Insect and Disease Control (1996) Barbara W. Ellis and Fern Marshall Bradley, eds. Rodale Press, Emmaus, PA. 534 pp.

Owens, Dave (2002) The Garden Guy. A seasonal guide to organic gardening in the desert Southwest. Poco Verde Landscape, Publ. Tempe, AZ. 155 pp.

Philbrick H, Philbrick J (1995) The Bug Book-Harmless Insect Controls. 3rd Ed. Garden Way Publishing, Charlotte, VT. 136 pp.

Robinson, Peter (1997) Complete Guide to Water Gardening. American Horticultural Society. DK Publications, New York, NY. 219 pp.

Rodale Book of Composting (1992) Rodale Press, Emmaus, PA. 278 pp.

Rodale's All-New Encyclopedia of Organic Gardening (1997) Rodale Press, Emmaus, PA. 690 pp.

Shuler, Carol (1993) Low-Water-Use-Plants for California and the Southwest. Fisher Books, Perseus Books, Cambridge, MA. 138 pp.

Sunset Water Gardens (2004) Sunset Press, Menlo Park, CA. 192 pp.

Sunset Western Garden Annual (2004) Sunset Publications, Menlo Park, CA. 384 pp.

The IPM Practitioner (2004) 2004 Directory of Least-Toxic Pest Control Products. (Quarles, W, ed.) Bio-Integral Resource Center (BIRC), Berkeley, CA. 52 pp.

Tumey HA, McIlveen G (1983) Insect control guide for organic gardeners. Texas A & M University Cooperative Extension Bull. College Station. 10 pp.

Worf GL, Heimann MF (1993) Young Greenhouse, Nursery Bedding, Garden, and Landscape Plants Disorder: Damping Off and Root Rot. University of Wisconsin Cooperative Extension Bull., A2700. Madison. 4 pp.

Index